W9-ANK-220

Algebraic Geometry
and
Commutative Algebra

Volume I

Algebraic Geometry
and
Commutative Algebra
in Honor of Masayoshi NAGATA

VOLUME I

Edited by
Hiroaki HIJIKATA
Heisuke HIRONAKA
Masaki MARUYAMA
Hideyuki MATSUMURA
Masayoshi MIYANISHI
Tadao ODA
Kenji UENO

1988

ACADEMIC PRESS
Harcourt Brace Jovanovich, Publishers
Tokyo Orlando San Diego New York
Austin Boston London Sydney Toronto

Typeset by LaTeX.
TeX is a trademark of the American Mathematical Society.

Published in the United States by Academic Press, Inc., Orland, Florida 32887

Library of Congress Catalog Card Number 88-082929

ISBN 0-12-348031-0

Mathematics Subject Classification(1980) 13-06, 14-06, 05-02, 12A55, 13C10, 13D03, 13E05, 13H05, 13H10, 13N05, 14C17, 14C20, 14D20, 14D22, 14H10, 14H45, 14H99, 14J15, 14J25, 14J30, 14J40, 14K99, 32G13, 32J25, 34A20.

Printed in JAPAN

The Drawing of Professor Masayoshi Nagata
by John Fogarty based on a photograph by Satoshi Hasui.

Foreword

February 9, 1987 was the sixtieth birthday of Professor Masayoshi Nagata. In addition, he received the Japan Academy Award in June, 1986 for his decisive contributions to commutative algebra.

To mark the occasion, his friends and students got together to dedicate the papers in this volume as well as those listed below, which are published elsewhere.

The editorial committee for this volume consisted of

Hiroaki Hijikata (Kyoto University),
Heisuke Hironaka (Harvard University and Kyoto University),
Masaki Maruyama (Kyoto University),
Hideyuki Matsumura (Nagoya University),
Masayoshi Miyanishi (Osaka University),
Tadao Oda (Tohoku University) and
Kenji Ueno (Kyoto University).

The publication of this volume was made possible thanks to the cooperation and effort of many people: the authors, the referees, those who converted the manuscripts to TeX files and those who read the proofs. We take this opportunity to thank all of them for their contribution. Special thanks go to Dr. Chiaki Tsukamoto (Kyoto University) whose technical assistance concerning TeX was particularly valuable. We are also grateful to Professor John Fogarty (University of Massachusetts) who produced the drawing of Professor Nagata in this volume from the photograph kindly provided by Professor Satoshi Hasui (Kyoto Sangyo University).

All the papers in this volume except that by Abhyankar are in final form and no version of any of them will be submitted for publication elsewhere.

April 20, 1988   Editors

The following papers, which are dedicated to Professor Masayoshi Nagata, are published elsewhere:

Y. Aoyama and S. Goto, Some special cases of a conjecture of Sharp, J. Math. Soc. Japan **26**(1986), 613–634.

D. E. Dobbs and T. Ishikawa, On seminormal underrings, Tokyo J. Math. **10**(1987), 157–159.

A. Fujiki, On the de Rham cohomology group of a compact Kähler symplectic manifold, in *Algebraic Geometry, Sendai, 1985* (T. Oda, ed.), Adv. Studies in Pure Math. **10**, Kinokuniya, Tokyo and North-Holland, Amsterdam, New York, Oxford, 1987, 105–165.

M.-N. Ishida, An elliptic surface covered by Mumford's fake projective plane, to appear in Tohoku Math. J. **40**(1988).

H. Ishii, On the coincidence of L-functions, Japan. J. Math. **12**(1986), 37–44.

T. Katsura and F. Oort, Families of supersingular Abelian surfaces, Compositio Math. **62**(1987), 107–185.

Y. Kawamata, Crepant blowing-ups of 3-dimensional canonical singularities and its application to degenerations of surfaces, Ann. of Math. **127**(1988), 93–163.

Y. Kobayashi, Enumeration of irreducible binary words, to appear in Discrete Applied Math.

V. Lakshmibai and C. S. Seshadri, Théorie monômiale standard pour $\widehat{SL_2}$, C. R. Acad. Sci. Paris. Sér. I Math. **305**(1987), 183–185.

Y. Miyaoka, On the Kodaira dimensions of minimal threefolds, to appear in Math. Ann. **281**(1988)

P. Monsky, Class numbers in $\mathbf{Z}_p^d$-extensions, III, Math. Z. **193**(1986), 491–514.

S. Mori, Flip theorem and the existence of minimal models for 3-folds, J. Amer. Math. Soc. **1**(1988), 117–253.

A. Moriwaki, Torsion freeness of higher direct images of canonical bundles, Math. Ann. **276**(1987), 385–398.

T. Oda, K. Saito's period map for holomorphic functions with isolated critical points, in *Algebraic Geometry, Sendai, 1985* (T. Oda, ed.), Adv. Studies in Pure Math. 10, Kinokuniya, Tokyo and North-Holland, Amsterdam, New York, Oxford, 1987, 591–648.

H. Saito, Elliptic units and Kummer's criterion for imaginary quadratic fields, J. Number Theory **25**(1987), 53–71.

M.-H. Saito, Y. Shimizu and S. Usui, Variation of mixed Hodge structure and the Torelli problem, in *Algebraic Geometry, Sendai, 1985* (T. Oda, ed.), Adv. Studies in Pure Math. 10, Kinokuniya, Tokyo and North-Holland, Amsterdam, New York, Oxford, 1987, 649–693.

T. Shioda, Algebraic cycles on hypersurfaces in $\mathbf{P}^n$, *ibid.*, 717–732.

H. Tanimoto, Normality, seminormality and quasinormality of $\mathbf{Z}[\sqrt[n]{m}\,]$, Hiroshima Math. J. **17**(1987), 29–40.

K. Toki, A remark on a homogeneous linear ordinary differential equation in characteristic $p > 0$ whose p-curvature is zero, Science Report Yokohama National Univ. Sec. I **34**(1987), 1–4.

P. M. H. Wilson, Fano fourfolds of index greater than one, J. Reine Angew. Math. **379**(1987), 171–181.

H. Yanagihara, On seminormal local rings and multicross singularities, Kobe J. Math. **4**(1987), 209–217.

H. Yoshida, On representation of finite groups in the space of Siegel modular forms and theta series, to appear in J. Math. Kyoto Univ. **28**(1988).

T. Yoshida, On a theorem of Benson and Parker, to appear in J. Algebra.

Table of Contents of Volume I

Table of Contents of Volume II

Algebraic Geometry and Commutative Algebra
in Honor of Masayoshi NAGATA
pp. 1-26 (1987)

Determinantal Loci and Enumerative Combinatorics of Young Tableaux*

Shreeram S. ABHYANKAR

§1. Introduction.

In the First Chapter we shall describe several formulae for enumerating certain types of objects. These objects may be certain tabular arrangements of integers called Young tableaux, or they may be certain types of monomials. Although what Alfred Young [5] introduced at the turn of the last century were mainly unitableaux, we shall also enumerate bitableaux as well as multitableaux of higher width; in fact our primary concern is with bitableaux and their relation with determinantal loci. The said enumerative formulae may very well be described as examples of what may be called determinantal polynomials in binomial coefficients.

In the Second Chapter we shall indicate how to establish these enumerative formulae. Here an important role is played by transformations of determinantal polynomials and recurrence relations satisfied by them.

In the Third Chapter we shall discuss a universal identity satisfied by the minors of any matrix.

In the Fourth Chapter we shall give several applications of the said enumerative formulae and universal identity. These applications will include enumerative proofs of the straightening law of Doubilet-Rota-Stein. They will also include enumerative proofs of the Second Fundamental Theorem of Invariant Theory and its generalizations. Finally they will include computations of Hilbert functions of polynomial ideals of certain determinantal loci.

This paper is essentially the text of a lecture which I gave on 23 July 1984 in Kyoto under the auspices of the joint U.S.-Japan conference on singularities. The material of that lecture has now been expanded into the monograph [2]. So this paper may serve as a brief preview of that monograph; in particular, the titles and contents of the various chapters of this paper follow the same pattern as that of the monograph.

---

*Supported by the National Science Foundation under grant umber DMS-8500491 and by the Office of Naval Research under University Research Initiative grant number N00014-86-0689.

Received March 4, 1987.

First Chapter: YOUNG TABLEAUX AND DETERMINANTAL POLYNOMIALS IN BINOMIAL COEFFICIENTS

§2. Tableaux and monomials.

To introduce the concept of a tableau, let q be a positive integer and let $m = (m(1), m(2), \ldots, m(q))$ be a sequence of positive integers. By a *multivector* a of *width* q we essentially mean a multisequence

$$a(k, 1) < a(k, 2) < \ldots < a(k, p)$$

of positive integers, for $k = 1, 2, \ldots, q$, where the nonnegative integer p is called the *length* of a. We say that a is *bounded* by m to mean that

$$a(k, i) \leq m(k) \text{ for } k = 1, 2, \ldots, q \text{ and } i = 1, 2, \ldots, p.$$

Given any other multivector b of width q and length p', we define $a \leq b$ to mean that $p \geq p'$ and

$$a(k, i) \leq b(k, i) \text{ for } k = 1, 2, \ldots, q \text{ and } i = 1, 2, \ldots, p'.$$

By a *tableau* T of *width* q we essentially mean a sequence

$$T[1]$$
$$T[2]$$
$$.$$
$$.$$
$$.$$
$$T[d]$$

of multivectors of width q where the nonnegative integer d is called the *depth* of T. We say that T is *standard* to mean that

$$T[1] \leq T[2] \leq \ldots \leq T[d]$$

and the length of $T[e]$ is nonzero for $e = 1, 2, \ldots, d$. We say that T is *bounded* by m to mean that $T[e]$ is bounded by m for $e = 1, 2, \ldots, d$. We say that T is *predominated* by a to mean that $a \leq T[e]$ for $e = 1, 2, \ldots, d$. The sum of the lengths

$$\text{length of } T[1] + \text{ length of } T[2] + \cdots + \text{ length of } T[d]$$

is called the *area* of T.

By a *bivector* we mean a multivector of width two, and by a *bitableaux* we mean a tableaux of width two.

Henceforth let a be a multivector of width q and positive length p, such that a is bounded by m.

As the first type of objects to be enumerated, by stab(q, m, p, a, V) we denote the set of all standard tableaux of width q and area V which are bounded by m and predominated by a. We also put

$$\text{Stab}(q, m, p, a, V) = \text{card}(\text{stab}(q, m, p, a, V))$$

where card denotes cardinal number.

Before introducing the second type of objects to be enumerated, which we shall denote by mon(q, m, p, a, V), as general notation to be used, we put

$$\mathbf{Q} = \text{ the set of all rational numbers}$$

and

$$\mathbf{Z} = \text{ the set of all integers}$$

and

$$\mathbf{N} = \text{ the set of all nonnegative integers}$$

and

$$\mathbf{Z}(p) = \text{ the set of all sequences } y = (y(1), y(2), \ldots, y(p))$$
$$\text{of integers } y(1), y(2), \ldots, y(p) \text{ of length } p$$

and

$$\mathbf{N}(p) = \{ y \in \mathbf{Z}(p) : y(i) \in \mathbf{N} \text{ for } i = 1, 2, \ldots, p \}$$

and, for any two integers A and B, by $[A, B]$ we denote the *closed integral segment* between A and B, i.e.,

$$[A, B] = \{ D \in \mathbf{Z} : A \leq D \leq B \}.$$

For every finite subset Y of $\mathbf{Z}(q)$ we introduce the *index* of Y which we denote by ind(Y) and which we define by putting

$$\text{ind}(Y) = \text{ the largest nonnegative integer } j \text{ for which there exist}$$
$$\text{elements } y_1, y_2, \ldots, y_j \text{ in } Y \text{ such that}$$
$$y_i(k) < y_{i+1}(k) \text{ for } i = 1, 2, \ldots, j-1 \text{ and } k = 1, 2, \ldots, q$$

and we note that then ind(Y) is a nonnegative integer, and we also note that: ind$(Y) = 0$ if and only if Y is empty.

We consider the *q-dimensional positive integral cube bounded by* m which we denote by cub(q, m) and which we define by putting

$$\text{cub}(q, m) = \text{ the set of all } y \text{ in } \mathbf{Z}(q) \text{ such that}$$
$$1 \leq y(k) \leq m(k) \text{ for all } k \in [1, q].$$

For every $k \in [1, q]$ and $i \in [1, p]$, firstly we define tru$[q, m, p, a, k, i]$ to be the element of $\mathbf{Z}(q)$ obtained by putting

$$\text{tru}[q, m, p, a, k, i](\hat{k}) = \begin{cases} m(k) & \text{if } \hat{k} = 1, \ldots, k-1, k+1, \ldots, q \\ a(k, i) - 1 & \text{if } \hat{k} = k \end{cases}$$

and secondly, we put

$$\text{truc}(q, m, p, a, k, i) = \text{cub}(q, \text{tru}[q, m, p, a, k, i]);$$

here $\text{tru}[q, m, p, a, k, i]$ may be called the $(a, k, i)^{\text{th}}$ *truncation of m*, whereas $\text{truc}(q, m, p, a, k, i)$ may be called the $(a, k, i)^{\text{th}}$ *truncation of* $\text{cub}(q, m)$.

By a *protomonomial on* $\text{cub}(q, m)$ we mean a map from $\text{cub}(q, m)$ into the set of all nonnegative integers $\mathbf{N}$. Here the word protomonomial is meant to suggest the exponent system of a monomial; for example, if $(X_y)_{y \in \text{cub}(q,m)}$ is a family of elements X_y in a ring R and if t is a protomonomial on $\text{cub}(q, m)$ then

$$\prod_{y \in \text{cub}(q,m)} X_y^{t(y)}$$

is the corresponding monomial in the said family.

We put

$$\text{mon}(q, m) = \text{the set of all protomonomials on } \text{cub}(q, m)$$

and for every $t \in \text{mon}(q, m)$ we define

$$\text{supp}(t) = \{\, y \in \text{cub}(q, m) \,:\, t(y) \neq 0 \}$$

and

$$\text{abs}(t) = \sum_{y \in \text{cub}(q,m)} t(y)$$

where supp and abs are meant to suggest *support* and *absolute value* respectively. We also put

$$\text{mon}(q, m, p) = \{\, t \in \text{mon}(q, m) \,:\, \text{ind}(\text{supp}(t)) \leq p \}$$

and we define

$$\text{mon}(q, m, p, a) = \{\quad t \in \text{mon}(q, m, p) : \\ \text{ind}(\text{supp}(t) \cap \text{truc}(q, m, p, a, k, i)) \leq i - 1 \\ \text{for } k = 1, 2, \ldots, q \text{ and } i = 1, 2, \ldots, p \quad \}$$

and for every $V \in \mathbf{Z}$ we put

$$\text{mon}(q, m, p, a, V) = \{\, t \in \text{mon}(q, m, p, a) \,:\, \text{abs}(t) = V \}$$

and

$$\text{Mon}(q, m, p, a, V) = \text{card}(\text{mon}(q, m, p, a, V)).$$

§3. Determinantal polynomials of any width.

Having described the two types of objects to be enumerated, let us now describe the relevant enumeration formulae. These formulae will be labelled as

$$F^{(1k)}, F^{(1*k)}, \ldots, F^{(4*k)}, F^{(5k)}, F^{(6)}, F^{(7*)}, F^{(7)}, F^{(8)}$$

and they may be thought of as examples of what may be called determinantal polynomials in binomial coefficients. These determinantal polynomials will be defined in terms of certain apparently infinite summations because formal manipulations with summations are easier to handle when the summation indices range freely. It can be seen that the relevant summations are essentially finite, i.e., all except a finite number of summands in these summations are zero.

Out of the above determinantal polynomials, $F^{(5k)}, F^{(6)}, \ldots, F^{(8)}$ make sense for tableaux of any width, and they will be introduced in this section. The remaining ones, i.e., $F^{(1k)}, F^{(1*k)}, \ldots, F^{(4*k)}$ make sense only for bitableaux, and they will be introduced in the next section.

Let us recall that, for any A in $\mathbf{Z}$ and for any V in any overring of $\mathbf{Q}$, the *ordinary binomial coefficient* is obtained by putting

$$\binom{V}{A} = \begin{cases} \dfrac{V(V-1)\ldots(V-A+1)}{A!} & \text{if } A \geq 0 \\ 0 & \text{if } A < 0 \end{cases}$$

and let us introduce the *twisted binomial coefficient* by putting

$$\begin{bmatrix} V \\ A \end{bmatrix} = \begin{cases} \dfrac{(V+1)(V+2)\ldots(V+A)}{A!} & \text{if } A \geq 0 \\ 0 & \text{if } A < 0 \end{cases}$$

Observe that $\binom{V}{A}$ and $\begin{bmatrix} V \\ A \end{bmatrix}$ may be regarded as polynomials in an indeterminate V with coefficients in $\mathbf{Q}$; thus for example, in the polynomial ring $\mathbf{Q}[V]$ we have the equation $\begin{bmatrix} V \\ A \end{bmatrix} = \binom{V+A}{A}$.

For every $V \in \mathbf{Z}$ let us put

$$\mathbf{Z}(p, V) = \{ v \in \mathbf{Z}(p) : v(1) + v(2) + \ldots + v(p) = V \}$$

and

$$\mathbf{N}(p, V) = \mathbf{Z}(p, V) \cap \mathbf{N}(p)$$

and

$$\mathbf{Z}(p, V, 8) = \{ v \in \mathbf{N}(p) : v(1) + 2v(2) + \ldots + pv(p) = V \}$$

and

$$\mathbf{Z}(p, V, 7*) = \mathbf{Z}(p, V + [p(p-1)/2])$$

and

$$\mathbf{Z}(p, V, 7) = \{ v \in \mathbf{Z}(p, V, 7*) : v(1) > v(2) > \ldots > v(p) \geq 0 \}.$$

Now, firstly for every $k \in [1,q]$ and $i \in [1,p]$ and $j \in [1,p]$ and $v \in \mathbf{Z}(p)$ we define

$$G_{ij}^{(8k)}(q,m,p,a,v) = \begin{bmatrix} m(k) - a(k,j) \\ v(i) + v(i+1) + \ldots + v(p) + j - i \end{bmatrix}$$

and

$$G_{ij}^{(7k)}(q,m,p,a,v) = \begin{bmatrix} m(k) - a(k,j) \\ v(i) + j - p \end{bmatrix}$$

and

$$G_{ij}^{(6k)}(q,m,p,a,v) = \begin{bmatrix} m(k) - a(k,j) \\ v(i) + j - i \end{bmatrix}$$

and secondly for every $L \in [6,8]$ and $k \in [1,q]$ and $v \in \mathbf{Z}(p)$ we define

$$G^{(Lk)}(q,m,p,a,v) = \text{the } p \text{ by } p \text{ matrix whose } (i,j)^{\text{th}} \text{ entry is}$$
$$G_{ij}^{(Lk)}(q,m,p,a,v)$$

and thirdly for every $L \in [6,8]$ and $v \in \mathbf{Z}(p)$ we define

$$H^{(L)}(q,m,p,a,v) = \prod_{k=1}^{q} \det G^{(Lk)}(q,m,p,a,v)$$

and fourthly for every $k \in [1,q]$ and $v \in \mathbf{Z}(p)$ we define

$$G^{[5k]}(q,m,p,a,v) = \prod_{i=1}^{p} \begin{bmatrix} m(k) - a(k,i) \\ v(i) \end{bmatrix}$$

and

$$H^{(5k)}(q,m,p,a,v) = G^{[5k]}(q,m,p,a,v) \prod_{n \in [1,q] \setminus \{k\}} \det G^{(6n)}(q,m,p,a,v)$$

and fifthly for every $V \in \mathbf{Z}$ we define

$$F^{(8)}(q,m,p,a,V) = \sum_{v \in \mathbf{Z}(p,V,8)} H^{(8)}(q,m,p,a,v)$$

and

$$F^{(7)}(q,m,p,a,V) = \sum_{v \in \mathbf{Z}(p,V,7)} H^{(7)}(q,m,p,a,v)$$

and

$$F^{(7*)}(q,m,p,a,V) = (1/p!) \sum_{v \in \mathbf{Z}(p,V,7*)} H^{(7)}(q,m,p,a,v)$$

and

$$F^{(6)}(q,m,p,a,V) = (1/p!) \sum_{v \in \mathbf{Z}(p,V)} H^{(6)}(q,m,p,a,v)$$

and sixthly for every $k \in [1, q]$ and $V \in \mathbf{Z}$ we define

$$F^{(5k)}(q, m, p, a, V) = \sum_{v \in \mathbf{Z}(p, V)} H^{(5k)}(q, m, p, a, v).$$

§4. Determinantal polynomials of width two.

If $q = 2$, then for every $k \in [1, 2]$ we define

$$C[m, p, a, k] = \sum_{i=1}^{p} [m(k) - a(k, i)]$$

and we define

$$C(m, p, a) = C[m, p, a, 1] + C[m, p, a, 2] + p - 1.$$

Again assuming $q = 2$, given any $k \in [1, 2]$, upon letting

$$k' = \begin{cases} 2 & \text{if } k = 1 \\ 1 & \text{if } k = 2, \end{cases}$$

firstly for every $i \in [1, p]$ and $j \in [1, p]$ and $e \in \mathbf{Z}(p)$ we define

$$G_{ij}^{(1k)}(m, p, a, e) = \binom{m(k) - a(k, j) + j - i}{m(k) - a(k, j) - e(i)}$$

and secondly for every $e \in \mathbf{Z}(p)$ we define

$$G^{(1k)}(m, p, a, e) = \begin{array}{l} \text{the } p \text{ by } p \text{ matrix whose } (i, j)^{\text{th}} \text{ entry} \\ \text{is } G_{ij}^{(1k)}(m, p, a, e) \end{array}$$

and

$$G^{[1k]}(m, p, a, e) = \prod_{i=1}^{p} \left[\begin{matrix} m(k') - a(k', i) \\ e(i) \end{matrix} \right]$$

and

$$H^{(1k)}(m, p, a, e) = G^{[1k]}(m, p, a, e) \det G^{(1k)}(m, p, a, e)$$

and thirdly for every $i \in [1, p]$ and $j \in [1, p]$ and $e \in \mathbf{Z}(p)$ we define

$$G_{ij}^{(1*k)}(m, p, a, e) = \left[\begin{matrix} m(k') - a(k', i) \\ e(j) \end{matrix} \right] \binom{m(k) - a(k, j) + j - i}{m(k) - a(k, j) - e(j)}$$

and fourthly for every $e \in \mathbf{Z}(p)$ we define

$$G^{(1*k)}(m, p, a, e) = \begin{array}{l} \text{the } p \text{ by } p \text{ matrix whose } (i, j)^{\text{th}} \text{ entry} \\ \text{is } G_{ij}^{(1*k)}(m, p, a, e) \end{array}$$

and

$$H^{(1*k)}(m, p, a, e) = \det G^{(1*k)}(m, p, a, e)$$

and fifthly for every $L \in \{1, 1*\}$ and $E \in \mathbf{Z}$ we define

$$H_E^{(Lk)}(m, p, a) = \sum_{e \in \mathbf{Z}(p, E)} H^{(Lk)}(m, p, a, e)$$

and sixthly for every $L \in \{1, 1*\}$ and $D \in \mathbf{Z}$ and $E \in \mathbf{Z}$ we define

$$H_E^{[Lk]}(m, p, a, D) = (-1)^{C[m,p,a,k]-E} \begin{pmatrix} E \\ D + E - C[m, p, a, k] \end{pmatrix}$$

and seventhly for every $L \in \{1, 1*\}$ and $D \in \mathbf{Z}$ we define

$$F_D^{(Lk)}(m, p, a) = \sum_{E \in \mathbf{Z}} H_E^{[Lk]}(m, p, a, D) H_E^{(Lk)}(m, p, a)$$

and eighthly for every $i \in [1, p]$ and $j \in [1, p]$ and $e \in \mathbf{Z}(p)$ we define

$$G_{ij}^{(2k)}(m, p, a, e) = \begin{pmatrix} m(k) - a(k, i) + i - j \\ m(k) - a(k, i) - e(i) \end{pmatrix} \begin{pmatrix} m(k') - a(k', j) + j - i \\ e(i) \end{pmatrix}$$

and

$$G_{ij}^{(2*k)}(m, p, a, e) = \begin{pmatrix} m(k) - a(k, i) + i - j \\ m(k) - a(k, i) - e(j) \end{pmatrix} \begin{pmatrix} m(k') - a(k', j) + j - i \\ e(j) \end{pmatrix}$$

and ninthly for every $L \in \{2, 2*\}$ and $e \in \mathbf{Z}(p)$ we define

$$G^{(Lk)}(m, p, a, e) = \begin{array}{l} \text{the } p \text{ by } p \text{ matrix whose } (i, j)^{\text{th}} \text{ entry} \\ \text{is } G_{ij}^{(Lk)}(m, p, a, e) \end{array}$$

and

$$H^{(Lk)}(m, p, a, e) = \det G^{(Lk)}(m, p, a, e)$$

and tenthly for every $L \in \{2, 2*\}$ and $E \in \mathbf{Z}$ we define

$$H_E^{(Lk)}(m, p, a) = \sum_{e \in \mathbf{Z}(p, E)} H^{(Lk)}(m, p, a, e)$$

and eleventhly for every $L \in \{2, 2*\}$ and $D \in \mathbf{Z}$ we define

$$F_D^{(Lk)}(m, p, a) = \sum_{E \in \mathbf{Z}} \begin{pmatrix} E \\ D \end{pmatrix} H_E^{(Lk)}(m, p, a)$$

and twelfthly for every $i \in [1, p]$ and $j \in [1, p]$ and $d \in \mathbf{Z}(p)$ we define

$$G_{ij}^{(3k)}(m, p, a, d) = \begin{bmatrix} m(k') - a(k', i) - d(i) \\ m(k) - a(k, j) \end{bmatrix} \begin{pmatrix} m(k) - a(k, j) + j - i \\ d(i) \end{pmatrix}$$

and

$$G_{ij}^{(3*k)}(m,p,a,d) = \left[\begin{array}{c} m(k') - a(k',i) - d(j) \\ m(k) - a(k,j) \end{array}\right] \left(\begin{array}{c} m(k) - a(k,j) + j - i \\ d(j) \end{array}\right)$$

and thirteenthly for every $L \in \{3, 3*\}$ and $d \in \mathbf{Z}(p)$ we define

$$G^{(Lk)}(m,p,a,d) = \text{the } p \text{ by } p \text{ matrix whose } (i,j)^{\text{th}} \text{ entry}$$
$$\text{is } G_{ij}^{(Lk)}(m,p,a,d)$$

and

$$H^{(Lk)}(m,p,a,d) = \det G^{(Lk)}(m,p,a,d)$$

and fourteenthly for every $L \in \{3, 3*\}$ and $D \in \mathbf{Z}$ we define

$$F_D^{(Lk)}(m,p,a) = \sum_{d \in \mathbf{Z}(p,D)} H^{(Lk)}(m,p,a,d)$$

and fifteenthly for every $i \in [1,p]$ and $j \in [1,p]$ and $v \in \mathbf{Z}(p)$ we define

$$G_{ij}^{(4k)}(m,p,a,v) = \left[\begin{array}{c} m(k') - a(k',j) \\ v(i) + j - i \end{array}\right]$$

and sixteenthly for every $v \in \mathbf{Z}(p)$ we define

$$G^{(4k)}(m,p,a,v) = \text{the } p \text{ by } p \text{ matrix whose } (i,j)^{\text{th}} \text{ entry}$$
$$\text{is } G_{ij}^{(4k)}(m,p,a,v)$$

and

$$G^{[4k]}(m,p,a,v) = \prod_{i=1}^{p} \left[\begin{array}{c} m(k) - a(k,i) \\ v(i) \end{array}\right]$$

and

$$H^{(4k)}(m,p,a,v) = G^{[4k]}(m,p,a,v) \det G^{(4k)}(m,p,a,v)$$

and seventeenthly for every $i \in [1,p]$ and $j \in [1,p]$ and $v \in \mathbf{Z}(p)$ we define

$$G_{ij}^{(4*k)}(m,p,a,v) = \left[\begin{array}{c} m(k) - a(k,i) \\ v(j) \end{array}\right] \left[\begin{array}{c} m(k') - a(k',j) \\ v(j) + j - i \end{array}\right]$$

and eighteenthly for every $v \in \mathbf{Z}(p)$ we define

$$G^{(4*k)}(m,p,a,v) = \text{the } p \text{ by } p \text{ matrix whose } (i,j)^{\text{th}} \text{ entry}$$
$$\text{is } G_{ij}^{(4*k)}(m,p,a,v)$$

and

$$H^{(4*k)}(m,p,a,v) = \det G^{(4*k)}(m,p,a,v)$$

and ninteenthly for every $L \in \{4, 4*\}$ and $V \in \mathbf{Z}$ we define

$$F^{(Lk)}(m, p, a, V) = \sum_{v \in \mathbf{Z}(p,V)} H^{(Lk)}(m, p, a, v)$$

and twentiethly for every $L \in \{1, 1*, 2, 2*, 3, 3*\}$ and $V \in \mathbf{Z}$ we define

$$F^{(Lk)}(m, p, a, V) = \sum_{D \in \mathbf{Z}} (-1)^D F_D^{(Lk)}(m, p, a) \begin{bmatrix} C(m, p, a) - D \\ V \end{bmatrix}.$$

Second Chapter: ENUMERATION OF YOUNG TABLEAUX

§5. Counting tableaux of any width.

In 1982 I proved the following

Theorem 1. *For all $V \in \mathbf{Z}$ we have*

$$\mathrm{Stab}(q, m, p, a, V) = F^{(8)}(q, m, p, a, V) = F^{(7)}(q, m, p, a, V).$$

If q is even then for all $k \in [1, q]$ and $V \in \mathbf{Z}$ we have

$$\begin{aligned} F^{(7)}(q, m, p, a, V) = F^{(7*)}(q, m, p, a, V) &= F^{(6)}(q, m, p, a, V) \\ &= F^{(5k)}(q, m, p, a, V). \end{aligned}$$

All except the first equation in the above theorem are obtained by transformations of determinantal polynomials. To explain the strategy of proving the first equation, given any tableau T of width q and given any $v \in \mathbf{N}(p)$, let us say that (p, v) is a *shape* of T to mean that, upon letting d to be the depth of T, we have

$$\text{length of } T[j] \le p \text{ for } j = 1, 2, \dots, d$$

and

$$\mathrm{card}(\{\, j \in [1, d] \ : \ \text{the length of } T[j] \text{ is } i \,\}) = v(i) \text{ for } i = 1, 2, \dots, p.$$

Upon letting

$$\begin{aligned} \mathrm{stas}(q, m, p, a, v) \ = \ \ &\text{the set of all standard tableaux of width } q \text{ and} \\ &\text{shape } (p, v) \text{ which are bounded by } m \text{ and} \\ &\text{predominated by } a \end{aligned}$$

we get a disjoint partition

$$\mathrm{stab}(q, m, p, a, V) = \coprod_{v \in \mathbf{Z}(p, V, 8)} \mathrm{stab}(q, m, p, a, v)$$

and therefore to establish the first equation in the above theorem it suffices to prove that

$$\text{card}(\text{stas}(q, m, p, a, v)) = H^{(8)}(q, m, p, a, v).$$

The above equation is proved by showing that $\text{stas}(q, m, p, a, v)$ and $H^{(8)}(q, m, p, a, v)$ satisfy a common recurrence relation.

The above theorem was included in the Lectures which I gave at the University of Nice in France in May 1983 and the notes of which by Andre Galligo have appeared in [1]. Note that the above entities $H^{(8)}$ and $F^{(8)}$ respectively correspond to psi and phi on page 70 of [1]. So $F^{(8)}$ may be called the (general) Nice Formula.

§6. Bitableaux.

Henceforth let us assume that $q = 2$.

§7. Counting bitableaux.

In 1982, by transformations of determinantal polynomials, I proved the following

Theorem 2. *Given any $k \in [1, 2]$ and $L \in \{2, 2*, 3, 3*, 4, 4*\}$, for all $V \in \mathbf{Z}$ we have*

$$F^{(Lk)}(m, p, a, V) = F^{(6)}(2, m, p, a, V)$$

and moreover if

$$p = \min(m(1), m(2)) \text{ and } a(1, i) = i = a(2, i) \text{ for } i = 1, 2, \ldots, p$$

then for all $V \in \mathbf{Z}$ we have

$$F^{(Lk)}(m, p, a, V) = \begin{bmatrix} m(1)m(2) - 1 \\ V \end{bmatrix}.$$

The above theorem was also included in [1] where the above entity $F^{(2k)}$ was explicitly given on page 68 and it was called H. So $F^{(2k)}$ may be called the (special) Nice Formula.

§8. Counting monomials.

In the spring of 1984 I proved the following

Theorem 3. *Given any $k \in [1, 2]$, for all $V \in \mathbf{Z}$ we have*

$$F^{(1k)}(m, p, a, V) = F^{(2k)}(m, p, a, V)$$

and

$$\text{Mon}(2, m, p, a, V) = F^{(1k)}(m, p, a, V).$$

Here the first equation was proved by making some more transformations of determinantal polynomials, whereas the second equation was proved by showing that $\mathrm{mon}(2, m, p, a, V)$ and $F_D^{(1k)}(m, p, a)$ satisfy a common recurrence relation.

§9. Bitableaux and monomials.

By Theorems 1 to 3 we get the following

Theorem 4. $\mathrm{Stab}(2, m, p, a, V) = \mathrm{Mon}(2, m, p, a, V)$ *for all* $V \in \mathbf{N}$.

In view of the above theorem we may pose the

Problem. Prove Theorem 4 directly by finding a "natural" one-to-one correspondence between $\mathrm{stab}(2, m, p, a, V)$ and $\mathrm{mon}(2, m, p, a, V)$. (I can do this for $p = 1$).

The above Problem in turn leads to the

Meta-Problem. If two naturally defined finite sets have the same cardinality, does there necessarily exist a *concrete* (= constructive = algorithmic) one-to-one correspondence between them?

In view of the said recurrence relation satisfied by $F_D^{(Lk)}$ and in view of elementary properties of binomial coefficients, by Theorems 1 to 3 we also get the following

Theorem 5. *Given any* $k \in [1, 2]$ *and* $L \in \{1, 1*, 2, 2*, 3, 3*\}$, *we have*

$$\mathrm{Stab}(2, m, p, a, V) = F^{(Lk)}(m, p, a, V) \text{ for all } V \in \mathbf{Z}$$

and

$$F^{(Lk)}(m, p, a, V) = \sum_{D \in \mathbf{Z}} (-1)^D F_D^{(Lk)}(m, p, a) \left[\begin{matrix} V \\ C(m, p, a) - D \end{matrix} \right]$$

$$\text{for all } V \in \mathbf{N}$$

and

$$F_D^{(Lk)}(m, p, a) \in \mathbf{N} \text{ for all } D \in \mathbf{Z}$$

and

$$F_0^{(Lk)}(m, p, a) \neq 0$$

and

$$F_D^{(Lk)}(m, p, a) = 0 \text{ whenever } D \notin [0, C(m, p, a)],$$

and so in particular $F^{(Lk)}(m, p, a, V)$ *may be regarded as a polynomial of degree* $C(m, p, a)$ *in* V *with coefficients in* $\mathbf{Q}$.

Upon letting

$$\text{stib}(2, m, V) = \text{the set of all standard bitableaux of area } V \\ \text{which are bounded by } m$$

and

$$\text{mon}[[2, m, V]] = \{ t \in \text{mon}(2, m) : \text{abs}(t) = V \}$$

we see that if

$$p = \min(m(1), m(2)) \text{ and } a(1, i) = i = a(2, i) \text{ for } i = 1, 2, \ldots, p$$

then

$$\text{stib}(2, m, V) = \text{stab}(2, m, p, a, V)$$

and

$$\text{mon}[[2, m, V]] = \text{mon}(2, m, p, a, V)$$

and hence upon letting

$$\text{Stib}(2, m, V) = \text{card}(\text{stib}(2, m, V))$$

and

$$\text{Mon}[[2, m, V]] = \text{card}(\text{mon}[[2, m, V]])$$

by Theorem 4 we get the following

Theorem 6. *For all* $V \in \mathbf{N}$ *we have* $\text{Stib}(2, m, V) = \text{Mon}[[2, m, V]]$.

Third Chapter: UNIVERSAL DETERMINANTAL IDENTITY

§10. Preamble.

We shall now formulate a universal identity satisfied by the minors of any matrix. In the general case of this identity, the minors may be of different sizes, and so we call it the mixed size case.

§11. The mixed size case.

To describe the mixed size case, let x be an $m(1)$ by $m(2)$ matrix with entries in any given field. As usual, for all $i \in [1, m(1)]$ and $j \in [1, m(2)]$, the $(i, j)^{\text{th}}$ entry of x is denoted by x_{ij}. With the bivector a we associate the p by p submatrix $\text{sub}(x, a)$ of x obtained by putting

$$\text{sub}(x, a)_{ij} = x_{a(1,i), a(2,j)} \text{ for all } i \text{ and } j \text{ in } [1, p].$$

With a we also associate the ath minor of X obtained by putting

$$\text{mor}(x, a) = \det \text{sub}(x, a).$$

Let there be given any finite sequence of minors of x, and let d be the length of that sequence. This is essentially equivalent to giving a bitableau T of depth d bounded by m; note that now $\text{mor}(x, T[e])_{1 \leq e \leq d}$ is the said sequence of minors of x. With the bitableaux T we associate the Tth *monomial in the minors* of x obtained by putting

$$\text{mom}(x, T) = \prod_{e \in [1, d]} \text{mor}(x, T[e]).$$

Let $\text{sim}(T)[e](1)$ and $\text{sim}(T)[e](2)$ be the row numbers and column numbers of $T[e]$, i.e., of the minor associated with $T[e]$; more precisely, for each $k \in [1, 2]$ let

$$\text{sim}(T)[e](k) = \{ \, T[e](k, i) \, : \, i \in [1, \text{length of } T[e]] \, \};$$

here we are thinking of T as a mapping, and sim is meant as an abbreviation for *semiimage*.

Now let $d = 2$. In other words, let there be given an arbitrary pair of minors of x. More precisely, let there be given a bitableau T of depth 2 bounded by m. For $e = 1, 2$, let us mark some rows and columns $B[e](1)$ and $B[e](2)$ of $T[e]$, i.e., let there be given any

$$B[e](1) \subset \text{sim}(T)[e](1) \text{ and } B[e](2) \subset \text{sim}(T)[e](2).$$

Let $\text{nex}(T, B)$ be the set of all "numerically exchangeable subsets" A of B, i.e., let $\text{nex}(T, B)$ be the set of all quartets

$$A = (A[1](1), A[1](2), A[2](1), A[2](2))$$

such that, for all $e \in [1, 2]$ we have

$$A[e](1) \subset B[e](1) \text{ and } A[e](2) \subset B[e](2)$$

and such that upon exchanging $A[1]$ and $A[2]$ between $T[1]$ and $T[2]$ we get a bitableau $\text{sat}(T, A)$ of depth 2 bounded by m, whereby, for all $e \in [1, 2]$ and $k \in [1, 2]$, upon letting

$$e' = \begin{cases} 2 \text{ if } e = 1 \\ 1 \text{ if } e = 2, \end{cases}$$

we have

$$\text{sim}(\text{sat}(T, A))[e](k) = (\text{sim}(T)[e](k) \setminus A[e](k)) \cup A[e'](k);$$

here sat is meant as an abbreviation of satellite. Also let $\text{pex}(T, B)$ be the set of all "properly exchangeable subsets" of A, i.e., let $\text{pex}(T, B)$ be the set obtained by deleting the empty quartet from $\text{nex}(T, B)$.

Given any $k \in [1, 2]$, let

$$k' = \begin{cases} 2 \text{ if } k = 1 \\ 1 \text{ if } k = 2 \end{cases}$$

and consider the *disjointness condition*

$$B[1](k') \cap B[2](k') = \emptyset$$

and the *cardinality condition*

$$\text{uni}(m, k, T, B) > m(k')$$

where $\text{uni}(m, k, T, B)$ is a nonnegative integer which we call the *numerically flipped unification* of (m, k, T, B) and which we shall define in a moment. Now the mixed size case of the universal determinantal identity asserts the following

Theorem 7. *If the above disjointness condition and cardinality condition are satisfied, then*

$$\text{mom}(x, T) = \sum_{A \in \text{pex}(T, B)} \pm \text{mom}(x, \text{sat}(T, A))$$

with suitable plus minus signs.

Note that this is a homogeneous quadratic equation in the minors of x.

§12. The cardinality condition.

To describe $\text{uni}(m, k, T, B)$, first we introduce the *image of the tabular complementation* $\text{im}(\text{com}(m, k, T)[e])$ of (m, k, T) at e by putting

$$\text{im}(\text{com}(m, k, T)[e]) = [1, m(k')] \setminus \text{sim}(T)[e](k')$$

and then we introduce the *flipping unification* $\text{ufi}(m, k, T, B)[e]$ of (m, k, T, B) *at e* by putting

$$\text{ufi}(m, k, T, B)[e] = B[e'](k') \cap \text{im}(\text{com}(m, k, T)[e])$$

and finally we define $\text{uni}(m, k, T, B)$ by putting

$$\begin{aligned} \text{uni}(m, k, T, B) = \quad & \text{card}(B[1](k)) + \text{card}(B[2](k)) \\ & + \text{card}(\bigcap_{e \in [1,2]}[\text{sim}(T)[e](k) \setminus B[e](k)]) \\ & + \text{card}(\text{ufi}(m, k, T, B)[1]) + \text{card}(\text{ufi}(m, k, T, B)[2]) \\ & + \text{card}(\bigcap_{e \in [1,2]}[\text{im}(\text{com}(m, k, T)[e]) \setminus \text{ufi}(m, k, T, B)[e]]) \end{aligned}$$

§13. The maximal size case.

The mixed size case is proved by reducing it to the maximal size case by which we mean the case of two $m(1)$ by $m(1)$ minors of the $m(1)$ by $m(2)$ matrix x, i.e., the case where: $k = 2$, and length of $T[1] = m(1) =$ the length of $T[2]$, and the disjointness condition is replaced by the stronger condition

$$B[1](1) = \emptyset = B[2](1).$$

§14. The basic case.

The maximal size case is in turn proved by reducing it to the basic case by which we mean the maximal size case when $m(2) = 2m(1)$ and the two minors form the left half and the right half of x.

§15. Laplace development.

The basic case is proved by making the Laplace development of the $2m(1)$ by $2m(1)$ determinant which is obtained by first doubling x and then putting zeroes in the bottom left quarter outside of $B[1](2)$ and in the top right quarter outside of $B[2](2)$.

§16. The full depth case.

By repeated applications of the mixed size case we obtain what may be called the full depth case. To describe the full depth case, let d be any positive integer, and let T be any bitableau of depth d, such that T is bonded by m. Given any other bitableau T' of any depth d', such that T' is bounded by m, let us say that T and T' have the same *content* to mean that for all $k \in [1,2]$ and $j \in \mathbf{N}$ we have

$$\sum_{e \in [1,d]} \operatorname{card}(\{\, i \in [1,m(k)] \,:\, T[e](k,i) = j\,\})$$
$$= \sum_{e \in [1,d']} \operatorname{card}(\{\, i \in [1,m(k)] \,:\, T'[e](k,i) = j\,\}).$$

Note that if T and T' have the same content, then obviously they have the same area. Finally let $\operatorname{sit}(m,d)$ denote the set of all standard bitableaux bounded by m whose depth is at most d. Now the full depth case asserts the following

Theorem 8. *We have*

$$\operatorname{mom}(x,T) = \sum_{S \in \operatorname{sit}(m,d)} \operatorname{fin}(m,T,S)\operatorname{mom}(x,S) \ \text{ with } \ \operatorname{fin}(m,T,S) \in \mathbf{Z}$$

where the numerical function fin *has the property which says that: if* $S \in \operatorname{sit}(m,d)$ *is such that* $\operatorname{fin}(m,T,S) \neq 0$, *then* S *and* T *have the same content and* $S[1] \leq T[1]$.

§17. Deduction of the full depth case.

To indicate how to deduce the full depth case from the mixed size case, we note that, in case T is not standard, there exists a unique $e*$ in $[1, d-1]$ such that

$$T[1] \leq T[2] \leq \ldots \leq T[e* - 1] \leq T[e*] \not\leq T[e* + 1].$$

Let t be the bitableau of depth 2 obtained by taking $t[1] = T[e*]$ and $t[2] = T[e*+1]$. For each $e \in [1,2]$ let $r[e]$ be the length of $t[e]$. Consider the condition

$$(1) \quad \begin{cases} r[1] \geq r[2] \text{ and } t[1](1,i) \leq t[2](1,i) \text{ for all } i \in [1, r[2]] \\ \text{and there exists } v \in [1, r[2]] \text{ such that } t[1](2,v) > t[2](2,v) \\ \text{and } t[1](2,j) \leq t[2](2,j) \text{ for all } j \in [1, v-1]. \end{cases}$$

Now $t[1] \not\leq t[2]$ and hence it follows that if condition (1) is not satisfied then the set

$$\{\, i \in [1, r[2]] \ : \ \text{either } i \leq r[1] \text{ and } t[1](1,i) > t[2](1,i), \text{ or } i > r[1] \,\}$$

is nonempty, and we let v to be the smallest element of the above set. Let $k = 2$ if condition (1) is satisfied, and $k = 1$ if condition (1) is not satisfied. Let us define the quartet

$$B = (B[1](1), B[1](2), B[2](1), B[2](2))$$

by putting

$$B[1](k') = \emptyset$$

and

$$B[2](k') = \text{sim}(t)[2](k') \setminus \text{sim}(t)[1](k')$$

and

$$B[1](k) = \{\, t[1](k,i) \ : \ i \in [v, r[1]] \,\}$$

and

$$B[2](k) = \{\, t[2](k,i) \ : \ i \in [1, v] \,\}.$$

Now it is easy to check that, with t replacing T, the disjointness condition and the cardinality condition are satisfied, and $\text{sat}(t, A)[1] < t[1]$ for all $A \in \text{pex}(t, B)$. By Theorem 7 we get

$$\text{mom}(x, T) = \sum_{A \in \text{pex}(t, B)} \pm \text{mom}(x, s(T, A))$$

where for every $A \in \text{pex}(t, B)$ we have let $s(T, A)$ to be the bitableau of depth d obtained by putting

$$s(T, A) = \begin{cases} \text{sat}(t, A)[e - e* + 1] & \text{if } e \in [e*, e* + 1] \\ T[e] & \text{if } e \in [1, d] \setminus [e*, e* + 1]. \end{cases}$$

Now apply Theorem 7 to all bitableaux $s(T, A)$ which are not standard. And so on. This process must stop after a finite number of steps, and so we get Theorem 8.

§18. The straightening law.

The full depth case, i.e., Theorem 8, subsumes the existence part of the straightening law of Doubilet-Rota-Stein [3]. We may now pose the following problem.

§19. Problem.

Determine the number of steps required to straighten T, i.e., determine the number of times we have to apply Theorem 7 to get Theorem 8. Also, determine the coefficients $\mathrm{fin}(m, T, S)$ occurring in Theorem 8. In particular, for what values of S is $\mathrm{fin}(m, T, S) \neq 0$?

Fourth Chapter: APPLICATIONS TO IDEAL THEORY

§20. Determinantal loci.

Let K be a field. Let $\mathrm{rec}(m(1), m(2))$ be the *positive integral rectangle* of width $m(1)$ and length $m(2)$, i.e., let $\mathrm{rec}(m(1), m(2))$ be the set of all pairs (i, j) with $i \in [1, m(1)]$ and $j \in [1, m(2)]$. Let X be an $m(1)$ by $m(2)$ matrix such that the $m(1)m(2)$ elements X_{ij}, with (i, j) ranging over $\mathrm{rec}(m(1), m(2))$, are independent indeterminates over K. Let $K[X]$ be the ring of polynomials in the $m(1)m(2)$ indeterminates X_{ij} with coefficients in K, and let $K(X)$ be the quotient field of $k[X]$. For every $V \in \mathbf{N}$ let us put

$K[X]_V = $ the set of all homogeneous polynomials of degree V in the $m(1)m(2)$ indeterminates X_{ij}, together with the zero polynomial

and let us note that

$$K[X] = \sum_{V \in \mathbf{N}} K[X]_V$$

where the sum is direct, and let us also note that

$$K[X]_0 = K.$$

Given any subset Y of $\mathrm{rec}(m(1), m(2))$, let $K[Y]$ be the ring of polynomials in the indeterminates X_{ij} with (i, j) ranging over Y and with coefficients in K, and let $K(Y)$ be the quotient field of $K[Y]$, and for every $V \in \mathbf{N}$ let

$$K[Y]_V = K[Y] \cap K[X]_V.$$

Note that now

$$K[X] = K[\mathrm{rec}(m(1), m(2))] \text{ and } K(X) = K(\mathrm{rec}(m(1), m(2)))$$

and, in the usual sense which we shall recall in a moment, $K[X]$ is a homogeneous ring whose V^{th} homogenous component is $K[X]_V$, and $K[Y]$ is homogenous subring of $K[X]$, $K(Y)$ is a subfield of $K(X)$, and the Vth homogeneous component of $K[Y]$ is $K[Y]_V$.

We shall now introduce several *determinantal ideals* in $K[X]$, i.e., ideals in $K[X]$ generated by various size minors of X; the zero sets of determinantal ideals may be called *determinantal loci*; such determinantal ideals are clearly homogeneous ideals in $K[X]$ and hence the corresponding determinantal loci may be viewed as algebraic varieties in the projective space over K of dimension $m(1)m(2) - 1$. We shall also consider determinantal ideals in subrings $K[Y]$ of $K[X]$; we are particularly interested in subrings $K[Y]$ generated by saturated subsets Y, where by a *saturated subset* of $\text{rec}(m(1), m(2))$ we mean a subset Y of $\text{rec}(m(1), m(2))$ having the property which says that: if b is any bivector bounded by m such that

$$(b(1,i), b(2,i)) \in Y \text{ for all } i \text{ in } [1, \text{length of } b]$$

then

$$(b(1,i), b(2,j)) \in Y \text{ for all } i \text{ and } j \text{ in } [1, \text{length of } b].$$

The determinantal ideals which we shall discuss will all turn out to be prime ideals; by calculating their hilbert function, it will turn out that many of them are hilbertian; finally it will also turn out that for many of them, the corresponding varieties are rational; the terms *hilbert function*, *hilbertian ideal*, and *rational variety* will be explained in a moment.

To start with let $\hat{G}[u]$ be the set of all u by u minors of X and let $I[u]$ be the ideal in $K[X]$ generated by $\hat{G}[u]$. More precisely, given any $u \in \mathbf{N}$, firstly we let

$$G[u] = \text{the set of all bivectors of length } u \text{ which are bounded by } m$$

and

$$\hat{G}[u] = \{ \text{mor}(X, b) : b \in G[u] \}$$

and

$$I[u] = \hat{G}[u] K[X]$$

and

$$I[u]_V = I[u] \cap K[X]_V \text{ for all } V \in \mathbf{N}$$

and we note that, in the usual sense which we shall recall in a moment, $I[u]$ is a homogeneous ideal in $K[X]$ and its Vth homogeneous component is $I[u]_V$, and secondly for every subset Y of $\text{rec}(m(1), m(2))$ we let

$$G*[u, Y] = \{ b \in G[u] : (b(1,i), b(2,j)) \in Y \text{ for all } i \text{ and } j \text{ in } [1, u] \}$$

and

$$\hat{G}*[u, Y] = \{ \text{mor}(X, b) : b \in G*[u, Y] \}$$

and

$$I*[u,Y] = \hat{G}*[u,Y]K[Y]$$

and

$$I**[u,Y,X] = \hat{G}*[u,Y]K[X]$$

and

$$I*[u,Y]_V = I*[u,Y] \cap K[Y]_V \text{ for all } V \in \mathbf{N}$$

and

$$I**[u,Y,X]_V = I**[u,Y,X] \cap K[X]_V \text{ for all } V \in \mathbf{N}$$

and we note that $I*[u,Y]$ and $I**[u,Y,X]$ are homogeneous ideals in $K[Y]$ and $K[X]$ respectively and their V^{th} homogenous components are $I*[u,Y]_V$ and $I**[u,Y,X]_V$ respectively.

Now with the given bivector a of length p, we shall associate certain determinantal ideals generated by different size minors of X. In greater detail, firstly for every $k \in [1,2]$ and $u \in [1,p]$ we let

$$G(p,a,k,u) = \{ b \in G[u] : b(k,u) < a(k,u) \}$$

and

$$\hat{G}(p,a,k,u) = \{ \text{mor}(X,b) : b \in G(p,a,k,u)\}$$

·and

$$I(p,a,k,u) = \hat{G}(p,a,k,u)K[X]$$

and

$$I(p,a,k,u)_V = I(p,a,k,u) \cap K[X]_V \text{ for all } V \in \mathbf{N}$$

and we note that $I(p,a,k,u)$ is a homogeneous ideal in $K[X]$ and its Vth homogeneous component is $I(p,a,k,u)_V$, and secondly we let

$$G(p,a) = G[p+1] \cup [\bigcup_{k\in[1,2]} \bigcup_{u\in[1,p]} G(p,a,k,u)]$$

and

$$\hat{G}(p,a) = \{ \text{mor}(X,b) : b \in G(p,a) \}$$

and

$$I(p,a) = \hat{G}(p,a)K[X].$$

and

$$I(p,a)_V = I(p,a) \cap K[X]_V \text{ for all } V \in \mathbf{N}$$

and we note that $I(p,a)$ is a homogeneous ideal in $K[X]$ and its Vth homogeneous component is $I(p,a)_V$, and thirdly for every $k \in [1,2]$ and $u \in [1,p]$ and for every subset Y of $\text{rec}(m(1),m(2))$ we let

$$G*(p,a,k,u,Y) = \{ b \in G(p,a,k,u) : \quad (b(1,i),b(2,j)) \in Y$$
$$\text{for all } i \text{ and } j \text{ in } [1,u] \}$$

and

$$\hat{G}*(p, a, k, u, Y) = \{ \, \mathrm{mor}(X, b) \; : \; b \in G*(p, a, k, u, Y) \, \}$$

and

$$I*(p, a, k, u, Y) = \hat{G}*(p, a, k, u, Y)K[Y]$$

and

$$I**(p, a, k, u, Y, X) = \hat{G}*(p, a, k, u, Y)K[X]$$

and

$$I*(p, a, k, u, Y)_V = I*(p, a, k, u, Y) \cap K[Y]_V \text{ for all } V \in \mathbf{N}$$

and

$$I**(p, a, k, u, Y, X)_V = I**(p, a, k, u, Y, X) \cap K[X]_V \text{ for all } V \in \mathbf{N}$$

and we note that $I*(p, a, k, u, Y)$ and $I**(p, a, k, u, Y, X)$ are homogeneous ideals in $K[Y]$ and $K[X]$ respectively and their V^{th} homogeneous components are $I*(p, a, k, u, Y)_V$ and $I**(p, a, k, u, Y, X)_V$ respectively, and fourthly for every subset Y of $\mathrm{rec}(m(1), m(2))$ we let

$$G*(p, a, Y) = \{ \, b \in G(p, a) \; : \; \begin{array}{l} (b(1, i), b(2, j)) \in Y \\ \text{for all } i \text{ and } j \text{ in } [1, \text{length of } b] \end{array} \}$$

and

$$\hat{G}*(p, a, Y) = \{ \, \mathrm{mor}(X, b) \; : \; b \in G*(p, a, Y) \, \}$$

and

$$I*(p, a, Y) = \hat{G}*(p, a, Y)K[Y]$$

and

$$I**(p, a, Y, X) = \hat{G}*(p, a, Y)K[X]$$

and

$$I*(p, a, Y)_V = I*(p, a, Y) \cap K[Y]_V \text{ for all } V \in \mathbf{N}$$

and

$$I**(p, a, Y, X)_V = I**(p, a, Y, X) \cap K[Y]_V \text{ for all } V \in \mathbf{N}$$

and we note that $I*(p, a, Y)$ and $I**(p, a, Y, X)$ are homogeneous ideals in $K[Y]$ and $K[X]$ respectively and their V^{th} homogeneous components are $I*(p, a, Y)_V$ and $I**(p, a, Y, X)_V$ respectively.

Before proceeding further, let us recall the relevant definitions concerning

§21. Vector spaces and homogeneous rings.

Given any map $w : M \to H$, where M and H are any sets, and given any subset $M*$ of M, by $w|M*$ we shall denote the restriction of w to $M*$, i.e., by $w|M*$ we shall denote the map $M* \to H$ induced by w.

Given any set M, by a map $v : M \to K$ with *finite support* we mean a map $v : M \to K$ such that

$$\text{card}(\{ s \in M \ : \ v(s) \neq 0 \}) < \infty.$$

Given any map $w : M \to H$, where M is a set and H is a K-vector-space, we say that w is K-*independent* to mean that w has the property which says that if $v : M \to K$ is any map with finite support such that

$$\sum_{s \in M} v(s)w(s) = 0$$

then $v(s) = 0$ for all $s \in M$. Given any map $w : M \to H$, where M is a set and H is a K-vector-space, and given any K-vector-subspace $H*$ of H, we say that w is a K-*generator* of $H*$ to mean that $w(s) \in H*$ for all $s \in M$, and for every $x \in H*$ there exists a map $x* : M \to K$ with finite support such that

$$x = \sum_{s \in M} x*(s)w(s).$$

Given any map $w : M \to H$, where M is a set and H is a K-vector-space, and given any K-vector-subspace $H*$ of H, we say that w is a K-*basis* of $H*$ to mean that w is K-independent and w is a K-generator of $H*$. Given any K-vector-space H, by $[H : K]$ we denote the cardinality of any set M for which there exists a map $w : M \to H$ such that w is a K-basis of H; (it is well known that the said cardinality is independent of M). Given any map $w : M \to H$, where M is a set and H is a K-vector-space, and given any K-vector-subspace $H*$ of H, and given any K-vector-subspace H' of $H*$, we say that w is K-*basis* of $H*$ *modulo* H' to mean that the composition of w followed by the canonical map $H \to H/H'$ is a K-basis of the K-vector-space which is the image of $H*/H'$ under the natural injective map of $H*/H'$ into H/H'.

Let us now recall the terminology concerning homogeneous rings. By a *homogeneous ring* we mean a ring H together with a family $(H_V)_{V \in \mathbf{N}}$ of additive subgroups of H such that the underlying additive group of H is the direct sum of the said family, and such that for all V and $V*$ in $\mathbf{N}$ and for all $x \in H_V$ and $x* \in H_{V*}$ we have $xx* \in H_{V+V*}$, and such that H_0 is a field, and such that $H = H_0[H_1]$, and finally such that $0 < [H_1 : H_0] < \infty$. Here H_V is called the Vth *homogeneous component* of H. By a *homogeneous subring* of H we mean a homogeneous ring $\tilde{H}$ such that $\tilde{H}$ is a subring of H, and $\tilde{H}_0 = H_0$, and for every $V \in \mathbf{N}$ we have $\tilde{H}_V = \tilde{H} \cap H_V$. By a *homogeneous ideal* in H we mean an ideal $\hat{H}$ in H such that $\hat{H}$ is generated by

$$\bigcup_{V \in \mathbf{N}} (\hat{H} \cap H_V);$$

we put $\hat{H}_V = \hat{H} \cap H_V$ and we call $\hat{H}_V$ the Vth *homogeneous component* of $\hat{H}$.

Finally recall that, given any homogeneous ideal $\hat{H}$ in the polynomial ring H in a finite number of indeterminates over K, the map $\mathbf{N} \to \mathbf{N}$ which sends V to $[H_V/\hat{H}_V : K]$ is called the *hilbert function* of $\hat{H}$ in H, and by a theorem of Hilbert there exists a unique polynomial in $h(V)$ in an indeterminate V with rational coefficients such that for all large enough nonnegative integers V we have $h(V) = [H_V/\hat{H}_V : K]$; note that $h(V)$ is called the *hilbert polynomial* of $\hat{H}$ in H; this motivates the definition according to which we say that $\hat{H}$ is *hilbertian* to mean that *for all nonnegative integers V we have* $h(V) = [H_V/\hat{H}_V : K]$.

§22. Standard basis.

Let us now return to the ring $K[X]$. Let $\text{tab}(2, m)$ be the set of all bitableaux bounded by m. Let $q : \text{tab}(2, m) \to K[X]$ be defined by putting $q(T) = \text{mom}(X, T)$ for all $T \in \text{tab}(2, m)$. Note that now a is a bivector, and q is no more the width of a. Let $\text{stab}(2, m)$ be the set of all standard bitableaux bounded by m. Let $q* : \text{mon}(2, m) \to K[X]$ be the map obtained by putting $q*(t) = X^t$ for all $t \in \text{mon}(2, m)$ where

$$X^t = \prod_{y \in \text{cub}(2, m)} X_{y(1), y(2)}^{t(y)}.$$

From Theorems 6 and 8 we get a proof of the following Theorem 9 which, among other things, says that $q|\text{stab}(2, m)$ is K-basis of $K[X]$; we may call this the *standard basis* of $K[X]$; from Theorems 4 and 8 we can deduce the following Theorem 10 which, among other things, says that this basis partitions well for various ideals in $K[X]$; Theorem 10 also says that $q*$ provides bases modulo the said ideals. The fact that $q|\text{stab}(2, m)$ is a K-basis of $K[X]$ is equivalent to the straightening law of Doubilet-Rota-Stein [3].

Theorem 9. *We have that $q|\text{stab}(2, m)$ is a K-basis of $K[X]$, and for every $V \in \mathbf{N}$ we have that $q|\text{stib}(2, m, V)$ is a K-basis of $K[X]_V$. In particular, the expansion given in Theorem 8 is unique, i.e., if $T' \in \text{tab}(2, m)$ is such that $\text{mom}(X, T') = \text{mom}(X, T)$ then for all $S \in \text{stab}(2, m)$ we have $\text{fin}(m, T', S) = \text{fin}(m, T, S)$ where we put $\text{fin}(m, T, S) = 0$ whenever the depth of S is greater than d. It follows that if $\text{fin}(m, T, S) \neq 0$ then $S[1] \leq T[e]$ for all $e \in [1, d]$.*

Theorem 10. (1) *We have that $q|[\text{stab}(2, m) \setminus \text{stab}(2, m, p, a)]$ is a K-basis of $I(p, a)$, whereas $q|\text{stab}(2, m, p, a)$ is a K-basis of $K[X]$ modulo $I(p, a)$.*

(2) *For every $V \in \mathbf{N}$ we have that $q|[\mathrm{stib}(2, m, V) \setminus \mathrm{stab}(2, m, p, a, V)]$ is a K-basis of $I(p, a)_V$, whereas $q|\mathrm{stab}(2, m, p, a, V)$ is a K-basis of $K[X]_V$ modulo $I(p, a)_V$.*

(3) *For every $V \in \mathbf{N}$ we have*

$$\mathrm{Stab}(2, m, p, a, V) = [K[X]_V / I(p, a)_V : K] = \mathrm{Mon}(2, m, p, a, V).$$

(4) *We have that $q*|\mathrm{mon}(2, m, p, a)$ is a K-basis of $K[X]$ modulo $I(p, a)$.*

(5) *For every $V \in \mathbf{N}$ we have that $q*|\mathrm{mon}(2, m, p, a, V)$ is a K-basis of $K[X]_V$ modulo $I(p, a)_V$.*

§23. Second fundamental theorem of invariant theory.

From Theorems 1, 2 and 8 we can get a new proof of the following Theorem 11 which was originally proved by E. Pascal [4] in 1888 as part of the second fundamental theorem of invariant theory.

Theorem 11. *For every $u \in \mathbf{N}$ we have that $I[u + 1]$ is a prime ideal in $K[X]$.*

§24. Generalized second fundamental theorem of invariant theory.

From Theorems 4 to 10 we can deduce the following Theorem 12 which may be called a generalized second fundamental theorem of invariant theory.

Theorem 12. (1) *There exists a subset $y(p, a, p)$ of $\mathrm{rec}(m(1), m(2))$ together with a ring homomorphism $f(p, a) : K[X] \to K(y(p, a, p))$ such that $\mathrm{card}(y(p, a, p)) = 1 + C(m, p, a)$, and $f(p, a)$ is identity on $K[y(p, a, p)]$, and $I(p, a) = \mathrm{Ker}(f(p, a))$, and such that for every saturated subset Y of $\mathrm{rec}(m(1), m(2))$ we have $I*(p, a, Y) = \mathrm{Ker}(f(p, a)|K[Y])$; (here, as usual, Ker denotes kernel).*

(2) *$I(p, a)$ is the sum of the $2p + 1$ ideals*

$$I(p, a, 1, 1), I(p, a, 2, 1), \ldots, I(p, a, 1, p), I(p, a, 2, p), I[p + 1]$$

each of which is a homogeneous prime ideal in $K[X]$.

(3) *$I(p, a)$ is a homogeneous prime ideal in $K[X]$, and the quotient field of the residue class ring $K[X]/I(p, a)$ is a pure transcendental extension of (the image of) K whose transcendence degree equals $1 + C(m, p, a)$.*

(4) *For all $V \in \mathbf{N}$ and $L \in \{1, 1*, 2, 2*, 3, 3*\}$ and $k \in [1, 2]$ we have*

$$[K[X]_V / I(p, a)_V : K] = F^{(Lk)}(m, p, a, V)$$

with $F^{(Lk)}(m, p, a, V)$ as in Theorem 5, and hence $I(p, a)$ is hilbertian and its hilbert function, as well as its hilbert polynomial, in $K[X]$ is $F^{(Lk)}(m, p, a, V)$.

(5) *For every saturated subset Y of* rec$(m(1), m(2))$, *we have that* $I*(p, a, Y)$ *is the sum of the* $2p + 1$ *ideals*

$$\begin{cases} I*(p, a, 1, 1, Y), I*(p, a, 2, 1, Y), \dots, \\ I*(p, a, 1, p, Y), I*(p, a, 2, p, Y), I*[p + 1, Y] \end{cases}$$

each of which is a homogeneous prime ideal in $K[Y]$.

(6) *For every saturated subset Y of* rec$(m(1), m(2))$, *we have that* $I**(p, a, Y, X)$ *is the sum of the* $2p + 1$ *ideals*

$$\begin{cases} I**(p, a, 1, 1, Y, X), & I**(p, a, 2, 1, Y, X), & \dots, \\ I**(p, a, 1, p, Y, X), & I**(p, a, 2, p, Y, X), & I**[p + 1, Y, X] \end{cases}$$

each of which is a homogeneous prime ideal in $K[X]$.

(7) *For every saturated subset Y of* rec$(m(1), m(2))$, *we have that* $I*(p, a, Y)$ *and* $I**(p, a, Y, X)$ *are homogeneous prime ideals in* $K[Y]$ *and* $K[X]$ *respectively.*

Remark 1. Given any $L \in \{1, 1*, 2, 2*, 3, 3*\}$ and $k \in [1, 2]$, in view of Theorem 5 we know that $F^{(Lk)}(m, p, a, V)$ is a polynomial of degree $C = C(m, p, a)$ in V with rational coefficients, and $C!$ times the coefficients of V^C in $F^{(Lk)}(m, p, a, V)$ equals the positive integer $F_0^{(Lk)}(m, p, a)$; now parts (3) and (4) of Theorem 12 give a reconfirmation of these assertions because of the theorem of Hilbert which says that, if $\hat{H}$ is a homogeneous prime ideal in a polynomial ring in a finite number, say $e + 1$, of indeterminates over K, and if $d + 1$ is the transcendence degree of the quotient field of $H/\hat{H}$ over (the image of) K, and if $h(V)$ is the hilbert polynomial of $\hat{H}$ in H, then d is the degree of $h(V)$ in V and $d!$ times the coefficient of V^d in $h(V)$ is a positive integer which equals the order of the variety defined by $\hat{H}$ in the e dimensional projective space over K, where we recall that the said order equals the number of points in which the said variety is met by a complementary dimensional (which means $e - d$ dimensional) linear space in the said projective space. Note that in our case of $\hat{H} = I(p, a)$, the quotient field of $H/\hat{H}$ is a pure transcendental extension of K, and so the said variety is rational, i.e., it is birationally equivalent to the d dimensional projective space over K.

Remark 2. Given any $u \in [1, \min(m(1), m(2))]$, let $E[u]$ be the bivector of length u obtained by putting $E[u](1, U) = U = E[u](2, U)$ for all $U \in [1, u]$. Now clearly $I[u+1] = I(u, E[u])$ and for every saturated subset Y of rec$(m(1), m(2))$ we have $I*[p + 1, Y] = I*(p, E[p], Y)$ and $I**[p + 1, Y, X] = I**(p, E[p], Y, X)$. Therefore the relevant parts of Theorem 12 and Remark 1 are applicable to the ideals $I[u + 1]$, $I*[p + 1, Y]$ and $I**[p + 1, Y, X]$. In particular, Theorem 11 becomes a special case of part (3) of Theorem 12.

Remark 3. Given any saturated subset Y of rec$(m(1), m(2))$, in view of Theorem 12 we may ask whether the ideals $I*(p, a, Y)$ and $I**(p, a, Y, X)$ are hilbertian.

Remark 4. I was led to conjecture part (7) of Theorem 12 in studying the problem of finding the singularities of Schubert varieties of flag manifolds.

References

[1] S. S. Abhyankar : Combinatoire des tableaux de Young, varietes determi-nantielles et calcul de fonctions de Hilbert, Rend. Sem. Mat. Univ. Torino, **42**(1984), 65–88.

[2] S. S. Abhyankar : Enumerative Combinatorics of Young Tableaux, Marcel Dekker, New York, forthcoming.

[3] P. Doubilet - G. C. Rota - J. Stein : Foundations of combinatrics IX: Foundations of combinatorial methods in invariant theory, Stud. Appl. Math., **53**(1974), 185–216.

[4] E. Pascal : Mem. del R. Acc. dei Lincei, Series V, **4a**(1888).

[5] A. Young : On quantitative substitutional analysis I, Proc. London Math. Soc., **33**(1901) 97–146.

Shreeram S. ABHYANKAR

Department of Mathematics
Purdue University
West Lafayette, IN 47907
U.S.A.

Poona University
Pune 411007
India

Bhaskaracharya Pratishthana
Pune 411004
India

Algebraic Geometry and Commutative Algebra
in Honor of Masayoshi NAGATA
pp. 27–34 (1987)

A Conjecture of Sharp —
The Case of Local Rings with dim nonCM ≤ 1 or
dim ≤ 5

Yoichi AOYAMA and Shiro GOTO*

§1. Introduction.

We continue to discuss a conjecture of Sharp on the existence of dualizing complexes from [3]. In the previous paper [3] we have shown that the conjecture is affirmative for local rings in the following cases: (A denotes a local ring and K_A is the canonical module of A.) (1) A is (FLC) ([3, Theorem 2.1]); (2) $\dim A \leq 4$ ([3, Theorem 3.2]); (3) A is (S_{d-2}) ($d = \dim A$), depth $A \geq d-1$ and depth $K_A \geq 3$ ([3, Theorem 4.11]).

In this paper we will prove that the conjecture is affirmative for local rings with dim nonCM ≤ 1 or dim ≤ 5, namely,

Theorem 1.1. *Let A be a local ring. If A has a dualizing complex and $\dim \mathrm{nonCM}(A) \leq 1$, then A is a homomorphic image of a Gorenstein ring.*

Theorem 1.2. *Let A be a local ring. If A has a dualizing complex and $\dim A \leq 5$, then A is a homomorphic image of a Gorenstein ring.*

In order to prove Theorem 1.1, we make use of the Cohen-Macaulayfication due to Faltings ([7]) and the theory of (FLC) local rings ([5], [8] and [12], cf. [3, §1]). If we had Theorem 1.1, Theorem 1.2 can be proved by a similar method to one given in [3, §3].

Throughout this paper a *ring* means a commutative noetherian ring with unit.

§2. Sharp's Conjecture.

Let A be a ring. For a finitely generated A-module M of finite dimension, we put

$$\mathrm{Assh}_A(M) = \{\mathfrak{p} \in \mathrm{Ass}_A(M) \mid \dim A/\mathfrak{p} = \dim M\}.$$

*The authors were patially supported by Grant-in-Aid for Co-operative Reserch.
Received January 30, 1987.

Let $\mathfrak{a}$ be an ideal of A and N an A-module. $E_A(N)$ denotes the injective envelope of N and $H_\mathfrak{a}^i(N)$ is the i-th local cohomology module of N with respect to $\mathfrak{a}$. We denote by $R(A, \mathfrak{a})$ the *Rees algebra* of A with respect to $\mathfrak{a}$, i.e.,

$$R(A, \mathfrak{a}) = \bigoplus_{n \geq 0} \mathfrak{a}^n \cong A[\mathfrak{a}T] \subseteq A[T]$$

with an indeterminate T. We put

$$V(\mathfrak{a}) = \{\mathfrak{p} \in \operatorname{Spec}(A) \mid \mathfrak{p} \supseteq \mathfrak{a}\},$$

$$\operatorname{CM}(A) = \{\mathfrak{p} \in \operatorname{Spec}(A) \mid A_\mathfrak{p} \text{ is Cohen-Macaulay}\}$$

and

$$\operatorname{nonCM}(A) = \operatorname{Spec}(A) \setminus \operatorname{CM}(A).$$

We denote by $\widehat{}$ the *maximal-ideal-adic completion* over a local ring.

Definition 2.1. *Let t be an integer. A finitely generated module M over a ring A is said to be (S_t) if*

$$\operatorname{depth} M_\mathfrak{p} \geq \min\{t, \dim M_\mathfrak{p}\}$$

for every $\mathfrak{p} \in \operatorname{Supp}_A(M)$.

Definition 2.2. *Let A be a local ring with the maximal ideal $\mathfrak{m}$. A finitely generated A-module M is said to be (FLC) if $H_\mathfrak{m}^i(M)$ is finitely generated for $i \neq \dim M$.*

Definition 2.3. ([11, 5.6]). *Let A be a d-dimensional local ring with the maximal ideal $\mathfrak{m}$. An A-module K is called the canonical module of A if*

$$K \otimes_A \hat{A} \cong \operatorname{Hom}_A(H_\mathfrak{m}^d(A), E_A(A/\mathfrak{m})).$$

The canonical module of A is usually denoted by K_A if it exists.

Definition 2.4. ([13, 2.4], cf. [10, p.258]). *A complex $I^\bullet$ over a ring A is called a dualizing complex of A if it satisfies the following four conditions:*
(D1) *$I^\bullet$ is bounded.*
(D2) *$H^i(I^\bullet)$ is finitely generated for every i.*
(D3) *Each I^i is an injective A-module.*
(D4) *Whenever $X^\bullet$ is a complex over A satisfying (D1) and (D2) for $X^\bullet$, the map*

$$\theta(X^\bullet, I^\bullet) \; : \; X^\bullet \longrightarrow \operatorname{Hom}_A([\operatorname{Hom}_A(X^\bullet, I^\bullet)], I^\bullet)$$

defined in [13, §2] (cf. [3, §1]) is a quasi-isomorphism, i.e., θ induces isomorphisms on their cohomology modules.

If a ring A has a dualizing complex, then A has a dualizing complex $I^\bullet$ such that

$$\bigoplus_{i \in \mathbf{Z}} I^i \cong \bigoplus_{\mathfrak{p} \in \operatorname{Spec}(A)} E_A(A/\mathfrak{p}),$$

which we call a *fundamental dualizing complex* ([9]). If a local ring A has a dualizing complex $I^\bullet$, then A has the canonical module and

$$K_A \cong H^s(I^\bullet)$$

where

$$s = \inf\{i \in \mathbf{Z} \mid H^i(I^\bullet) \neq 0\}$$

(cf. [4,2.24 and 13]). A ring which is a homomorphic image of a finite-dimensional Gorenstein ring has a dualizing complex (cf. [13,3.7 and 9] and [10, V.2.4]), and it is not known whether there is a ring with dualizing complexes which is not a homomorphic image of a Gorenstein ring. Sharp showed that a Cohen-Macaulay ring with dualizing complexes is a homomorphic image of a finite-dimensional Gorenstein ring ([14, 4.3]), and he posed the following conjecture ([14, 4.4]).

(SC) **Sharp's Conjecture:** If a ring A has a dualizing complex, then A is a homomorphic image of a finite-dimensional Gorenstein ring.

We refer the reader to [4] for a summary of the elements of the theory of dualizing complexes.

§3. Proofs of Theorem 1.1 and Theorem 1.2.

Let A be a $d+1$-dimensional semi-local ring with the maximal ideals $\mathfrak{m}_1, \ldots,$ $\mathfrak{m}_s$ and $\mathfrak{m} = \mathfrak{m}_1 \cdots \mathfrak{m}_s$ $(d \geq 1)$. We assume that the following two conditions are satisfied:

(a) A has a fundamental dualizing complex

$$D^\bullet : 0 \longrightarrow D^0 \longrightarrow \cdots \longrightarrow D^{d+1} \longrightarrow 0$$

such that

$$D^0 \cong \bigoplus_{\mathfrak{p} \in \operatorname{Ass}(A)} E_A(A/\mathfrak{p})$$

and

$$D^{d+1} \cong \bigoplus_{i=1}^{s} E_A(A/\mathfrak{m}_i).$$

(b) $\dim \operatorname{nonCM}(A) \leq 1$.

We first note that the length of every maximal chain of prime ideals in A is equal to $d+1$ by the assumption (a) ([14, 2.7 or 3.1]). Let $\mathfrak{a}$ be an ideal such that $\mathrm{nonCM}(A) = V(\mathfrak{a})$. (Note that $\mathrm{nonCM}(A)$ is a closed set (cf. [14, 2.10]).) We have height $\mathfrak{a} \geq d$ by the assumption (b). Take elements $x_1, \ldots, x_d$ from $\mathfrak{a} \cap \mathfrak{m}$ such that $\mathrm{height}(x_1, \ldots, x_d)A = d$, and put $I = (x_1, \ldots, x_d)A$. For a minimal prime ideal $\mathfrak{p}$ of I, $A_\mathfrak{p}$ is (FLC) by [3, 1.17]. Taking powers of $x_1, \ldots, x_d$ if necessary, we may assume that, for every minimal prime ideal $\mathfrak{p}$ of I, all of $x_1, \ldots, x_d (\in A_\mathfrak{p})$ are contained in a $\mathfrak{p}A_\mathfrak{p}$-primary ideal $\mathfrak{b}$ such that, for every system of parameters $a_1, \ldots, a_d$ for $A_\mathfrak{p}$ contained in $\mathfrak{b}$,

$$(a_1, \ldots, a_i)A_\mathfrak{p} : a_{i+1} = (a_1, \ldots, a_i)A_\mathfrak{p} : \mathfrak{b}$$

holds for $0 \leq i < d$ (cf. [3, 1.18]). For every prime ideal $\mathfrak{p} \not\supseteq I$, we have

$$H^i(D^\bullet)_\mathfrak{p} = 0$$

for $i > 0$ as $A_\mathfrak{p}$ is Cohen-Macaulay. Hence there is a positive integer t such that $I^t H_\mathfrak{m}^i(A) = 0$ for $i \leq d$. Let $\mathfrak{p}$ and $\mathfrak{q}$ be prime ideals such that $\mathfrak{p} \not\supseteq I$ and $\mathfrak{q} \supseteq I + \mathfrak{p}$. Since $A_\mathfrak{p}$ is Cohen-Macaulay and height $\mathfrak{q} \geq d$,

$$\mathrm{depth}\, A_\mathfrak{p} + \dim A_\mathfrak{q}/\mathfrak{p}A_\mathfrak{q} = \dim A_\mathfrak{p} + \dim A_\mathfrak{q}/\mathfrak{p}A_\mathfrak{q} = \dim A_\mathfrak{q} \geq d.$$

Hence there is a positive integer u such that $I^u H_I^i(A) = 0$ for $i < d$ by virtue of [6]. Taking powers of $x_1, \ldots, x_d$ again if necessary, we may assume that all of $x_1, \ldots, x_d$ are contained in

$$\prod_{i \leq d}(\mathrm{Ann}(H_\mathfrak{m}^i(A)))^{2^{d-1}} \quad \text{and} \quad \prod_{i < d}(\mathrm{Ann}(H_I^i(A)))^{2^{d-1}}$$

(Note that $H_{\underline{x}}^i(A) = H_{\underline{x}^n}^i(A)$ for $n \geq 1$ where $\underline{x}^n = (x_1^n, \ldots, x_d^n)$.) Put $C = R(A, I^{d-1})$. For a non-maximal prime ideal $\mathfrak{p}$ of A, $C_\mathfrak{p}$ is Cohen-Macaulay. In fact, if $\mathfrak{p} \not\supseteq I$ then $C_\mathfrak{p} \cong A_\mathfrak{p}[T]$ is Cohen-Macaulay, and if $\mathfrak{p} \supseteq I$ then $C_\mathfrak{p} \cong R(A_\mathfrak{p}, I_\mathfrak{p}^{d-1})$ is Cohen-Macaulay by [3, 1.19]. We put

$$C_i = C\left[\frac{1}{x_i^{d-1}T}\right]_0 \cong A\left[\frac{x}{x_i} \mid x \in I\right]$$

for $i = 1, \ldots, d$. Let $\mathfrak{p}$ and $\mathfrak{q}$ be prime ideals of C_i such that $\mathfrak{p} \not\supseteq \mathfrak{m}C_i$ and $\mathfrak{q} \supseteq \mathfrak{m}C_i + \mathfrak{p}$. If $\mathfrak{p}$ is a minimal prime ideal, we have $\dim C_{i\mathfrak{q}}/\mathfrak{p}C_{i\mathfrak{q}} \geq 2$ as $\dim C_i/\mathfrak{m}_iC_i \leq d - 1$ and $\dim C_i/\mathfrak{p} = d + 1$. If $\mathfrak{p}$ is not minimal, we put $\mathfrak{n} = \mathfrak{p} \cap A$. Then $C_\mathfrak{n}$ is Cohen-Macaulay as $\mathfrak{n}$ is not a maximal ideal, whence $C_{i\mathfrak{p}}$ is Cohen-Macaulay. Therefore $\mathrm{depth}\, C_{i\mathfrak{p}} + \dim C_{i\mathfrak{q}}/\mathfrak{p}C_{i\mathfrak{q}} \geq 1 + 1 = 2$. Hence there is an integer $r \geq d$ such that $\mathfrak{m}^r H_\mathfrak{m}^1(C_i) = 0$ for $i = 1, \ldots, d$ by virtue of [6]. Take an element y from $\mathfrak{m}^r$ such that $\mathrm{height}(I + yA) = d + 1$, and put

$$J = I^r(I^r + yA).$$

Let $R = R(A, J)$, $\mathfrak{N}_i = \mathfrak{m}_i R + R_+$ for $i = 1, \ldots, s$, and $\mathfrak{N} = \mathfrak{N}_1 \cap \cdots \cap \mathfrak{N}_s = \mathfrak{N}_1 \cdots \mathfrak{N}_s = \mathfrak{m} R + R_+$. Note that R has a dualizing complex (as R is a finitely generated A-algebra).

Theorem 3.1. *Under the situation above, $H_{\mathfrak{N}}^i(R)$ is finitely generated for $i \neq d + 2$.*

Proof. It is sufficient to show that $R_{\mathfrak{P}}$ is Cohen-Macaulay for every homogeneous prime ideal $\mathfrak{P} \neq \mathfrak{N}_1, \ldots, \mathfrak{N}_s$ (cf. [3, 1.17] and [15, §1]). We put $\mathfrak{p} = \mathfrak{P} \cap A$. First suppose $\mathfrak{p}$ is not a maximal ideal. If $\mathfrak{p} \not\supseteq I$, then $R_{\mathfrak{p}} \cong A_{\mathfrak{p}}[T]$ is Cohen-Macaulay. If $\mathfrak{p} \supseteq I$, then $R_{\mathfrak{p}} \cong R(A_{\mathfrak{p}}, I_{\mathfrak{p}}^r)$ is Cohen-Macaulay by [3, 1.19]. Hence $R_{\mathfrak{P}}$ is Cohen-Macaulay. Now suppose $\mathfrak{p}$ is a maximal ideal. Considering $A_{\mathfrak{p}}$ and $R_{\mathfrak{p}}$ instead of A and R respectively, we may assume that A is a local ring. Since $J^r = (x_1^{2r}, \ldots, x_d^{2r}, yx_1^r, \ldots, yx_d^r) J^{r-1}$ and $\mathfrak{P} \not\supseteq R_+$, $\mathfrak{P} \not\ni x_i^{2r} T$ or $\mathfrak{P} \not\ni yx_i^r T$ for some i. Without loss of generality we may assume $\mathfrak{P} \not\ni x_1^{2r} T$ or $\mathfrak{P} \not\ni yx_1^r T$. Suppose $\mathfrak{P} \not\ni x_1^{2r} T$. We put $U = x_1^{2r} T$, $S = R[1/U]$, $B = S_0$ and $\mathfrak{Q} = \mathfrak{P} S \cap B$ ($\supseteq \mathfrak{m} B$). Since $S = B[U, 1/U]$ and U is algebraically independent over B, $S_{\mathfrak{P} S}$ is Cohen-Macaulay if and only if so is $B_{\mathfrak{Q}}$. The Cohen-Macaulayness of $B_{\mathfrak{Q}}$ follows from [7, Satz 3] because the elements $x_1, \ldots, x_d, y$ satisfy the conditions of [7, Satz 3] and $B = S_0 \cong A[x/x_1^{2r} \mid x \in J] = A[x_2/x_1, \ldots, x_d/x_1, y/x_1^r]$ is the coordinate ring of an affine chart of Y, the Cohen-Macaulay scheme given in [7, Satz 3] (cf. [7, Bemerkung b) S.190]). In the case where $\mathfrak{P} \not\ni yx_1^r T$, the proof is the same as the above (with $B \cong A[x_2/x_1, \ldots, x_d/x_1, x_1^r/y]$).

By Theorem 3.1, [3, 1.20] and [14, 4.3], we have the validity of Theorem 1.1 not only for a local ring A with $\mathrm{Ass}(A) = \mathrm{Assh}(A)$ but also for a semi-local ring which satisfies the condition (a).

Proof of Theorem 1.1. We proceed by induction on $d = \dim A$. Let $A \supset (0) = \mathfrak{q}_1 \cap \cdots \cap \mathfrak{q}_t$ be a primary decomposition of the zero ideal such that $\dim A/\mathfrak{q}_i = d$ if and only if $i \leq s$ $(1 \leq s \leq t)$. If $s = t$, the proof is already done. Let $s < t$. Put $\mathfrak{a} = \mathfrak{q}_1 \cap \cdots \cap \mathfrak{q}_s$ and $\mathfrak{b} = \mathfrak{q}_{s+1} \cap \cdots \cap \mathfrak{q}_t$. Let $\mathfrak{p}$ be a prime ideal with $\dim A/\mathfrak{p} > 1$. Then $A_{\mathfrak{p}}$ is Cohen-Macaulay, especially $\mathrm{Ass}(A_{\mathfrak{p}}) = \mathrm{Assh}(A_{\mathfrak{p}})$. Hence we have $\mathfrak{p} \not\supseteq \mathfrak{a} + \mathfrak{b}$ and $\dim A/(\mathfrak{a} + \mathfrak{b}) \leq 1$. If $\mathfrak{p} \supseteq \mathfrak{a}$, $(A/\mathfrak{a})_{\mathfrak{p}} \cong A_{\mathfrak{p}}$ is Cohen-Macaulay. If $\mathfrak{p} \supseteq \mathfrak{b}$, $(A/\mathfrak{b})_{\mathfrak{p}} \cong A_{\mathfrak{p}}$ is Cohen-Macaulay. Therefore $\dim \mathrm{nonCM}(A/\mathfrak{a}) \leq 1$ and $\dim \mathrm{nonCM}(A/\mathfrak{b}) \leq 1$. Since $\mathrm{Ass}(A/\mathfrak{a}) = \mathrm{Assh}(A/\mathfrak{a})$, $A/\mathfrak{a}$ is a homomorphic image of a Gorenstein local ring R. Since $\dim A/\mathfrak{b} < d$, $A/\mathfrak{b}$ is a homomorphic image of a Gorenstein local ring S by the induction hypothesis. We may assume $\dim R = \dim S = d$. Let f be the surjective homomorphism from $R \oplus S$ to $A/\mathfrak{a} \oplus A/\mathfrak{b}$, and put $B = f^{-1}(A)$. We have a commutative diagram of B-modules with exact rows

$$
\begin{array}{ccccccccc}
0 & \longrightarrow & B & \longrightarrow & R \oplus S & \longrightarrow & R \oplus S/B & \longrightarrow & 0 \\
& & \downarrow & & \downarrow f & & \downarrow \iota & & \\
0 & \longrightarrow & A & \longrightarrow & A/\mathfrak{a} \oplus A/\mathfrak{b} & \longrightarrow & A/(\mathfrak{a} + \mathfrak{b}) & \longrightarrow & 0.
\end{array}
$$

By the same argument as in [3, Proof of Proposition 2.7 (d) $\Rightarrow$ (a)], it is known that B is a d-dimensional local ring with dualizing complexes. Since $\dim(R \oplus S/B) \leq 1$, we have $\dim \operatorname{nonCM}(B) \leq 1$. Therefore B is a homomorphic image of a Gorenstein ring as $\operatorname{Ass}(B) = \operatorname{Assh}(B)$, whence so is A.

Corollary 3.2. (to Theorem 1.1). *Let A be a d-dimensional local ring. If A has a dualizing complex and A is (S_{d-2}), then A is a homomorphic image of a Gorenstein ring.*

Before proving Theorem 1.2, we recall the following

Lemma 3.3. ([3, Lemma 3.1]). *Assume that (SC) is affirmative for local rings of dimension $< d$ and (S_2) local rings of dimension d. Then (SC) is affirmative for local rings of dimension d.*

Proof of Theorem 1.2.

See [3, Theorem 3.2] for the case of $\dim A \leq 4$. (We also have a proof for the case of $\dim A = 4$ by Corollary 3.2 and Lemma 3.3.) Let $\dim A = 5$. Suppose that the assertion is false. Then, by virtue of Lemma 3.3, there is a five-dimensional local ring A such that A has a dualizing complex, is not a homomorphic image of a Gorenstein ring and is (S_2). We note that $\operatorname{Ass}(A) = \operatorname{Assh}(A)$ (cf. [3, 1.4]). A is not (S_3) by Corollary 3.2. We put

$$T(A) = \{\mathfrak{p} \in \operatorname{Spec}(A) \mid \operatorname{depth} A_{\mathfrak{p}} = 2 < \dim A_{\mathfrak{p}}\}.$$

Then $T(A)$ is not empty. Let $\mathfrak{a}$ be an ideal such that $V(\mathfrak{a}) = \operatorname{nonCM}(A)$. As A is (S_2), height $\mathfrak{a} \geq 3$. There is an A-regular sequence x, y in $\mathfrak{a}$. Then we have $T(A) \subset \operatorname{Ass}(A/(x,y))$ and $T(A)$ is a finite set. We put

$$s(A) = \max\{\dim A_{\mathfrak{p}} \mid \mathfrak{p} \in T(A)\},$$

$$T_0(A) = \{\mathfrak{p} \in T(A) \mid \dim A_{\mathfrak{p}} = s(A)\}$$

and

$$T_1(A) = T(A) \setminus T_0(A).$$

Consider all such local rings as stated above, and take a local ring A from them whose $s(A)$ is the smallest. If a local ring $(R, \mathfrak{n})$ has a dualizing complex, R is (S_2) and $\dim R \geq 3$, then $H_{\mathfrak{n}}^2(R)$ is of finite length. Hence there is a non-zero divisor $a \in \left(\bigcap_{\mathfrak{p} \in T_0(A)} \mathfrak{p} \right) \setminus \left(\bigcup_{\mathfrak{p} \in T_1(A)} \mathfrak{p} \right)$ such that $a H_{\mathfrak{p}A_{\mathfrak{p}}}^2(A_{\mathfrak{p}}) = 0$ for every $\mathfrak{p}$ in $T_0(A)$. Let $C = \operatorname{Hom}_{A/aA}(K_{A/aA}, K_{A/aA})$. As A is (S_2), we have $\operatorname{Ass}(A/aA) = \operatorname{Assh}(A/aA)$ and $A/aA \subseteq C$ (cf. [2, 1.8]). Then it is known by [2, Theorem 3.2] that C is a semi-local ring which satisfies the conditions (a) and (b) stated before (cf. [4, p.9]). Hence, by the fact mentioned before the

Proof of Theorem 1.1, there exists a Gorenstein semi-local ring G such that $\text{Max}(G) = \{\mathfrak{n} \cap G \mid \mathfrak{n} \in \text{Max}(C)\}$, every maximal chain of prime ideals in G has length five, the length of a fundamental dualizing complex is equal to five and C is a homomorphic image of G. Let B be the fibre product of the ring homomorphism $A \to C$ and the epimorphism $G \to C$. We have an exact sequence of B-modules $0 \to B \to A \oplus G \to C \to 0$. By the same argument as that in [3, Proof of Theorem 3.2], it is known that B is a five-dimensional local ring with dualizing complexes. Since A is a homomorphic image of B, B is not a homomorphic image of a Gorenstein ring and therefore not (S_3). Since A is S_2), G is Gorenstein and C is (S_2), B is (S_2) by the exact sequence above. Hence we have $T(B) \neq \emptyset$ and $s(B) \geq s(A)$ by the choice of A. Take a prime ideal $\mathfrak{P}$ of B from $T_0(B)$. We have depth $B_{\mathfrak{P}} = 2$. If $C_{\mathfrak{P}} = 0$, $B_{\mathfrak{P}} \cong A_{\mathfrak{P}}$ as G is Gorenstein. Hence $\mathfrak{P}A$ is in $T_0(A)$, which contradicts $\mathfrak{P}A \not\supseteq a$. Hence we have $C_{\mathfrak{P}} \neq 0$. Put $\dim C_{\mathfrak{P}} = r$. Then $\dim B_{\mathfrak{P}} = \dim A_{\mathfrak{P}} = \dim G_{\mathfrak{P}} = r + 1 = s(B) \geq s(A) > 2$. From the exact sequence

$$0 \longrightarrow B_{\mathfrak{P}} \longrightarrow A_{\mathfrak{P}} \oplus G_{\mathfrak{P}} \longrightarrow C_{\mathfrak{P}} \longrightarrow 0,$$

we have depth $A_{\mathfrak{P}} = 2$ as depth $B_{\mathfrak{P}} = 2$, depth $G_{\mathfrak{P}} = r + 1 > 2$ and depth $C_{\mathfrak{P}} \geq 2$. Therefore $\mathfrak{P}A$ is in $T_0(A)$ and $s(B) = s(A)$. Hence $aH^2_{\mathfrak{P}A_{\mathfrak{P}}}(A_{\mathfrak{P}}) = 0$ and the map $H^2_{\mathfrak{P}A_{\mathfrak{P}}}(A_{\mathfrak{P}}) \to H^2_{\mathfrak{P}A_{\mathfrak{P}}}(A_{\mathfrak{P}}/aA_{\mathfrak{P}})$ is injective. Next we show that $\text{Coker}((A/aA)_{\mathfrak{P}} \to C_{\mathfrak{P}})$ is of finite length. Let $\mathfrak{q}$ be a prime ideal of A with $a \in \mathfrak{q} \leftrightarrow \mathfrak{P}A$. If height $\mathfrak{q}/aA < 2$, then height $\mathfrak{q} = 5 - \dim A/\mathfrak{q} = 5 - (4 - \text{height } \mathfrak{q}/aA) \leq 2$. Hence $A_{\mathfrak{q}}$ is Cohen-Macaulay and so is $(A/aA)_{\mathfrak{q}}$. If height $\mathfrak{q}/aA \geq 2$, then height $\mathfrak{q} \geq 3$. Hence depth $A_{\mathfrak{q}} \geq 3$ as $\mathfrak{q} \notin T(A)$, and $\text{depth}(A/aA)_{\mathfrak{q}} \geq 2$. Hence the map $(A/aA)_{\mathfrak{q}} \to C_{\mathfrak{q}}$ is an isomorphism by [1, Proposition 2]. Therefore $\text{Coker}((A/aA)_{\mathfrak{P}} \to C_{\mathfrak{P}})$ is of finite length, whence the map $H^2_{\mathfrak{P}A_{\mathfrak{P}}}(A_{\mathfrak{P}}/aA_{\mathfrak{P}}) \to H^2_{\mathfrak{P}A_{\mathfrak{P}}}(C_{\mathfrak{P}})$ is injective. Hence the composition map $H^2_{\mathfrak{P}A_{\mathfrak{P}}}(A_{\mathfrak{P}}) \to H^2_{\mathfrak{P}A_{\mathfrak{P}}}(A_{\mathfrak{P}}/aA_{\mathfrak{P}}) \to H^2_{\mathfrak{P}A_{\mathfrak{P}}}(C_{\mathfrak{P}})$ is injective. From the exact sequence

$$0 = H^1_{\mathfrak{P}B_{\mathfrak{P}}}(C_{\mathfrak{P}}) \longrightarrow H^2_{\mathfrak{P}B_{\mathfrak{P}}}(B_{\mathfrak{P}}) \longrightarrow$$
$$H^2_{\mathfrak{P}B_{\mathfrak{P}}}(A_{\mathfrak{P}} \oplus G_{\mathfrak{P}}) \cong H^2_{\mathfrak{P}A_{\mathfrak{P}}}(A_{\mathfrak{P}}) \longrightarrow H^2_{\mathfrak{P}B_{\mathfrak{P}}}(C_{\mathfrak{P}}),$$

we have $H^2_{\mathfrak{P}B_{\mathfrak{P}}}(B_{\mathfrak{P}}) = 0$, which contradicts depth $B_{\mathfrak{P}} = 2$. Now the proof is completed.

References

[1] Y. Aoyama, On the depth and the projective dimension of the canonical module, Japan. J. Math. (N.S.), **6** (1980), 61-66.

[2] Y. Aoyama, Some basic results on canonical modules, J. Math. Kyoto Univ., **23** (1983), 85-94.

[3]   Y. Aoyama and S. Goto, Some special cases of a conjecture of Sharp, J. Math. Kyoto Univ., **26** (1986), 613-634.

[4]   Y. Aoyama and S. Goto, A brief summary of the elements of the theory of dualizing complexes and Sharp's conjecture, The Curves Seminar at Queen's, vol. 4, Queen's Papers in Pure and Appl. Math. **76**, 1986.

[5]   M. Brodmann, Local cohomology of certain Rees- and form-rings I, J. Algebra, **81** (1983), 29-57.

[6]   G. Faltings, Über die Annulatoren lokaler Kohomologiegruppen, Arch. Math.(Basel), **30** (1978), 473-476.

[7]   G. Faltings, Über Macaulayfizierung, Math. Ann., **238** (1978), 175-192.

[8]   S. Goto and K. Yamagishi, The theory of unconditioned strong d-sequences and modules of finite local cohomology, Preprint.

[9]   J. E. Hall, Fundamental dualizing complexes for commutative noetherian rings, Quart. J. Math. Oxford Ser.(2), **30** (1979), 21-32.

[10] R. Hartshorne, Residues and duality, Lect. Notes in Math. **20**, Springer Verlag, 1966.

[11] J. Herzog, E. Kunz et al., Der kanonische Modul eines Cohen-Macaulay-Rings, Lect. Notes in Math. **238**, Springer Verlag, 1971.

[12] P. Schenzel, Standard systems of parameters and their blowing-up rings, J. Reine Angew. Math., **344** (1983), 201-220.

[13] R. Y. Sharp, Dualizing complexes for commutative Noetherian rings, Math. Proc. Cambridge Philos. Soc., **78** (1975), 369-386.

[14] R. Y. Sharp, Necessary conditions for the existence of dualizing complexes in commutative algebra, Sém. Algèbre P. Dubreil 1977/78, Lect. Notes in Math. **740**, Springer Verlag, 1979, 213-229.

[15] G. Valla, Certain graded algebras are always Cohen-Macaulay, J. Algebra, **42** (1976), 537-548.

Yoichi Aoyama

Department of Mathematics
Faculty of Science
Ehime University
Matsuyama, 790
Japan

Shiro Goto

Department of Mathematics
College of Humanities and Sciences
Nihon University
Setagaya-ku, Tokyo, 156
Japan

Algebraic Geometry and Commutative Algebra
in Honor of Masayoshi NAGATA
pp. 35–44 (1987)

A Structure Theorem for Power Series Rings

Michael ARTIN and Christel ROTTHAUS

The object of this paper is to prove the following result, which was announced in [3] for the case that R is a field of characteristic zero.

Theorem 1: *Let R be an excellent discrete valuation ring and let $X = (X_1, \ldots, X_n)$ be variables. Then the power series ring $\hat{R}[[X]]$ is a direct limit of smooth $R[X]$-algebras.*

This theorem has an important application to the proof [14] by the second author of the approximation property for excellent henselian rings containing **Q**. Though the result is the subject of the other papers [9], [13], a complete proof does not seem to be available up to now. Our proof is based on the Weierstrass Preparation Theorem, and uses arguments similar to those in [2], where the approximation property for henselizations of finite type algebras is proved. We refer the reader to [10] for the foundations of the theory of henselian rings, and to [3], [4], [13] for a discussion of the general context in which this theorem should be placed.

0. Theorem 1 follows from Néron's p-desingularization [2] in the case $n = 0$. We will review the results we need here.

A local morphism $\tau : R \to R'$ of discrete valuation rings is *regular* if it has the following properties:

(i) If p is a local parameter for R, then $\tau(p)$ is a local parameter for R'.

(ii) The residue class field of R' is separable over the residue class field of R.

(iii) The field of fractions of R' is separable over the field of fractions of R.

If $\tau : R \to R'$ is regular, then R' is a direct limit of finite type smooth R-algebras. More precisely the following can be shown:

Theorem 2: ([2] §4): *Let $\tau : R \to R'$ be a regular local morphism of discrete valuation rings and suppose that τ factors: $R \to B \xrightarrow{\sigma} R'$, where B is a finite type R-algebra and σ is injective. Then there is a factorization of σ:*

Received December 3, 1986.

$$
\begin{array}{ccc}
R & \xrightarrow{\ \tau\ } & R' \\
\downarrow & \nearrow{\sigma} & \uparrow{\varphi} \\
B & \longrightarrow & C
\end{array}
$$

satisfying the following conditions:
(i) C is a finite type smooth R-algebra,
(ii) φ is injective,
(iii) $C \subseteq B[p^{-1}]$.

Theorem 2 implies the following well-known statement which we will use for the proof of Theorem 1:

Theorem 3: *Let R be an excellent discrete valuation ring, $\hat{R}$ its completion, $X = (X_1, \ldots, X_n)$ variables, and let $\tau : R[X] \to \hat{R}[[X]]$ be the canonical embedding. Moreover suppose that there is given a commutative diagram:*

$(*)$
$$
\begin{array}{ccc}
R[X] & \xrightarrow{\ \tau\ } & \hat{R}[[X]] \\
\downarrow & \nearrow{\sigma} & \\
B & &
\end{array}
$$

where B is a $R[X]$-algebra of finite type and σ is injective. Then $(*)$ can be embedded in a commutative diagram

$(**)$
$$
\begin{array}{ccc}
R[X] & \xrightarrow{\ \tau\ } & \hat{R}[[X]] \\
\downarrow & \nearrow{\sigma} & \uparrow{\varphi} \\
B & \longrightarrow & C
\end{array}
$$

such that
(a) C is a $R[X]$-algebra of finite type, of the form $R[X,Y]/(f_1, \ldots, f_r)$, where $(Y) = (Y_1, \ldots, Y_N)$ are variables,
(b) φ is injective and $\varphi(Y_i) = \bar{y}_i \in \mathfrak{m}_{\hat{R}[[X]]}$,
(c) $(f) = (f_1, \ldots, f_r)$ is a prime ideal in $R[X,Y]$, with $\mathrm{ht}(f) = N - d$, where $d = \dim C - \dim R[X]$,
(d) There is a $(N-d) \times (N-d)$-minor of the Jacobian matrix, say

$$
\delta(X,Y) = \det\left(\frac{\partial f_i}{\partial Y_j}\right)_{i,j=1,\ldots,N-d} \,,
$$

such that $\delta(X, \bar{y}) \notin \mathfrak{p}\hat{R}[[X]]$.

Theorem 3 follows from Theorem 2 by considering the localization of τ at the height one prime ideals $\mathfrak{p}R[X]$ and $\mathfrak{p}\hat{R}[[X]]$.

1. We suppose that there is given a commutative diagram:

(1.a)

$$\begin{array}{ccc} R[X] & \xrightarrow{\ \tau\ } & \hat{R}[[X]] \\ \big\downarrow & \nearrow_{\sigma} & \\ B & & \end{array}$$

where B is a finitely generated $R[X]$-algebra and σ is injective. To prove Theorem 1, we have to show that (1.a) can be embedded in a commutative diagram

(1.b)

$$\begin{array}{ccc} R[X] & \xrightarrow{\ \tau\ } & \hat{R}[[X]] \\ \big\downarrow & \nearrow_{\sigma} & \big\uparrow \\ B & \longrightarrow & S \end{array}$$

where S is a smooth $R[X]$-algebra of finite type. By theorem 3 we may assume that B satisfies the following conditions:

(1.1) $B = R[X, Y]/(f_1, \ldots, f_r)$, where $Y = (Y_1, \ldots, Y_N)$ is a set of variables; $(f) = (f_1, \ldots, f_r)$ is a prime ideal in $R[X, Y]$, with $\operatorname{ht}(f) = N - d = m$, where $d = \dim B - \dim R[X]$.

(1.2) Let $\sigma(Y_i) = \bar{y}_i \in \mathfrak{m}_{\hat{R}[[X]]}$. Then there is a $m \times m$ minor of the Jacobian matrix of (f), say

$$\delta(X, Y) = \det\left(\frac{\partial f_i}{\partial Y_j}\right)_{i,j=1,\ldots,m}$$

such that $\delta(X, \bar{y}) \notin \mathfrak{p}\hat{R}[[X]]$.

2. We may replace B by $C = R[X, Y]/(f_1, \ldots, f_m)$.

Proof: Suppose that the map $\sigma' : C \to \hat{R}[[X]]$ induced by σ factors through a smooth $R[X]$-algebra S':

$$\begin{array}{ccc} C & \xrightarrow{\ \sigma'\ } & \hat{R}[[X]] \\ \psi'\big\downarrow & \nearrow_{\varphi'} & \\ S' & & \end{array}$$

Let $\mathfrak{m} = \varphi'^{-1}(\mathfrak{m}_{\hat{R}[[X]]}) \in \operatorname{Spec}(S')$ be the preimage of the maximal ideal in S'. Then $S'_{\mathfrak{m}}$ is a regular local domain and φ' factors through $S'_{\mathfrak{m}}$. By inverting an element of S' we get a smooth domain over which σ' factors and which is still of finite type over $R[X]$. This implies:

(2.1) $\operatorname{Ker} \psi'$ is a prime ideal in C contained in the minimal prime ideal $\operatorname{Ker} \sigma' = (f)$ of C.

This shows that ψ' factors through B.

3.   The proof of the theorem is by induction on the number of variables $X_1, \ldots, X_n$. The theorem is known for excellent discrete valuation rings (Theorem 2), and we suppose now that it has been proved for rings of the type $R[X_1, \ldots, X_{n-1}]$. Under this assumption we prove the following lemma:

Lemma: *Let* $\sigma : R[X, Y] \to \hat{R}[[X]]$ $(X = (X_1, \ldots, X_n); Y = (Y_1, \ldots, Y_N))$ *be the homomorphism defined by* $\sigma(Y_i) = \bar{y}_i \in \mathfrak{m}\hat{R}[[X]]$, *and let* $f_1, \ldots, f_m, g \in R[X, Y]$ *be polynomials satisfying the following conditions:*
  (i) $g(X, \bar{y}) \notin \mathfrak{p}\hat{R}[[X]]$,
  (ii) $g(X, \bar{y})$ *divides* $f_i(X, \bar{y})$ *in* $\hat{R}[[X]]$ *for all* $i = 1, \ldots, m$,
  (iii) $\partial g / \partial Y_j$ *has no constant terms for all* $j = 1, \ldots, N$.
Then there is a smooth $R[X]$-*algebra* S' *of finite type and a diagram*

$$
\begin{array}{ccc}
R[X, Y] & \overset{\sigma}{\longrightarrow} & \hat{R}[[X]] \\
{\scriptstyle \tau} \downarrow & \nearrow {\scriptstyle \psi} & \\
S' & &
\end{array}
$$

such that (a) $\sigma = \psi \circ \tau$ *and* (b) $g(X, \tau(Y))$ *divides* $f_i(X, \tau(Y))$ *in* S' *for all* $i = 1, \ldots, m$.

Proof of the lemma:

(3.1)   Because of condition (i), we get, after performing a Weierstrass transformation if necessary:

$$g(X, \bar{y}) = \bar{a}(X_n)\varepsilon,$$

where $\varepsilon \in \hat{R}[[X]]^*$ is invertible and $\bar{a}(X_n) = \sum_{j=0}^{r-1} \bar{a}_j X_n^j + X_n^r$ is a monic polynomial with coefficients $\bar{a}_j \in \hat{R}[[X_1, \ldots, X_{n-1}]]$.

(3.2)   By the Weierstrass preparation theorem, we find presentations:

$$\bar{y}_\nu = \bar{a}(X_n)\bar{w}_\nu + \sum_{j=0}^{r-1} \bar{y}_{\nu j} X_n^j,$$

with $\bar{y}_{\nu j} \in \hat{R}[[X_1, \ldots, X_{n-1}]]$. Define:

$$\bar{y}_\nu^* = \sum_{j=0}^{r-1} \bar{y}_{\nu j} X_n^j.$$

(3.3) Let $Y_{\nu j}$; A_j $(j = 0, \ldots, r-1; \nu = 1, \ldots, N)$ be new variables, and define

$$Y_\nu^* = \sum_{j=0}^{r-1} Y_{\nu j} X_n^j, \qquad A(X_n) = X_n^r + \sum_{j=0}^{r-1} A_j X_n^j.$$

Division of the polynomials

$$f_i(X, Y^*); \ g(X, Y^*) \in R[X_1, \ldots, X_n; Y_{\nu j}; A_j]$$

by $A(X_n)$ gives us

(3.4) $\quad f_i(X, Y^*) = A(X_n) Q_i + \sum_{j=0}^{r-1} F_{ij} X_n^j \quad$ for $i = 1, \ldots, m,$

$$g(X, Y^*) = A(X_n) Q + \sum_{j=0}^{r-1} G_j X_n^j,$$

where $F_{ij}, G_j \in R[X_1, \ldots, X_{n-1}; Y_{\nu j}; A_j]$, $Q_i, Q \in R[X_1, \ldots, X_n; Y_{\nu j}; A_j]$.

(3.5) Note that $\bar{a}(X_n)$; $g(X, \bar{y})$ and $g(X, \bar{y}^*)$ differ only by unit factors of $\hat{R}[[X]]$.

For, by Taylor's formula, $g(X, \bar{y}^*)$ has the form

$$g(X, \bar{y}^*) = g(X, \bar{y}) + \sum_{k=1}^{N} \frac{\partial g}{\partial Y_k}(X, \bar{y}) \bar{a}(X_n) \bar{w}_k + \bar{a}(X_n)^2 \bar{h}$$

where $\bar{w}_k$ is as in (3.2), and $\bar{h} \in \hat{R}[[X]]$. Assertion (3.5) follows by combining this formula with assumption (iii) of the lemma.

This implies in particular:

(3.6) $\quad g(X, \bar{y}^*)$ divides $f_i(X, \bar{y}^*)$ for all $i = 1, \ldots, m.$

Therefore, by (3.4):

(3.7) $\quad \bar{y}_{\nu j}$ and $\bar{a}_j \in \hat{R}[[X_1, \ldots, X_{n-1}]]$ are solutions of the equations for $G_j = F_{ij} = 0$ in $\hat{R}[[X_1, \ldots, X_{n-1}]][Y_{\nu j}, A_j]$. Put $T_0 = R[X_1, \ldots, X_{n-1}, Y_{\nu j}, A_j]$ and define

$$\rho : T_0 \longrightarrow \hat{R}[[X_1, \ldots, X_{n-1}]]$$

by $\rho(Y_{\nu j}) = \bar{y}_{\nu j}$ and $\rho(A_j) = \bar{a}_j$. Then ρ induces a morphism

$$\bar{\rho} : T_0 / (G_j, F_{ij}) \longrightarrow \hat{R}[[X_1, \ldots, X_{n-1}]].$$

(3.8) By the induction hypothesis $\bar{\rho}$ factors through a smooth $R[X_1, \ldots, X_{n-1}]$-algebra S_1 of finite type:

$$
\begin{array}{ccc}
T_0/(G_j, T_{ij}) & \xrightarrow{\;\bar{\rho}\;} & \hat{R}[[X_1, \ldots, X_{n-1}]] \\
\downarrow & \nearrow \varphi_1 & \\
S_1 & &
\end{array}
$$

Let $Y'_{\nu j}$ and A'_j denote the images of $Y_{\nu j}$ and A_j in S_1.

(3.9) Let $S' = S_1[X_n, W_1, \ldots, W_N]$, and define morphisms:

$$\psi : S' \longrightarrow \hat{R}[[X]]$$

by $\psi|S_1 = \varphi_1$; $\psi(X_n) = X_n$; $\psi(W_\nu) = \bar{w}_\nu$ and

$$\mu : R[X, Y] \longrightarrow S'$$

by $\mu(Y_\nu) = Y'_\nu = A'W_\nu + \sum_{j=0}^{r-1} Y'_{\nu j} X_n^j$ where $A' = X_n^r + \sum_{j=0}^{r-1} A'_j X_n^j$. Then σ factors over the smooth $R[X]$-algebra S' by $\sigma = \psi \circ \mu$.

(3.10) **Claim:** There is an element $u \in S'$ such that $\psi(u)$ is invertible in $\hat{R}[[X]]$ and $g(X, Y')$ divides $f_i(X, Y')$ in $S'[u^{-1}]$, for all $i = 1, \ldots, m$.

Proof: Setting

$$Y'^*_\nu = \sum_{j=0}^{r-1} Y'_{\nu j} X_n^j \in S',$$

Taylor's formula implies that $f_i(X, Y') = f_i(X, Y'^*) + A'H_i$, where H_i is some element of S'. From (3.4) and (3.8) we know that $f_i(X, Y'^*)$ is divisible by A' in S', hence so is $f_i(X, Y')$. Similarly, $g(X, Y')$ is divisible by A' in S', say that $g(X, Y') = A'u$. Since $\psi(g(X, Y')) = g(X, \bar{y})$ and $\psi(A') = \bar{a}(X_n)$ differ by a unit in $\hat{R}[[X]]$, $\psi(u)$ is invertible in $\hat{R}[[X]]$. This proves the claim. The lemma follows by setting $S = S'[u^{-1}]$.

4. Proof of the Theorem.

We use the assumptions and notations of Sections 1 and 2.

(4.1) Let

$$J = \left(\frac{\partial f_i}{\partial Y_j} \right)_{\substack{i=1,\ldots,m \\ j=1,\ldots,N}}$$

be the Jacobian matrix of $(f) = (f_1, \ldots, f_m)$. Denote by δ_ℓ $(\ell = 1, \ldots, r)$ the $m \times m$ minors of J, where

$$\delta_1 = \delta = \det \left(\frac{\partial f_i}{\partial Y_j} \right)_{i,j=1,\ldots,m}$$

Then for all $\ell = 1, \ldots, r$ there are $N \times m$ matrices N_ℓ with $JN_\ell = \delta_\ell I$, where I is the $m \times m$ identity matrix.

Define $g(X,Y) = \delta(X,Y)^2 p(X,Y)$, where $p(X,Y) \in R[X,Y]$ is suitably chosen so that conditions (i), (ii) and (iii) in the lemma are satisfied. Since $f_i(x, \bar{y}) \neq 0$, we know from Section 3 that there is a factorization:

$$R[X,Y] \xrightarrow{\ \sigma\ } \hat{R}[[X]]$$

$$\mu \downarrow \quad \Big/ \psi$$

$$S'$$

where S' is a smooth $R[X]$-algebra of finite type and $g(X, \mu(Y))$ divides $f_i(X, \mu(Y))$ in S' for all $i = 1, \ldots, m$. Set $Y_\nu' = \mu(Y_\nu)$. The proof now follows the argument of [2] (5.11).

(4.2) For $i = 1, \ldots, m$, we get in S':

$$
\begin{aligned}
f_i(X, Y') &= g(X, Y')W_i' \\
&= \delta_1(X, Y')^2 p(X, Y')W_i',
\end{aligned}
$$

for some elements $W_i' \in S'$. Since $g(X, \bar{y}) \neq 0$ but $f_i(x, \bar{y}) = 0$, we have $\psi(W_i') = \bar{w}_i = 0$.

(4.3) Introducing new variables $U_{\ell\nu}$; $\ell = 1, \ldots, r_j$, $\nu = 1, \ldots, N$ we define:

$$Z_\nu = Y_\nu' + \sum_{l=1}^{r} \delta_\ell(X, Y')U_{\ell\nu} \in S'[U]$$

and consider for $j = 1, \ldots, m$:

$$
\begin{aligned}
f_j(X, Z) = f_j(X, Y') \ &+\ \sum_{k=1}^{N} \frac{\partial f_j}{\partial Y_k}(X, Y')\Big(\sum_{\ell=1}^{r} \delta_\ell U_{\ell k}\Big) \\
&+\ \sum_{\ell, k=1}^{r} \delta_\ell \delta_k Q_{j\ell k}
\end{aligned}
$$

where $Q_{j\ell k}$ are polynomials in $U_{\mu\nu}$ whose monomials have degree at least 2 in $U_{\mu\nu}$.

(4.4) By performing elementary operations on the matrix equation

$$
\begin{pmatrix} f_1(X, Z) \\ \vdots \\ f_m(X, Z) \end{pmatrix} = \begin{pmatrix} f_1(X, Y') \\ \vdots \\ f_m(X, Y') \end{pmatrix} + J \begin{pmatrix} \sum_1^r \delta_\ell U_{\ell 1} \\ \vdots \\ \sum_1^r \delta_\ell U_{\ell N} \end{pmatrix} + \sum_{\ell=1}^{r} \delta_\ell \begin{pmatrix} \sum_{k=1}^r \delta_k Q_{1\ell k} \\ \vdots \\ \sum_{k=1}^r \delta_k Q_{m\ell k} \end{pmatrix}
$$

and by using the equations in (4.1) and (4.2) we get:

$$(*) \quad \begin{pmatrix} f_1(X,Z) \\ \vdots \\ f_m(X,Z) \end{pmatrix} = J \left[\delta_1 \begin{pmatrix} P_1(X,Y',W') \\ \vdots \\ P_N(X,Y',W') \end{pmatrix} + \sum_{\ell=1}^{r} \delta_\ell \begin{pmatrix} U_{\ell 1} + H_{\ell 1} \\ \vdots \\ U_{\ell N} + H_{\ell N} \end{pmatrix} \right],$$

such that

(a) $P_i(X,Y',W') \in (W_1',\ldots,W_m') \subseteq S'$. In particular, $\psi(P_i(X,Y',W')) = 0$ for all $i = 1,\ldots,N$.

(b) The $H_{\ell i}$ are polynomials in $U_{\nu\mu}$ whose monomials have degree at least 2.

(4.5) Let $\mathfrak{a} \subseteq S'[U_{\ell j}]$ be the ideal generated by

$$P_j(X,Y',W') + U_{1j} + H_{1j} \qquad j = 1,\ldots,N$$

and

$$U_{ij} + H_{ij} \qquad j = 1,\ldots,N; \ i = 2,\ldots,r.$$

Let $\tilde{\varphi} : S'[U_{ij}] \longrightarrow \hat{R}[[X]]$ be the morphism defined by $\tilde{\varphi}|_{S'} = \psi$ and $\tilde{\varphi}(U_{ij}) = 0$. because of (4.4)(a),(b): $\mathfrak{a} \subset \mathrm{Ker}\tilde{\varphi}$. Let $\varphi : S_1 = S'[U_{ij}]/\mathfrak{a} \to \hat{R}[[X]]$ be the morphism induced by $\tilde{\varphi}$. Since S' is smooth over $R[X]$, the defining equations show that S_1 is smooth over $R[X]$ in a neighborhood of the locus $\{U_{ij} = 0\}$.

(4.6) Consider the morphism $\tilde{\gamma} : R[X,Y] \to S'[U_{ij}]$ defined by $\tilde{\gamma}(Y_\nu) = Z_\nu$. $\tilde{\varphi}(U_{ij}) = 0$ implies that $\sigma : R[X,Y] \to \hat{R}[[X]]$ factors through $\gamma : \sigma = \tilde{\varphi} \circ \tilde{\gamma}$. Because of $(*)$ in (4.4) the $\tilde{\gamma}(f_i)$ are contained in $\mathfrak{a}$. Hence we have found the desired factorization of $\bar{\sigma}$ through the smooth $R[X]$-algebra S:

$$
\begin{array}{ccc}
R[X,Y]/(f_1,\ldots,f_m) & \xrightarrow{\ \bar{\sigma}\ } & \hat{R}[[X]] \\
\Big\downarrow{\scriptstyle \gamma} & \nearrow & \\
S & \varphi &
\end{array}
$$

5. Corollary: *Let R be an excellent discrete valuation ring; $X = (X_1,\ldots, X_n)$ variables over R. Then $\hat{R}[[X]]$ is a direct limit of $R[X]$-algebras of the form $R^h\langle X,Y\rangle$ where $Y = (Y_1,\ldots,Y_N)$ are variables and $R^h\langle X,Y\rangle$ denotes the henselization of the polynomial ring $R[X,Y]$ with respect to the maximal ideal $(\mathfrak{m}_R, X, Y)$.*

Proof: By theorem 1, $\bar{R}[[X]]$ is a direct limit of smooth $R[X]$-algebras C. Since $\hat{R}[[X]]$ is henselian, it is also the direct limit of the hensilizations C^h. Localizing C suitably, we may assume that it is an etale extension of some polynomial ring $R[X,Y]$ $(Y = (Y_1,\ldots,Y_N))$ ([8] Chap.III, Prop.3.1.) since the residue class field of $R[[X]], C$, and $R[X,Y]$ are the same, we may assume that the maximal ideal of $\hat{R}[[X]]$ lies over the ideal (p, X, Y) of $R[X, Y]$. then $R^h\langle X, Y \rangle \approx C^h \subset \hat{R}[[X]]$.

References

[1] M. Artin: On the solutions of analytic equations, Invent. Math. 5(1968) 277-291.

[2] M. Artin: Algebraic approximation of structures over complete local rings, Pub. Math. Inst. Hautes Études Sci. 36(1969) 23-58.

[3] M. Artin: Algebraic Structure of power series rings, Contemp. Math. 13, Amer. Math. Soc., Providence (1982) 223-227.

[4] M. Artin and J. Denef: Smoothing of a ring homomorphism along a section, Arithmetic and Geometry. Vol.II, Birkhäuser, Boston 1983.

[5] J. Becker, J. Denef and L. Lipshitz: The approximation property for some 5-dimensional henselian rings, Trans. of the AMS 276(1983) 301-309.

[6] J. Becker, J. Denef, L. Lipshitz and L. van den Dries: Ultraproducts and approximation in local rings I, Invent. Math. 51(1979) 189-203.

[7] R. Elkik: Solutions d'équations à coefficients dans un anneau hensélian, Ann. Sci. École Normal Sup. 4ᵉ sér. 6(1973) 553-601.

[8] B. Iversen: Generic local structure of the morphisms in commutative algebra, Lec. Notes in Math. 310, Springer Verlag, Berlin 1973.

[9] H. Kurke, T. Mostowski, G.Pfister, G. Popescu and M. Roczen: Die Approximationseigenshaft lokaler Ringe, Lec. Notes in Math. 634, Springer Verlag, Berlin 1978.

[10] M. Nagata: On the theory of henselian rings I, II, Nagoya Math. J. 5(1953) 45-57; 7(1954) 1-19.

[11] A. Néron: Modeles minimaux des variétés abéliennes sur les corps locaux et globaux, Pub. Math. Inst. Hautes Études Sci. 21, 1964.

[12] A. Ploski: Note on a theorem of M. Artin, Bull. Acad. Polonaise Sci., ser. Math. 22(1974) 1107-1109.

[13] D. Popescu: General Néron Desingularization, Nagoya Math. J. 100(1985) 97-126.

[14] C. Rotthaus: On the approximation property for excellent rings, Invent. Math. 88 (1987) 39-63.

Michael Artin

Department of Mathematics
Massachusetts Institute of Technology
Cambridge, MA 02139
U.S.A.

Christel Rotthaus

Department of Mathematics
Michigan State University
East Lansing, MI 48824-1027
U.S.A.

Algebraic Geometry and Commutative Algebra
in Honor of Masayoshi NAGATA
pp. 45–58 (1987)

On Rational Plane Sextics with Six Tritangents

Wolf BARTH* and Ross MOORE

§0. Introduction.

A rational plane sextic curve in general has ten nodes. It is the aim of this note to describe a family of such sextics S_x (parametrized by $x \in \mathbf{P}_3$) admitting six tritangents. The double plane X branched over S_x is the Kummer surface of an abelian surface carrying a polarization of type $(1, 5)$. Surprisingly enough the double plane Y branched over the union of the six tritangents is also the Kummer surface of such an abelian surface.

Each curve S_x is the projection from $x \in \mathbf{P}_3$ of the same sextic space curve

$$S: (\lambda^6 - 2\lambda : 2\lambda^5 + 1 : \lambda^4 : \lambda^2).$$

All the properties needed to control the double cover X (miraculously) drop out of some polynomial identities. One of these facts is the existence of the six tritangents L_i; another is the possibility of choosing rational curves $M_i \subset X$ over L_i which are mutually disjoint. (This allows us to apply Nikulin's theorem [N] to the sixteen disjoint rational curves M_i, $i = 1, \ldots, 6$, and the ten curves in X over the nodes of S_x.)

The reader will realize that the curve S is invariant under an action of the icosahedral group on $\mathbf{P}_3$. Although elementary, this fact is not so easy to check, cf. [BHM, Lemma 4.1]. Indeed the parameter λ on S is the "Ikosaeder Transzendente" of F. Klein parametrizing elliptic curves with a level-5 structure. The polynomial identities needed are related to the modular equation (the polynomial P_3 in Section 1 below) describing 2-torsion quotients of such curves.

The generic abelian surface with $(1, 5)$-polarization can be embedded in $\mathbf{P}_4$ as a Horrocks-Mumford surface [HM]. It was the study of these surfaces, and of the vector bundle on $\mathbf{P}_4$ related to them, that gave rise to the observations collected in this note. This note however is intended to be self-contained. It should be considered as an attempt to describe, by simple equations, abelian surfaces with $(1, 5)$-polarization in terms of their Kummer surfaces.

The search for equations of the sextic plane curves described above was begun in 1985 by the first author during his stay at Kyoto University. He is

*Supported by DFG Research Grant Ba 423/2-1
Received December 13, 1986.

indebted to Professor M. Nagata and to JSPS for having made possible this visit. He is also indebted to D. Morrison for pointing out to him the crucial condition (4.1 b) below.

Convention: The base field always is $\mathbf{C}$.

§1. Some Polynomials.

The polynomial

$$P_3(\lambda, \mu; s, t) = s^2 t \lambda^3 + t^3 \lambda^2 \mu + s^3 \lambda \mu^2 - st^2 \mu^3$$

is homogeneous of bidegree 3, 3 in the two variables $\lambda : \mu$ and $s : t \in \mathbf{P}_1$. It has the symmetries

$$P_3(\lambda, \mu; s, t) = P_3(s, t; -\mu, \lambda)$$
$$P_3(\varepsilon^k \lambda, \mu; \varepsilon^{2k} s, t) = \varepsilon^{2k} P_3(\lambda, \mu; s, t), \quad \varepsilon = e^{2\pi i/5}.$$

Putting

$$\Delta(s, t) = st(s^{10} + 11 s^5 t^5 - t^{10})$$

we find $4\Delta(s, t)$ for the discriminant of P_3 considered as a polynomial in $\lambda : \mu$ and $-4\Delta(\lambda, \mu)$ for the discriminant when considered as a polynomial in $s : t$. The twelve roots of $\Delta(s, t)$ are the vertices of the icosahedron

$$s : t = 0, \infty, \varepsilon^k \eta, \varepsilon^k \eta' \quad (k = 0, \ldots, 4)$$

with

$$\eta = \varepsilon + \varepsilon^4, \qquad \eta' = \varepsilon^2 + \varepsilon^3.$$

The roots of $\Delta(s, t)$ and $\Delta(\lambda, \mu)$ correspond to each other under the relation defined by $P_3 = 0$ as shown in Figure 1. This follows easily using the identity

$$P_3(\eta, 1; s, t) = \eta(s - \eta t)^2(s - \eta' t)$$

and the symmetries above.

We further use

$$P_3^2 = s^4 t^2 \lambda(\lambda^5 - 2\mu^5) + s^2 t^4 \mu(2\lambda^5 + \mu^5)$$
$$+ t(2s^5 + t^5)\lambda^4 \mu^2 + s(s^5 - 2t^5)\lambda^2 \mu^4$$

$$\frac{1}{2}\partial_\lambda P_3^2 = s^4 t^2(3\lambda^5 - \mu^5) + 5 s^2 t^4 \lambda^4 \mu$$
$$+ 2t(2s^5 + t^5)\lambda^3 \mu^2 + s(s^5 - 2t^5)\lambda \mu^4$$

$$\frac{1}{2}\partial_s P_3^2 = 2 s^3 t^2 \lambda(\lambda^5 - 2\mu^5)$$
$$+ st^4 \mu(2\lambda^5 + \mu^5) + 5 s^4 t \lambda^4 \mu^2 + (3s^5 - t^5)\lambda^2 \mu^4.$$

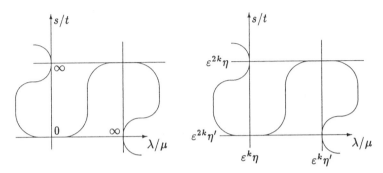

Figure 1.

In P_3^2 appear the sextic polynomials

$$\lambda(\lambda^5 - 2\mu^5), \quad \mu(2\lambda^5 + \mu^5), \quad \lambda^4\mu^2, \quad \lambda^2\mu^4.$$

One easily checks:

(1.1) *These four polynomials have no common zero* $\lambda : \mu$.

(1.2) *For* $\lambda_1 : \mu_1 \neq \lambda_2 : \mu_2 \in \mathbf{P}_1$ *the matrix*

$$\begin{pmatrix} \lambda_1(\lambda_1^5 - 2\mu_1^5) & \mu_1(2\lambda_1^5 + \mu_1^5) & \lambda_1^4\mu_1^2 & \lambda_1^2\mu_1^4 \\ \lambda_2(\lambda_2^5 - 2\mu_2^5) & \mu_2(2\lambda_2^5 + \mu_2^5) & \lambda_2^4\mu_2^2 & \lambda_2^2\mu_2^4 \end{pmatrix}$$

has rank two.

(1.3) *For all* $(\lambda, \mu) \neq (0, 0)$ *the matrix of derivatives*

$$\begin{array}{l} (1/2)\partial_\lambda: \\ (1/2)\partial_\mu: \end{array} \begin{pmatrix} 3\lambda^5 - \mu^5 & 5\lambda^4\mu & 2\lambda^3\mu^2 & \lambda\mu^4 \\ -5\lambda\mu^4 & \lambda^5 + 3\mu^5 & \lambda^4\mu & 2\lambda^2\mu^3 \end{pmatrix}$$

has rank two.

§2. The sextic space curve S.

The four polynomials from Section 1 parametrize in $\mathbf{P}_3$ the curve

$$S: (\lambda : \mu) \mapsto x(\lambda, \mu) = (\lambda(\lambda^5 - 2\mu^5) : \mu(2\lambda^5 + \mu^5) : \lambda^4\mu^2 : \lambda^2\mu^4).$$

This parametrization is injective (1.2) and smooth (1.3). So $S \subset \mathbf{P}_3$ is a smooth rational curve of degree six.

(2.1) *For fixed* $s : t$ *with* $\Delta(s, t) \neq 0$ *the three roots* $\lambda_i : \mu_i$ *of* $P_3(\lambda, \mu; s, t)$ *define collinear points* $x(\lambda_i, \mu_i)$ *on* S, *i.e., three points on a trisecant* $L(s, t)$ *to* S.

Proof. $P_3(\lambda, \mu; s, t) = 0$ implies $P_3^2 = \partial_s P_3^2 = 0$. A glance at the explicit form of these polynomials (Section 1) shows that the three points $x(\lambda_i, \mu_i)$ lie on the two planes with dual coordinates

$$
\begin{array}{ccccccc}
s^4 t^2 & : & s^2 t^4 & : & t(2s^5 + t^5) & : & s(s^5 - 2t^5) \\
2s^3 t^2 & : & st^4 & : & 5s^4 t & : & 3s^5 - t^5.
\end{array}
$$

One easily checks their independence.

(2.2) *For fixed* $s : t$ *with* $\Delta(s, t) \neq 0$ *the tangents to the three points* $x(\lambda_i, \mu_i) \in L(s, t)$ *are dependent. In fact, together with* $L(s, t)$ *they span the plane* $W(s, t)$ *with dual coordinates*

$$
s^4 t^2 \quad : \quad s^2 t^4 \quad : \quad t(2s^5 + t^5) \quad : \quad s(s^5 - 2t^5).
$$

Proof. $P_3(\lambda_i, \mu_i; s, t) = 0$ implies $P_3^2 = \partial_\lambda P_3^2 = 0$.

(2.3) *Any trisecant to* S *is of the form* $L(s, t)$, $s : t \in \mathbf{P}_1$.

Proof. Collinear triplets of points are described in $S \times S \times S$ by the condition that a 3×4-matrix drops its rank. The determinantal variety defined by this condition has codimension ≤ 2 [F, Thm. 14.4 (b)], the trisecant variety therefore has no discrete components. S generating $\mathbf{P}_3$, its projection S_x into $\mathbf{P}_3$ from $x = x(\lambda, \mu) \in S$ is an irreducible plane quintic. There are genus drops $\delta(p)$ attached to the singularities $p \in S_x$ such that $\sum \delta(p) = 6$.

If $\Delta(\lambda, \mu) \neq 0$, there are three distinct values $s_i : t_i$ satisfying $P_3(\lambda, \mu; s_i, t_i) = 0$. As $\Delta(s_i, t_i) \neq 0$, each trisecant $L(s_i, t_i)$ meets S in three distinct points. By (2.2) the plane $W(s_i, t_i)$ touches S in these three points, hence $L(s_i, t_i)$ will not meet S in a fourth point. If $L(s_i, t_i) = L(s_j, t_j)$, then $W(s_i, t_i) = W(s_j, t_j)$ and (1.2) applied to s_i, t_i instead of λ_i, μ_i shows $i = j$. So for all points $x(\lambda, \mu) \in S$ but finitely many, the three trisecants $L(s_i, t_i)$ through x are distinct. On the projected curve S_x they define three distinct singularities p_i. At each p_i two branches of S_x meet, and by (2.2) they meet not transversally. This implies $\delta(p_i) \geq 2$; and in view of $\sum \delta(p) = 6$, these points p_i are the only singularities on S_x. The lines $L(s_i, t_i)$ are the only trisecants through x.

By the dimension argument above, this also holds for the points $x \in S$ where $\Delta(\lambda, \mu) = 0$.

(2.4) *Any pair of coplanar tangents to* S *belongs to the three tangents of some plane* $W(s, t)$, $s : t \in \mathbf{P}_1$.

Proof. Pairs of points $x \in S$ with coplanar tangents are described in $S \times S$ by the condition that a 4×4-determinant vanishes. The variety of coplanar pairs of tangents therefore has no discrete components. Projecting S to $\mathbf{P}_1$ from some tangent $T = T_x(S)$, $x \in S$, we obtain a covering $S \to \mathbf{P}_1$ of degree ≤ 4. Its number of branch points is ≤ 6.

If $x = x(\lambda, \mu)$ with $\Delta(\lambda, \mu) \neq 0$ there are three distinct planes $W(s_i, t_i)$ containing T. As $\Delta(s_i, t_i) \neq 0$, the plane $W(s_i, t_i)$ touches S in three distinct points and there are two tangents T_i, T'_i in this plane meeting T. Together we count six distinct tangents meeting T and causing six different branch points on the projection $S \to \mathbf{P}_1$ from T. So T_i, T'_i, $i = 1, 2, 3$, are the only tangents to S meeting T.

This count works for all but twelve points $x \in S$. By the dimension argument above the assertion holds for all tangents to S.

The planes $W(s, t)$ parametrize a curve

$$S': (s, t) \mapsto x^*(s, t) = (s^4 t^2 : s^2 t^4 : t(2s^5 + t^5) : s(s^5 - 2t^5))$$

in $\mathbf{P}_3^*$. The curves S and S' enjoy a strange sort of duality:

(2.5) *The trisecants of S are the lines dual to the tangents to S', and conversely: the tangents to S are the duals to the trisecants of S'.*

Proof. $x \in L(s, t)$ means $x^* \in \mathbf{P}_3^*$ lies on the line spanned by $x^*(s, t)$ and $\partial_s x^*(s, t)$, i.e., the tangent to S' at $x^*(s; t)$. The second assertion follows from the observation that the coordinate transformation

$$x_0^* = x_2, \quad x_1^* = x_3, \quad x_2^* = x_1, \quad x_3^* = x_0$$

$$s = \lambda, \quad t = \mu$$

interchanges the role of S and S'.

(2.6) *The curve S admits precisely six double tangents, joining pairs of points $x(\lambda, \mu)$ with $\Delta(\lambda, \mu) = 0$. These double tangents lie on D, the surface swept out by the trisecants $L(s, t)$. Any other tangent to S meets the curve only in its point of contact, and it meets D in ≤ 4 points outside of S.*

Proof. Let L be tangent to S at $x(\lambda, \mu)$ and meet S again at another point $x(\lambda', \mu')$. In (2.2) we observed $L \subset W(s_i, t_i)$ for the three solutions $s_i : t_i$ of $P_3(\lambda, \mu; s_i, t_i) = 0$. This implies $P_3(\lambda', \mu'; s_i, t_i) = 0$. Using $\partial_s P_3^2(\lambda, \mu; s_i, t_i) = \partial_s P_3^2(\lambda', \mu'; s_i, t_i) = 0$ we see $L = L(s_i, t_i)$, $i = 1, 2, 3$. Then necessarily

$$0 = \frac{1}{2} \partial_\lambda (\partial_s P_3^2(\lambda, \mu; s_i, t_i)) = \partial_\lambda P_3(\lambda, \mu; s_i, t_i) \cdot \partial_s P_3(\lambda, \mu; s_i, t_i).$$

So $\Delta(\lambda, \mu)$ or $\Delta(s_i, t_i) = 0$. In fact $\Delta(\lambda, \mu) = \Delta(s_i, t_i) = 0$, cf. Section 1. And conversely: If $\Delta(\lambda, \mu) = 0$ the tangent at $x(\lambda, \mu)$ touches S again at $x(\lambda', \mu')$ corresponding to the other root (λ', μ') of $P_3(-, -; s_i, t_i)$.

Assume L to be a tangent of S lying totally on D. Under the duality (2.5) it corresponds to a trisecant L^* of S' on the tangent scroll of S'. Since S' generates $\mathbf{P}_3$, L^* coincides with one tangent to S'. L is thus a trisecant, in fact a double tangent, to S.

Finally, let us count the intersections of a tangent L, not on D, with D outside of S. Under (2.5) these intersections correspond to intersections of L^* with tangents to S'. Projecting S' onto $\mathbf{P}_1$ from L^* defines a covering of degree ≤ 3, hence with ≤ 4 branch points. Each tangent of S' meeting L^* causes one of these.

(2.7) *Any pair L, M of coplanar tangents meets in a point not on D (unless L or M is a bitangent).*

Proof. Under (2.5) the assertion dualizes as follows: In $W^* \in S'$—the point dual to the plane $W(s,t)$ containing L and M according to (2.4)—the trisecants L^* and M^* of S' meet. They span a plane which should not contain any tangent of S'. If $\Delta(s,t) \neq 0$, the trisecants L^* and M^* meet S' transversally in together five points, so there can be no sixth point in which S' touches the plane.

§3. The projected curves S_x.

For $x \in \mathbf{P}_3$, by $S_x \subset \mathbf{P}_2$ we denote the *projection of S from x*. As a curve in $\mathbf{P}_2$ it is determined up to automorphisms of $\mathbf{P}_2$ only. Whenever $x \notin S$, we have $\deg S_x = 6$.

We denote by $D \subset \mathbf{P}_3$ the (closure of the) *surface swept out by all the trisecants* to S. Whenever $x \notin D$, the singularities of S_x are double points only. Let $T \subset \mathbf{P}_3$ be the *tangent scroll* to S. For $x \notin D \cup T$ each singularity of S_x consists of two smooth branches. If these branches would touch, their corresponding points on S would have coplanar tangents, and by (2.4) this would imply $x \in D$. So for $x \notin D \cup T$ the curve S_x has ordinary double points only. By the genus formula their number is ten.

(3.1) *Assume $x \notin D \cup T$. Then S_x has exactly six tritangents L_i. None of these L_i contains a node of S_x nor do three L_i have a point in common.*

Proof. If $L \subset \mathbf{P}_2$ is a tritangent to S_x, the plane $W \subset \mathbf{P}_3$ through x determined by L touches S in three points, hence W contains coplanar tangents.

By (2.4) we have $W = W(s,t)$ for some $(s:t) \in \mathbf{P}_1$, and L is the projection of $L(s,t)$. In this way the tritangents to S_x correspond one to one with the trisecants $L(s,t)$ satisfying $x \in W(s,t)$.

A general point $x \in \mathbf{P}_3$ lies on six distinct planes $W(s,t)$. Two of these come together if the plane $x^* \in \mathbf{P}^*$ contains a tangent to the curve S' from (2.5), i.e., only if $x \in D$.

A tritangent cannot pass through a node of S_x, because then $L.S_x > 6$.

If three tritangents L_i have a point in common, x lies on a line M common to three planes $W(s_i, t_i)$, i.e., M^* is a trisecant to S' and M a tangent to S by (2.5).

(3.2) *Assume $x \notin D \cup T$ as above. Then each tritangent L_i to S_x touches S_x in three distinct points, unless $x \in W(s,t)$ for some $(s:t) \in \mathbf{P}_1$ with*

$\Delta(s,t) = 0$. *In this case the projection L_i of $L(s,t)$ has fourfold contact with S_x in one point and touches S_x in another one.*

Proof. If $\Delta(s,t) \neq 0$, the three intersections of $L(s,t)$ with S are distinct and project to three distinct points. If $\Delta(s,t) = 0$ two intersections come together. The plane $W(s,t)$ has intersection multiplicity 4 with S at this point and its projection, when x lies in this plane, is a point where $L(s,t)$ and S_x have intersection number 4.

Denote by $W \subset \mathbf{P}_3$ the union of the twelve planes mentioned in (3.2). In the sequel we choose the center of projection $x \in \mathbf{P}_3$ general, i.e., *not on D, T, or W*.

(3.3) *Assume $x \notin D \cup W \cup T$. Except for the six tritangents L_i the curve S_x has precisely six double tangents: the projections of the six double tangents to S from (2.6).*

Proof. If a double tangent of S_x is not the projection of a double tangent to S, it determines a plane $W(s,t)$ containing two tangents of S, cf. (2.4). So the double tangent is a tritangent.

§4. The double plane X.

Here we describe the surface $X = X_x$, the double cover of $\mathbf{P}_2$ branched over S_x for $x \in \mathbf{P}_3$ general. To be precise: This double cover has ten nodes over the ten double points of S_x; by X we mean its minimal desingularization. X is a K3-surface containing ten smooth rational curves N_i, one over each double point $p_i \in S_x$.

If $L \subset \mathbf{P}_2$ is a tritangent to S_x, the double cover over L decomposes into two rational curves M, $M' \subset X$. By (3.1) M and M' are disjoint from the ten N_i. For any collection $\{L_i\}$ of tritangents to S_x we define the *reduced double cover* for $\cup L_i$ to be $\cup(M_i \cup M_i')$ where M_i and M_i' are torn apart at the three points $M_i \cap M_i'$. This reduced double cover of $\cup L_i$ is unramified.

(4.1) *The following properties are equivalent:*
a) *The reduced double cover over $\cup_1^6 L_i$ is trivial.*
b) *The eighteen points of contact of the six tritangents are cut out on S_x by some cubic curve in $\mathbf{P}_2$.*
c) *Over each triangle L_i, L_j, L_k of tritangents the reduced double cover is trivial.*

Proof. Properties a) and b) are equivalent by the definition of branched double covers. Obviously a) implies c). If c) holds, fix one tritangent L_1 with M_1, $M_1' \subset X$ the two curves over it. For $i = 2, \ldots, 6$ let M_i (resp. M_i') be the curve over L_i that meets M_1 (resp. M_1'). Condition c) guarantees $M_i \cap M_j' = \emptyset$ for $i \neq j$ and this is equivalent to a).

(4.2)   *The reduced double cover over $\cup_1^6 L_i$ is not trivial.*

Proof.   The triviality of this double cover is a topological fact, so it does not change under small deformations. If it is trivial for one $x \in \mathbf{P}_3 \setminus (D \cup T \cup W)$, then it is trivial for all these x. Condition (4.1 b) then implies there is a cubic cone in $\mathbf{P}_3$ with vertex x cutting out on S the eighteen points where the six trisecants $L(s,t)$ meet S, $x \in W(s,t)$. By continuity this then holds for all $x \in \mathbf{P}_3$.

Consider, e.g., $x = (0:0:0:1)$. Then $x \in W(s,t)$ for

$$(s:t) = (1:0) \quad \text{and} \quad (2^{1/5}:\varepsilon^k), \ 0 \le k \le 4.$$

These six values $(s:t)$ determine the eighteen zeros of

$$f(\lambda,\mu) = \lambda^2 \mu \cdot \prod_{k=0}^{4} (2^{2/5}\varepsilon^k\lambda^3 + \varepsilon^{3k}\lambda^2\mu + 2^{3/5}\lambda\mu^2 - 2^{1/5}\varepsilon^{2k}\mu^3).$$

We observe

$$f(\varepsilon\lambda,\mu) = \varepsilon^2\lambda^2\mu \prod_{k=0}^{4}(2^{2/5}\varepsilon^{k+3}\lambda^3 + \varepsilon^{3k+2}\lambda^2\mu + 2^{3/5}\varepsilon\lambda\mu^2 - 2^{1/5}\varepsilon^{2k}\mu^3)$$

$$= \varepsilon^3\lambda^2\mu \prod_{k=0}^{4}(2^{2/5}\varepsilon^{k+2}\lambda^3 + \varepsilon^{3(k+2)}\lambda^2\mu + 2^{3/5}\lambda\mu^2 - 2^{1/5}\varepsilon^{2(k+2)}\mu^3)$$

$$= \varepsilon^3 f.$$

If this is cut out by a cubic on S, this f is a cubic polynomial $Q(f_0,\ldots,f_4)$ in

$$f_0 = \mu(2\lambda^5 + \mu^5), \quad f_1 = \lambda(\lambda^5 - 2\mu^5), \quad f_2 = \lambda^2\mu^4, \quad f_4 = \lambda^4\mu^2.$$

These polynomials are labelled such that $f_i(\varepsilon\lambda,\mu) = \varepsilon^i f(\lambda,\mu)$. It follows that Q is a linear combination of f_i, f_j, f_k with $i+j+k \equiv 3$ (5), i.e., of

$$f_0 f_1 f_2, \quad f_0 f_4^2, \quad f_1^3, \quad f_2^2 f_4.$$

These four polynomials are divisible by λ^3, whereas λ^3 does not divide f. Hence there is no cubic in $\mathbf{P}_3$ cutting out the eighteen roots of f.

(4.3)   *Over any triangle L_i, L_j, L_k, the reduced double cover is non-trivial.*

Proof.   The assertion is topological. If the cover is trivial over one triangle for one x, it is trivial over all triangles into which the first triangle deforms, when $x \in \mathbf{P}_3 \setminus (D \cup T \cup W)$ moves. It suffices to show that in the 3-parameter family obtained when x moves, the monodromy is transitive on triangles. The triviality over all triangles would then contradict (4.1 c) and (4.2).

Triangles of tritangents for S_x correspond to triplets of roots of the polynomial

$$f(x;s,t) = x_0 s^4 t^2 + x_1 s^2 t^4 + x_2 t(2s^5 + t^5) + x_3 s(s^5 - 2t^5).$$

To show that the monodromy acts transitively on these triplets, we consider the variety

$$I = \{ (x; s_1 : t_1, s_2 : t_2, s_3 : t_3) \in \mathbf{P}_3 \times \mathbf{P}_1^3 :$$
$$f(x; s_1, t_1) = f(x; s_2, t_2) = f(x; s_3, t_3) = 0 \}.$$

I contains a component projecting surjectively onto $\mathbf{P}_1^3$. This shows that any triplet can be deformed into any other one.

Assertion (4.3) is visualized as in Figure 2.

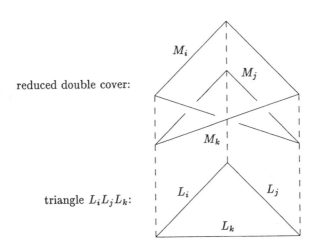

reduced double cover:

triangle $L_i L_j L_k$:

Figure 2.

So over each line L_i, L_j, L_k it is possible to choose one curve M_i, M_j, M_k with $M_i.M_j = M_i.M_k = M_j.M_k = 0$.

(4.4) *It is possible to choose curves* $M_1, \ldots, M_6 \subset X$ *over* $L_1, \ldots, L_6$ *such that* $M_i.M_j = 0$ *for* $1 \leq i \neq j \leq 6$.

Proof. Fix the first curve $M_1 \subset X$ over L_1 arbitrarily. For $i = 2, \ldots, 6$ let $M_i \subset X$ be the curve over L_i uniquely determined by $M_i.M_1 = 0$. ($L_i \cap L_j \cap S_x = \emptyset$ whenever $i \neq j$, cf. (3.1).) Then $M_i.M_j = 0$ for any pair $i \neq j > 1$ by the above observation applied to the triangle L_1, L_i, L_j.

So on X we identified sixteen disjoint rational curves: $M_1, \ldots, M_6$ and $N_1, \ldots, N_{10}$. By Nikulin's theorem [N], X is a *Kummer surface*, i.e., there is an abelian surface A, the blow up $\tilde{A} \to A$ of sixteen half-periods, and a double cover $\tilde{A} \to X$ branched over $M_1, \ldots, N_{10}$, such that these branch curves are precisely the images of the sixteen exceptional curves in $\tilde{A}$.

(4.5) *The line bundle $\mathcal{O}_{\mathbf{P}_2}(1)$ induces on $A = A_x$ a polarization of type $(1, 5)$.*

Proof. Let $\tilde{H}$ be the pullback of $\mathcal{O}_{\mathbf{P}_2}(1)$ to $\tilde{A}$. Then $\tilde{H}^2 = \deg(\tilde{A} \to \mathbf{P}_2) = 4$. The pullback to $\tilde{A}$ of a general line in $\mathbf{P}_2$ avoids the ten exceptional curves over $N_1, \ldots, N_{10}$ meeting transversally in one point each of the six exceptional curves over $M_1, \ldots, M_6$. When passing from $\tilde{A}$ to A the latter are blown down. This yields $4 + 6 = 10$ as the self-intersection of the line bundle induced by $\tilde{H}$ on A.

§5. The double plane Y.

Here, for $x \in \mathbf{P}_3$ general, we consider the double cover $Y \to \mathbf{P}_2$ branched over the six tritangents $L_1, \ldots, L_6$ to the curve S_x. To be precise, this double cover has fifteen nodes over the intersections $L_i \cap L_j$. By $Y = Y_x$ we mean its minimal desingularization. Y is a K3-surface containing fifteen disjoint smooth rational curves N_{ij}, one over each intersection $L_i \cap L_j$, $1 \leq i < j \leq 6$.

Over S_x the covering splits as $S' \cup S''$ with two rational curves S', $S'' \subset Y$. By the *reduced cover* $S' \cup S'' \to S_x$ we mean the double cover, where the eighteen intersections of S' with S'' over the tritangents L_i are separated. The curve $S' \cup S''$ thus has twenty nodes, two each over each node $p \in S_x$.

(5.1) *The reduced cover $S' \cup S'' \to S_x$ is nontrivial, i.e., the induced maps S', $S'' \to S_x$ are not isomorphisms.*

Proof. Triviality of the cover is equivalent to the existence of a cubic curve in $\mathbf{P}_2$ cutting out on S_x the eighteen points of contact with its tritangents L_i. But on $\cup_1^6 L_i$ this cubic would cut out the same points, in conflict with (4.1 b) and (4.2).

Next we specify generators for the fundamental group $\pi_1(S_x)$: Fix a smooth base point $p_0 \in S_x$. Let $\nu : \mathbf{P}_1 \to S_x$ be the normalization and $q_0 = \nu^{-1} p_0 \in \mathbf{P}_1$. For each node $p_i \in S_x$ let $\nu^{-1} p_i = \{q_i, q_i'\} \subset \mathbf{P}_1$. Let γ_i, γ_i' be paths in $\mathbf{P}_1$ joining q_0 to q_i, q_i' (they are unique up to homotopy). Put finally $\omega_i := (-\nu \gamma_i') \circ \nu \gamma_i$. The class $[\omega_i] \in \pi_1(S_x)$ depends on the node p_i only (and on the ordering of q_i, q_i', which influences the sign of $[\omega_i]$). The ten classes $[\omega_i]$ generate $\pi_1(S_x)$.

(5.2) *The reduced cover $S' \cup S''$ is nontrivial over each loop ω_i.*

Proof. Consider the varieties

$$V = \{\, (p, x) \in \mathbf{P}_2 \times \mathbf{P}_3 \setminus (D \cup T \cup W) : p \text{ node on } S_x \,\}$$
$$V' = \{\, (L, x) \in \mathrm{Sec}(S) \times \mathbf{P}_3 : x \in L \}.$$

Under projection from x, an open dense set of the irreducible variety V' is isomorphic to V. So V is irreducible. The projection $V \to \mathbf{P}_3 \setminus (D \cup T \cup W)$

is finite and unramified of degree ten, cf. the beginning of Section 3, with the monodromy acting transitively on its fibres by the connectedness of V. If the reduced cover would be trivial over one ω_i, it would therefore be trivial over all ω_i, hence trivial over S_x. This contradicts (5.1).

Assertion (5.2) means that both curves S', $S'' \subset Y$ are smooth over each node $p_i \in S_x$. So both of them are smooth rational curves on Y. Picking one of them at random and calling it N_0, we have sixteen disjoint smooth rational curves N_0, N_{ij} on Y. Again by Nikulin's theorem [N] the surface Y is a *Kummer surface*. Let $\tilde{A} \to Y$ be the double cover, E_0, $E_{ij} \subset \tilde{A}$ the exceptional curves over N_0, N_{ij} and $\sigma\colon \tilde{A} \to A$ their blow-down.

(5.3) *The line bundle* $\mathcal{O}_{\mathbf{P}_2}(1)$ *induces on* $A = A_x$ *a polarization of type* $(2, 10)$.

Proof. Let $\tilde{H}$ be the pullback of $\mathcal{O}_{\mathbf{P}_2}(1)$ to $\tilde{A}$. Then $\tilde{H}^2 = 4$, $\tilde{H}.E_{ji} = 0$, and $\tilde{H}.E_0 = 6$. The line bundle $\tilde{H} \otimes \mathcal{O}_{\tilde{A}}(6E_0)$ descends to a line bundle H on A with

$$H^2 = 4 + 72 - 36 = 40.$$

So H defines on A a polarization of type $(1, 20)$ or $(2, 10)$. To see that the latter is the case, consider one tritangent, say L_1. Its pullback to Y is of the form $2M_1 + \sum_2^6 N_{1j}$. Hence $\tilde{H}$ on $\tilde{A}$ is divisible by two; so also must H be, on A.

§6. Moduli.

As in Section 4 let A_x be the abelian surface with Kummer surface X_x $(x \in \mathbf{P}_3 \setminus (D \cup T \cup W))$. Let $\pi\colon \tilde{A}_x \to \mathbf{P}_2$ be the map of degree four factoring over the double plane X_x. Let $\mathcal{L}$ be the line bundle on A_x induced by $\pi^* \mathcal{O}_{\mathbf{P}_2}(1)$ on $\tilde{A}_x$. We observe that $\mathcal{L}$ is of type $(1, 5)$.

Let ι on A_x be the involution with fixed points the sixteen half-periods e_1, ..., e_{16} (corresponding to $M_1, \ldots, N_{10}$). By construction $\iota^* \mathcal{L} = \mathcal{L}$. We can lift ι to $\mathcal{L}$ such that it acts by $+1$ on the vector space $\pi^* \mathcal{O}_{\mathbf{P}_2}(1)$. By Riemann-Roch $h^0(\mathcal{L}) = 5$. We put $h^0(\mathcal{L})^{\pm} = \dim H^0(\mathcal{L})^{\pm}$ with $H^0(\mathcal{L})^{\pm}$ the eigenspaces for ι.

(6.1) $h^0(\mathcal{L})^+ = 3$.

Proof. Since $H^0(\mathcal{L})^+$ contains $\alpha_* \pi^* H^0(\mathcal{O}_{\mathbf{P}_2}(1))$ with $\alpha\colon \tilde{A}_x \to A_x$, we must exclude the possibility $h^0(\mathcal{L})^+ \geq 4$. The pullback of the generic line in $\mathbf{P}_2$ does not contain any $e_7, \ldots, e_{16}$, so ι acts by $+1$ over $e_7, \ldots, e_{16}$, hence by -1 over e_1, ..., e_6 and any section in $H^0(\mathcal{L})^+$ vanishes at $e_1, \ldots, e_6$. If $h^0(\mathcal{L})^+ \geq 4$, the sections $s \in H^0(\mathcal{L})^+$ vanishing (doubly) at e_7 would form a vector space of dimension ≥ 3. Counting intersections of the curves $\{s = 0\}$ this is possible only if all these curves are reducible. However, the pullback of the general line through the node of S_x corresponding to e_7 is easily seen to be irreducible, a contradiction.

(6.2) *For x, $y \in \mathbf{P}_3 \setminus (D \cup T \cup W)$ assume $A_x = A_y$ (as polarized surfaces).
Then the branch curves S_x, $S_y \subset \mathbf{P}_2$ are projectively equivalent.*

Proof. The decomposition $H^0(\mathcal{L}) = H^0(\mathcal{L})^+ \oplus H^0(\mathcal{L})^-$ depends only on A.
By (6.1) the map $\tilde{A} \to \mathbf{P}_2$ is given by the linear system $H^0(\mathcal{L})^+$. This map
factors through $X_x = \operatorname{Km} A_x$ and determines the branch curve of the double
plane X_x.

(6.3) *Given $x \in \mathbf{P}_3 \setminus (D \cup T \cup \dot{W})$ there are at most sixty different points
$y \in \mathbf{P}_3 \setminus (D \cup T \cup W)$ with $S_y \subset \mathbf{P}_2$ isomorphic to S_x.*

Proof. Let $\phi : \mathbf{P}_2 \to \mathbf{P}_2$ be an isomorphism inducing an isomorphism
$\varphi : S_x \to S_y$ which lifts to the smooth curve as $\tilde{\varphi} : S \to S$. Now ϕ maps double
tangents of S_x to those of S_y, hence by (3.3) the vertices of the icosahedron
$\Delta(\lambda, \mu) = 0$ (cf. Section 1) are permuted under $\tilde{\varphi}$. So $\tilde{\varphi}$ is one of the sixty
elements φ_i, $i = 1, \ldots, 60$, in the icosahedral group.
  The ten nodes on S_x determine ten secants of S in $\mathbf{P}_3$ and those determine x,
the center of projection, as their common point of intersection. $\tilde{\varphi}$ maps the ten
pairs of points where these secants meet S to ten pairs on secants corresponding
to the nodes of S_y. The latter secants determine Y. As there are ≤ 60 choices
φ_i for $\tilde{\varphi}$, there are at most sixty points y.

Assertion (6.3) means that varying x we obtain a three-dimensional family
of abelian surfaces A_x. This family will cover an open set in the moduli space
of these surfaces. In particular, for a dense set of points x the surface A_x will
have Picard number $\rho(A_x) = 1$. This allows us to identify the abelian surfaces
from Section 4 with those from Section 5.

(6.4) *Let A_x be the abelian surface, birationally equivalent to the double
cover $\tilde{A}_x \to X_x$ from Section 4 with $\mathcal{L}_x$ on A_x the induced $(1, 5)$-polarization;
and similarly let B_x, with $(2, 10)$-polarization $\mathcal{M}_x$, be the abelian surface for Y_x.
Then A_x and B_x are isomorphic with $\mathcal{M}_x = \mathcal{L}_x^{\otimes 2}$.*

Proof. We trace a tritangent L to S_x under the fourfold covers

$$\tilde{A}_x \longrightarrow X_x \longrightarrow \mathbf{P}_2, \qquad \tilde{B}_x \longrightarrow Y_x \longrightarrow \mathbf{P}_2.$$

The inverse image of L splits as $M \cup M'$ on X_x. On $\tilde{A}_x$ the curve M becomes
an exceptional curve blowing down to a half-period $e \in A_x$. The inverse image
$\tilde{C} \subset \tilde{A}_x$ of M' is smooth of genus three. The hyperelliptic curve $\tilde{C}$ is uniquely
determined by the eight branch points for $\tilde{C} \to M'$ which lie over the three
points of contact of L with S_x and the five points where L intersects the other
tritangents. In A_x the image C of $\tilde{C}$ acquires a triple point at e.
  The inverse image $M \subset Y_x$ of L is a smooth (-2)-curve. The inverse image
of this in $\tilde{B}_x$ is another copy of $\tilde{C}$ and its image in B_x another copy of C. By
the universal property of jacobians the maps $\tilde{C} \to A_x$ and $\tilde{C} \to B_x$ factorize

$$\tilde{C} \longrightarrow \operatorname{Jac}(\tilde{C}) \longrightarrow A_x, \qquad \tilde{C} \longrightarrow \operatorname{Jac}(\tilde{C}) \longrightarrow B_x$$

inducing exact sequences

$$0 \longrightarrow E \longrightarrow \mathrm{Jac}(\tilde{C}) \longrightarrow A_x \longrightarrow 0,$$
$$0 \longrightarrow F \longrightarrow \mathrm{Jac}(\tilde{C}) \longrightarrow B_x \longrightarrow 0$$

with elliptic curves E, F. Dualizing we obtain embeddings

$$A_x^\vee, B_x^\vee \longrightarrow \mathrm{Jac}(\tilde{C}).$$

If the images of $A_x^\vee$ and $B_x^\vee$ are different subtori, their intersection is an elliptic curve on $A_x^\vee$ and $B_x^\vee$. For general x this is impossible, because we know $\rho(A_x^\vee) = 1$. This implies $A_x^\vee \cong B_x^\vee$ for all x. Under the induced map $A_x \to B_x$ the images of $\tilde{C}$ correspond. This proves

$$\mathcal{M}_x = \mathcal{O}_{B_x}(2C) = \mathcal{L}_x^{\otimes 2}.$$

§7. Explanations.

In this final section we discuss the relation of the preceding sections with [HM]. This shall explain some of the apparent coincidences described above.

The *vector bundle $\mathcal{F}$ of Horrocks-Mumford* on $\mathbf{P}_4$ has a 4-dimensional vector space $H^0(\mathcal{F})$ of sections. The general section $s \in H^0(\mathcal{F})$ vanishes on a smooth abelian surface $A \subset \mathbf{P}_4$. In this way $\mathbf{P}_3 = \mathbf{P}H^0(\mathcal{F})$ is birational to a compactification of the *moduli space of abelian surfaces with a polarization of type $(1,5)$ and a distinguished level structure*, cf. [HM, §6]. The icosahedral group operates on this $\mathbf{P}_3$ permuting the points corresponding to different level structures on the same surface.

The boundary of this moduli space was described in [BHM], i.e., the singular surfaces corresponding to its points were classified there. Most notably there appears a one-parameter family of non-reduced structures on quintic scrolls [HV]. These scrolls are defined by elliptic curves with level-5 structure. Using as parameter $\lambda : \mu \in \mathbf{P}_1$, F. Klein's "Ikosaeder Transzendente" parametrizing this curve, its image in $\mathbf{P}_3 = \mathbf{P}H^0(\mathcal{F})$, upon the right choice of coordinates, is just the curve S from Section 2. The equation $P_3(\lambda, \mu; s, t) = 0$ describes for given $\lambda : \mu$ the three parameters $s_i : t_i \in \mathbf{P}_1$ corresponding to the three 2-torsion quotients of the elliptic curve belonging to $\lambda : \mu$.

The essential new observation is that the abelian surface $A \subset \mathbf{P}_4$ corresponding to $x \in \mathbf{P}H^0(\mathcal{F})$ is reconstructed as a fourfold cover of the plane $\mathbf{P}_2$ starting with a double plane X ramified over the projected curve S_x. A proof of this is somewhat involved; beyond the scope of this paper. (In fact, the fourfold cover is not A, but the dual surface $A^\vee$.) Singular surfaces arise when x, the center of projection, belongs to the trisecant surface D. When $x \in (T \cup W) \setminus D$, the corresponding surface is smooth but special. We intend to describe these surfaces, as well as other interesting ones, in another context.

References

[BHM] W. Barth, K. Hulek and R. Moore, Degenerations of Horrocks-Mumford surfaces, Math. Ann. **277**(1987), 735–755.

[F] W. Fulton, *Intersection Theory*, Springer-Verlag, Berlin, Heidelberg, New York, Tokyo, 1984.

[HM] G. Horrocks and D. Mumford, A rank 2 vector bundle on $\mathbf{P}_4$ with 15,000 symmetries, Topology **12**(1973), 63–81.

[HV] K. Hulek and A. Van de Ven, The Horrocks-Mumford bundle and the Ferrand construction, Manuscr. Math. **50**(1986), 313–335.

[N] V. V. Nikulin, On Kummer surfaces, (English translation), Math. USSR-Izv. **9**(1975), 261–275.

Wolf BARTH

University of Erlangen-Nürnberg
Department of Mathematics
Bismarckstrasse $1\frac{1}{2}$, D-8520 Erlangen
Federal Republic of Germany

Ross MOORE

Macquarie University
School of Mathematics and Physics
North Ryde, Sydney
Australia

Algebraic Geometry and Commutative Algebra
in Honor of Masayoshi NAGATA
pp. 59–64 (1987)

On Rings of Invariants of Finite Linear Groups

Shizuo ENDO

Let k be a field. An element τ of finite order in $\mathrm{GL}_n(k)$ is called a pseudo-reflection if $\mathrm{rank}(\tau - I) = 1$. A finite subgroup of $\mathrm{GL}_n(k)$ is called small if it contains no pseudo-reflection.

The purpose of this note is to prove the following theorem.

Theorem A. *Let k be a field and let $S = k[X_1, X_2, \ldots, X_n]$ be the polynomial ring in indeterminates X_1, X_2, $\ldots$, X_n over k. Let G and G' be small finite subgroups of $\mathrm{GL}_n(k)$ acting linearly on S. Then the following conditions are equivalent:*

(1) $S^G \cong S^{G'}$ *as k-algebras.*

(2) G *and* G' *are conjugate in* $\mathrm{GL}_n(k)$.

Theorem A is viewed as a global version of the following theorem of D. Prill ([6]).

Let S be a regular complex-analytic algebra of dimension n. Let G and G' be small finite subgroups of $\mathrm{GL}_n(\mathbf{C})$ acting linearly on S. Then the following conditions are equivalent:

(1) $S^G \cong S^{G'}$ *as complex-analytic algebras.*

(2) G *and* G' *are conjugate in* $\mathrm{GL}_n(\mathbf{C})$.

It should be noted that in the local case, some generalizations of the theorem of Prill are given in [1], [4], etc.

§1. Fundamental groups.

We recall some basic properties of the fundamental groups of Noetherian normal domains (cf. [2], [3]). Let R be a Noetherian normal domain and let K be the quotient field of R. We denote by K_s the fixed separable closure of K, and consider only the extensions of K contained in K_s. Let L be a finite extension of K and let S be the integral closure of R in L. We say that S (resp. L) is a 1-unramified extension of R (resp. K) if every prime ideal of height 1 in S is unramified over R.

We easily see the following:

Received February 2, 1987.

(a) If L is 1-unramified over K and if F is 1-unramified over L, then F is also 1-unramified over K.

(b) If L and L' are 1-unramified over K, then the composite LL' of L and L' is also 1-unramified over K.

(c) If L is 1-unramified over K, then the smallest Galois extension of K containing L is also 1-unramified over K.

We define $K_1 = \bigcup_L L$ and $\pi(R) = \varprojlim_L \mathrm{Gal}(L/K)$, where L runs over all 1-unramified Galois extensions of K contained in K_s. Then K_1 is a (not always finite) Galois extension of K and $\pi(R)$ is the (profinite) Galois group of this extension. Let R_1 denote the integral closure of R in K_1. We say that R_1 (resp. K_1) is the universal 1-unramified extension of R (resp. K) and that $\pi(R)$ is the fundamental group of R.

Let $V = \mathrm{Spec}\,R$ and let $\mathrm{Sing}\,V$ denote the singular locus of V. Then, by the purity of branch locus ([5, (41.1)]), the group $\pi(R)$ coincides with the étale fundamental group $\pi^{et}(V - \mathrm{Sing}\,V)$ of the open subscheme $V - \mathrm{Sing}\,V$ of V. Especially, if R is regular, then $\pi(R) = \pi^{et}(V)$.

The following proposition is well known (e.g., [1], [3], [7]).

Proposition 1. *Let R be one of the following rings:*

(1) *a Henselian regular local rings whose residue class field is algebraically closed.*

(2) *a polynomial ring in a finite number of indeterminates over an algebraically closed field of characteristic 0.*

Then we have $\pi(R) = \{I\}$.

Let S be a Noetherian normal domain and let $\mathrm{Aut}\,S$ be the group of all automorphisms of S. An element τ ($\neq I$) of finite order in $\mathrm{Aut}\,S$ is called a pseudo-reflection if there exists a prime ideal $\mathfrak{p}$ of height 1 in S such that $(\tau - I)(S) \subseteq \mathfrak{p}$. A finite subgroup of $\mathrm{Aut}\,S$ is called small if it contains no pseudo-reflection. In the case of linear actions as in the title page, these definitions coincide with those given before.

Let G be a finite subgroup of $\mathrm{Aut}\,S$ and assume that S^G is Noeterian. Then, by ramification theory (e.g., [5, §41]), we see that S is 1-ramified over S^G if and only if G is small.

We here give a slight generalization of [1, Satz 6].

Proposition 2. *Let S and S' be Noetherian normal domains such that $\pi(S) = \pi(S') = \{I\}$, and let G and G' be small finite subgroups of $\mathrm{Aut}\,S$ and $\mathrm{Aut}\,S'$, respectively. Assume that S^G and $S'^{G'}$ are Noetherian. Then the following conditions are equivalent:*

(1) S^G *and* $S'^{G'}$ *are isomorphic.*

(2) *There exists an isomorphism $\rho\colon S \to S'$ such that $\rho G \rho^{-1} = G'$.*

Under the assumptions in Proposition 2, S and S' are the universal 1-unramified extensions of S^G and $S'^{G'}$, respectively, and so $\pi(S^G) = G$ and

$\pi(S'^{G'}) = G'$. Hence, Proposition 2 follows immediately from the uniqueness of universal 1-unramified extensions of a Noetherian normal domain.

§2. Proof of Theorem A

The implication (2) $\Rightarrow$ (1) is obvious. In order to prove (1) $\Rightarrow$ (2) we may assume by a standard result in representation theory that k is algebraically closed.

Set $\mathfrak{M} = \sum_{i=1}^{n} SX_i$ and $\mathfrak{m} = \mathfrak{M} \cap S^G$. Then $\mathfrak{m}$ is a maximal ideal of S^G and $\mathfrak{M}$ is the unique maximal ideal of S containing $\mathfrak{m}$.

Assume that there exists a k-(algebra) isomorphism $\sigma: S^G \to S^{G'}$. Set $\mathfrak{m}' = \sigma(\mathfrak{m})$ and let $\mathfrak{M}'_1, \mathfrak{M}'_2, \ldots, \mathfrak{M}'_t$ be all maximal ideals of S containing $\mathfrak{m}'$. Given a Noetherian ring R, a prime ideal $\mathfrak{p}$ of R and a finitely generated R-module T, let $T^*_\mathfrak{p}$ denote the completion of T at $\mathfrak{p}$. The k-isomorphism $\sigma: S^G \to S^{G'}$ induces a k-isomorphism $\sigma^*: (S^G)^*_\mathfrak{m} \to (S^{G'})^*_{\mathfrak{m}'}$. The groups G and G' act on $S^*_\mathfrak{m}$ and $S^*_{\mathfrak{m}'}$, respectively, and we have $(S^*_\mathfrak{M})^G = (S^*_\mathfrak{m})^G = (S^G)^*_\mathfrak{m}$ and $(S^*_{\mathfrak{m}'})^{G'} = (S^{G'})^*_{\mathfrak{m}'}$. Write $\mathfrak{M}' = \mathfrak{M}'_1$ and set $H' = \{ \tau' \in G' \mid \tau'(\mathfrak{M}') = \mathfrak{M}' \}$. Then it follows from ramification theory that $(S^*_\mathfrak{M})^{H'}$ and $(S^*_{\mathfrak{m}'})^{G'}$ are k-isomorphic. Therefore we get a k-isomorphism $\mu^*: (S^*_\mathfrak{M})^G \to (S^*_{\mathfrak{M}'})^{H'}$. Both $S^*_\mathfrak{M}$ and $S^*_{\mathfrak{M}'}$ are the formal power series rings over the algebraically closed field k, and hence, by Proposition 1, we have $\pi(S^*_\mathfrak{M}) = \pi(S^*_{\mathfrak{M}'}) = \{I\}$. We easily see that G and H' are small subgroups of Aut $S^*_\mathfrak{M}$ and Aut $S^*_{\mathfrak{M}'}$, respectively. Therefore, by Proposition 2, the k-isomorphism $\mu^*: (S^*_\mathfrak{M})^G \to (S^*_{\mathfrak{M}'})^{H'}$ can be extended to a k-isomorphism $\rho^*: S^*_\mathfrak{M} \to S^*_{\mathfrak{M}'}$ such that $\rho^* G \rho^{*-1} = H'$. Hence we have $|G| = |H'| \leq |G'|$. Similarly we can show that $|G'| \leq |G|$, and so it follows that $H' = G'$. Thus $\mathfrak{M}'$ is the unique maximal ideal of S containing $\mathfrak{m}'$, and we get a k-isomorphism $\rho^*: S^*_\mathfrak{M} \to S^*_{\mathfrak{M}'}$ such that $\rho^* G \rho^{*-1} = G'$.

Since k is algebraically closed, we can write $\mathfrak{M}' = \sum_{i=1}^{n} S(X_i - a_i)$, $a_1, a_2, \ldots, a_n \in k$. Then G' acts linearly on the k-vector space $\sum_{i=1}^{n} k(X_i - a_i)$, and the matrix representation of G' on $X_1 - a_1, X_2 - a_2, \ldots, X_n - a_n$ is the same as the one of G' on $X_1, X_2, \ldots, X_n$. Therefore we may assume that $\mathfrak{M} = \mathfrak{M}'$.

Under this assumption, G and G' act linearly on the formal power series ring $S^*_\mathfrak{M} = k[[X_1, X_2, \ldots, X_n]]$ and there exists a k-automorphism ρ^* of $S^*_\mathfrak{M}$ such that $\rho^* G \rho^{*-1} = G'$. Then, the actions of G and G' on $S^*_\mathfrak{M}$ induce those of G and G' on the k-vector space $V = \mathfrak{M} S^*_\mathfrak{M} / \mathfrak{M}^2 S^*_\mathfrak{M}$, and hence G and G' can be regarded as subgroups of $GL(V)$. Moreover, the k-automorphism ρ^* of $S^*_\mathfrak{M}$ induces a k-linear automorphism $\bar{\rho}$ of V such that $\bar{\rho} G \bar{\rho}^{-1} = G'$. This shows that G and G' are conjugate in $GL(V)$. However, V can be identified with $\sum_{i=1}^{n} kX_i$ ($\subseteq S \subseteq S^*_\mathfrak{M}$) as representation spaces of G and G'. Thus we conclude that G and G' are conjugate in $GL_n(k)$, which completes the proof of (1) $\Rightarrow$ (2).

Remark 1. In the case where char $k = 0$, Theorem A can be generalized as follows:

Let S be a polynomial ring in a finite number of indeterminates over a field k of characteristic 0 and let $\mathrm{Aut}_k\, S$ denote the group of all k-automorphisms of S. Let G and G' be small finite subgroups of $\mathrm{Aut}_k\, S$. Then the following conditions are equivalent:

(1) $S^G \cong S^{G'}$ as k-algebras.

(2) G and G' are conjugate in $\mathrm{Aut}_k\, S$.

This is an immediate consequence of Proposition 1, (2) and Proposition 2. However, it should be noted that in the case where char $k > 0$, this does not always hold.

§3. Additional results.

In this section, we give two propositions as supplements to Theorem A.

Proposition 3. *Let k be an algebraically closed field and let $S = k[X_1, X_2, \ldots, X_n]$ be the polynomial ring in indeterminates X_1, X_2, $\ldots$, X_n over k. Let R be a normal k-subalgebra of S such that S is a finitely generated R-module. Assume further that R is a graded subalgebra of S with respect to the natural grading of S. Then the following conditions are equivalent:*

(1) *S is 1-unramified over R.*

(2) *There exists a small finite subgroup G of $\mathrm{GL}_n(k)$ acting linearly on S such that $R = S^G$.*

Proof. Clearly, it is enough to prove (1) $\Rightarrow$ (2). Suppose that S is 1-unramified over R. In the case where char $k = 0$, S is the universal 1-unramified extension of R, because $\pi(S) = \{I\}$. Therefore the group $G = \pi(R)$ acts on S and we have $S^G = R$. Since R is a graded subalgebra of S, it follows that the action of G on S is linear.

Consider, next, the case where char $k = p > 0$. As before, set $\mathfrak{M} = \sum_{i=1}^n SX_i$ and $\mathfrak{m} = \mathfrak{M} \cap R$. Then $\mathfrak{M}$ is the unique maximal ideal of S containing the maximal ideal $\mathfrak{m}$ of R. For brevity, denote by S^* and R^* the completions $S^*_{\mathfrak{M}}$ and $R^*_{\mathfrak{m}}$, respectively. Since S is 1-unramified over R, S^* is also 1-unramified over R^* (e.g. [2, Satz 2.5]). But, we have $\pi(S^*) = \{I\}$ by Proposition 1. Therefore S^* is the universal 1-unramified extension of R^*, and so we have $(S^*)^G = R^*$, where $G = \pi(R^*)$. It is noted that R^* consists of all elements F of $S^* = k[[X_1, X_2, \ldots, X_n]]$ such that every homogeneous part of F is contained in R.

An element σ of G is expressible as follows:

$$\sigma(X_i) = \sum_{j=1}^n a_{ij}(\sigma)X_j + F_i, \quad a_{ij}(\sigma) \in k,\ F_i \in \mathfrak{M}^2 S^*,$$

for each $1 \le i \le n$. Define the element σ' of $\mathrm{Aut}_k S^*$ by

$$\sigma'(X_i) = \sum_{j=1}^{n} a_{ij}(\sigma)X_j, \quad 1 \le i \le n,$$

and set $G' = \{\, \sigma' \mid \sigma \in G \,\}$. Then G' is a subgroup of $\mathrm{GL}_n(k) \subseteq \mathrm{Aut}_k S^*$. From the fact noted above it follows that $G' \subseteq G$. Further, define the subgroup H of G to be the kernel of the natural homomorphism $G \to \mathrm{GL}(\mathfrak{M}S^*/\mathfrak{M}^2 S^*)$. Then we have $H \lhd G$ and $G = HG'$, $H \cap G' = \{I\}$.

Now, let P be a homogeneous element of $(S^*)^{G'}$, and let d denote the degree of P with respect to $X_1, X_2, \ldots, X_n$. Set $P_0 = \prod_{\tau \in H} \tau(P)$. Then it is easy to see that $P_0 \in (S^*)^G = R^*$. For each element τ of H, we can write

$$\tau(P) = P + Q_\tau, \qquad Q_\tau \in \mathfrak{M}^{d+1} S^*.$$

Then we have $P_0 = P^h + Q$, $Q \in \mathfrak{M}^{dh+1} S^*$, where $|H| = h$, and so $P^h \in R^*$. Therefore, for each element τ of H, we have $P^h = \tau(P^h) = (P + Q_\tau)^h = P^h + hP^{h-1}Q_\tau + \ldots + Q_\tau^h$, and hence it follows that $Q_\tau = 0$, i.e., $\tau(P) = P$. This shows that H acts trivially on $(S^*)^{G'}$. Accordingly we have $(S^*)^{G'} = R^*$, and so $G' = G$. Therefore G acts linearly on $S^* = k[[X_1, X_2, \ldots, X_n]]$. Thus G acts linearly on $S = k[X_1, X_2, \ldots, X_n]$ and we have $S^G = R$, which completes the proof of $(1) \Rightarrow (2)$.

Remark 2. In the case where char $k = 0$, Proposition 3 can be generalized easily as follows:

Let S be a polynomial ring in a finite number of indeterminates over an algebraically closed field k of characteristic 0. Let R be a normal (not always graded) k-subalgebra of S such that S is a finitely generated R-module. Then the following conditions are equivalent:

(1) S is 1-unramified over R.
(2) There exists a small finite subgroup of $\mathrm{Aut}_k S$ such that $R = S^G$.

It is also noted that in the case where char $k > 0$, this does not always hold.

Proposition 4. *Let S be a polynomial ring in n indeterminates, $n \ge 4$, over a field k of caharacteristic $p > 0$. Then there exists a normal graded k-subalgebra R of S satisfying the following conditions:*

(1) *S is a finitely generated R-module.*
(2) *There is no small finite subgroup H of $\mathrm{Aut}_k S$ such that $R \cong S^H$ as k algebras.*

Proof. It is well known (e.g. [8]) that there exists a finite subgroup G of $\mathrm{GL}_n(k)$ genarated by pseud-reflections acting linearly on S such that S^G is not a polynomial ring. We shall show that such $R = S^G$ is as desired.

It is clear that R satisfies the condition (1). Therefore we only need to show that R satisfies the condition (2). In order to show this we may assume that k is algebraically closed.

Let $\mathfrak{M}$ denote the maximal graded ideal of S and set $\mathfrak{m} = \mathfrak{M} \cap R$. Then G acts on $S_{\mathfrak{M}}^*$ and we have $(S_{\mathfrak{M}}^*)^G = R_{\mathfrak{m}}^*$. We see that G is also generated by pseudo-reflections as a subgroup of $\mathrm{Aut}_k S_{\mathfrak{M}}^*$. Since $\pi(S_{\mathfrak{M}}^*) = \{I\}$, it follows that $\pi(R_{\mathfrak{m}}^*) = \{I\}$.

Suppose that there is a small finite subgroup H of $\mathrm{Aut}_k S$ such that $R \cong S^H$ as k-algebras. We denote by $\mathfrak{n}$ the image of $\mathfrak{m}$ under the k-isomorphism $R \to S^H$. Let $\mathfrak{N}$ be a maximal ideals of S containing $\mathfrak{n}$, and set $D = \{\tau \in H \mid \tau(\mathfrak{N}) = \mathfrak{N}\}$. Then D is a small subgroup of $\mathrm{Aut}_k S_{\mathfrak{N}}^*$, and therefore $D = \pi((S_{\mathfrak{N}}^*)^D)$. However, we have $\pi(R_{\mathfrak{m}}^*) = \{I\}$ and $R_{\mathfrak{m}}^* = (S^H)_{\mathfrak{n}}^* \cong (S_{\mathfrak{n}}^*)^H \cong (S_{\mathfrak{N}}^*)^D$. Thus we must have $D = \{I\}$, and so $R_{\mathfrak{m}}^*$ is regular. Since $\mathfrak{m}$ is the unique maximal graded ideal of R, it follows that R is a polynomial ring, which is a contradiction.

References

[1] D. Denneberg and O. Riemenschneider, Verzweigung bei Galois-erweiterungen und Quotienten regulärer analytischer Raumkeime, Invent. Math. 7 (1969), 111–119.

[2] D. Denneberg, Universell-endliche Erweiterungen analytischer Algebren, Math. Annalen 200 (1973), 307–326.

[3] A. Grothendieck, Revêtements Étales et Groupe Fondamental (SGA 1), Lecture Notes in Math. 224, Springer, Berlin-Heidelberg-New York, 1971.

[4] G. Müller, Endliche Automorphismengruppen analytischer C-Algebren und ihre Invarianten, Math. Annalen 260 (1982), 375–396.

[5] M. Nagata, Local rings, Interscience, New York-London, 1962.

[6] D. Prill, Local classification of quotients of complex manifolds by discontinuous groups, Duke Math. J. 34 (1967), 375–386.

[7] T. Kambayashi and V. Srinivas, On étale coverings of the affine space, Lecture Notes in Math. 1008, Springer, Berlin-Heidelberg-New York-Tokyo, 1983, 75–82.

[8] H. Nakajima, Regular rings of invariants of unipotent groups, J. Algebra 85 (1983), 253–286.

Shizuo ENDO

Department of Mathematics
Tokyo Metropolitan University
Tokyo, 158
Japan

Algebraic Geometry and Commutative Algebra
in Honor of Masayoshi NAGATA
pp. 65–72 (1987)

Invariant Differentials

John FOGARTY

§1. Introduction.

The present note discusses a differential criterion for smoothness of stable geometric quotients in characteristic zero. Specifically, let x be a stable point for the action of a reductive group G on a smooth affine variety X, let $\pi: X \to Y$ be the quotient and let $y = \pi(x)$. Let $\Omega_{X/Y}$ be the module of relative Kähler differentials of X over Y. G operates on $\Omega_{X/Y}$ compatibly with its action on X. We show that there is a canonical homomorphism

$$\theta_y: \Omega^G_{X/Y}/m_y\Omega^G_{X/Y} \longrightarrow \Omega^G_G$$

of the 'fibre' at y of the sheaf of invariant differentials into the space of invariant differentials on G. The latter may be thought of as the "pre-dual" of the Lie algebra of G. The criterion will be:

Y is smooth at y if and only if θ_y is an isomorphism.

The latter condition is equivalent, in turn, to $\Omega_{X/Y}$ being free of rank g ($= \dim G$) in a neighborhood of y. In this paper, we prove the necessity of the differential condition, and the sufficiency in the case where G is a torus.

Using the étale slice theorem ([3], [4]), the proof comes down to the following. Let E be the normal space in X to the orbit $G \cdot x$ at x. The stabilizer $H = G_x$ of x (a finite group, by the stability assumption) operates on E and Y is smooth at y if and only if $k[E]^H$ is itself a polynomial ring. On the other hand, setting $R = k[E]$, and

$$\Omega_E = \Omega_{R/R^H},$$

one shows that θ_y is an isomorphism if and only if $\Omega^H_E = (0)$. Thus one wants to prove that if G is a finite group, and E is a finite dimensional G-module, the following are equivalent.

(i) $k[E]^G$ is a polynomial ring.

(ii) $\Omega^G_E = (0)$.

For the present proof that (ii) implies (i), we need to assume that G is abelian.

Received January 22, 1987.
Revised April 17, 1987.

§2. Use of the étale slice theorem.

Let k be a field of characteristic zero. This is a standing assumption and no further mention will be made of it. Let G be a reductive group over k and let X be a smooth affine G-variety over k. Let x be a stable point for the action of G. This means that the orbit $G \cdot x$ is closed and the isotropy group $H = G_x$ is finite. (See [4, App. 1].) By the étale slice theorem, the following are true.

(i)   There is an H-invariant locally closed smooth subvariety W of X, containing x, such that $U = GW$ is open in X.

(ii)   There is a strongly étale map $\alpha \colon G \times_H W \to U$ such that, in the diagram:

$$
\begin{array}{ccccc}
G \times W & & G \times_H W & & \\
\| & & \| & & \\
T & \xrightarrow{\ \gamma\ } & V & \xrightarrow{\ \alpha\ } & U \\
{\scriptstyle p_2}\downarrow & & {\scriptstyle \psi}\downarrow & & \downarrow{\scriptstyle \eta} \\
W & \xrightarrow{\ \delta\ } & Z & \xrightarrow{\ \beta\ } & Y
\end{array}
$$

(1) $\beta = \alpha/G$ is étale and the right square is cartesian, where ψ and η are the quotient maps for the actions of G. (This is actually the definition of "strongly étale" (see [4]).)

(2) $G \times_H W$ is the quotient of $G \times W$ by the action of H given by $(g \cdot w)^h = (gh, h^{-1}w)$, γ is the quotient map for this action and $\delta = \gamma/G$.

Now, quite generally, when one has maps of schemes, $X \xrightarrow{f} Y \to Z \to S$, there is an exact sequence (Hurwitz sequence)

$$(X, Y, Z) \qquad 0 \longrightarrow \Upsilon_{X/Y/Z} \longrightarrow f^*\Omega_{Y/Z} \xrightarrow{u} \Omega_{X/Z} \longrightarrow \Omega_{X/Y} \longrightarrow 0,$$

where $\Omega_{\cdot/\cdot}$ stands for the sheaf of relative Kähler differentials, and $\Upsilon_{X/Y/Z} = \ker u$. There is also an exact sequence

$$\Upsilon_{X/Z/S} \longrightarrow \Upsilon_{X/Y/S} \longrightarrow \Upsilon_{X/Y/Z} \longrightarrow 0.$$

For such matters, and other untapped resources, see [2].

In our situation, since H acts freely on T, γ is étale, and $\Upsilon_{T/V/k} = \Omega_{T/V} = (0)$. Thus the sequence (T, V, Z) yields an isomorphism

(a) $\Omega_{V/Z} \otimes_{\mathcal{O}_V} \mathcal{O}_T \xrightarrow{\ \sim\ } \Omega_{T/Z}.$

Since $\mathcal{O}_T = \mathcal{O}_G \otimes_k \mathcal{O}_W$, we see that $\Omega_{W/Z} \otimes_{\mathcal{O}_W} \mathcal{O}_T = \Omega_{W/Z} \otimes_k \mathcal{O}_G$, and (T, W, Z) yields an exact sequence

(b) $0 \longrightarrow \mathcal{O}_G \otimes_k \Omega_{W/Z} \longrightarrow \Omega_{T/Z} \longrightarrow \Omega_{T/W} \longrightarrow 0.$

Taking H-invariants, (a) gives an isomorphism $\Omega_{V/Z} \xrightarrow{\sim} \Omega^H_{T/Z}$. Applying $(\ )^{G \times H}$ to (b), and taking account of the facts that $\Omega_{T/W} = \Omega_G \otimes_k \mathcal{O}_W$ and $\mathcal{O}^H_W = \mathcal{O}_Z$, we get the fundamental exact sequence:

$$0 \longrightarrow \mathcal{O}^G_G \otimes_k \Omega^H_{W/Z} \longrightarrow \Omega^G_{V/Z} \longrightarrow \Omega^G_G \otimes_k \mathcal{O}_Z \longrightarrow 0.$$

We note in passing that $\mathrm{Hom}_k(\Omega^G_G, k) = \mathrm{Lie}(G) = \mathrm{Der}_k(\mathcal{O}_G, \mathcal{O}_G)^G$ and that $\mathcal{O}^G_G = k$.

Proposition. *Set $y = \eta(x)$. Then $\Omega^G_{X/Y} \otimes_k \kappa(y) \to \Omega^G_G \otimes_k \kappa(y)$ is an isomorphism if and only if $\Omega^H_{W/Z} = (0)$.*

Proof. By étale descent and the fundamental exact sequence, together with the observation that, α being étale, it follows from (V, U, Y) that $\Omega_{U/Y} \otimes_{\mathcal{O}_U} \mathcal{O}_V = \Omega_{V/Y}$. Also, $\Omega_{V/Z} = \Omega_{U/Y} \otimes_{\mathcal{O}_Y} \mathcal{O}_Z$, from which it follows, since G is reductive, that $\Omega^G_{V/Z} = \Omega^G_{U/Y} \otimes_{\mathcal{O}_Y} \mathcal{O}_Z$. For the last assertion, see [4, p. 28, (1)]. The argument given there for rings of invariants is formal in the Reynolds operators, and works as well for modules of invariants.

§3. The finite group case.

A further consequence of the étale slice theorem allows us to replace W in the remaining analysis with the H-module $N(X, G \cdot x, x)$, the normal space at x to the orbit $G \cdot x$ in X. This reduces matters to the following question. Let E be a finite dimensional vector space over k and let G be a finite subgroup of $\mathrm{GL}(E)$. Let $R = k[E]$ be the ring of polynomial functions on E and let $\Omega_E = \Omega_{R/R^G}$. Are the following equivalent?

(1)
    (i) R^G is a polynomial ring.

    (ii) $\Omega^G_E = (0)$.

Of course, it is well known (see [1]) that (i) is equivalent to either of the following.

    (iii) G is reflective, i.e., is generated by reflections.
         ($\sigma \in \mathrm{GL}(E)$ is a reflection if $1 - \sigma$ has rank 1.)

    (iv) R is a free R^G-module.

First we show that (i) implies (ii) in (1). Recall that if R, m is a local ring and M is a finite R-module, the depth, $\mathrm{dp}_R M$, of M is the length r of a maximal sequence $f_1, \ldots, f_r$ with

$$\left(\sum_{i=1}^{j-1} f_i M : f_j R \right) = \sum_{i=1}^{j-1} f_i M, \qquad 1 \leq j \leq r.$$

Lemma 1. *Let R, m be a local ring, and let G be a finite group of automorphisms of R such that the order of G is invertible in R. Let M be a finite R-G-module. If $M^G \neq (0)$, then $\mathrm{dp}_R M \leq \mathrm{dp}_{R^G} M^G$.*

Proof. If M has depth 0, there is nothing to prove, so assume that $\mathrm{dp}_R M > 0$, and let $f \in m$ be a non-zero-divisor in M. Let $F = N_G(f) = \prod_{\sigma \in G} f^\sigma$. Then $F \in R^G$ and is a non-zero-divisor in M, hence a non-zero-divisor in M^G. Now, since the order of G is invertible in R, we have $(M/FM)^G = M^G/(FM)^G = M^G/FM^G$, the second equality holding because F is a non-zero-divisor. Thus $\mathrm{dp}_{R^G} M^G > 0$. Replacing M with M/FM, the lemma follows by induction on the depth.

A local k-algebra R is said to be algebraic over k if R is a ring of franctions of a k-algebra of finite type.

Lemma 2. *If R is an algebraic local k-algebra and G is a finite group of k-automorphisms of R, then R^G is also algebraic. If R and R^G are regular, then Ω^G_{R/R^G} has homological dimension ≤ 1 as R^G-module.*

Proof. The first assertion is easily proved, and we omit the details. If R is regular, then $\Omega_{R/k}$ is a free R-module, and so it has depth $d = \dim R$ as R-module. By Lemma 1, $\Omega^G_{R/k}$ has depth $\geq d$ as R^G-module. Since R^G is regular, we see that $\Omega^G_{R/k}$ has homological dimension ≤ 0 over R^G, i.e., $\Omega^G_{R/k}$ is a free R^G-module (of rank d). If K is the fraction field of R, then K^G is the fraction field of R^G since G is a finite group. Since K is separable over K^G,

$$\Upsilon_{R/R^G/k} \otimes_R K = \Upsilon_{K/K^G/k} = (0),$$

i.e., $\Upsilon_{R/R^G/k}$ is a torsion module. But $\Upsilon_{R/R^G/k}$ is a submodule of the free module $\Omega_{R^G/k} \otimes_{R^G} R$, so that $\Upsilon^G_{R/R^G/k} = (0)$, and we have the exact sequence:

$$(*) \qquad \begin{array}{ccccccccc} 0 & \longrightarrow & \Omega_{R^G/k} & \longrightarrow & \Omega^G_{R/k} & \to & \Omega^G_{R/R^G} & \to & 0 \\ & & \| & & & & & & \\ & & (\Omega_{R^G/k} \otimes_{R^G} R)^G & & & & & & \end{array}$$

Since the first two terms in the sequence are free, the lemma follows.

Proposition. *Let R be a regular algebraic semilocal domain over k, with fraction field K and let G be a finite group of k-automorphisms of R such that R^G is local. If $\dim R = 1$, then $\Omega^G_{R/R^G} = (0)$.*

Proof. R^G is a discrete valuation ring with fraction field K^G and R is the integral closure of R^G in K. Passing to completions, the following facts are easily verified.

(i) $$\widehat{R^G} = \hat{R}^G.$$

(ii) $$\Omega_{\hat{R}/\hat{R}^G} = \Omega_{R/R^G} \otimes_R \hat{R}.$$

(iii) $$\Omega^G_{R/R^G} \otimes_{R^G} \hat{R}^G = \Omega^G_{\hat{R}/\hat{R}^G}.$$

By faithfully flat descent, it suffices to show that $\Omega^G_{\hat{R}/\hat{R}^G} = (0)$. Thus we may assume that R and R^G are complete. In that case, $R = \bigoplus_i R_{m_i}$ is the direct sum of its local components, and $\Omega_{R/R^G} = \bigoplus_i \Omega_{R_{m_i}/R^G}$. G permutes the summands transitively, and if $G_j = \{\sigma \in G \;;\; m_j^\sigma = m_j\}$, then R^{G_j} is étale over R^G and $\Omega^G_{R/R^G} = \Omega^{G_j}_{R_{m_j}/R^G}$ for each and every j. Thus we may assume that R is local with maximal ideal m. Let $G^\circ = \{\sigma \in G \;;\; \sigma \equiv 1 \bmod m\}$, $R^\circ = R^{G^\circ}$ and $m^\circ = m \cap R^\circ$. Then by Hilbert ramification theory, $m^G R^\circ = m^\circ$, and we have the standard:

Lemma 3. *If R, m is a local ring such that R/m has characteristic zero, and if G is a finite group of automorphisms of R such that $m^G R = m$, then $\Omega_{R/R^G} = (0)$.*

Proof. One has the exact sequence:

$$m/m^2 \overset{\delta}{\longrightarrow} \Omega_{R/R^G}/m\Omega_{R/R^G} \longrightarrow \Omega_{(R/m)/(R^G/m^G)} \longrightarrow 0,$$

δ being induced by $d: R \to \Omega_{R/R^G}$. Since R/m has characteristic 0, $R^G/m^G = (R/m)^G$ and the last term of the sequence is (0). Thus

$$\Omega_{R/R^G} = m\Omega_{R/R^G} + R(\operatorname{im} \delta).$$

By Nakayama's lemma, it follows that $\Omega_{R/R^G} = R(\operatorname{im} \delta)$. Since $m = m^G R$, we see that Ω_{R/R^G} is generated over R by δm^G. But $dm^G = (0)$, and the lemma follows.

Lemma 4. *Let R, m be a complete local valuation ring such that $K = R/m$ has characteristic zero and let G be a finite group of automorphisms of R acting trivially on K. Then $\Omega^G_{R/R^G} = (0)$.*

Proof. Taking a field of representatives of K, which we also denote by K, from the elements of R^G, we may write $R = k[[t]]$. Note that R is generated as R^G-algebra by t, whence $\Omega_{R/R^G} = R\,dt$. Now if $\omega \in \Omega_{R/R^G}$, write $\omega = f(t)dt$, with

$$f(t) = \sum_{i=0}^{\infty} a_i t^i \in R, \qquad a_i \in K,$$

and let $F(t) = \sum_{i=0}^{\infty}(a_i/(i+1))t^{i+1}$. Then $\omega = dF$, i.e., $\Omega_{R/R^G} = d_{R/R^G}(R)$. Furthermore, if $\omega \in \Omega^G_{R/R^G}$, then $dF = dF^\sigma$ for all $\sigma \in G$, i.e., $F^\sigma = F + C_\sigma$, with $C_\sigma \in \ker(d_{R/R^G}) = S$. Also, $C_{\sigma\tau} = C_\sigma^\tau + C_\tau$. Since the characteristic is zero, we have $H^1(G, S) = (0)$, and there is an element $g \in S$ with $C_\sigma = g^\sigma - g$. Replacing F with $F - g$, we see that $\Omega^G_{R/R^G} = d_{R/R^G}(R^G) = (0)$.

The proposition follows directly from Lemmas 3 and 4. To show that (i) implies (ii) in (1) we note that a graded k-algebra of finite type is a polynomial ring if and only if its local ring at the "vertex" is regular. Replacing R in (i) with the latter ring, we see that, by the proposition, Ω^G_{R/R^G} has support of codimension ≥ 2 in $\operatorname{Spec} R^G$. On the other hand, by Lemma 2, there is an exact sequence:

$$0 \longrightarrow F_1 \xrightarrow{A} F_0 \longrightarrow \Omega^G_{R/R^G} \longrightarrow 0$$

with F_i free R^G-modules. Since Ω^G_E is a torsion module, $\operatorname{rank} F_1 = \operatorname{rank} F_0$. Then $\det A \notin P$ for all prime ideals of height 1 in R^G. Hence $\det A$ is a unit of R^G and $\Omega^G_{R/R^G} = (0)$. Since Ω^G_{R/R^G} is now the localization of Ω^G_E at the vertex, $\Omega^G_E = (0)$.

To prove (ii) implies (i), we assume, as noted above, that G is abelian.

Lemma 5. *Let G be a finite group and let E be a finite dimensional k-G-module. If G is abelian, the following are equivalent.*

(a) *G is generated by reflections.*

(b) *There is a basis $x_1, \ldots, x_n$ of E and positive integers $a_1, \ldots, a_n$ such that $k[E]^G = k[x_1^{a_1}, \ldots, x_n^{a_n}]$.*

(c) *G is the direct sum of cyclic groups G_i, the i-th summand G_i acting on E via*

$$\begin{pmatrix} 1 & & & & & \\ & 1 & & & & \\ & & \ddots & & & \\ & & & \chi_i & \cdots\cdots\cdots & \\ & & & & \ddots & \\ & & & & & 1 \end{pmatrix} \cdots i$$

Proof. (c) implies (b) is evident and (b) implies (a) is a classical result (see [1]). To see that (a) implies (c), write the action:

$$\begin{pmatrix} \chi_1 & & & \\ & \chi_2 & & \\ & & \ddots & \\ & & & \chi_n \end{pmatrix}$$

with $\chi_j \in \operatorname{Hom}(G, G_m)$. Clearly, if $\sigma \in G$ is a reflection, then $\chi_j(\sigma) = 1$ for all but one j. Let $G_j = \chi_j(G)$. If τ_j generates the cyclic group G_j, then $\tau_j = \chi_j(\rho_1 \ldots \rho_h)$, where $\rho_1, \ldots, \rho_h$ are reflections with $\chi_j(\rho_k) \neq 1$ for all k. Therefore $\rho_1 \ldots \rho_h$ is a reflection of the same order as τ_j. This shows that the homomorphism $\chi_j: G_j \to G$ has a right inverse. The result then follows by induction, since $\ker \chi_j$ is also generated by reflections.

NOTATION: $\alpha = (\alpha_1, \ldots, \alpha_n)$, $\epsilon_j = (0, \ldots, \overset{j}{1}, \ldots, 0)$, $\alpha \cdot \beta = (\alpha_1 \beta_1, \ldots, \alpha_n \beta_n)$ and $|\alpha| = \alpha_1 + \cdots + \alpha_n$.

Now, for each j, choose a_j minimal so that $x_j^{a_j} \in R^G$. Then, by Lemma 5, R^G is a polynomial ring if and only if every invariant monomial is a monomial in the $x_j^{a_j}$, i.e., if $A = (a_1, \ldots, a_n)$, then every invariant monomial is of the form $X^{A \cdot \beta}$ for some β. If this is *not* the case, choose a monomial X^α in R^G of minimum positive degree which is not of the form $X^{A \cdot \beta}$. Note that, by the choice just made, α cannot be of the form $a\epsilon_j$ for any j.

Using the division algorithm for integers, we may write $\alpha = A \cdot \beta + \lambda$ where $0 \leq \lambda_j < a_j$. Then by the minimality of $|\alpha|$, we see immediately that $\alpha_j < a_j$ for each j. Now write:

$$dX^\alpha = \omega_1 + \cdots + \omega_n, \qquad \omega_j = \alpha_j X^{\alpha - \epsilon_j} dx_j.$$

Observe that ω_j is G-invariant.

If $\Omega_E^G = (0)$, then $\Omega_{R^G/k} \to \Omega_{R/k}^G \to 0$ is exact, and so we can write

$$(2) \qquad \omega_j = \sum_{k=1}^{n} f_{jk} x_k^{a_k - 1} dx_k + \sum_p c_p dv_p,$$

where $f_{jk} \in R_{|\alpha|-a_j}^G$, $c_p \in R_{|\alpha|-b_p}^G$ and each v_p is a monomial in $R_{b_p}^G$ which is not of the form $X^{A \cdot \beta}$. By the division process, we may assume that each v_p is of the form X^γ with $\gamma_j < a_j$. By the minimality of $|\alpha|$ again, we have $b_p \geq |\alpha|$, whence $b_p = |\alpha|$, and the c_p are constants. Thus $\sum c_p dv_p$ is an exact differential of degree $|\alpha|$.

Finally, write $v_p = X^\beta$, $dv_p = \sum \beta_j X^{\beta - \epsilon_j} dx_j$, and note that the coefficient of dx_j in the first sum on the right in (2) is of degree $\geq a_j - 1$ in x_j. In the second sum on the right, the corresponding coefficient has degree $\beta_j - 1$ in x_j. Therefore, since there can be no cancellation of terms between the two sums on the right, all f_{jk} are zero. Thus ω_j is an exact differential. This, however, implies that $\alpha = a\epsilon_j$ for some j, contradicting the choice of α. Therefore, in view of the exact sequence $(*)$, $\Omega_E^G \neq (0)$.

References

[1] Bourbaki, N., Groupes et algèbres de Lie, Ch. VI, Hermann, Paris.

[2] Grothendieck, A. and Dieudonné, J., Éléments de géométrie algébrique, Ch. 0$_{III}$, Publ. Math. de l'IHES, No. 20.

[3] Luna, D., Slice étales, Bull. Soc. Math. France, Memoire 33, (81), 1973.

[4] Mumford, D. and Fogarty, J., Geometric invariant theory, 2nd ed., Springer, 1982.

John FOGARTY

Department of Mathmatics and Statistics
University of Massachusetts
Amherst, MA 01003
U.S.A.

Algebraic Geometry and Commutative Algebra
in Honor of Masayoshi NAGATA
pp. 73–98 (1987)

Classification of Polarized Manifolds
of Sectional Genus Two

Takao FUJITA

Introduction

Let L be an ample (not necessarily very ample) line bundle on a compact complex manifold M with $\dim M = n$. The sectional genus $g(M, L)$ of the polarized manifold (M, L) is defined by the formula $2g(M, L) - 2 = (K + (n-1)L)L^{n-1}$, where K is the canonical bundle of M. We have a complete classification of polarized manifolds with $g(M, L) \leq 1$ (see [F8]). In this paper we consider the case $g(M, L) = 2$.

The problem is trivial when $n = 1$. The case $n = 2$ was studied in [BLP]. When $n \geq 3$, we first show that one of the following conditions is satisfied (see (1.10)).

1) K is numerically equivalent to $(3-n)L$ and $L^n = 1$.

2) M is a double covering of $\mathbf{P}^n$ with branch locus being a smooth hypersurface of degree 6, and L is the pull-back of $\mathcal{O}(1)$.

2') M is the blowing-up at a point p of another polarized manifold (M', L') of the above type 2) and $L = L'_M - E$, where E is the exceptional divisor. In this case $n = 3$.

3) There is a surjective morphism $f : M \to C$ onto a smooth curve C, $\rho(M) = 2$ and any general fiber F of f is a hyperquadric in $\mathbf{P}^n$ with $L_F = \mathcal{O}(1)$.

4) (M, L) is a scroll over a smooth surface.

5) (M, L) is a scroll over a smooth curve of genus two.

In §2, we study the above case 1). The case 3) is studied in §3. For technical reasons, the case 4) will be studied in a forthcoming paper [F10]. Combining these results we get an almost complete classification of the case $n \geq 3$ (see [F9]; for remaining problems, see (2.14)). In §4, for the sake of completeness, we give our classification theory of the case $n = 2$ in a slightly different form from [BLP], since we use these results in [F10].

The author heartily thanks Professor Y. Miyaoka for many helpful comments during the preparation of this paper.

Received August 28, 1987.

Notation, Convention and Terminology

We use the notation in [F8] and [F9], and usually follow the customary notation in algebraic geometry today. Line bundles and invertible sheaves are used interchangeably, and are identified with the linear equivalence classes of Cartier divisors. The tensor products of line bundles are denoted additively, while we use multiplicative notation for intersection products in Chow rings. The numerical equivalence of line bundles is denoted by $\sim$, while we write $=$ for linear equivalence. The linear equivalence classes are denoted by [ ], while we use { } for homology classes of an albegraic cycle. Given a morphism $f : X \to Y$ and a line bundle A on Y, we denote f^*A by A_X, or sometimes by A for short when there is no danger of confusion. The $\mathcal{O}(1)$'s of projective spaces $\mathbf{P}_\alpha$, $\mathbf{P}_\beta$, ... will be denoted by H_α, H_β, ... The canonical bundle of a manifold X is denoted by K_X.

§1. Classification, first step

From now on, throughout this paper, let (M, L) be a polarized manifold with $n = \dim M \geq 3$, $d = d(M, L)$ and $g(M, L) = 2$. The invertible sheaf $\mathcal{O}_M[L]$ will be denoted by $\mathcal{L}$.

Theorem (1.1). *If $K + (n-1)L$ is not nef, then (M, L) is a scroll over a curve of genus two.*

Proof. Apply [F8; Theorem 2] and use the method in [F8; (2.8)].

(1.2). From now on, we assume that $K + (n-1)L$ is nef. For the moment, until (1.5), we further assume that $K + (n-2)L$ is nef. Then, since $(K + (n-1)L)L^{n-1} = 2$, we have $0 \leq (K + (n-2)L)L^{n-1} = 2 - d$. So $d = 1$ or 2.

(1.3). When $d = 2$, M is a double covering of $\mathbf{P}^n$ with branch locus being a smooth hypersurface of degree six, and L is the pull-back of $\mathcal{O}(1)$.

To see this recall that $\mathrm{Bs}|m(K + (n-2)L)| = \emptyset$ for some $m > 0$ since $K + (n-2)L$ is nef (cf. [KMM]). Take a member D of $|m(K + (n-2)L)|$. Then $L^{n-1}D = 0$ since $d = 2$. Hence $D = 0$ because L is ample. So $K + (n-2)L \sim 0$, and $-K$ is ample. Therefore the numerical equivalence implies the linear equivalence. Thus $K + (n-2)L = 0$ in $\mathrm{Pic}(M)$. Moreover $\Delta(M, L) = 1$ by [F6;(1.11)], so the theory in [F4] applies.

(1.4). When $d = 1$, we claim that $K \sim (3 - n)L$.

To see this, we will first show that $F^2L^{n-2} > 0$ for $F = K + (n-2)L$.

Indeed, otherwise, by the base point free theorem (cf. [KMM]) and [F7; Appendix], there is a morphism $f : M \to C$ onto a curve C such that $F = f^*A$ for some ample line bundle A on C. Since $1 = FL^{n-1} = aL^{n-1}X$ for $a = \deg A$ and any general fiber X of f, we have $L^{n-1}X = 1 = a$. So $L^{n-1}F_x = 1$ for the fiber F_x over every point x on C. Hence F_x is irreducible and reduced since L is ample. Let L_x be the restriction of L to F_x. Then (F_x, L_x) is a polarized variety with $d(F_x, L_x) = 1$. On the other hand, since $K_x + (n-2)L_x = 0$, any general fiber is a del Pezzo manifold. So $\Delta(F_x, L_x) = 1$ for every $x \in C$ by the lower semi-continuity of the Δ-genus. Hence $\mathrm{Bs}|L_x|$ is at most finite. Since $L_x^{n-1} = 1$ and $h^0(F_x, L_x) = n-1$, there are $n-1$ members of $|L_x|$ meeting at only one point, and the intersection is transverse. Now we infer that $\mathcal{E} = f_* \mathcal{L}$ is a locally free sheaf on C of rank $n-1$, and $B = \mathrm{Supp}(\mathrm{Coker}(f^*\mathcal{E} \to \mathcal{L}))$ is a section of f such that $B \cap F_x = \mathrm{Bs}|L_x|$ for every $x \in C$. Furthermore $B = \mathrm{Bs}|L + 2A|$, where we write A instead of f^*A. Indeed, $H^1(L + 2A - [F_x]) = 0$ by Kodaira's vanishing theorem since $L + 2A - F_x \sim K + (n-1)L$. So the restriction mapping $H^0(M, L + 2A) \to H^0(F_x, L_x)$ is surjective for every x. This implies $F_x \cap \mathrm{Bs}|L + 2A| = \mathrm{Bs}|L_x|$, proving the assertion. Now, let M_1 be the blowing-up of M along B. By the above observation we infer $\mathrm{Bs}|L + 2A - E| = \emptyset$ on M_1, where E is the exceptional divisor over B. Let $\mathcal{N}$ be the normal bundle of B in M and let q be the genus of C. Since $B \simeq C$, we have $c_1(\mathcal{N}) = c_1(\Omega_B) - KB = 2q - 2 - (A - (n-2)L)B = 2q - 3 + (n-2)LB$. Since $(E, \mathcal{O}[-E]) \simeq (\mathbf{P}(\mathcal{N}), \mathcal{O}(1))$, we have $(-E)^{n-1}\{E\} = -c_1(N) = 3 - 2q - (n-2)LB$. So $0 \le (L + 2A - E)^n = (L + 2A)^n + n(L + 2A)(-E)^{n-1} - (-E)^{n-1}E = 2q - 2 - 2LB$ since $(L + 2A)^n = L^n + 2nL^{n-1}A = 1 + 2n$ and $(L + 2A)(-E)^{n-1} = -(L + 2A)B = -LB - 2$. Hence $q \ge 1 + LB \ge 2$. So $0 < h^1(C, [x]) \le h^1(M, [F_x])$, which contradicts the vanishing theorem since $F_x \sim K + (n-2)L$. Thus we conclude $F^2 L^{n-2} > 0$.

Now we set $I = K + (n-3)L$. Then $IL^{n-1} = 0$ and $1 \le F^2 L^{n-2} = (I + L)^2 L^{n-2} = I^2 L^{n-2} + 1$. So the problem is reduced to the following:

Proposition (1.5). *Let L be an ample line bundle on a projective variety V with $\dim V = n$ and let I be a line bundle on V such that $IL^{n-1} = 0$ and $I^2 L^{n-2} \ge 0$. Then $I \sim 0$.*

Proof. We use the induction on n. The assertion follows from the index theorem when $n = 2$. We will derive a contradiction assuming $n \ge 3$ and $IC \ne 0$ for some curve C in V. Clearly we may assume that V is normal. Take a large integer t and let Λ be the linear subsystem of $|tL|$ consisting of members containing C. If t is large enough, we have $\mathrm{Bs}\Lambda = C$ and any general member D of Λ is irreducible and reduced. Then $IL^{n-2}\{D\} = 0$ and $I^2 L^{n-3}\{D\} \ge 0$, so $I_D \sim 0$ by the induction hypothesis. This contradicts $IC \ne 0$ and $C \subset D$.

(1.6). Now we consider the case in which $K + (n-2)L$ is not nef. By [F8; Theorem 3'], one of the following conditions is satisfied.

a) There is an effective divisor E on M such that $(E, L_E) \simeq (\mathbf{P}^{n-1}, \mathcal{O}(1))$ and $\mathcal{O}[E]_E = \mathcal{O}(-1)$.

b0) (M, L) is isomorphic to either $(\mathbf{P}^3, \mathcal{O}(j))$ with $j = 2$ or 3, $(\mathbf{P}^4, \mathcal{O}(2))$ a hyperquadric in $\mathbf{P}^4$ with $L = \mathcal{O}(2)$, or a del Pezzo variety.

b1) There is a surjective morphism $f : M \to C$ onto a smooth curve C and $\rho(M) = 2$ for the Picard number ρ. Moreover, any general fiber F is a hyperquadric in $\mathbf{P}^n$ with $L_F = \mathcal{O}_F(1)$, unless f is a $\mathbf{P}^2$-bundle with $(F, L_F) \simeq (\mathbf{P}^2, \mathcal{O}(2))$ for every fiber F.

b2) (M, L) is a scroll over a smooth surface.

(1.7). In the above case b0), we easily see that $g(M, L) \neq 2$. So this case is ruled out.

(1.8). Assume that f is a $\mathbf{P}^2$-bundle in case b1). Then, if we set $H = K + 2L$, we have $(F, H_F) \simeq (\mathbf{P}^2, \mathcal{O}(1))$. So $(M, H) \simeq (\mathbf{P}_C(\mathcal{E}), \mathcal{O}(1))$ for the locally free sheaf $\mathcal{E} = f_*(\mathcal{O}_M[H])$. Since $L = 2H + f^*B$ for some $B \in \mathrm{Pic}(C)$, we have $HL^2 = 4H^3 + 4H^2B \equiv 0$ modulo 4. Hence $g(M, L) \neq 2$.

(1.9). In the above case a), M is the blowing-up of another manifold M' at a point p and E is the exceptional divisor. Moreover, $L + E \in \mathrm{Pic}(M)$ is the pull-back of an ample line bundle L' on M' (cf. [F3, §5]). We easily see that $g(M', L') = 2$. $K + (n-1)L$ is the pull-back of $K' + (n-1)L'$, where K' is the canonical bundle of M'. Hence $K' + (n-1)L'$ is nef. If $K' + (n-2)L'$ is nef, then (M', L') is of the type (1.3) because $d' = d(M', L') = d+1 \geq 2$. If $K' + (n-2)L'$ is not nef, then (M', L') satisfies the condition a) in (1.6). Indeed, in both cases b1) and b2), by (1.8), there is a curve Z in a fiber of f such that $p \in Z$ and $L'Z = 1$. But $L'Z > LZ^\sharp > 0$ for the proper transform $Z^\sharp$ of Z on M. This yields a contradiction as desired.

Repeating this process of blowing-down if necessary, we obtain a polarized manifold (M'', L'') not satisfying the condition a). This must be of the type (1.3) by the above reasoning. Therefore $2 = d(M'', L'') = d + r$, where r is the number of blowing-down processes. So $d = r = 1$. Now we see $\Delta(M, L) = 1$. The structure of (M, L) is studied in [F4; III], and is sectionally hyperelliptic of type (∞). In particular $n \leq 3$.

(1.10). Summing up, we obtain the following.

Theorem. Let (M, L) be a polarized manifold with $n = \dim M \geq 3$, $d = d(M, L)$ and $g(M, L) = 2$. Then one of the following conditions is satisfied.

1) K is numerically equivalent to $(3 - n)L$ and $d = 1$.

2) M is a double covering of $\mathbf{P}^n$ with branch locus being a smooth hypersurface of degree 6, and L is the pull-back of $\mathcal{O}(1)$, $d = 2$.

2') M is the blowing-up at a point p of another polarized manifold (M', L') of the above type 2) and $L = L'_M - E$, where E is the exceptional divisor. In this case $d = 1$ and $n = 3$.

3) (M, L) is a scroll over a smooth surface.

4) There is a surjective morphism $f : M \to C$ onto a smooth curve C, $\rho(M) = 2$ and any general fiber F of f is a hyperquadric in $\mathbf{P}^n$ with $L_F = \mathcal{O}(1)$.

5) (M, L) is a scroll over a smooth curve of genus two.

Remark (1.11). We have $\Delta(M, L) \geq 2$ in most cases above. By the theory of Δ-genus, we infer $\Delta(M, L) > 0$ if $g = 2$. Moreover, if $\Delta = 1$, then (M, L)

is of the above type 2), 2$'$) or a weighted hypersurface of degree 10 in the weighted projective space $\mathbf{P}(5,2,1,\cdots,1)$. In the last case (M,L) is of the type 1). However, there are many other polarized manifolds of the type 1).

(1.12). The case 1) and 4) will be studied further in the following sections. In case 3), $(M,L) \simeq (\mathbf{P}(\mathcal{E}),\mathcal{O}(1))$ for some vector bundle $\mathcal{E}$ on a surface S. Moreover, $g(S,A) = 2$ for $A = \det\mathcal{E}$. Such vector bundles will be classified in [F10].

§2. The case $K \sim (3-n)L$

Throughout this section let (M,L) be a polarized manifold of the type (1.10; 1).

Theorem (2.1). $H^1(M,\mathcal{O}_M) = 0$ and $K = (3-n)L$ in $\mathrm{Pic}(M)$.

Proof. If $n \geq 4$, $-K$ is ample and the assertion follows from the vanishing theorem. So we consider the case $n = 3$.

Since $c_1(M)_{\mathbf{R}} = 0$ by virtue of [Y], there is a Kähler form representing $c_1(L)_{\mathbf{R}}$ with vanishing Ricci tensor. Then $c_2(M)L \geq 0$ by [CO]. So $\chi(M,L') = L^3/6 + c_2L/12 > 0$ for any $L' \sim L$ by the Riemann-Roch theorem. Hence $h^0(M,L') > 0$ since $h^j(M,L') = 0$ for $j > 0$ by Kodaira's vanishing theorem. So $|L'| \neq \emptyset$.

Let L_0 be any line bundle with $L_0 \sim L$ and let $D_0 \in |L_0|$. If L_0, L', L'' are numerically equivalent but not linearly equivalent, then $C' = D_0 \cap D'$ and $C'' = D_0 \cap D''$ are different curves on D_0 for $D' \in |L'|$ and $D'' \in |L''|$. Indeed, otherwise, $H^0(D_0,L'-L'') \neq 0$ while $H^0(M,L'-L'') = 0$ and $H^1(M,L'-L''-L_0) = 0$ by the vanishing theorem. This is impossible because we have an exact sequence $H^0(M,L'-L'') \to H^0(D_0,L'-L'') \to H^1(M,L'-L''-L_0)$. Note also that C' is irreducible and reduced since $LC' = L^3 = 1$.

Let $L_1 \sim L$ and $L_1 \neq L_0$. Take $D_1 \in |L_1|$ and set $C = D_0 \cap D_1$. Let $L' \sim L \sim L''$ and suppose that L_0, L_1, L', L'' are all different to each other in $\mathrm{Pic}(M)$. Take $D' \in |L'|$ and $D'' \in |L''|$. Then $p' = C \cap D' = C \cap D_0 \cap D'$ is a point on C. Moreover, the intersection is transverse since $D_0D_1D' = L^3 = 1$. Similarly $p'' = C \cap D''$ is a point on C and $p'' \neq p'$. Indeed, if $p'' = p'$, then $H^0(C,L'-L'') \neq 0$. Similarly as before, this is impossible because $h^0(D_0,L'-L'') = 0$ and $h^1(D_0,L'-L''-L_1) \leq h^1(M,L'-L''-L_1) + h^2(M,L'-L''-L_1-L_0) = 0$. Thus $p' \neq p''$ and hence $D_0 \cap D_1 \cap D' \cap D'' = \emptyset$.

Now, we will derive a contradiction assuming $h^1(M,\mathcal{O}_M) > 0$. Take four different line bundles N_0, N_1, N_2, N_3 such that $2N_j = 0$ in $\mathrm{Pic}(M)$. For any $L' \sim L$, let D_j be a member of $|L' + N_j|$. Then $D_0 \cap D_1 \cap D_2 \cap D_3 = \emptyset$ by the above observation. Since $2D_j \in |2L'|$ for any j, we infer $\mathrm{Bs}|2L'| = \emptyset$. The same reasoning shows $\mathrm{Bs}|2L''| = \emptyset$ for any $L'' \sim L$. There are infinitely many line bundles L'' with this property because $h^1(M,\mathcal{O}_M) > 0$. For most such L'' and $D'' \in |L''|$, $D_0 \cap D_1 \cap D'' = p''$ is a point on $C = D_0 \cap D_1$, and p'' moves if so does L''. Therefore, $\mathrm{Bs}|2p| = \emptyset$ for infinitely many simple points p on C. This is impossible because the arithmetic genus of C is $g(M, L) = 2$.

Thus we conclude $H^1(M, \mathcal{O}_M) = 0$. The Riemann-Roch theorem implies $\chi(M, \mathcal{O}_M) = 0$, hence $h^3(M, \mathcal{O}_M) = h^0(M, K) > 0$. This implies $K = 0$ because $K \sim 0$.

Remark (2.2). When $n = 3$, as we see in the above proof, we have $h^0(M, L) = \chi(M, L) > 0$ and $\Delta(M, L) \leq 3$. We conjecture that this is true even if $n > 3$.

We can use Miyaoka's inequality in [Mi2] to obtain $c_2 L \geq 0$. Moreover, when $h^1(M, \mathcal{O}_M) > 0$, we can use Kawamata's theory on Albanese mappings in order to show $\chi(M, L) > 0$. Thus there are several methods to prove (2.1). But in any case we need *BIG* theorems.

Proposition (2.3). *If $\Delta(M, L) = 1$ in addition, then (M, L) is a weighted hypersurface of degree 10 in the weighted projective space $\mathbf{P}(5, 2, 1, \cdots, 1)$.*

Proof. This follows from results in [F4;III]. Indeed, since $g(M, L) = 2$, (M, L) is sectionally hyperelliptic. Moreover $K = (3 - n)L$ only if (M, L) is of type $(-)$. So [F4;(16.7)] applies.

Proposition (2.4). *If $\Delta(M, L) = 2$ in addition, then (M, L) is a weighted complete intersection of type $(6, 6)$ in the weighted projective space $\mathbf{P}(3, 3, 2, 2, 1, \cdots, 1)$.*

Proof. Let $D_1, \ldots, D_{n-1}$ be general members of $|L|$, and let $V_j = D_1 \cap \cdots \cap D_{n-j}$ for $1 \leq j \leq n - 1$. Since $\dim \mathrm{Bs}|L| < \Delta(M, L)$, we have $\dim V_j = j$. Moreover V_j is irreducible and reduced because $L^j V_j = L^n = 1$.

We claim $H^i(M, tL) = 0$ for any $0 < i < j$, $t \in \mathbf{Z}$. Indeed, if $n = 3$ and $t = 0$, this follows from (2.1). Otherwise we can apply Kodaira's vanishing theorem.

Next we claim $H^i(V_j, tL) = 0$ for any $0 < i < n$, $t \in \mathbf{Z}$. To prove this, we use the induction on j from above. If we set $V_n = M$, this is true for $j = n$ by the above claim. For smaller j, we use the exact sequence $H^i(V_j, tL) \to H^i(V_{j-1}, tL) \to H^{i+1}(V_j, (t - 1)L)$.

This claim implies that $H^0(V_j, tL) \to H^0(V_{j-1}, tL)$ is surjective for any $j \geq 2$, $t \in \mathbf{Z}$. So $\Delta(V_1, L) = \cdots \Delta(V_j, L) = \cdots = \Delta(M, L) = 2$, hence $h^0(V_1, L) = 0$. Therefore (V_1, L) is a weighted complete intersection of type $(6, 6)$ in $\mathbf{P}(3, 3, 2, 2)$ (see the Appendix). Now, using [F2; Prop. 2.2 & 2.4], we infer that (V_j, L) is a weighted complete intersection of type $(6, 6)$ in $\mathbf{P}(3, 3, 2, 2, 1, \cdots, 1)$ by induction on j. Thus we prove (2.4).

Remark. This result can be viewed as a higher dimensional version of [C].

Proposition (2.5). *#(torsion part of $\mathrm{Pic}(M)$) ≤ 5.*

Proof. If $n \geq 4$, M is a Fano manifold and hence simply connected. So we may assume $n = 3$. Let T be the torsion subgroup of $\mathrm{Pic}(M)$ and assume $\tau = \#T \geq 6$. Take a Galois étale covering $\pi : \tilde{M} \to M$ such that $\mathrm{Gal}(\tilde{M}/M) \simeq T$ and $\pi^* N = 0$ for any $N \in T$. Let $N_1, \ldots, N_\tau$ be the elements of T and take a member D_α of $|L + N_\alpha|$ for each $\alpha = 1, \cdots, \tau$. By the method in (2.1) we infer $D_1 \cap \cdots \cap D_\tau = \emptyset$. Setting $\pi^* L = \tilde{L}$, we have $\pi^* D_\alpha \in |\tilde{L}|$. So $\mathrm{Bs}|\tilde{L}| = \emptyset$. The canonical bundle of $\tilde{M}$ is trivial and $\tilde{L}$ is ample. So, by vanishing theorem and Riemann-Roch theorem, we have $h^0(\tilde{M}, \tilde{L}) = \chi(\tilde{M}, \tilde{L}) = \tau\chi(M, L) = \tau h^0(M, L) \geq \tau$. On the other hand, we have $\tilde{d} = d(\tilde{M}, \tilde{L}) = \tau d(M, L) = \tau$ and $\tilde{\Delta} = \Delta(\tilde{M}, \tilde{L}) = 3 + \tau - h^0(\tilde{M}, \tilde{L}) \leq 3$. Note also that $2g(\tilde{M}, \tilde{L}) - 2 = 2\tilde{L}^3 = 2\tau$ and hence $\tilde{g} = g(\tilde{M}, \tilde{L}) = \tau + 1$. If $\tilde{d} > 2\tilde{\Delta}$, then $\tilde{g} \leq \tilde{\Delta}$ by [F2]. Therefore $\tilde{d} \leq 2\tilde{\Delta}$. When $\tau \geq 6$, this is possible only when $\tilde{d} = \tau = 6$ and $\tilde{\Delta} = 3$. Then $(\tilde{M}, \tilde{L})$ is hyperelliptic in the sense [F6] by [F6; (1.10)]. Moreover, using results in [F6] (especially Table II in §6, (5.6.2) and (5.22.4)), we infer that $(\tilde{M}, \tilde{L})$ is of the type $(\Sigma^3(3)_{3,-1}^+)$. This means that $\tilde{M}$ is a double covering of $W \simeq \mathbf{P}_\xi^1 \times \mathbf{P}_\zeta^2$ with branch locus B being a smooth connected divisor in $|4H_\xi + 6H_\zeta|$ and $\tilde{L} = H_\xi + H_\zeta$, where H_ξ and H_ζ denote pull-backs of $\mathcal{O}(1)$'s of $\mathbf{P}_\xi^1$ and $\mathbf{P}_\zeta^2$ respectively.

We claim that $\mathrm{Pic}(\tilde{M}) \simeq H^2(\tilde{M}; \mathbf{Z})$ is generated by H_ξ and H_ζ. Indeed, $\tilde{M}$ can be embedded in the ambient space of the line bundle $2H_\xi + 3H_\zeta$ over $\mathbf{P}_\xi^1 \times \mathbf{P}_\zeta^2$. So [L; Theorem 2.1] applies.

$T \simeq \mathbf{Z}/6\mathbf{Z}$ since $\tau = 6$. Let σ be a generator of T. Then $\sigma^* H_\xi \in \mathrm{Pic}(\tilde{M})$ is nef, $(\sigma^* H_\xi)^2 = 0$ in the Chow ring of $\tilde{M}$ and $\mathrm{Pic}(\tilde{M})/\langle\sigma^* H_\xi\rangle \simeq \mathbf{Z}$. From these we infer $\sigma^* H_\xi = H_\xi$. Therefore H_ξ comes from $\mathrm{Pic}(M)$ and $\chi(\tilde{M}, H_\xi)$ is a multiple of $\tau = 6$. However, we have $\chi(M, H_\xi) = \chi(W, H_\xi) + \chi(W, -H_\xi - 3H_\zeta) = 2$. Thus we get a contradiction, as desired.

Proposition (2.6). *If $\#(\text{torsion part of } \mathrm{Pic}(M)) = 5$, then $\pi_1(M) \simeq \mathbf{Z}/5\mathbf{Z}$ and the universal covering $\tilde{M}$ is a hypersurface of degree 5 in $\mathbf{P}^4$*

Proof. Let T, $\pi : \tilde{M} \to M$ and $\tilde{L}$ be as in (2.5). Then $\mathrm{Bs}|\tilde{L}| = \emptyset$ as before. Since $\tilde{L}^3 = 5$ and $h^0(\tilde{M}, \tilde{L}) = \chi(\tilde{M}, \tilde{L}) = 5\chi(M, L) = 5$, $|\tilde{L}|$ gives a morphism $\rho : \tilde{M} \to \mathbf{P}^4$ such that $\tilde{L} = \rho^* \mathcal{O}(1)$. Let $W = \mathrm{Im}(\rho)$ and set $w = \deg(W)$. Then $5 = \tilde{L}^3 = w \cdot \deg(\rho)$. We have $w > 1$ since $\rho^* : H^0(\mathbf{P}^4, \mathcal{O}(1)) \to H^0(\tilde{M}, \tilde{L})$ is bijective and factors through $H^0(W, \mathcal{O}_W(1))$. Hence $w = 5$ and ρ is a birational morphism onto W. Now, for $s \gg 0$, we have $h^0(\tilde{M}, s\tilde{L}) \geq h^0(W, \mathcal{O}_W(s)) = h^0(\mathbf{P}^4, \mathcal{O}(s)) - h^0(\mathbf{P}^4, \mathcal{O}(s - 5)) = (5/6)(s^3 + 5s)$. On the other hand, by the Riemann-Roch theorem and by $\chi(M, L) = 1$, we infer $\chi(M, sL) = (1/6)(s^3 + 5s)$. Hence $h^0(\tilde{M}, s\tilde{L}) = h^0(W, \mathcal{O}_W(s))$ for $s \gg 0$. This implies $\rho_* \mathcal{O}_{\tilde{M}} = \mathcal{O}_W$ and hence W is normal. Therefore $\tilde{M} \simeq W$ by Zariski's Main Theorem. Thus $\tilde{M}$ is a smooth hypersurface and hence simply connected by Lefschetz Theorem. So π is the universal covering of M, as desired.

(2.7). In case (2.6), we can further describe the structure of M as follows. Let N be a generator of T. Then $\pi_* \mathcal{O}_{\tilde{M}} \simeq \oplus_{j=0}^{4} \mathcal{O}_M[jN]$. So $H^0(\tilde{M}, \tilde{L}) \simeq \oplus_j H^0(M, L_j)$ for $L_j = L + jN$ and $h^0(M, L_j) = \chi(M, Lj) = \chi(M, L) = 1$. If $\zeta_j \in H^0(M, L_j)$ is a base and if $\sigma \in \text{Gal}(\tilde{M}/M)$ is the automorphism of $\tilde{M}$ corresponding to $N \in T \simeq \text{Gal}(\tilde{M}/M)$, then $\sigma^* \tilde{\zeta}_j = \exp(2\pi j \sqrt{-1}/5) \tilde{\zeta}_j$ for $\tilde{\zeta}_j = \pi^* \zeta_j \in H^0(\tilde{M}, \tilde{L})$. Consider $(\tilde{\zeta}_0 : \cdots : \tilde{\zeta}_4)$ as a homogeneous coordinate system of $\mathbf{P}^4$ in the natural way. Then σ extends to a linear automorphism of $\mathbf{P}^4$ by the above formula. Note that $h^0(M, 5L) = 25$ while there are 26 monomials of $\tilde{\zeta}_0, \cdots, \tilde{\zeta}_4$ contained in $H^0(M, 5L_0)$. Hence there is a relation among them, which gives the equation $\phi(\tilde{\zeta}) = 0$ of $\tilde{M}$ in $\mathbf{P}^4$.

Now it is obvious how to construct examples of the type (2.6). Let $G = \mathbf{Z}/5\mathbf{Z}$ act on $\mathbf{P}^4$ as before. Take a G-invariant polynomial ϕ in $\tilde{\zeta}$'s of degree five. $W = \{\phi = 0\}$ is smooth if ϕ is general. G acts on W and there is no fixed point. So $M = W/G$ is a manifold. Since $\mathcal{O}_W(1)$ is G-invariant, $\mathcal{O}_W(1) = \pi^* L$ for some $L \in \text{Pic}(M)$. Then (M, L) is a polarized manifold of the type (2.6). A simplest example is obtained by setting $\phi(\tilde{\zeta}) = \tilde{\zeta}_0^5 + \cdots + \tilde{\zeta}_4^5$. This is a three-dimensional version of Godeaux surface.

Proposition (2.8). *If #(torsion part of* $\text{Pic}(M)$*) $= 4$, then* $\pi_1(M) \simeq \mathbf{Z}/4\mathbf{Z}$ *and the universal covering* $\tilde{M}$ *is a weighted complete intersection of type* (4, 4) *in the weighted projective space* $\mathbf{P}(2, 2, 1, 1, 1, 1)$.

Proof. We employ similar method and notation as in (2.6). There we have $\tilde{L}^3 = 4$ and $h^0(\tilde{M}, \tilde{L}) = 4$. So $|\tilde{L}|$ gives a morphism $\rho : \tilde{M} \to \mathbf{P}^3$. $T \simeq \text{Gal}(\tilde{M}/M)$ is isomorphic to either $\mathbf{Z}/4\mathbf{Z}$ or $\mathbf{Z}/2\mathbf{Z} \oplus \mathbf{Z}/2\mathbf{Z}$.

Assume that $T \simeq \mathbf{Z}/2\mathbf{Z} \oplus \mathbf{Z}/2\mathbf{Z}$. Then, if $\zeta_N \in H^0(M, L + N)$ for $N \in T$, we have $\zeta_N^2 \in H^0(M, 2L)$. Since $h^0(M, 2L) = \chi(M, 2L) = 3$, there is a relation among ζ_N^2's. This implies that $\rho(\tilde{M})$ is contained in a hyperquadric. But then $\tilde{L}^3 = 0$, contradiction.

Thus $T \simeq \mathbf{Z}/4\mathbf{Z}$. Let N be a generator of T, set $\Gamma(s, j) = H^0(M, sL + jN)$ and let $\zeta_j (j = 0, 1, 2, 3)$ be a base of $\Gamma(1, j)$. Then $\tilde{\zeta}_j = \pi^* \zeta_j$ form a basis of $H^0(\tilde{M}, \tilde{L})$. Since ρ is surjective, ζ_j's are algebraically independent in the graded algebra $\Gamma = \oplus_{s,j} \Gamma(s, j)$.

There are exactly two monomials of ζ_j's contained in $\Gamma(2, 1)$, namely ζ_{01} and ζ_{23}, where ζ_{ij} denotes $\zeta_i \zeta_j$. On the other hand $\dim \Gamma(2, 1) = \chi(M, 2L) = 3$. Hence we have $\eta_1 \in \Gamma(2, 1)$ such that ζ_{01}, ζ_{23} and η_1 form a linear basis of $\Gamma(2, 1)$. Similarly we have $\eta_3 \in \Gamma(2, 3)$ such that η_3, ζ_{03}, ζ_{12} form a basis of $\Gamma(2, 3)$. Note also that $\Gamma(2, 0) = \langle \zeta_{00}, \zeta_{13}, \zeta_{22} \rangle$ and $\Gamma(2, 2) = \langle \zeta_{02}, \zeta_{11}, \zeta_{33} \rangle$.

We claim that Γ is generated by ζ_j's and η_i's. To prove this, let R be the subalgebra generated by them and we will show $\Gamma(s, j) \subset R$ by the induction on s. By the above observation this is true if $s \leq 2$.

In order to consider the case $s \geq 3$, let D_j be the divisor $\{\zeta_j = 0\}$ for $j = 0, 1, 2, 3$. They are all irreducible and reduced, different to each other, and $L^2 D_j = 1$. For any $j \neq k$, $C_{jk} = D_j \cap D_k$ is an irreducible reduced curve with

$LC_{jk} = 1$.

We claim $H^1(M, jN) = 0$ for any j. To see this, assume $h^1(M, N) > 0$ for example. Then, using the exact sequence $H^0(M, N) \to H^0(M, L) \to H^0(D_3, L) \to H^1(M, N) \to H^1(M, L) = 0$, we get $h^0(D_3, L) > h^0(M, L) = 1$. We have $h^0(D_3, 2N) \leq h^0(M, 2N) + h^1(M, 3N - L) = 0$ by the vanishing theorem. Therefore $h^0(C_{23}, L) \geq h^0(D_3, L) - h^0(D_3, 2N) \geq 2$. Since $LC_{23} = 1$, this implies $C_{23} \simeq \mathbf{P}^1$, which is clearly absurd. Thus we conclude $h^1(M, N) = 0$. Similarly we prove $H^1(M, jN) = 0$ for any j.

Using Serre duality and vanishing theorem, we infer $H^q(M, sL + jN) = 0$ for any $0 < q < 3$, $s \in \mathbf{Z}$, j from this claim. So $H^1(D_\alpha, sL + jN) = 0$ and the restriction maps $\Gamma(s, j) \to H^0(D_\alpha, sL + jN)$ and $H^0(D_\alpha, sL + jN) \to H^0(C_{\alpha\beta}, sL+jN)$ are surjective for every α, β, s, j. Therefore $\oplus_{s,j} H^0(C_{\alpha\beta}, sL + jN) \simeq \Gamma/(\zeta_\alpha = \zeta_\beta = 0)$. Hence, in order to show $\Gamma(s, j) \subset R$ by induction on s, it suffices to find some $C_{\alpha\beta}$ such that $R \cap \Gamma(s, j) \to H^0(C_{\alpha\beta}, sL + jN)$ is surjective.

For $(s, j) = (3, 0)$, take C_{13}. Then $p_0 = D_0 \cap C_{13}$ and $p_2 = D_2 \cap C_{13}$ are different points on C_{13}. So $\zeta_{000} = \zeta_0^3$ and $\zeta_{022} = \zeta_0 \zeta_2^2$ are linearly independent in $H^0(C_{13}, 3L)$. Since $h^0(C_{13}, 3L) = 2$ by Riemann-Roch theorem, they form a base. Thus C_{13} is a desired one. Similarly, take C_{02}, ζ_{113} and ζ_{333} for $\Gamma(3, 1)$. For $\Gamma(3, 2)$, take C_{13} and ζ_{002}, ζ_{222}. For $\Gamma(3, 3)$, take C_{02} and ζ_{111}, ζ_{133}.

For $(s, j) = (4, 0)$, take C_{02}. $p_1 = D_1 \cap C_{02}$ and $p_3 = D_3 \cap C_{02}$ are different points. So ζ_{11} and ζ_{33} are linearly independent on C_{02}. Hence ζ_{1111}, ζ_{1133} and ζ_{3333} are linearly independent in $H^0(C_{02}, 4L)$. So they generate $H^0(C_{02}, 4L)$ and C_{02} is a desired one. We remark that $\eta_1 \eta_3 \in H^0(M, 4L)$ and there is a relation of the form $\eta_1 \eta_3 + \tau_0 \zeta_0 + \tau_2 \zeta_2 + a_0 \zeta_{1111} + a_1 \zeta_{1133} + a_2 \zeta_{3333} = 0$, where a_i's are constants and τ_j's are polynomials in η's and ζ's of degree three.

For $(s, j) = (4, 1)$, we take C_{13}. If $\eta_1(p_2) = 0$, then $\{\eta_1 = 0\}_{C_{13}} = p_2 + x$ for some $x \in |(2L + N)_C - (L + 2N)_C| = |(L + 3N)_{C_{13}}|$, while $H^0(C_{13}, L + 3N) = 0$ because $\Gamma(1, 3) \to H^0(C_{13}, L + 3N)$ is surjective. This contradiction shows $\eta_1(p_2) \neq 0$. Therefore $\eta_1 \zeta_{00} \notin \langle \eta_1 \zeta_{22}, \eta_3 \zeta_{02} \rangle$ in $H^0(C_{13}, 4L + N)$. Similarly we infer $\eta_1(p_0) \neq 0$ and $\eta_1 \zeta_{22} \notin \langle \eta_1 \zeta_{00}, \eta_3 \zeta_{02} \rangle$. Hence $\eta_1 \zeta_{00}$, $\eta_1 \zeta_{22}$ and $\eta_3 \zeta_{02}$ are linearly independent and generate $H^0(C_{13}, 4L + N)$. Thus C_{13} is a desired one.

We take C_{13} for $(s, j) = (4, 3)$ too. Quite similarly as above, $\eta_3 \zeta_{00}$, $\eta_3 \zeta_{22}$ and $\eta_1 \zeta_{02}$ generate $H^0(C_{13}, 4L + 3N)$.

For $\Gamma(4, 2)$, take C_{02}. We have $\eta_1(p_1) \neq 0$ and $\eta_1(p_3) \neq 0$ similarly as above. Therefore η_1^2, ζ_{1113}, ζ_{1333} are linearly independent on C_{02} and generate $H^0(C_{02}, 4L + 2N)$. We remark that $\eta_3^2 \in \Gamma(4, 2)$ and hence there is a relation of the form $c_1 \eta_1^2 + c_2 \eta_3^2 + \sigma_2 \zeta_0 + \sigma_0 \zeta_2 + b_1 \zeta_{1113} + b_2 \zeta_{1333} = 0$, where b_1, b_2, c_1, c_2 are constants ($c_1 \neq 0$ and $c_2 \neq 0$) and σ_0, σ_2 are polynomials in η's and ζ's of degree three.

In order to treat the case $s = 5$, we use the following

Lemma (2.9). *Let C be an irreducible reduced curve with $h^1(C, \mathcal{O}_C) = 2$ and let T be a line bundle on C such that $\deg(T) = 3$ and $\mathrm{Hom}(\omega, \mathcal{O}[T]) = 0$,* *where ω is the dualizing sheaf of C. Then the mapping $H^0(\omega) \otimes H^0(\mathcal{O})[T] \to H^0(\omega[T])$ is surjective.*

Proof. We have $h^0(T) = 2$ by the Riemann-Roch theorem. Let $\mathcal{F}$ be the subsheaf of $\mathcal{O}[T]$ generated by global sections. If $\mathcal{F} \neq \mathcal{O}[T]$, then $\deg(\mathcal{F}) \leq 2$ and $h^0(\mathcal{F}) = 2$. This is possible only when $\mathcal{F} \simeq \omega$. Then $\text{Hom}(\omega, \mathcal{O}[T]) \neq 0$, contradicting the assumption. Hence $\text{Bs}|T| = \emptyset$. So we can apply [F6;(A7)].

(2.10). *Proof of* (2.8), *continued.* As for $\Gamma(5,0)$, we let $C = C_{02}$ and set $T = 3L + 2N$. Since $\omega \simeq \mathcal{O}_C[2L + 2N]$ and $H^0(C, L) = 0$, the mapping $H^0(C, 2L+2N) \otimes H^0(C, T) \to H^0(C, 5L)$ is surjective by (2.9). Since $H^0(C, 2L+2N)$ and $H^0(C, T)$ come from $\Gamma(2,2)$ and $\Gamma(3,2)$ respectively, they are generated by the restrictions of monomials of ζ's and η's. From this we infer that $R \cap \Gamma(5,0) \to H^0(C_{02}, 5L)$ is surjective, as desired.

For $\Gamma(5,1)$, we let $C = C_{13}$, $T = 3L + N$ and use the above method. For $\Gamma(5,2)$, we let $C = C_{02}$ and $T = 3L$. For $\Gamma(5,3)$, we let $C = C_{13}$ and $T = 3L + 3N$.

In order to consider the case $s = 6$, we use the following

Lemma (2.11). *Let C be an irreducible reduced curve with $h^1(C, \mathcal{O}_C) = 2$ and let Q, T be line bundles on C such that $\deg(T) = 3$, $\deg(Q) = 4$, $\text{Hom}(\omega, \mathcal{O}_C[T]) = 0$ and $h^0(Q - T) \neq 0$. Then the mapping $H^0(\omega) \otimes H^0(Q) \to H^0(\omega[Q])$ is surjective.*

Proof. We have $\text{Bs}|Q| = \emptyset$. Moreover, if p is a member of $|Q - T|$, then $\text{Bs}|Q - p| = \text{Bs}|T| = \emptyset$. This implies that the rational mapping $C \to \mathbf{P}^2$ defined by $|Q|$ is a birational morphism. Hence we can apply [F6;(A7)].

(2.12). *Proof of* (2.8), *continued.* As for $\Gamma(6,0)$, we let $C = C_{01}$, $T = 3L + N$ and $Q = 4L + 3N$. Then $H^0(\omega) \otimes H^0(Q_C) \to H^0(C, 6L)$ is surjective by (2.11). So $R \cap \Gamma(6,0) \to H^0(C, 6L)$ is surjective, as desired.

For $\Gamma(6,1)$, we let $C = C_{01}$, $T = 3L + N$ and $Q = 4L$. For $\Gamma(6,2)$, let $C = C_{01}$, $T = 3L+2N$ and $Q = 4L+N$. For $\Gamma(6,3)$, let $C = C_{03}$, $T = 3L+3N$ and $Q = 4L$.

As for $\Gamma(s,j)$ with $s \geq 7$, any $C = C_{\alpha\beta}$ has the desired property. Indeed, for $F = (s-2)L+(j-\alpha-\beta)N$, we see that F_C is very ample. So $H^0(\omega) \otimes H^0(F) \to H^0(\omega[F]) = H^0(C, sL + jN)$ is surjective by [F6;(A7)]. Hence $R \cap \Gamma(s,j) \to H^0(C, sL + jN)$ is surjective.

Now we have proved that $R = G$. So, if $\tilde{\zeta}_j = \pi^*\zeta_j \in H^0(\tilde{M}, \tilde{L})$ and $\tilde{\eta}_i = \pi^*\eta_i \in H^0(\tilde{M}, 2\tilde{L})$, then the graded algebra $\oplus_{s \geq 0} H^0(\tilde{M}, s\tilde{L})$ is generated by $\tilde{\zeta}_j$'s and $\tilde{\eta}_j$'s. Hence we have an embedding $\tilde{M} \subset \mathbf{P}(2,2,1,1,1,1)$.

As we have observed in case $(s,j) = (4,0)$ and $(4,2)$, there are two relations ψ_1 and ψ_2 of degree four among ζ_j's and $\tilde{\eta}_j$'s. The zeros of them give irreducible divisors D_1, D_2 in $\mathbf{P}(2, 2, 1, 1, 1, 1)$ such that $\tilde{M} \subset D_1 \cap D_2$. Comparing dimensions and degrees we infer $\tilde{M} = D_1 \cap D_2$. So $\tilde{M}$ is simply connected. Thus we complete the proof of (2.8).

Remark. Alternately, one can use also the method in the Appendix.

Remark (2.13). If (M, L) is of the type (2.8), we can describe the structure of (M, L) more precisely similarly as in (2.7). Moreover, one can construct examples explicitly. We also see that all the polarized threefolds of this type are deformations to each other. Details of the proof are left to the reader. Our result may be viewed as a polarized version of a result in [Mi1].

(2.14). At present, following problems remain unsolved.

1) Find examples and classify polarized manifolds (M, L) such that $n = \dim M \geq 4$, $K = (3 - n)L$, $L^n = 1$ and $\Delta(M, L) > 2$.

2) Find examples and classify polarized threefolds (M, L) with $L^3 = 1$, $K = \mathcal{O}$, $\Delta(M, L) = 3$ and #(torsion part of $\mathrm{Pic}(M)$) ≤ 3.

§3. The case of a hyperquadric fibration over a curve

In this section we study (M, L) of the type (1.10; 4). So we have a surjective morphism $f : M \to C$ onto a smooth curve C such that any general fiber F of f is a hyperquadric in $\mathbf{P}^n$ with $L_F = \mathcal{O}_F(1)$.

Claim (3.1). *Every fiber of f is irreducible and reduced.*

This fact was pointed out by Ionescu[12] and is proved by the method in [Mo].

(3.2). For every $x \in C$ let F_x be the fiber of f over x and let L_x be the restriction of L to F_x. Then $\Delta(F_x, L_x) = 0$ by the lower-semicontinuity of the Δ-genus. Hence F_x is a hyperquadric. In particular $h^0(F_x, L_x) = n + 1$ and $\mathrm{Bs}|L_x| = \emptyset$. Therefore $\mathcal{E} = f_*\mathcal{L}$ is a locally free sheaf of rank $n + 1$ and the natural homomorphism $f^*\mathcal{E} \to \mathcal{L}$ is surjective. This yields an embedding $\iota : M \to P = \mathbf{P}(\mathcal{E})$ such that $\iota^*H = L$, where H is the tautological line bundle on P. M is a divisor on P and is a member of $|2H + \pi^*B|$ for some $B \in \mathrm{Pic}(C)$, where π is the projection $P \to C$.

(3.3). $2e + (n + 1)b \geq 0$ *for $b = \deg(B)$ and $e = c_1(\mathcal{E})$.*

Indeed, for all x, the restriction of the equation defining M in P is a polynomial of degree two in homogeneous coordinates of $P_x = \pi^{-1}(x) \simeq \mathbf{P}^n$. Taking the determinant of them, we get a section of $2\det(\mathcal{E}) + (n + 1)B$ over C such that its zeros are exactly the points x over which F_x are singular. In particular we have the inequality above.

(3.4). One easily sees $d(M, L) = H^n(2H + B)\{P\} = 2e + b$.

(3.5). $b + e + 2q = 3$ *for the genus q of C.*

Indeed, the canonical bundle K^P of P is $-(n+1)H + \pi^*(K^C + \det(\mathcal{E}))$. So $K = (1-n)H + f^*A$ for $A = K^C + \det(\mathcal{E}) + B$ and $K + (n-1)L = f^*A$. Since $(K + (n-1)L)L^{n-1} = 2$, we have $1 = \deg(A) = 2q - 2 + e + b$. This gives the above equality.

Claim (3.6). $q = 0$ *or* 1.

Indeed, for any fiber F over $x \in C$, we have $h^1(C, x) \le h^1(M, F) = 0$ since $F \sim K + (n-1)L$. This implies $q \le 1$.

(3.7). For the moment, until (3.14), we assume $q = 1$. In view of (3.3), (3.5) and $d = 2e + b \ge 1$, we infer that there are only the following three possibilities:

1) $b = 1$, $e = 0$ and $d = 1$.
2) $b = 0$, $e = 1$ and $d = 2$.
3) $b = -1$, $e = 2$, $n = 3$ and $d = 3$.

(3.8). In case (3.7; 1), we have $\chi(M, L) = \chi(P, H) = \chi(C, \mathcal{E}) = 0$. Indeed, from the exact sequence $0 \to \mathcal{O}_P[-H - B] \to \mathcal{O}_P[H] \to \mathcal{O}_M[L] \to 0$, we obtain $H^i(M, L) \simeq H^i(P, H) \simeq H^i(C, \mathcal{E})$ for every i.

We have $\deg(\mathcal{Q}) \ge 0$ for any quotient bundle $\mathcal{Q}$ of $\mathcal{E}$ of rank one. Indeed, $\deg(\mathcal{Q}) = HZ$ for the section Z of π induced by $\mathcal{Q}$. If $HZ < 0$, then $MZ = 2HZ + b < 0$. So $Z \subset M$. Then we should have $0 < LZ = HZ$, a contradiction.

Thus we can describe the structure of (M, L). As we see below, polarized manifolds of this type do really exist in arbitrary dimension.

Example (3.9). Let C be any smooth elliptic curve and let N_0, N_1, ..., N_n be line bundles of degree zero on C. We choose them generically so that $N_i - N_j$ is not a torsion in $\mathrm{Pic}(C)$ for any $i \ne j$.

Let $\mathcal{E} = N_0 \oplus \cdots \oplus N_n$, $P = \mathbf{P}(\mathcal{E})$, $H = \mathcal{O}_P(1)$ and let B be any line bundle of degree one on C. Then any general member of $|2H + \pi^*B|$ is smooth, where π is the projection $P \to C$.

To see this, let D_i be the member of $|H - \pi^*N_i|$ induced by the subbundle N_i of $\mathcal{E}$. Then $C_i = D_0 \cap \cdots \cap D_{i-1} \cap D_{i+1} \cap \cdots \cap D_n$ is the section of π induced by the quotient bundle N_i of $\mathcal{E}$. Of course $C_i \cap D_i = \emptyset$. On the other hand, $h^0(C, B + N_i + N_j) = 1$ for any i, j. Let $x_{ij} \in |B + N_i + N_j|$ and $F_{ij} = \pi^{-1}(x_{ij})$. By the assumption on N_i's, x_{ij}'s are *different* $(n+1)^2$ points. Since $2D_i + F_{ii} \in |2H + \pi^*B|$ for every i, we have $\mathrm{Bs}|2H + \pi^*B| \subset \cap_{i=0}^n (D_i \cup F_{ii}) = \cup_i (F_{ii} \cap (\cap_{j \ne i} D_j)) = \{p_0, \cdots, p_n\}$ where $p_i = F_{ii} \cap C_i$. We have also $D_i + D_j + F_{ij} \in |2H + \pi^*B|$ for $i \ne j$, which is non-singular at p_i since $p_i \notin D_j$ and $p_i \notin F_{ij}$. Therefore any general member of $|2H + \pi^*B|$ is non-singular at each point p_i. This is enough by Bertini's theorem.

Thus, any general member M of $|2H + \pi^*B|$ is smooth. Moreover, we may assume $C_i \not\subset M$ for any i. Let L be the restriction of H to M. Then we claim $LZ > 0$ for any curve Z in M.

Indeed, $HZ \geq 0$ since $\mathcal{E}$ is semipositive. If $HZ = 0$, then the pull-back $\mathcal{E}_Z$ of $\mathcal{E}$ to Z has a quotient bundle Q of rank one with $\deg(Q) = 0$. So $h^0(Z, Q - N_i) > 0$ for some i. Comparing the degrees we get $Q = [N_i]_Z$. On the other hand $N_i - N_j \neq 0$ in $\mathrm{Pic}(Z)$ for $i \neq j$ since it is not a torsion in $\mathrm{Pic}(C)$. Hence $\mathrm{Hom}(\mathcal{E}_Z, Q) \simeq \mathbf{C}$ and Q must be the quotient bundle N_i of $\mathcal{E}$. This implies $Z = C_i$, contradicting $C_i \not\subset M$.

For any subvariety Y of M, we have $Y \not\subset D_i$ for some i. By the above claim we have $Y \cap D_i \neq \emptyset$ since $D_i \sim L$. From this observation we infer $L^r Y > 0$ for $r = \dim(Y)$ by induction on r. Hence L is ample on M by Nakai's criterion. It is now obvious that (M, L) is a polarized manifold of the type (3.8).

Problem. Find an example (M, L) of the type (3.8) such that $\mathcal{E}$ is indecomposable. Perhaps there are many.

(3.10). In case (3.7; 2), we have $\chi(M, L) = 1$ similarly as in (3.8). Moreover $\mathcal{E}$ is numerically semipositive. Indeed, if $HZ < 0$ for a curve Z in P, then $MZ = (2H + B)Z = 2HZ < 0$. So $Z \subset M$, which is impossible because $L = H_M$ is ample.

Polarized manifolds of this type do really exist in arbitrary dimension. To see this, we recall the following

Lemma (3.11). *Let $\mathcal{E}$ be any ample vector bundle on an elliptic curve C. Then $h^1(C, \mathcal{E}) = 0$ and $h^0(C, \mathcal{E}) = c_1(\mathcal{E})$.*

Proof is easy and well-known (cf., e. g., [A]).

Example (3.12). Let C, $\mathcal{E}$ be as in (3.11) and assume that $c_1(\mathcal{E}) = 1$. Note that $\mathcal{E}$ is indecomposable. Let $P = \mathbf{P}(\mathcal{E})$ and $H = \mathcal{O}_P(1)$. Then $\mathrm{Bs}|2H + \pi^* B| = \emptyset$ for any $B \in \mathrm{Pic}(C)$ with $\deg(B) = 0$, where π is the projection $P \to C$.

In order to prove this, we use the induction on $r = \mathrm{rank}(\mathcal{E})$. The assertion is obvious if $r = 1$. So we assume $r > 1$. Suppose that we have $x \in \mathrm{Bs}|2H + \pi^* B|$. For any $N \in \mathrm{Pic}(C)$ with $\deg(N) = 0$, we have $h^0(P, H + \pi^* N) = h^0(C, \mathcal{E} \otimes N) = 1$ by (3.11). Let D_N be the unique member of $|H + \pi^* N|$. Since D_N moves as N varies in $\mathrm{Pic}^0(C)$, we find $x \in D_N$ for some N. This $D_N = D$ corresponds to an injection $\mathcal{O}_C[-N] \to \mathcal{E}$, the cokernel $\mathcal{E}'$ of which is again an ample vector bundle with $c_1(\mathcal{E}') = 1$. Moreover $D \simeq \mathbf{P}(\mathcal{E}')$ with $\mathcal{O}(1)$ being the restriction of H. Therefore $\mathrm{Bs}|[2H + \pi^* B]_D| = \emptyset$ by the induction hypothesis. On the other hand, we have an exact sequence $0 \to \mathcal{O}_P(H + \pi^*[B - N]) \to \mathcal{O}_P[2H + \pi^* B] \to \mathcal{O}_D[2H + \pi^* B] \to 0$ and $H^1(P, H + \pi^*[B - N]) = H^1(C, \mathcal{E} \otimes [B - N]) = 0$ by (3.11). So $H^0(P, 2H + \pi^* B) \to H^0(D, 2H + \pi^* B)$ is surjective. This implies $\mathrm{Bs}|[2H + \pi^* B]_D| = D \cap \mathrm{Bs}|2H + \pi^* B| \ni x$. This yields a contradiction, as desired. Thus we prove the claim.

By this claim any general member M of $|2H + \pi^* B|$ is smooth. The restriction $L = H_M$ is ample since so is $\mathcal{E}$. The polarized manifold (M, L) is clearly of the type (3.7; 2).

(3.13). In case (3.7; 3), we have $\chi(M, L) = \chi(P, H) = \chi(C, \mathcal{E}) = 2$ similarly as before. Moreover $\mathcal{E}$ is ample.

To see this, let Y be any subvariety of P. If $Y \subset M$, then $H_Y = L_Y$ is ample. If $Y \not\subset M$, then $2H_Y = [M]_Y + A_Y$ for some $A \in \mathrm{Pic}(C)$ with $\deg(A) = 1$. Hence, in either case, $|mH_Y|$ contains a non-zero member for some $m > 0$. This implies the ampleness of H by Nakai's criteion (cf.[F3; Appendix B]).

Furthermore, since $2e + (n + 1)b = 0$, every fiber of f is smooth by (3.3). Thus f is a $(\mathbf{P}^1 \times \mathbf{P}^1)$-bundle over C.

Example (3.14). Let $p : \tilde{C} \to C$ be an étale double covering of an elliptic curve C and let ι be the sheet changing involution such that $\tilde{C}/\iota \simeq C$. Let $\mathcal{E}_1$ be an ample vector bundle on $\tilde{C}$ such that $\mathrm{rank}(\mathcal{E}_1) = 2$, $c_1(\mathcal{E}_1) = 1$. Set $\mathcal{E}_2 = \iota^*\mathcal{E}_1$, $\tilde{\mathcal{E}} = \mathcal{E}_1 \otimes \mathcal{E}_2$ and let $\tilde{P} = \mathbf{P}(\tilde{\mathcal{E}})$, $\tilde{P}_1 = \mathbf{P}(\tilde{\mathcal{E}}_1)$ and $\tilde{P}_2 = \mathbf{P}(\tilde{\mathcal{E}}_2)$. Let $\tilde{M}$ be the fiber product $\tilde{P}_1 \times_{\tilde{C}} \tilde{P}_2$ over $\tilde{C}$. Then we have a natural embedding $\tilde{M} \subset \tilde{P}$ such that each fiber $\tilde{M}_X$ of $\tilde{M} \to \tilde{C}$ over $x \in \tilde{C}$ is a smooth quadric in $\tilde{P}_x \simeq \mathbf{P}^3$. Now, since $\iota^*\tilde{\mathcal{E}} = \iota^*\mathcal{E}_1 \otimes \iota^*\mathcal{E}_2 \simeq \mathcal{E}_2 \otimes \mathcal{E}_1$, ι induces a factor changing involution of $\tilde{\mathcal{E}}$ and lifts to an involution $\tilde{\iota}$ of $\tilde{P}$ such that $\tilde{\iota}(\tilde{M}) = \tilde{M}$. Let $P = \tilde{P}/\tilde{\iota}$ and $M = \tilde{M}/\tilde{\iota}$ be the quotients. Then $P = \mathbf{P}(\mathcal{E})$ for some vector bundle $\mathcal{E}$ on C such that $p^*\mathcal{E} \simeq \tilde{\mathcal{E}}$. The tautological line bundle H on P is ample since so is $\tilde{\mathcal{E}}$. Setting $L = H_M$, we easily see that (M, L) is a polarized threefold of the type (3.7; 3).

(3.15). From now on, throughout this section, we study the case $q = 0$. So $C \simeq \mathbf{P}_\xi^1$. The pull-backs of $\mathcal{O}_C(1)$ will be denoted by H_ξ. We have $b + e = 3$ by (3.5) and $d = b + 2e$ by (3.4). So $b = 6 - d$, $e = d - 3$ and $6n \geq (n - 1)d$ by (3.3). Hence $d \leq 9$ if $n \geq 3$ and $d \leq 8$ if $n \geq 4$.

(3.16). Similarly as in (3.8), we have $H^i(M, L) \simeq H^i(P, H) \simeq H^i(C, \mathcal{E})$ for every i. So $h^0(M, L) \geq \chi(C, \mathcal{E}) = n + d - 2$. We claim $h^1(M, L) = 0$, $h^0(M, L) = n + d - 2$ and $\Delta(M, L) = 2$.

Indeed, otherwise, we would have $\Delta(M, L) \leq 1$. This is impossible by (1.11).

(3.17). Since $C \simeq \mathbf{P}^1$, $\mathcal{E}$ is a direct sum of $n + 1$ line bundles of degrees $e_0, \cdots, e_n$, which will be denoted by $\mathcal{O}(e_0, \cdots, e_n)$. Clearly we may assume $e_0 \leq \cdots \leq e_n$. We have $e_0 \geq -1$ since $h^1(C, \mathcal{E}) = h^1(M, L) = 0$ by (3.16).

Lemma (3.18). $e_1 \geq 0$ *if $d \geq 2$.*

Proof. Suppose that $e_0 = e_1 = -1$. Then the surjection $\mathcal{E} \to \mathcal{O}(e_0, e_1)$ yields an embedding of $Z = \mathbf{P}(\mathcal{O}(e_0, e_1))$ in P. Then $Z \simeq \mathbf{P}_\sigma^1 \times C$ and $H_Z = H_\sigma - H_\xi$. So $Z \not\subset M$ since H_Z is not ample. On the other hand $Z \cap M \neq \emptyset$ since each fiber of Z over $x \in C$ is a line and meets M_x. Hence $Z \cap M$ is a curve in M. Therefore $0 < L\{Z \cap M\} = H(2H + bH_\xi)\{Z\} = b - 4 = 2 - d$. This is impossible if $d \geq 2$.

Lemma (3.19). $e_0 \geq 0$ *if $d \geq 5$.*

Proof. Suppose that $e_0 = -1$. Then $\mathcal{E} \to \mathcal{O}(e_0)$ gives a section Z of $P \to C$ such that $HZ = -1$. So $Z \not\subset M$. Hence $0 \leq MZ = (2H + bH_\xi)Z = b - 2 = 4 - d$, contradicting $d \geq 5$.

Lemma (3.20). $e_0 \geq 1$ *if $d \geq 7$.*

Proof is similar as above.

(3.21). For each case $d = 1, \cdots, 9$, we will study the structure of (M, L). First we suppose $d = 1$. Then $e = -2$ and $b = 5$. We claim $e_2 \geq 0$.

Indeed, if $e_0 = e_1 = e_2 = -1$, then we have an embedding of $Z = \mathbf{P}(e_0, e_1, e_2) \simeq \mathbf{P}^2 \times C$ in P such that $H_Z = H_\sigma - H_\xi$. Since $Z \subset \mathrm{Bs}|H|$ and $H^0(P, H) \simeq H^0(M, L)$, we infer $Z \cap M \subset \mathrm{Bs}|L|$. So $\dim(\mathrm{Bs}|L|) \geq \dim(Z \cap M) \geq 2$ since M is an ample divisor on P. This is impossible because $\Delta(M, L) = 2$.

Now we conclude $e_0 = e_1 = -1$ and $e_2 = \cdots = e_n = 0$. So $P = \{(\xi_0 : \xi_1) \times (\sigma_0 : \sigma_1 : \sigma_{20} : \sigma_{21} : \cdots : \sigma_{n0} : \sigma_{n1}) \in \mathbf{P}_\xi^1 \times \mathbf{P}_\sigma^{2n-1} \mid \xi_0 : \xi_1 = \sigma_{20} : \sigma_{21} = \cdots = \sigma_{n0} : \sigma_{n1}\}$ and $H = H_\sigma - H_\xi$. We let $M \in |2H_\sigma + 3H_\xi|$ be defined by $q_0(\sigma)\xi_0^3 + q_1(\sigma)\xi_0^2\xi_1 + \cdots + q_3(\sigma)\xi_1^3 = 0$, where $q_j(\sigma)$'s are quadric polynomials in σ_0, σ_1, σ_{20}, σ_{21}, ..., σ_{n0}, σ_{n1}.

We claim $n \leq 3$. To prove this, we substitute $\sigma_0 = a_{00}\xi_0 + a_{01}\xi_1$, $\sigma_1 = a_{10}\xi_0 + a_{11}\xi_1$, $\sigma_{20} = a_2\xi_0$, $\sigma_{21} = a_2\xi_1$, ..., $\sigma_{n0} = a_n\xi_0$, $\sigma_{n1} = a_n\xi_1$ in the above equation defining M, where a_{00}, a_{01}, a_{10}, a_{11}, a_2, ..., a_n are constants. Then we get an equation of the form $Q_0(a)\xi_0^5 + Q_1(a)\xi_0^4\xi_1 + \cdots + Q_5(a)\xi_1^5 = 0$, where Q_j's are quadric polynomials in a's. When $n \geq 4$, the homogeneous equation $Q_0(a) = \cdots = Q_5(a) = 0$ has a non-trivial solution. Taking such a solution, we define a rational map $\alpha : \mathbf{P}_\xi^1 \to \mathbf{P}_\sigma^{2n-1}$ by setting

$$\alpha(\xi_0 : \xi_1) = (a_{00}\xi_0 + a_{01}\xi_1 : a_{10}\xi_0 + a_{11}\xi_1 : a_2\xi_0 : a_2\xi_1 : \cdots : a_n\xi_1 : a_n\xi_1).$$

If α is not a morphism, then $a_{00}a_{11} - a_{01}a_{10} = a_2 \cdots = a_n = 0$. Since (a) ia non-trivial, either the ratio $a_{00} : a_{10}$ or $a_{01} : a_{11}$ is well-defined, and these are equal if both are defined. So the equations $\sigma_0 : \sigma_1 = a_{00} : a_{10} = a_{01} : a_{11}$, $\sigma_{20} = \sigma_{21} = \cdots = \sigma_{n0} = \sigma_{n1} = 0$ determines a point z on $\mathbf{P}_\sigma^{n-1}$. Let Z be the fiber of $\mathbf{P}_\xi^1 \times \mathbf{P}_\sigma^{2n-1} \to \mathbf{P}_\sigma^{2n-1}$ over z. Then $Z \subset M$ by the choice of z. On the other hand, $HZ = -1$ since $H = [H_\sigma - H_\xi]_P$, while $HZ = LZ > 0$. Thus this possibility is ruled out.

Now we see that α is a morphism. Let Γ be its graph. Then $\Gamma \subset M$ by the definition of α. So we must have $H\Gamma = L\Gamma > 0$. However, $H\Gamma = (H_\sigma - H_\xi)\Gamma = 0$ since $H_\sigma\Gamma = H_\xi\Gamma = 1$.

Thus we get a contradiction, proving the claim.

(3.22). Suppose that $d = 2$. Then $e = -1$ and $b = 4$. Using (3.18) we infer $\mathcal{E} = \mathcal{O}(-1, 0, \cdots, 0)$. So, similarly as in (3.21), $P = \{(\xi_0 : \xi_1), (\sigma_0 : \sigma_{10} : \sigma_{11} : \cdots : \sigma_{n0} : \sigma_{n1}) \in \mathbf{P}_\xi^1 \times \mathbf{P}_\sigma^{2n} \mid \xi_0 : \xi_1 = \sigma_{10} : \sigma_{11} = \cdots = \sigma_{n0} : \sigma_{n1}\}$ and $H = $

$H_\sigma - H_\xi$. Let $M \in |2H_\sigma + 2H_\xi|$ be defined by $q_0(\sigma)\xi_0^2 + q_1(\sigma)\xi_0\xi_1 + q_2(\sigma)\xi_1^2 = 0$, where q_i's are quadric polynomials in σ's. We claim $n \leq 3$.

To prove the claim, we substitute $\sigma_0 = a_{00}\xi_0 + a_{01}\xi_1$, $\sigma_{10} = a_1\xi_0$, $\sigma_{11} = a_1\xi_1$, $\ldots$, $\sigma_{n1} = a_n\xi_1$ in the above equation defining M, where a_{00}, a_{01}, a_1, $\ldots$, a_n are constants. Then we get an equation of the form $Q_0(a)\xi_0^4 + \cdots Q_4(a)\xi_1^4 = 0$, where Q_j's are quadric polynomials in a's. When $n \geq 4$, the homogeneous equation $Q_0(a) = \cdots = Q_4(a) = 0$ has a non-trivial solution. Taking such a solution, we define a map $\alpha : \mathbf{P}_\xi^1 \to \mathbf{P}_\sigma^{2n}$ by setting $\alpha(\xi_0 : \xi_1) = (a_{00}\xi_0 + a_{01}\xi_1 : a_1\xi_0 : a_1\xi_1 : \cdots : a_n\xi_0 : a_n\xi_1)$. If this is not a morphism, then $a_1 = \cdots = a_n = 0$ and the fiber Z of $\mathbf{P}_\xi^1 \times \mathbf{P}_\sigma^{2n} \to \mathbf{P}_\sigma^{2n}$ over $(1 : 0 : \cdots : 0)$ is contained in M. This is absurd since $HZ = -1$. Hence α is a morphism. Let Γ be its graph. Then $\Gamma \subset M$ by the definition of α. This is impossible since $H\Gamma = (H_\sigma - H_\xi)\Gamma = 0$. Thus we prove the claim.

(3.23). Suppose that $d = 3$. Then $e = 0$ and $b = 3$. Using (3.18), we infer $\mathcal{E} = \mathcal{O}(-1, 0, \cdots, 0, 1)$ or $\mathcal{O}(0, \cdots, 0)$.

1) If $\mathcal{E} = \mathcal{O}(0, \cdots, 0)$, then $P \simeq \mathbf{P}_\xi^1 \times \mathbf{P}_\eta^n$ and $H = H_\eta$. Let $M \in |2H_\eta + 3H_\xi|$ be defined by $q_0(\eta)\xi_0^3 + \cdots + q_3(\eta)\xi_1^3 = 0$, where q_i's are quadric polynomials in the homogeneous coordinate $(\eta_0 : \cdots : \eta_n)$ of $\mathbf{P}_\eta^n$. Then we claim $n \leq 3$.

Indeed, if $n \geq 4$, $q_0(\eta) = \cdots = q_3(\eta) = 0$ would have a non-trivial solution. This gives a fiber Z of $P \to \mathbf{P}_\eta^n$ such that $Z \subset M$. This is absurd since $HZ = 0$.

Thus $P \simeq \mathbf{P}_\xi^1 \times \mathbf{P}_\eta^3$. Note that $\mathrm{Bs}|L| = \emptyset$ and $\rho_{|L|}$ is the finite morphism $M \to \mathbf{P}_\eta^3$ of degree three.

Conversely, if we let $P = \mathbf{P}_\xi^1 \times \mathbf{P}_\eta^3$ and take a general member M of $|2H_\eta + 3H_\xi|$, then M contains no fiber of $P \to \mathbf{P}_\eta^3$. Hence $L = [H\eta]_M$ is ample and (M, L) is a polarized manifold of the above type.

2) If $\mathcal{E} = \mathcal{O}(-1, 0, \cdots, 0, 1)$, then $P \simeq \{(\xi_0 : \xi_1) \times (\sigma_0 : \sigma_{10} : \sigma_{11} : \cdots : \sigma_{n-1,0} : \sigma_{n-1,1} : \sigma_{n0} : \sigma_{n1} : \sigma_{n2}) \in \mathbf{P}_\xi^1 \times \mathbf{P}_\sigma^{2n+1} \mid \xi_0 : \xi_1 = \sigma_{10} : \sigma_{11} = \cdots = \sigma_{n-1,0} : \sigma_{n-1,1} = \sigma_{n0} : \sigma_{n1} : \sigma_{n2}\}$ and $H = H_\sigma - H_\xi$. Let $M \in |2H_\sigma + H_\xi|$ be defined by $q_0(\sigma)\xi_0 + q_1(\sigma)\xi_1 = 0$, where q_i's are quadric polynomials in σ.'s. Then we claim $n \leq 3$.

To see this, we substitute $\sigma_0 = a_{00}\xi_0 + a_{01}\xi_1$, $\sigma_{10} = a_1\xi_0$, $\sigma_{11} = a_1\xi_1$, $\ldots$, $\sigma_{n-1,0} = a_{n-1}\xi_0$, $\sigma_{n-1,1} = a_{n-1}\xi_1$, $\sigma_{n,0} = \sigma_{n1} = \sigma_{n2} = 0$ in the above equation defining M. Then we get an equation $Q_0(a)\xi_0^3 + \cdots + Q_3(a)\xi_1^3 = 0$, where Q_j's are quadric polynomials in a's. When $n \geq 4$, $Q_0(a) = \cdots = Q_3(a) = 0$ has a non-trivial solution. Taking such a solution, we define a rational mapping $\alpha : \mathbf{P}_\sigma^1 \to \mathbf{P}_\sigma^{2n+1}$ by setting $\alpha(\xi_0 : \xi_1) = (a_{00}\xi_0 + a_{01}\xi_1 : a_1\xi_0 : a_1\xi_1 : \cdots a_{n-1}\xi_0 : a_{n-1}\xi_1 : 0 : 0 : 0)$. If α is not a morphism, then the fiber Z of $\mathbf{P}_\xi^1 \times \mathbf{P}_\sigma \to \mathbf{P}_\sigma$ over $(1 : 0 : \cdots : 0)$ is contained in M. This is absurd since $HZ = -1$. Hence α is a morphism. Let Γ be its graph. Then $\Gamma \subset M$ by the definition of α. This is impossible since $H\Gamma = 0$.

Note that $\mathrm{Bs}|H|$ is $\{\sigma_{10} = \sigma_{11} = \sigma_{20} = \sigma_{21} = \sigma_{30} = \sigma_{31} = \sigma_{32} = 0\}$, the section of $P \to C$ induced by the surjection $\mathcal{E} \to \mathcal{O}(-1)$. So $\mathrm{Bs}|L| = M \cap \mathrm{Bs}|H|$

is a point. Let $\tilde{M}$ be the blowing-up of M at this point. Then $|L|$ gives a morphism $\tilde{M} \to \mathbf{P}^3$ of mapping degree two.

(3.24). Suppose that $d = 4$. Then $e = 1$ and $b = 2$. Using (3.18), we infer $\mathcal{E} \simeq \mathcal{O}(0, \cdots, 0, 1)$, $\mathcal{O}(-1, 0, \cdots, 0, 1, 1)$ or $\mathcal{O}(-1, 0, \cdots, 0, 2)$.

This last case is impossible. Indeed, if not, the injection $\mathcal{O}(2) \to \mathcal{E}$ gives a member D of $|H - 2H_\xi|$ on P. Then $D \cap M \neq \emptyset$. So $0 < L^{n-1}\{D \cap M\} = (H - 2H_\xi)(2H + 2H_\xi)H^{n-1} = 2e - 2 = 0$, a contradiction.

If $\mathcal{E} \simeq \mathcal{O}(-1, 0, \cdots, 0, 1, 1)$, the injection $\mathcal{O}(1, 1) \to \mathcal{E}$ yields an embedding $S = \mathbf{P}(\mathcal{O}(-1, 0, \cdots, 0))$ in P. Clearly $S \cap M \neq \emptyset$, while $S \not\subset M$ since H_S is not ample. So $H^{n-2}\{S \cap M\} > 0$. On the other hand, since S is a complete intersection of two members of $|H - H_\xi|$, we have $H^{n-2}\{S \cap M\} = H^{n-2}(H - H_\xi)^2(2H + 2H_\xi) = 2e - 2 = 0$. Thus this case is ruled out.

Now we conclude $\mathcal{E} \simeq \mathcal{O}(0, \cdots, 0, 1)$. So $P \simeq \{((\xi_0 : \xi_1) \times (\eta_0 : \cdots : \eta_{n-1} : \eta_{n0} : \eta_{n1})) \in \mathbf{P}^1_\xi \times \mathbf{P}^{n+1}_\eta \mid \xi_0 : \xi_1 = \eta_{n0} : \eta_{n1}\}$ and $H = H_\eta$. Let $M \in |2H_\eta + 2H_\xi|$ be defined by $q_0(\eta)\xi_0^2 + q_1(\eta)\xi_0\xi_1 + q_2(\eta)\xi_1^2 = 0$, where q_i's are quadric polynomials in η's. We claim $n \leq 3$. Indeed, if $n \geq 4$, $q_0(\eta) = q_1(\eta) = q_2(\eta) = \eta_{n0} = \eta_{n1} = 0$ would have a non-trivial solution. So there is a fiber Y of $\mathbf{P}^1_\xi \times \mathbf{P}^{n+1}_\eta \to \mathbf{P}^{n+1}_\eta$ such that $Y \subset M$. This is absurd since $HY = 0$.

Thus $\mathcal{E} \simeq \mathcal{O}(0, 0, 0, 1)$. The morphism $\rho_\eta : P \to \mathbf{P}^4_\eta$ makes P the blowing-up of $\mathbf{P}^4_\eta$ along the line $l = \{\eta_{30} = \eta_{31} = 0\}$. The exceptional divisor E is the member $\{\eta_{30} = \eta_{31} = 0\}$ of $|H_\eta - H_\xi|$. Since $M \in |4H_\eta - 2E|$, $\rho_\eta(M)$ is a hypersurface of degree four having double points along l. The morphism $M \to \rho_\eta(M)$ is finite since $L = [H_\eta]_M$ is ample. Hence M is the normalization of $\rho_\eta(M)$.

(3.25). Suppose that $d = 5$. Then $e = 2$ and $b = 1$. Using (3.19), we infer $\mathcal{E} \simeq \mathcal{O}(0, \cdots, 0, 2)$ or $\mathcal{O}(0, \cdots, 0, 1, 1)$.

We claim $e_2 \geq 1$. Indeed, otherwise, we have an embedding of $T = \mathbf{P}(\mathcal{O}(0, 0, 0)) \simeq \mathbf{P}^1_\xi \times \mathbf{P}^2_\eta$ in P such that $H_T = H_\eta$. Then $T \not\subset M$ since H_η is not ample. So $M_T \in |2H_\eta + H_\xi|$. Similarly as before we find a fiber Y of $T \to \mathbf{P}^2_\eta$ such that $Y \subset M$. This is impossible since $HY = 0$. Thus we prove the claim.

From this claim we obtain $\mathcal{E} \simeq \mathcal{O}(0, 0, 1, 1)$.

(3.26). Suppose that $d = 6$. Then $e = 3$ and $b = 0$. We claim $e_1 > 0$. Indeed, otherwise, we have an embedding of $S = \mathbf{P}(\mathcal{O}(0, 0)) \simeq \mathbf{P}^1_\xi \times \mathbf{P}^1_\eta$ in P such that $H_S = H_\eta$. Since $M_S \in |2H_\eta|$, we get a contradiction similarly as in (3.25).

From this claim and (3.19) we infer that $\mathcal{E} \simeq \mathcal{O}(0, 1, 1, 1)$. In particular $n = 3$. M is a smooth member of $|2H|$.

We have $h^0(M, L - H_\xi) = h^0(P, H - H_\xi) = 3$. $\mathrm{Bs}|H - H_\xi|$ is the section Z of $P \to C$ induced by the surjection $\mathcal{E} \to \mathcal{O}$. $Z \not\subset M$ since H_Z is not ample.

So $\text{Bs}|L - H_\xi| = Z \cap M = \emptyset$ since $ZM = 2HZ = 0$. Let $h : M \to \mathbf{P}_\tau^2$ be the morphism defined by $|L - H_\xi|$. Combining f and h we get a morphism $p : M \to \mathbf{P}_\xi^1 \times \mathbf{P}_\tau^2$. Since $L = p^*[H_\xi + H_\tau]$ is ample, p is a finite morphism. Moreover, $6 = d = L^3 = \deg(p) \cdot (H_\xi + H_\tau)^3 \{\mathbf{P}_\xi^1 \times \mathbf{P}_\tau^2\}$ implies $\deg(p) = 2$. By [F6; (2.6)], the branch locus B of p is a smooth divisor on $\mathbf{P}_\xi^1 \times \mathbf{P}_\tau^2$. From $K + 2L = H_\xi$ we infer $B \in |2H_\xi + 2H_\tau|$.

The existence of polarized manifolds of this type is obvious.

(3.27). Suppose that $d = 7$. Then $e = 4$ and $b = -1$. From (3.20) we infer $\mathcal{E} \simeq \mathcal{O}(1,1,1,1)$. So $P \simeq \mathbf{P}_\xi^1 \times \mathbf{P}_\tau^3$, $H = H_\xi + H_\tau$ and $M \in |2H_\tau + H_\xi|$. We easily see that the morphism $M \subset P \to \mathbf{P}_\tau^3$ makes M the blow-up of $\mathbf{P}_\tau^3$ along a curve Z, which is a complete intersection of two hyperquadrics in $\mathbf{P}_\tau^3$.

(3.28). Suppose that $d = 8$. Then $e = 5$ and $b = -2$. From (3.20) we infer $\mathcal{E} \simeq \mathcal{O}(1,1,1,2)$ or $\mathcal{O}(1,1,1,1,1)$.

When $\mathcal{E} \simeq \mathcal{O}(1,1,1,2)$, we have $h^0(M, L - H_\xi) = h^0(P, H - H_\xi) = 5$. Moreover $|H - H_\xi|$ gives a birational morphism $P \to \mathbf{P}_\tau^4$. In fact P is the blowing-up of $\mathbf{P}_\tau^4$ with center Z being linear plane $\mathbf{P}^2$. Since $M \in |2H - 2H_\xi| = |2H_\tau|$, M is the total transform of a hyperquadric Q in $\mathbf{P}_\tau^4$. So M is the blowing-up of Q with center $Q \cap Z$, which is a conic curve. We easily see that both Q and $Q \cap Z$ are smooth.

When $\mathcal{E} \simeq \mathcal{O}(1,1,1,1,1)$, we have $P \simeq \mathbf{P}_\xi^1 \times \mathbf{P}_\tau^4$, $H = H_\xi + H_\tau$ and $M \in |2H_\tau|$. Hence $M \simeq \mathbf{P}_\xi^1 \times Q$ for some hyperquadric Q.

(3.29). Suppose that $d = 9$. Then $e = 6$ and $b = -3$. Moreover $n = 3$ by (3.15). So $\mathcal{E} \simeq \mathcal{O}(1,1,1,3)$ or $\mathcal{O}(1,1,2,2)$.

If $\mathcal{E} \simeq \mathcal{O}(1,1,1,3)$, the injection $\mathcal{O}(3) \to \mathcal{E}$ gives an embedding $T = \mathbf{P}(\mathcal{O}(1,1,1)) \simeq \mathbf{P}_\xi^1 \times \mathbf{P}_\tau^2$ in P such that $H_T = H_\xi + H_\tau$. Then $T \subset M$ since $[M]_T = 2H - 3H_\xi$ is not effective. This is absurd since $\dim(T) = 3$.

Now we conclude $\mathcal{E} \simeq \mathcal{O}(1,1,2,2)$. So $P \simeq \{((\xi_0 : \xi_1) \times (\tau_0 : \tau_1 : \tau_{20} : \tau_{21} : \tau_{30} : \tau_{31})) \in \mathbf{P}_\xi^1 \times \mathbf{P}_\tau^5 \mid \xi_0 : \xi_1 = \tau_{20} : \tau_{21} = \tau_{30} : \tau_{31}\}$, $H = H_\xi + H_\tau$ and $M \in |2H_\tau - H_\xi|$. Let $S = \{\tau_{20} = \tau_{21} = \tau_{30} = \tau_{31} = 0\} \simeq \mathbf{P}_\xi^1 \times \mathbf{P}_\tau^1$. Then $S \subset M$ since $[2H_\tau - H_\xi]_S$ is not effective. The normal bundle of S in P is $[H_\tau - H_\xi] \oplus [H_\tau - H_\xi]$ since $\{\tau_{i0} = \tau_{i1} = 0\} \in |H_\tau - H_\xi|$ for $i = 2, 3$. Hence the normal bundle of S in M is $-H_\xi$. So S can be blown down to the direction of $S \to \mathbf{P}_\tau^1$. M is the blowing-up of another manifold M' with center Z being canonicallly identified with $\mathbf{P}_\tau^1$. Moreover H_τ and $H_\xi + S$ come from $\text{Pic}(M')$ since their restrictions to fibers of $S \to Z$ are trivial. $L' = L + S$ is the pull-back of an ample line bundle on M', which will be denoted by L' by abuse of notation.

Clearly we have $\text{Bs}|[H_\xi + S]_M| \subset S$. Hence, in view of the exact sequence
$$0 \to H^0(M, H_\xi) \to H^0(M, H_\xi + S) \to H^0(S, H_\xi + S) \to H^1(M, H_\xi) = 0,$$
we infer $h^0(M, H_\xi + S) = 3$ and $\text{Bs}|H_\xi + S| = \emptyset$. So we have a morphism $M' \to \mathbf{P}_\sigma^2$. On the other hand, since S is an ideal-theoretical intersection of

members of $|H_\tau - H_\xi|$ on P, we have $|H_\tau - H_\xi|_M = S + \Lambda$ for some linear system Λ such that $\text{Bs}\Lambda = \emptyset$ and $\dim \Lambda = 1$. Since the restriction of Λ to any fiber of $S \to Z$ is trivial, Λ gives a morphism $M' \to \mathbf{P}^1_\lambda$. thus we get a morphism $h : M' \to \mathbf{P}^2_\sigma \times \mathbf{P}^1_\lambda$ such that $h^*(H_\sigma + H_\lambda) = H\tau$ and $L' = h^*(2H_\sigma + H_\lambda)$. So h is finite since L' is ample. Moreover, calculating the degrees, we see that h is birational. Hence h is an isomorphism by Zariski's Main Theorem. Since $[H_\sigma]_S = 0$, the center Z of the blowing-up $M \to M'$ is a fiber of $M' \to \mathbf{P}^2_\sigma$ over a point x. This implies that $M \simeq \mathbf{P}^1_\lambda \times \Sigma_1$, where Σ_1 is the blowing-up of $\mathbf{P}^2_\sigma$ at x. We have $L = H_\lambda + 2H_\sigma - S$, $H_\xi = H_\sigma - S$ and $H_\tau = H_\lambda + H_\sigma$. Thus, $f : M \to C$ is given by the ruling $\Sigma_1 \to \mathbf{P}^1$, not by the projection $M \to \mathbf{P}^1_\lambda$.

(3.30). Summarizing we get the following

Theorem. *Polarized manifolds of the type* (3.15) *are classified as follows:*
1) $d = 1$, $\mathcal{E} \simeq \mathcal{O}(-1, -1, 0, 0)$ *and* $M \in |2H + 5H_\xi|$.
2) $d = 2$, $\mathcal{E} \simeq \mathcal{O}(-1, 0, 0, 0)$ *and* $M \in |2H + 4H_\xi|$.
3) $d = 3$, $\mathcal{E} \simeq \mathcal{O}(0, 0, 0, 0)$ *and* $M \in |2H + 3H_\xi|$. $\text{Bs}|L| = \emptyset$ *and* $|L|$ *makes M a triple corvering of* $\mathbf{P}^3$.
   3') $d = 3$, $\mathcal{E} \simeq \mathcal{O}(-1, 0, 0, 1)$ *and* $M \in |2H + 3H_\xi|$. $\text{Bs}|L|$ *is a point.*
4) $d = 4$, $\mathcal{E} \simeq \mathcal{O}(0, 0, 0, 1)$ *and* $M \in |2H + 2H_\xi|$. *M is the normalization of a hypersurface of degree four in* $\mathbf{P}^4$, *which has double points along a line.*
5) $d = 5$, $\mathcal{E} \simeq \mathcal{O}(0, 0, 1, 1)$ *and* $M \in |2H + H_\xi|$.
6) $d = 6$, $\mathcal{E} \simeq \mathcal{O}(0, 1, 1, 1)$ *and* $M \in |2H|$. *M is a double covering of* $\mathbf{P}^1 \times \mathbf{P}^2$ *with branch locus being a smooth divisor of bidegree* $(2, 2)$.
7) $d = 7$, $\mathcal{E} \simeq \mathcal{O}(1, 1, 1, 1)$ *and* $M \in |2H - H_\xi|$. *M is the blowing-up of* $\mathbf{P}^3$ *along a curve which is a complete intersection of two hyperquadrics.*
8) $d = 8$, $\mathcal{E} \simeq \mathcal{O}(1, 1, 1, 2)$ *and* $M \in |2H - 2H_\xi|$. *M is the blowing-up of a hyperquadric in* $\mathbf{P}^4$ *along a conic curve.*
   8*) $d = 8$, $\mathcal{E} \simeq \mathcal{O}(1, 1, 1, 1, 1)$ *and* $M \in |2H - 2H_\xi|$. *$M \simeq \mathbf{P}^1 \times Q$ for a hyperquadric Q in* $\mathbf{P}^4$.
9) $d = 9$, $\mathcal{E} \simeq \mathcal{O}(1, 1, 2, 2)$ *and* $M \in |2H - 3H_\xi|$. *$M \simeq \mathbf{P}^1 \times \Sigma_1$ for the blowing-up Σ_1 of* $\mathbf{P}^2$ *at a point.*

Remark (3.31). If $d \geq 5$, we have $\text{Bs}|L| = \emptyset$ by (3.19). So L is very ample by [F2; Theorem 4.1; c)]. Therefore we can use also the argument in [I1] in this case. His method is essentially different from ours.

§4. Polarized surfaces of sectional genus two

Here we classify polarized surfaces of sectional genus two by the same principle as in (1.10). Thus, throughout this section, let (S, L) be a polarized manifold with $\dim(S) = 2 = g(S, L)$. So $(K + L)L = 2$ for the canonical bundle K of S.

(4.1). If $K + L$ is not nef, then (S, L) is a scroll over a curve of genus two (see(1.11)). So, from now on, we assume that $K + L$ is nef.

(4.2). Suppose that K is nef. Then $0 \le KL = 2 - L^2$. Hence $d = L^2 = 1$ or 2.

If $d = 2$, then $KL = 0$, which implies $K \sim 0$ by the index theorem since $K^2 \ge 0$.

If $d = 1$, we have $(K - L)L = 0$. Therefore, unless $K \sim L$, we have $0 > (K - L)^2 = K^2 - 1$ by the index theorem. This implies $K^2 = 0$. So S is a minimal elliptic surface.

(4.3). From now on, we suppose that K is not nef. Obviously S is not $\mathbf{P}^2$ since $g(S, L) = 2$. So, as is well-known, one of the following conditions is valid.

a) S is a $\mathbf{P}^1$-bundle over a curve C.

b) There is a rational smooth curve E with $E^2 = -1$.

(4.4). In the above case a), $S \simeq \mathbf{P}(\mathcal{E})$ for some vector bundle $\mathcal{E}$ of rank two on C. Let H be the tautological line bundle on it and set $L = \delta H + \pi^* B$ for some $B \in \mathrm{Pic}(C)$. Note that $K = -2H + \pi^*(\det(\mathcal{E}) + K^C)$, where K^C is the canonical bundle of C. Set $q = h^1(C, \mathcal{O}_C)$, $e = c_1(\mathcal{E})$ and $b = \deg(B)$. Then $d = L^2 = \delta(\delta e + 2b)$ and $2 = (K + L)L = \delta^{-1}(\delta - 1)d + 2\delta(q - 1) \ge 2\delta(q - 1)$. We have $\delta \ge 2$ since $K + L$ is nef. From this we infer $q \le 1$.

(4.5). When $q = 1$, we have $(\delta - 1)d = 2\delta$. So $(d, \delta) = (3, 3)$ or $(4, 2)$.

In case $d = \delta = 3$, replacing $\mathcal{E}$ by $\mathcal{E} \otimes \mathcal{F}$ for an appropriate invertible sheaf $\mathcal{F}$ if necessary, we may assume that $H = K + L$. Then $b + e = 0$. Moreover $1 = \delta e + 2b = 3e + 2b$. So $e = 1$ and $b = -1$. Hence H is ample since so is $3H = L - \pi^* B$. Thus $\mathcal{E}$ is an indecomposable vector bundle with $c_1(\mathcal{E}) = 1$.

In case $d = 4$ and $\delta = 2$, we have $b + e = 1$. Moreover $K + L = \pi^*(B + \det(\mathcal{E}))$. Replacing $\mathcal{E}$ by $\mathcal{E} \otimes \mathcal{F}$ if necessary, we may assume $e = 0$ or 1.

(4.6). When $q = 0$, $\mathcal{E}$ is decomposable since $C \simeq \mathbf{P}^1$. We may assume $\mathcal{E} = \mathcal{O}(e, 0)$ for some $e \ge 0$. Then $b > 0$ since L is ample. So $2 = (K + L)L = \delta(\delta - 1)e + 2(\delta - 1)b - 2\delta \ge \delta(\delta - 1)e + 2(\delta - 1) - 2\delta = \delta(\delta - 1)e - 2$. Hence $\delta = 2$ if $e > 0$. If $e = 0$, we have $2 = 2(\delta - 1)b - 2\delta$ and hence $(b, \delta) = (2, 3)$ or $(3, 2)$. Since $S \simeq \mathbf{P}^1 \times \mathbf{P}^1$ in this case, we may assume $\delta = 2$ in any case by changing the role of these two ruling if necessary.

Assuming $\delta = 2$, we have $b + e = 3$. So $(b, e) = (3, 0)$, $(2, 1)$ or $(1, 2)$.

(4.7). We now study the case (4.3; b). In this case S is the blowing-up of another surface S' at a point p and E is the exceptional curve over p.

Definition (4.8). (S, L) is said to be *half-minimal* if $2K + L$ is nef. Note that $LE \ge 2$ for any exceptional curve E on S if (S, L) is half-minimal.

(4.9). For the moment, until (4.12), we further assume that (S, L) is half-minimal. We first show that S is ruled.

Indeed, set $L' = L + mE$ for $m = LE \ge 2$. Then L' is the pull-back of an ample line bundle on S', which is denoted by L' by abuse of notation. Then

$K'L' = KL - m = 2 - L^2 - m < 0$ for the canonical bundle K' of S'. So K' is not pseudo-effective, and hence S' is ruled.

Let q be the irregularity of S.

(4.10). Suppose that $q > 0$. Then $K^2 = (K')^2 - 1 \leq 8(1 - q) - 1$ by (4.9). On the other hand $0 \leq (2K + L)^2 = 4K^2 + 8 - 3d$ by half-minimality. From them we infer that $q = 1$, $K^2 = -1$ and $d = 1$. In particular S' is a $\mathbf{P}^1$-bundle over an elliptic curve C. So $S' \simeq \mathbf{P}_C(\mathcal{E})$ for a vector bundle $\mathcal{E}$ on C with $e = c_1(\mathcal{E}) = 0$ or 1. Letting H be the tautological line bundle on S', we set $L' = \delta H + \pi^* B$ for some $B \in \mathrm{Pic}(C)$. Let E' be the other exceptional curve on S such that $E + E'$ is a fiber of $S \to C$. Then $LE' = \delta - m$. We may assume $\delta - m \geq m$ by replacing E by E' if necessary. Setting $b = \deg(B)$, we obtain $\delta^2 e + 2\delta b - m^2 = L^2 = 1$ and $1 = KL = m - \delta e - 2b$. So $2b = m - \delta e - 1$ and $m(\delta - m) = \delta + 1$. This implies $m = 2$ and $\delta = 5$. Hence $2b = 1 - 5e$. So e is odd and hence $e = 1$, $b = -2$.

Thus S is the blowing-up of $S' = \mathbf{P}_C(\mathcal{E})$ at a point, $c_1(\mathcal{E}) = 1$, $L = 5H - 2A - 2E$ for some $A \in \mathrm{Pic}(C)$ with $\deg(A) = 1$, where E is the exceptional curve. Moreover, we easily see that $\mathcal{E}$ is indecomposable.

(4.11). Suppose that $q = 0$. Then S is rational by (4.9). We have $0 \leq (2K + L)L = 4 - d$ by the half-minimality.

In case $d = 4$, we have $(2K + L)L = 0$. Since $2K + L$ is nef, this implies $2K + L \sim 0$ by the index theorem. So $-K$ is ample and S is a del Pezzo surface. Moreover $L = -2K$. Hence $K^2 = 1$. Therefore S is the blowing-up of $\mathbf{P}^2$ at eight points, and $L = 6H - 2E_1 - \cdots - 2E_8$, where H is the pull-back of $\mathcal{O}(1)$ of $\mathbf{P}^2$ and E_j's are the exceptional curves.

In case $d = 3$, we have $KL = -1$ and $(3K + L)L = 0$. Hence $0 \geq (3K + L)^2 = 9K^2 - 3$ by the index theorem. So $K^2 \leq 0$. On the other hand $0 \leq (2K + L)^2 = 4K^2 - 1$. Combining them we get a contradiction and this case is ruled out.

In case $d = 2$, we have $KL = 0$ and $K^2 < 0$ by the index theorem. So $(2K + L)(K + L) = 2K^2 + 2 \leq 0$. Since $2K + L$ is nef and $K + L$ is ample, this implies $2K + L = 0$. Then $2 = L^2 = 4L^2$, which is absurd. Thus this case is impossible.

(4.12). We consider the remaining case $d = 1$. We have $KL = 1$ and $0 \leq (2K + L)^2 = 4K^2 + 5$. So $K^2 \geq -1$. We claim $K^2 = -1$.

Indeed, if $K^2 \geq 0$, we would have $\chi(S, tK + L) > 0$ for any $t \gg 0$ by the Riemann-Roch Theorem. Moreover $h^2(S, tK + L) = h^0(S, (1 - t)K - L) = 0$ since $((1 - t)K - L)L = -t < 0$. So $|tK + L| \neq \emptyset$ for $t \gg 0$ and K must be pseudo-effective, which is absurd.

Now we set $L_\mathbb{Q} = K + L$. Then $(S, L_\mathbb{Q})$ is a polarized surface such that $L_\mathbb{Q}^2 = 2$, $KL_\mathbb{Q} = 0$ and hence $g(S, L_\mathbb{Q}) = 2$. Note that $(S, L_\mathbb{Q})$ is not half-minimal since $(2K + L_\mathbb{Q})^2 = (3K + L)^2 = -2$.

(4.13). Now we study the case in which (S, L) is not half-minimal, assuming that S is not a $\mathbf{P}^1$-bundle. Then $(2K + L)E < 0$ for some extremal rational curve E. Since S is not a $\mathbf{P}^1$-bundle, E must be an exceptional curve. So $KE = -1$ and $LE = 1$. Setting $L' = L + E$, we get a polarized surface (S', L') as in (4.7). Since $(K' + L')_S = K + L$, we have $g(S', L') = 2$ and moreover $K' + L'$ is nef, where K' is the canonical bundle of S'. Now there are several possibilities:

a) K' is nef.

b) S' is a $\mathbf{P}^1$-bundle over a curve.

c) (S', L') is half-minimal and b) is not the case.

d) $L'E' = 1$ for some exceptional curve E' on S'.

In case d), we contract E' to a smooth point on another surface S'' and set $L'' = L' + E'$. Then $g(S'', L'') = 2$ and $K'' + L''$ is nef. We continue this process until we get a polarized surface (S_0, L_0) which satisfies one of the above conditions a), b) or c). Such a pair (S_0, L_0) will be called a relatively half-minimal model of (S, L).

(4.14). We have $L^2 = (L')^2 - 1$. Hence $d(S, L) = d(S_0, L_0) - k$ if (S_0, L_0) is obtained by k-times of contractions. In particular $d(S_0, L_0) = L^2 + k \geq 2$.

Corollary (4.15). If K_0 is nef, then $L_0^2 = 2$, $k = 1$ and $K_0 \sim 0$.

For a proof, recall (4.2).

(4.16). Suppose that S_0 is a $\mathbf{P}^1$-bundle over a smooth curve C of genus g. If $g > 0$, then (S_0, L_0) is of the type (4.5). If $g = 0$, then $S_0 \simeq \mathbf{P}(\mathcal{O}_C(e, 0))$ and $L_0 = 2H_\zeta + bH_\xi$ with $(e, b) = (0, 3)$, $(1, 2)$ or $(2, 1)$, where H_ζ is the tautological line bundle and H_ξ is the pull-back of $\mathcal{O}_C(1)$. Let (S_1, L_1) be the polarized surface just before the final contraction to get (S_0, L_0). So S_1 is the blowing-up of S_0 at a point p. In case $e = 0$ or 2, we contract the strict transform on S_1 of the fiber of $S_0 \to C$ passing p. Replacing half-minimal model in this way if necessary, we may assume $e = 1$. Then S_0 is the blowing-up of $\mathbf{P}^2$ at a point p_0. Thus S is obtained from $\mathbf{P}^2$ by $(k + 1)$ times blowing-ups, and $L = 4H - 2E_0 - E_1 - \cdots - E_k$, where H is the pull-back of $\mathcal{O}_{\mathbf{P}^2}(1)$ and E_j's are (the total transforms of) the exceptional curves.

(4.17). Suppose that (S_0, L_0) satisfies the condition (4.13; c). (S_0, L_0) is not of the type (4.10) since $L_0^2 > 1$. Hence, by (4.11), we see $L_0^2 = 4$ and $L_0 = -2K_0$. In particular S_0 is a del Pezzo surface and is the blowing-up of $\mathbf{P}^2$ at eight points.

(4.18). Finaly we study the case (4.12). Let (S_0, L_0) be a relatively half-minimal model of $(S, L_{\mathbb{Q}})$. Since $(K_0 + L_0)_S = K + L_{\mathbb{Q}} = 2K + L$, we have $K_0(K_0 + L_0) = K(2K + L) = -1$. From this we infer that (S_0, L_0) is of the type (4.17) since S is rational. Thus S_0 is the blowing-up of $\mathbf{P}^2$ at eight points $p_1, \cdots, p_8$ and S is obtained from S_0 by two times of blowing-ups. Moreover $L = 9H - 3E_1 - \cdots - 3E_8 - 2E_1' - 2E_2'$, where H is the pull-back of $\mathcal{O}_{\mathbf{P}^2}(1)$, E_i is the exceptional curve over p_i and E_j' is the exceptional curve of one of the last two blowing-ups.

(4.19). Thus, summing up, we obtain the following

Theorem. *Let* (S, L) *be a polarized manifold with* $\dim S = 2 = (K + L)L$, *where* K *is the canonical bundle. Then one of the following conditions is satisfied:*

0) *There is another polarized surface* (S', L') *such that* S *is the blowing-up of* S' *at a point* p *on* S' *and that* $L = L'_S - E_p$, *where* E_p *is the exceptional curve over* p.

1) K *ia numerically equivalent to* L *and* $L^2 = 1$.

1?) S *is a minimal elliptic surface and* $KL = L^2 = 1$.

2) K *is numerically trivial and* $L^2 = 2$.

3) *There is an indecomposable vector bundle* $\mathcal{E}$ *on an elliptic curve* C *with* $c_1(\mathcal{E}) = 1$ *such that* $S \simeq \mathbf{P}_C(\mathcal{E})$. $L^2 = 3$ *and* $L = 3H - \pi^* A$ *for some* $A \in \mathrm{Pic}(C)$ *with* $\deg(A) = 1$, *where* H *is the tautological line bundle and* π *is the projection* $S \to C$.

4) *There is a vector bundle* $\mathcal{E}$ *on an elliptic curve* C *with* $e = c_1(\mathcal{E}) = 0$ *or* 1 *such that* $S \simeq \mathbf{P}_C(\mathcal{E})$. $L = 2H + \pi^* B$ *for some* $B \in \mathrm{Pic}(C)$ *with* $e + \deg(B) = 1$, *where* H *and* π *are as in* 3). *In this case* $L^2 = 4$.

5) *There are a vector bundle* $\mathcal{E}$ *on an elliptic curve* C *with* $c_1(\mathcal{E}) = 1$ *and a point* p *on* $P = \mathbf{P}_C(\mathcal{E})$ *such that* S *is the blowing-up of* P *at* p. $L^2 = 1$ *and* $L = 5H - 2A - 2E_p$, *where* E_p *is the exceptional curve over* p *and* H *(resp.* A*) is the pull-back of the tautological line bundle of* P *(resp. a line bundle on* C *of degree one).*

6_0) $S \simeq \mathbf{P}^1 \times \mathbf{P}^1$ *and* L *is of bidegree* $(2, 3)$. $L^2 = 12$.

6_1) S *is the blowing-up of* $\mathbf{P}^2$ *at a point.* $L^2 = 12$ *and* $L = 4H - 2E$, *where* H *is the pull-back of* $\mathcal{O}_{\mathbf{P}^2}(1)$ *and* E *is the exceptional curve.*

6_2) $S \simeq \mathbf{P}(\mathcal{E})$ *for the vector bundle* $\mathcal{E} = \mathcal{O}(2, 0)$ *on* $\mathbf{P}^1$. $L^2 = 12$ *and* $L = 2H_\zeta + H_\xi$, *where* H_ξ *is the pull-back of* $\mathcal{O}_{\mathbf{P}^1}(1)$ *and* H_ζ *is the tautological line bundle.*

7) $-K$ *is ample,* $K^2 = 1$ *and* $L = -2K$. *In this case* S *is the blowing-up of* $\mathbf{P}^2$ *at eight points and* $L^2 = 4$.

8) *There is a polarized surface* (S_0, L_0) *of the above type* 7) *such that* S *is obtained by two times of blowing-ups from* S_0. $L^2 = 1$ *and* $L = -3K + E_1 + E_2$, *where* E_j *'s are (the total transform of) the exceptional curves (see* (4.18)).

9) *There is a vector bundle* $\mathcal{E}$ *on a curve* C *of genus two such that* $S \simeq \mathbf{P}_C(\mathcal{E})$ *and that* L *is the tautological line bundle on it.*

Appendix

Here we give a proof of the following fact.

Theorem. *Let* C *be an irreducible reduced curve and let* L *be a line bundle on* C *such that* $\deg(L) = 1$, $h^0(L) = 0$ *and* $\omega \simeq \mathcal{O}[2L]$, *where* ω *is the dualizing sheaf of* C. *Then* C *is a weighted complete intersection of type* $(6, 6)$ *in the weighted projective space* $\mathbf{P}(2, 2, 3, 3)$ *and* $L = \mathcal{O}(1)$.

Proof. Since $\deg(\omega) = 2$, we have $h^1(C, \mathcal{O}) = 2$. So $h^0(2L) = h^0(\omega) = 2$. Let η_0, η_1 be a linear base of $H^0(2L)$. For any $t > 2$, we have $h^0(tL) = t - 1$ by Riemann-Roch theorem. Let τ_0, τ_1 be a linear base of $H^0(3L)$. We first show that $G(C, L) = \oplus_{t \geq 0} H^0(tL)$ is generated by η_0, η_1, τ_0, τ_1 as a graded algebra.

Since $|2L|$ gives the canonical mapping $f : C \to \mathbf{P}^1$ of degree two, η_0 and η_1 are algebraically independent. In particular η_0^2, $\eta_0 \eta_1$ and η_1^2 are linearly independent. They generate $H^0(4L)$ since $h^0(4L) = 3$.

Assume that $\eta_0 \tau_0$, $\eta_0 \tau_1$, $\eta_1 \tau_0$ and $\eta_1 \tau_1$ are linearly dependent in $H^0(5L)$. Then $\eta_0 \tau_0' + \eta_1 \tau_1' = 0$ for some τ_0', $\tau_1' \in H^0(3L)$. Since the divisors $\{\eta_0 = 0\}$ and $\{\eta_1 = 0\}$ are disjoint, the divisor $\{\tau_1' = 0\}$ is of the form $\{\eta_0 = 0\} + x$, x being a point. Then $x \in |3L - 2L|$, contradicting $h^0(L) = 0$ Thus $\eta_i \tau_j$'s are linearly independent and generate $H^0(5L)$.

Note that $\dim\langle \eta_0^3, \eta_0^2 \eta_1, \eta_0 \eta_1^2, \eta_1^3 \rangle = 4$ in $H^0(6L)$. Take a general point z on $\mathbf{P}^1$ such that $f^{-1}(z)$ consists of two different points x and y. Take a linear combination τ of τ_0 and τ_1 such that $\tau(x) = 0$. If $\tau(y) = 0$, then $\{\tau = 0\} - x - y$ would be a member of $|L|$. So $\tau(y) \neq 0$. Hence $\tau^2 \notin \langle \eta_0^3, \eta_0^2 \eta_1, \eta_0 \eta_1^2, \eta_1^3 \rangle$. Therefore $\langle \eta_0^3, \eta_0^2 \eta_1, \eta_0 \eta_1^2, \eta_1^3, \tau_0^2, \tau_0 \tau_1, \tau_1^2 \rangle = H^0(6L)$.

By virtue of [F6;(A7)], the natural mapping $H^0(2L) \otimes H^0(tL) \to H^0((t+2)L)$ is surjective for any $t > 4$. Thus, combining these observations, we infer that $\eta_0, \eta_1, \tau_0, \tau_1$ generate $G(C, L)$.

Define a surjective homomorphism $R = \oplus_t H^0(\mathbf{P}(2, 2, 3, 3), \mathcal{O}(t)) \to G(C, L)$ by sending generators of R to η_0, η_1, τ_0, τ_1. We will study the kernel J of it. Let R_t be the part of R of degree t and set $J_t = J \cap R_t$. By the preceding observation we see $J_t = 0$ for $t < 6$ and $\dim(J_6) = 2$. Let ψ_1, ψ_2 be a base of J_6. Let J' be the ideal of R generated by them. Since they are irreducible as weighted homogeneous polynomials, we infer $\dim(J_t') = 2r_{t-6} - r_{t-12}$, where $r_t = \dim(R_t)$. Hence, by elementary computations, we get $r_t - \dim(J_t') = t - 1$. On the other hand we have $r_t - \dim(J_t) = h^0(tL) = t - 1$. Thus we conclude $J' = J$, proving the theorem.

References

[A]   M. F. Atiyah, Vector bundles over an elliptic curve, Proc. London Math. Soc. (3) **7** (1957), 414-452.

[BLP]   M. Beltrametti, A. Lanteri and M. Palleschi, Algebraic surfaces containing an ample divisor of arithmetic genus two, preprint, Univ. di Milano 1986.

[C]   F. Catanese, Surfaces with $K^2 = p_g = 1$ and their period mapping, in Proceedings on Algebraic Geometry at Copenhagen 1978, Lecture Notes in Math. **732**, Springer, 1979.

[CO]   B. Y. Chen and K. Ogiue, Some characterizations of complex space forms in terms of Chern classes, Quart. J. Math. Oxford **26** (1975), 459-464.

[F1]   T. Fujita, On the structure of polarized varieties with Δ-genera zero, J. Fac. Sci. Univ. of Tokyo **22** (1975) 103-115.

[F2]   T. Fujita, Defining equations for certain types of polarized varieties, in Complex Analysis and Algebraic Geometry (Baily and Shioda eds.), pp.165-173, Iwanami, Tokyo, 1977.

[F3]   T. Fujita, On the hyperplane section principle of Lefshetz, J. Math. Soc. Japan **32** (1980), 153-169.

[F4]   T. Fujita, On the strucrture of polarized manifolds with total deficiency one, part I and III, J. Math. Soc. Japan **32** (1980), 709-725 and **36** (1984), 75-89.

[F5]   T. Fujita, On polarized varieties of small Δ-genera, Tohoku Math. J. **34** (1982), 319-341.

[F6]   T. Fujita, On hyperelliptic polarized varieties, Tohoku Math. J. **35** (1983), 1-44.

[F7]   T. Fujita, On polarized manifolds of Δ-genus two; part I, J. Math. Soc. Japan **36** (1984), 709-730.

[F8]   T. Fujita, Polarized manifolds whose adjoint bundles are not semipositive, in Algebraic Geometry, Sendai 1985, pp. 167-178, Advanced Studies in Pure Math. 10, Kinokuniya, Tokyo, 1987

[F9]   T. Fujita, On polarized manifolds of sectional genus two, Proc. Japan Acad. **62** (1986), 69-72.

[F10]   T. Fujita, Vector bundles of small c_1-sectional genera, preprint.

[I1]   P. Ionescu, Embedded projective varieties of small invariants, in Proceedings on Algebraic Geometry at Bucharest 1982, pp. 142-186, Lecture Notes in Math. **1056**, Springer, 1984.

[I2]   P. Ionescu, Generalized adjunction and applications, INCREST Preprint Series No. 48/1985, Bucharest.

[KMM] Y. Kawamata, K. Matsuda and K. Matsuki, Introduction to the minimal model problem, in Algebraic Geometry, Sendai 1985, pp. 283-360, Advanced Studies in Pure Math. 10, Kinokuniya, Tokyo, 1987.

[L]   R. Lazarsfeld, A Barth-type theorem for branched coverings of projective space, Math. Ann. **249** (1980), 153-162.

[Mi1]   Y. Miyaoka, Tricanonical maps of numerical Godeaux surfaces, Inventiones math. **34** (1976), 99-111.

[Mi2]   Y. Miyaoka, the Chern classes and Kodaira dimension of a minimal variety, in Algebraic Geometry, Sendai 1985, pp. 449-476, Advanced Studies in Pure Math. 10, Kinokuniya, Tokyo, 1987.

[Mo]   S. Mori, Threefolds whose canonical bundles are not numerically effective, Ann. Math. **116** (1982), 133-176.

[Y]   S. T. Yau, On the Ricci curvature of a compact Kähler manifolds and the complex Monge-Ampère equations I, Comm. Pure Appl. Math. **31** (1978), 339-411.

Takao FUJITA

Department of Mathematics
College of Arts and Sciences
University of Tokyo
Komaba, Meguro, Tokyo 153
Japan

Algebraic Geometry and Commutative Algebra
in Honor of Masayoshi NAGATA
pp. 99–124 (1987)

Affine Surfaces with $\bar{\kappa} \leq 1$

R. V. Gurjar and M. Miyanishi*

Introduction.

We are mainly interested in the relationship between the divisor class group and the fundamental group of a non-singular affine surface defined over $\mathbf{C}$. Our naïve question is the following

(∗) Suppose the divisor class group $\mathrm{Pic}(S)$ of a non-singular affine surface S over $\mathbf{C}$ is trivial and the units in the coordinate ring $\Gamma(S)$ of S are just the elements of $\mathbf{C}^*$. Is $\pi_1(S) = (1)$?

It turns out that the answer to the above question is negative, but the surfaces for which π_1 is not trivial are rather special. If $\bar{\kappa}(S) = -\infty$, it is known [3; Chapter I] that $S \cong \mathbf{C}^2$ as an affine variety. If $\bar{\kappa}(S) = 0$, then we will verify that $\pi_1(S) = (1)$. Using the results of T. Fujita [1; §8], we will in fact show that there are only two types of surfaces in this case. See §2 for details.

If $\bar{\kappa}(S) = 1$, S has a $\mathbf{C}^*$-fibration $\pi: S \to C$, where C is isomorphic to $\mathbf{A}_{\mathbf{C}}^1$ or $\mathbf{P}_{\mathbf{C}}^1$. Then we shall show that the $\mathbf{C}^*$-fibration π is untwisted. In this case $C \cong \mathbf{P}^1$, any singular fiber of π is of the form

(i) $\mu\Delta$ with $\mu > 1$ and $\Delta \cong \mathbf{C}^*$, called a singular fiber of multiplicative type, or

(ii) $\mu\Delta$ with $\mu \geq 1$ and $\Delta \cong \mathbf{A}^1$, called a singular fiber of additive type.

In this case (i.e., when $C \cong \mathbf{P}^1$), $\pi_1(S) = (1)$ if and only if π has at most two singular fibers with multiplicity $\mu > 1$. Furthermore, when $\pi_1(S)$ is finite, S is obtained from a Platonic $\mathbf{C}^*$-fiber space $\mathbf{C}^2/G - \{0\}$ by applying "elementary" operations, where G is a binary icosahedral subgroup of $SL(2, \mathbf{C})$.

In §1, we will prove that if the fundamental group at infinity of S is finite then $\bar{\kappa}(S) = -\infty$. This result explains the resemblance between the results in [2, 6], obtained by topological methods and the results of M. Miyanishi, T. Sugie and T. Fujita [3] obtained by geometric methods. In §3, we will construct infinitely many contractible surfaces with $\bar{\kappa}(S) = 1$. In §4, we will give an example

*The first author was a JSPS fellow, and the second author was partly supported by the Grant-in-Aid for Scientific Research, The Ministry of Education, Science and Culture, Japan.
Received January 10, 1986.

with $\bar{\kappa} = 2$ for which the answer to ($*$) is affirmative. We shall also give a new contractible affine surface with $\bar{\kappa} = 2$ which is different from C. P. Ramanujam's example in [6].

For the definition of the logarithmic Kodaira dimension of a non-complete surface, see [3; Chapter I]. The examples of contractible surfaces with $\bar{\kappa} = 1$ in §3 show that some of the assertions in [1] are incorrect (for example, Corollary (7.16), Theorem (9.1)).

The authors would like to express their gratitude to T. Fujita and A. R. Shastri for pointing out some errors in the manuscript.

§1. Surfaces with $\bar{\kappa} = -\infty$.

In this section, we will prove the following

Theorem 1. *Let S be a nonsingular, quasi-projective surface defined over* **C**. *Assume that S is connected at infinity, the divisor at infinity for S does not have negative definite intersection form and the fundamental group at infinity of S is finite. Then $\bar{\kappa}(S) = -\infty$.*

Remark. The hypothesis is satisfied if S is affine and has finite fundamental group at infinity.

Before proving the Theorem, we need some preliminaries. Embed $S \subset \bar{S}$, where $\bar{S}$ is a nonsingular, projective surface such that $\bar{S} - S = \bigcup_{i=1}^{r} C_i$ is a divisor with normal crossings (C_i irreducible components of $\bar{S} - S$). Let $C = \bigcup_{i=1}^{r} C_i$. Then S is connected at infinity if C is connected. The fundamental group at infinity, $\pi_1^{\infty}(S)$, is defined as follows.

Choose a tubular neighbourhood T of C in $\bar{S}$ such that
(i) C is a strong deformation retract of T,
(ii) ∂T is a strong deformation retract of $T - C$, and
(iii) ∂T is a compact 3-manifold.
There exists a fundamental system of such neighbourhoods of C and the fundamental group $\pi_1(\partial T)$ is defined to be the fundamental group at infinity of S, denoted by $\pi_1^{\infty}(S)$. We will use the classification of the possible weighted dual graphs of C when $\pi_1^{\infty}(S)$ is finite. This was done by A. R. Shastri [7].

In order to state Shastri's result, first we will only assume that S is connected at infinity and $\pi_1^{\infty}(S)$ is finite.

For any integers λ, n with $0 < \lambda < n$ and $(\lambda, n) = 1$, let

$$\frac{n}{\lambda} = a_1 - \cfrac{1}{a_2 - \cfrac{1}{a_3 - \cfrac{\ddots}{ - \cfrac{1}{a_k}}}}$$

be the continued fraction expansion. Define $\langle n, \lambda \rangle$ to be the negative definite linear tree

$$
\overset{-a_1}{\underset{\circ}{}} \quad \overset{-a_2}{\underset{\circ}{}} \quad \cdots\cdots \quad \overset{-a_k}{\underset{\circ}{}}
$$

For $0 < \lambda_j < n_j$ with $(n_j, \lambda_j) = 1$, define a_{ji} using the continued fraction of n_j / λ_j. For any $a \in \mathbf{Z}$, let $\langle a; n_1, \lambda_1; n_2, \lambda_2; n_3, \lambda_3 \rangle$ denote the tree

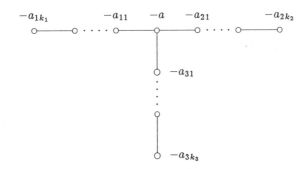

Finally, for any weighted tree T and a vertex v of T, let $T^{(v)}$ denote the tree obtained from T by adding two more vertices v_1, v_2 with weights at v_1, v_2 both 0 and two new links $[v, v_1]$ and $[v_1, v_2]$.

The result of Shastri mentioned earlier is the following

Theorem. *With the above notations and hypothesis, each $C_i \cong \mathbf{P}^1$ and the dual graph of C is equivalent (via blowing-up and blowing-down) to one of the following trees.*

(i) *The empty tree or* $\underset{0}{\circ}\text{------}\underset{0}{\circ}$.

(ii) $\langle n, \lambda \rangle$. *This is negative definite.*

(iii) $\langle a; 2, 1; n_2, \lambda_2; n_3, \lambda_3 \rangle$ *where $\{n_2, n_3\}$ is one of the pairs $\{2, n\}$ for any $n \ge 2$, $\{3, 3\}$, $\{3, 4\}$ or $\{3, 5\}$ and $a \ge 2$. This is negative definite.*

(iv) *The tree mentioned in* (iii) *except that $a \le 1$. This has exactly one positive eigenvalue, all the others being negative.*

(v) *The trees $T^{(v)}$ where T is one of the trees in* (ii) *or* (iii) *and $v \in T$ is any vertex. This tree also has one positive eigenvalue and others all negative eigenvalues.*

For the proof of Theorem 1, we will use the theory of Zariski decomposition of pseudo-effective divisors as discussed in Fujita's paper [1]. The definitions of rational twigs, abnormal rational club, bark of a tree, etc., will be used as in [1].

Now assume that S is as in the statement of Theorem 1. Using Shastri's result, suppose first that the dual graph of C is either $\underset{0}{\circ}\!\!-\!\!\!-\!\!\!-\!\!\underset{0}{\circ}$ or $T^{(v)}$. Then there exist curves C_1, C_2 in C such that $C_1^2 = 0 = C_2^2$, $C_1 \cdot C_2 = 1$ and C_1 meets no other curves in C ($C_i \cong \mathbf{P}^1$). Then clearly $(K + C) \cdot C_1 = -1$ and hence $|n(K + C)| = \emptyset$ for all $n \geq 1$.

Since C supports a divisor Δ with $\Delta^2 > 0$, we can now assume the dual graph of C to be as in (iv) of Shastri's Theorem. Assume $\bar{\kappa}(S) \geq 0$. Then $K + C$ is pseudo-effective. Let $K + C = H + N$ be the Zariski decomposition.

Case 1. Suppose every component of $N = (K + C)^-$ is a component of C. Then the Lemma 6.17 in [1] implies that the dual graph Γ of C is an abnormal rational club. But the intersection form on an abnormal rational club is negative definite. Hence this case cannot occur.

Case 2. Since Γ is connected and not an abnormal rational club, $\mathrm{Bk}(\Gamma) = \mathrm{Bk}^*(\Gamma)$ by definition. If $N = \mathrm{Bk}(\Gamma)$, then all the irreducible components of N are components of C, which is not possible by Case 1. Thus $N \neq \mathrm{Bk}^*(\Gamma)$.

Now Lemma (6.20) in [1] guarantees the existence of an exceptional curve of the first kind E on $\bar{S}$, not in C, satisfying one of the following conditions.

(1) $C \cap E = \emptyset$.

(2) $C \cdot E = 1$ and E meets a component of $\mathrm{Bk}(C)$.

If (1) occurs, we can blow-down E without changing the fundamental group at infinity or $\bar{\kappa}$. So we consider (2). We study the tree Γ more closely. $\Gamma = B + T_1 + T_2 + T_3$, where B is the unique curve which meets three other curves of Γ and $B^2 \geq -1$. We can completely list the possibilities for T_1, T_2, T_3. Blow down E to get a surface $\bar{W}$, $\pi: \bar{S} \to \bar{W}$ and $C' = \pi(C)$, $W = \bar{W} - C'$. Then C' looks like C (with weights changed). $W \subset S$, so it suffices to show that $\bar{\kappa}(W) = -\infty$.

We can blow down exceptional curves of the first kind in C' successively to obtain a minimal tree (retaining normal crossings) which is either linear or has exactly one curve which is a branch point of the new tree. This way we get a new compactification of W. If the new tree is linear or one of the branches at the branch point has a non-negative weight, then the Corollary (6.14) in Fujita's paper implies that $\bar{\kappa}(W) = -\infty$.

We now assume that the new tree Γ' has a unique branch point. It is easy to check that Γ' is again of the type (iv) in Shastri's Theorem. We can thus repeat the above argument for W and in finitely many steps reach a Zariski open subset of S with $\bar{\kappa} = -\infty$.

The proof of Theorem 1 is complete.

§2. The case $\bar{\kappa}(S) = 0$.

Let S denote a nonsingular, affine surface defined over $\mathbf{C}$. Assume that $\mathrm{Pic}\, S = (0)$ and $\Gamma(S)^* = \mathbf{C}^*$. In this section we will prove the following

Theorem 2. *With the above assumptions, if further* $\bar{\kappa}(S) = 0$, *then* S *is simply connected. In fact, S is isomorphic to one of the following surfaces.*

(1) *Let ℓ_1, ℓ_2, ℓ_3 be three non-concurrent lines in $\mathbf{P}^2$ and $P_1 \in \ell_1$, $P_2 \in \ell_2$ be points which are not points of intersection. Let $\sigma: V \to \mathbf{P}^2$ be the blowing-up with centers P_1, P_2 and $S = V - (\ell_1' \cup \ell_2' \cup \ell_3')$, where ℓ_i' is the proper transform of ℓ_i on V.*

(2) *Let C be a smooth conic in $\mathbf{P}^2$ and ℓ a line which is not tangent to C. Let P be a point of C, not lying on ℓ, and $\sigma: V \to \mathbf{P}^2$ be the blowing-up at P. Let $S = V - (C' \cup \ell')$, where C', ℓ' are the proper transforms.*

This result depends heavily on the Theorem (8.70) in Fujita's paper [1]. Embed $S \subset V$, where V is a non-singular, projective surface such that $D = V - S$ is a divisor with normal crossings. We call the pair (V, D) a normal completion of S.

For proving Theorem 2 and later results, we need some preliminaries.

Lemma 2.1. *Assume $\bar{\kappa}(S) \leq 1$, $\mathrm{Pic}\, S = (0)$ and S is a non-singular, affine surface over $\mathbf{C}$. Then S is rational. If $\bar{\kappa}(S) = 1$, then S has a $\mathbf{C}^*$-fibration $\pi: S \to C$, where C is a nonsingular rational curve. If further, $\Gamma(S)^* = \mathbf{C}^*$, then either $C \cong \mathbf{P}^1$ or $\mathbf{A}^1$.*

Proof. Since D supports an effective ample divisor, $|nK| \neq \emptyset$ for some $n \geq 1$ will imply that $\dim H^0(V, n(K + D))$ is bigger than λn^2 for some $\lambda > 0$ and n large. This contradicts $\bar{\kappa}(S) \leq 1$. Hence $|nK| = \emptyset$ for all n. $\mathrm{Pic}\, S = (0)$ implies that the irregularity of V is zero, since $\mathrm{Pic}\, V$ is finitely generated. Hence V is rational.

If $\bar{\kappa}(S) = 1$, then S has a $\mathbf{C}^*$-fibration $\pi: S \to C$, by the Theorem in Chapter II, Section 5 of [3]. It is clear that $C \cong \mathbf{P}^1$ or $\mathbf{A}^1$ if $\Gamma(S)^* = \mathbf{C}^*$.

Lemma 2.2. *Assume that $\mathrm{Pic}(S) = (0)$, $\Gamma(S)^* = \mathbf{C}^*$. Then $H_1(S, \mathbf{Z}) = (0)$. Assume, furthermore, that $p_g(V) = (0)$. Then $H_2(S, \mathbf{Z}) \cong H^1(D, \mathbf{Z})$, and hence, if every irreducible component of D is rational and the dual graph of D has no loop, then $H_2(S, \mathbf{Z}) = (0)$.*

Proof. Consider the natural homomorphism

$$i_*: H_2(D, \mathbf{Z}) \longrightarrow H_2(V, \mathbf{Z}).$$

Since $\Gamma(S)^* = \mathbf{C}^*$, the 2-cycles on V corresponding to the irreducible components of D are independent in $H_2(V, \mathbf{Z})$. Thus the above homomorphism is injective. We claim that the cokernel of i_* is torsion-free.

From the well known long exact sequence

$$\cdots \to H^1(V, \mathcal{O}_V) \to H^1(V, \mathcal{O}_V^*) \xrightarrow{\delta} H^2(V, \mathbf{Z}) \to H^2(V, \mathcal{O}_V) \to \cdots,$$

the cokernel of the map δ is torsion-free, since $H^2(V, \mathcal{O}_V)$ is a $\mathbf{C}$-vector space. $\mathrm{Pic}(S) = (0)$ implies that $H^1(V, \mathcal{O}_V^*)$ is generated by the line bundles $[C_1]$,

..., $[C_r]$, where C_i are the irreducible components of D. Thus the image of δ is generated by the cohomology classes of the curves $C_1, \ldots, C_r$. Any torsion element of the cokernel of i_* will correspond to an algebraic 2-cycle. From these observations, we see that i_* has torsion-free cokernel.

Now we have a short exact sequence

$$0 \longrightarrow H_2(D, \mathbf{Z}) \longrightarrow H_2(V, \mathbf{Z}) \longrightarrow K \longrightarrow 0,$$

with K a finitely generated torsion-free group. The corresponding long exact cohomology sequence gives a surjection

$$H^2(V, \mathbf{Z}) \longrightarrow H^2(D, \mathbf{Z}) \longrightarrow (0).$$

Finally, we use the long exact cohomology sequence of the pair (V, D) (with $\mathbf{Z}$-coefficients),

$$\cdots \to H^1(V) \to H^1(D) \to H^2(V, D)$$
$$\to H^2(V) \to H^2(D) \to H^3(V, D) \to H^3(V) \to \cdots$$

As remarked above, $H^1(V, \mathcal{O}_V) = (0)$ because Pic V is finitely generated. Hence $H_1(V, \mathbf{Z})$ is a torsion group. But by the arguments used above, $H^2(V, \mathbf{Z})$ is torsion-free, hence by the Universal Coefficient Theorem for cohomology, we see that $H_1(V, \mathbf{Z}) = (0)$. Now from the long exact sequence it follows that $H^3(V, D; \mathbf{Z}) = (0)$ since $H^2(V, \mathbf{Z}) \to H^2(D, \mathbf{Z})$ is a surjection. By Poincare duality, we have $H_1(V - D, \mathbf{Z}) = (0)$. If $p_g(V) = (0)$, then $K = (0)$ and hence $H^2(V) \to H^2(D)$ is injective. Then we have isomorphisms

$$H^1(D, \mathbf{Z}) \xrightarrow{\sim} H^2(V, D; \mathbf{Z}) \xrightarrow{\sim} H_2(V - D, \mathbf{Z}),$$

again from the same exact sequence and Poincare duality.

The remaining assertion in the Lemma is now easy to prove.

Remark. The Lemma (2.2) shows that if S has trivial Picard group and $\Gamma(S)^* = \mathbf{C}^*$, then $\pi_1(S)$ is a perfect group. This already imposes a strong restriction on a counterexample to the question (∗) in the introduction.

Let A be an irreducible component of D and write $D = A + D_1$. Let P be a point of A which is not a double point of D and $\sigma : V' \to V$ the blowing-up with center at P. Let $E = \sigma^{-1}(P)$ and D' the proper transform of D, $S' = V' - D'$. We say that S' is obtained from S by attaching a half-point $E - E \cap D'$, and that S is obtained from S' by detaching a half-point $E - E \cap D'$ (cf. Fujita [1]). Define an integer $\mu \geq 0$ by $\mu = 0$ if A is linearly independent from other components of D and

$$\mu = \min\{\, s > 0 \mid sA + \sum_i \alpha_i D_{1i} \sim 0, \ \alpha_i \in \mathbf{Z} \,\} \quad \text{otherwise,}$$

where $D_1 = \sum D_{1i}$ is the decomposition of D_1 into its irreducible components.

Lemma 2.3. *With the above notations, the following assertions hold true.*
(1) $h^0(V', m(D' + K_{V'})) = h^0(V, m(D + K_V))$ *for every* $m \geq 0$, *whence*
$\bar{\kappa}(S') = \bar{\kappa}(S)$.
(2) *There are exact sequences*

$$0 \longrightarrow \mathbf{Z}/\mu\mathbf{Z} \longrightarrow \operatorname{Pic}(S') \longrightarrow \operatorname{Pic}(S) \longrightarrow 0,$$

$$0 \longrightarrow \Gamma(S')^* \longrightarrow \Gamma(S)^* \longrightarrow \mu\mathbf{Z} \longrightarrow 0.$$

Proof. (1) is clear because $\sigma^*(D + K_V) = D' + K_{V'}$.

(2) $\operatorname{Pic}(S')$ is generated by the prime divisors in S and $E - E \cap D'$ (under
the usual equivalence relation). So clearly $\operatorname{Pic}(S') \to \operatorname{Pic}(S)$ is a surjection.
Letting $E_0 = E - E \cap D'$, the line bundle $\mathcal{L}_{E_0}$ corresponding to E_0 is clearly
trivial on S. Let λ be the order of $\mathcal{L}_{E_0}$ in $\operatorname{Pic}(S')$. If $\lambda > 0$ and finite, we get a
linear equivalence $\lambda E \sim \lambda' A' + \sum \beta_i D'_{1_i}$ on V'. Taking intersection with E, we
get $-\lambda = \lambda'$ which gives a relation $\lambda(E + A') \sim \sum \beta_i D'_{1_i}$. From the definitions
of λ, μ we see that $\lambda = \mu$ and this gives the first exact sequence (the case when
λ is infinite is treated similarly).

Clearly $\Gamma(S')^*$ is a subgroup of $\Gamma(S)^*$. The only possible extra units in $\Gamma(S)^*$
come from the torsionness of the line bundle $\mathcal{L}_{E_0}$ on S'. This gives the second
exact sequence.

This completes the proof of Lemma (2.3). It is clear that S is affine if S'
is affine. We also have a surjection $\pi_1(S) \to \pi_1(S')$ since S is a Zariski open
subset of S'.

We are now ready to prove Theorem 2.

Proof of Theorem 2. By Theorem (8.70) in [1], S is obtained by attaching
several half-points to S_0, where either
(1) $S_0 = \mathbf{P}^2 - (\ell_1 \cup \ell_2 \cup \ell_3)$, with ℓ_i non-concurrent lines, or
(2) $S_0 = \mathbf{P}^2 - C \cup \ell$, where C is a smooth conic and ℓ a line intersecting
transversally with C.

Consider the case (1). Since $\operatorname{rank}\Gamma(S_0)^*/\mathbf{C}^* = 2$, S is obtained from S_0 by
attaching two half points (using Lemma 2.3). Since $\Gamma(S)^* = \mathbf{C}^*$, S is obtained
in the way as described in the statement of Theorem 2.

In case (2), only one half-point is attached. If the center of blowing-up lies
on ℓ, we will get $\operatorname{Pic}(S) \cong \mathbf{Z}/2\mathbf{Z}$, using Lemma (2.3), which is absurd. Hence
the half-point is attached on C.

In case (1), $\pi_1(S_0) \cong \mathbf{Z} \times \mathbf{Z}$ and in case (2), $\pi_1(S_0) \cong \mathbf{Z}$. Because of the
surjection $\pi_1(S_0) \to \pi_1(S)$, $\pi_1(S)$ is also abelian and hence $\pi_1(S) \cong H_1(S, \mathbf{Z})$.
But by Lemma (2.2), $H_1(S, \mathbf{Z}) = (0)$. Hence S is simply connected.
This completes the proof of Theorem 2.

§3. The case $\bar{\kappa}(S) = 1$.

In this section, S will be an affine nonsingular surface defined over $\mathbf{C}$ with
$\bar{\kappa}(S) = 1$ (unless stated otherwise). By Lemma (2.1), there exists a $\mathbf{C}^*$-fibration

$\pi\colon S \to C$ with C a nonsingular rational curve. A general fiber of π is isomorphic to $\mathbf{C}^*$. A scheme-theoretic fiber of π which is not isomorphic to $\mathbf{C}^*$ is called *singular*. We may assume that S has a normal completion (V, D) with a surjective morphism $p\colon V \to B$ where C is a Zariski open subset of B and $p|_S = \pi$. An irreducible component of D is called a *vertical component* or a *fiber component* (resp. a *horizontal component*) if it is contained in a fiber of p (resp. not contained in any fiber of p). Then D has at most two horizontal components. We call the fibration $\pi\colon S \to C$ *twisted* (resp. *untwisted*) if there is only one (resp. two) horizontal components of D. (V, D) is said to be *minimal along fibers* if any vertical component of D which is an exceptional curve of the first kind meets at least three other components of D.

Lemma 3.1. *Let $\pi\colon S \to C$ be a $\mathbf{C}^*$-fibration on a nonsingular affine surface over $\mathbf{C}$ and F a singular fiber of π. Then $F = \Delta + \Gamma$ where*

(1) *either $\Gamma = \emptyset$, $\Gamma_{red} \cong \mathbf{C}^*$ or $\Gamma_{red} = A_1 + A_2$ with $A_1 \cong A_2 \cong \mathbf{A}^1$ and A_1, A_2 meet each other transversally in a single point*

(2) *Δ_{red} is a disjoint union of $\mathbf{A}^1$'s*

(3) *$\Gamma \cap \Delta = \emptyset$.*

Proof. We refer to [4] for an easy proof. See also the proof of Lemma (3.2) below.

The *multiplicity* of a singular fiber F is defined in the usual way as for fibrations of other kinds.

Lemma 3.2. *Let S be a nonsingular, affine surface such that $\bar{\kappa}(S) = 1$, $\operatorname{Pic} S = (0)$ and $\Gamma(S)^* = \mathbf{C}^*$. Then the following assertions hold true.*

(1) *The $\mathbf{C}^*$-fibration $\pi\colon S \to C$ is untwisted, C is isomorphic to $\mathbf{P}^1$ or $\mathbf{A}^1$. If $C \cong \mathbf{P}^1$, then all the fibers of π are irreducible. If $C \cong \mathbf{A}^1$, then all the fibers of π are irreducible and reduced except for one reducible fiber which consists of two irreducible components.*

(2) *Assume that $C \cong \mathbf{P}^1$. Let (V, D) be a normal completion as in the beginning of §3, and assume (V, D) is minimal along the fibers. Let H_1, H_2 be horizontal components of D. Let F be a singular fiber of π and write $F = mA$, where $m > 0$ and A is irreducible. Let $\tilde{F}$ be the fiber of p containing the closure $\bar{A}$ of A. Then the dual graph of $\tilde{F}$ is given as follows:*

Case $A \cong \mathbf{C}^*$. Then $\tilde{F}$ is a linear chain and $\bar{A}$ is the unique exceptional curve of the first kind in $\tilde{F}$;

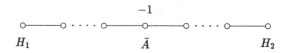

Case $A \cong \mathbf{A}^1$.   If $\tilde{F} \neq \bar{A}$, then $\tilde{F} - \bar{A}$ is connected;

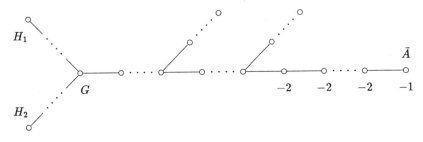

$$H_1 \qquad\qquad G \qquad\qquad -2 \quad -2 \quad -2 \quad -1 \qquad \bar{A}$$

$$H_2$$

Here all the slanted branches are supposed to be linear. $\bar{A}$ is the unique (-1) component except possibly for the component G when G meets H_1 and H_2. If $\tilde{F} = \bar{A}$, then H_1, H_2 and $\bar{A}$ meet each other in one point.

(3)   *Assume $C \cong \mathbf{P}^1$. Then π has at least one singular fiber F_i such that $(F_i)_{red} \cong \mathbf{A}^1$.*

(4)   *With $C \cong \mathbf{P}^1$, write $F_i = m_i A_i$ $(1 \leq i \leq n; A_i$ irreducible). Let Γ be the group defined by the following generators and relations*

$$\Gamma = \langle \gamma_1, \ldots, \gamma_n \mid \gamma_1^{m_1} = \cdots = \gamma_n^{m_n} = \gamma_1 \ldots \gamma_n = 1 \rangle.$$

Then we have an exact sequence

$$(1) \longrightarrow N \longrightarrow \pi_1(S) \longrightarrow \Gamma \longrightarrow (1)$$

and N is central in $\pi_1(S)$. Here N is the image of $\pi_1(F)$ in $\pi_1(S)$ for a general fiber F of π. Γ is a perfect group and N is finite provided Γ is finite.

Proof.   (1) $C \cong \mathbf{P}^1$ or $\mathbf{A}^1$ by Lemma (2.1). If π is twisted, then the unique horizontal component of D is a 2-section and the line bundle given by a section of p will be nontrivial on S. Let now H_1, H_2 be two horizontal components of D. We will give the details of proof when $C \cong \mathbf{P}^1$, the case $C \cong \mathbf{A}^1$ can be treated similarly. So now assume $C \cong \mathbf{P}^1$, F_1, ..., F_n all the singular fibers of π. Write

$$\tilde{F}_i = \sum_{j=1}^{r_i} \mu_{ij} C_{ij} + \sum_{j=r_i+1}^{s_i} \delta_{ij} D_{ij},$$

where C_{ij}'s and D_{ij}'s are irreducible components such that $C_{ij} \cap S \neq \emptyset$, $D_{ij} \cap S = \emptyset$.

We have a relation

$$H_2 - H_1 \sim \sum_{i,j} \alpha_{ij} C_{ij} + \left(\begin{array}{c} \text{a linear combination of} \\ \text{fiber components of } D \end{array} \right).$$

Therefore Pic(S) is the abelian group defined by the following generators and relations

$$\mathrm{Pic}(S) = \left\langle [C_{ij}] \;\; \begin{matrix} i = 1, \ldots, n \\ j = 1, \ldots, r_i \end{matrix} \;\middle|\; \begin{matrix} \sum_{i,j} \alpha_{ij}[C_{ij}] = 0 \\ \sum_{j=1}^{r_1} \mu_{1j}[C_{1j}] = \cdots = \sum_{j=1}^{r_n} \mu_{nj}[C_{nj}] \end{matrix} \right\rangle.$$

Here $\mathrm{rank}(\mathrm{Pic}(S)) \geq \sum_{i=1}^{n} r_i - (n-1) - 1 = \sum_{i=1}^{n} r_i - n$. Since $\mathrm{Pic}(S) = (0)$, $r_i = 1$ for all i. Hence each $F_i = m_i A_i$ with $A_i \cong \mathbf{A}^1$ or $\mathbf{C}^*$. When $C \cong \mathbf{A}^1$, we conclude that, say, $r_1 \leq 2$ and $r_2 = \cdots = r_n = 1$. Suppose $r_1 = 1$. Then it is easy to find an integer $N > 0$ such that

$$N(H_2 - H_1) \sim a\ell_\infty + \left(\begin{matrix} \text{a linear combination of} \\ \text{fiber components of } D \end{matrix} \right),$$

where ℓ_∞ is the complete fiber of p lying in D. But then $\Gamma(S)^* \neq \mathbf{C}^*$, a contradiction. Thus $r_1 = 2$. Write $F_1 = m_1 A_1 + m_1' A_1'$ and $F_i = m_i A_i$ for $2 \leq i \leq n$. Then $\mathrm{Pic}(S) = (0)$ implies that the following $(n+1) \times (n+1)$ matrix is unimodular

$$\begin{pmatrix} \alpha_{11} & \alpha_{12} & \alpha_{21} & \cdots & \alpha_{n1} \\ m_1 & m_1' & 0 & \cdots & 0 \\ 0 & 0 & m_2 & \cdots & 0 \\ \vdots & \vdots & \vdots & \ddots & \vdots \\ 0 & 0 & 0 & \cdots & m_n \end{pmatrix}.$$

From this we see that $m_2 = \cdots = m_n = 1$ and $\alpha_{11} m_1' - \alpha_{12} m_1 = \pm 1$. In particular, $(m_1, m_1') = 1$. This completes the proof of part (1) of Lemma (3.2). In the rest of the proof of Lemma (3.2), we assume $C \cong \mathbf{P}^1$.

(2)   Let F be a singular fiber of π and $\tilde{F} = p^{-1}(\pi(F))$.

Suppose $\tilde{F}$ is nonsingular. Then $A \cong \mathbf{A}^1$ and H_1, H_2 meet transversally at the point $H_1 \cap \tilde{F}$. Assume now that $\tilde{F}$ is reducible. Let G be a (-1) curve in $\tilde{F}$ other than $\bar{A}$. Since (V, D) is minimal along the fibers of p, the following is the unique possible case

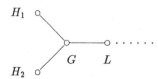

where L is a component of $\tilde{F}$. If $A \cong \mathbf{C}^*$, this is impossible because otherwise $\tilde{F}$ will contain a loop (for properties of singular fibers of a $\mathbf{P}^1$-fibration, see [3; Chapter I]). Hence if $A \cong \mathbf{C}^*$, $\bar{A}$ is the unique (-1) curve in $\tilde{F}$ and the dual graph of $\tilde{F}$ is as stated above.

If $A \cong \mathbf{A}^1$ and there exists a (-1) curve $G \neq \bar{A}$, then we have the situation in the diagram above. G occurs with multiplicity 1 in $\tilde{F}$ and hence there is a

(-1) curve in $\tilde{F}$ other than G ([3; Chapter I]). By the argument above, this curve must be $\bar{A}$. Also, we can blow down $\tilde{F}$ to a nonsingular, rational curve starting from $\bar{A}$. This implies easily that the configuration of $\tilde{F}$ is as stated in the Lemma.

(3) This is clear because D is connected.

(4) Let $P_i = \pi(F_i)$ and $U = C - \{P_1, \ldots, P_n\}$. Then $\pi^{-1}(U) \cong U \times \mathbf{C}^*$. Let $\gamma_1, \ldots, \gamma_n$ be small loops around $P_1, \ldots, P_n$ in $\mathbf{P}^1$. Then letting δ_i to be a small loop in S 'around' A_i, we see that δ_i maps onto $\gamma_i^{m_i}$ under the map $S \to \mathbf{P}^1$. These facts, together with a successive application of Van-Kampen's theorem gives the short exact sequence

$$(1) \longrightarrow N \longrightarrow \pi_1(S) \longrightarrow \Gamma \longrightarrow (1),$$

with Γ as in the statement of Lemma (3.2). N is the image of $\pi_1(F)$ into $\pi_1(S)$ for a general fiber of π. N is central in $\pi_1(S)$ because $\pi_1(\pi^{-1}(U))$ surjects onto $\pi_1(S)$ and $\pi_1(F)$ is central in $\pi_1(U \times \mathbf{C}^*)$.

Since $H_1(S, \mathbf{Z}) = (0)$ by Lemma (2.2), $\Gamma = [\Gamma, \Gamma]$, hence Γ is a perfect group. Suppose Γ is finite. If $N \cong \mathbf{Z}$, we get a central extension

$$(1) \longrightarrow \mathbf{Z} \longrightarrow \pi_1(S) \longrightarrow \Gamma \longrightarrow (1).$$

This extension corresponds to a class $\xi \in H^2(\Gamma, \mathbf{Z})$. Since Γ is finite, we have $H^2(\Gamma, \mathbf{Z}) \cong H^1(\Gamma, \mathbf{Q}/\mathbf{Z})$. See [8; Chapter 3]. Γ operates trivially on $\mathbf{Q}/\mathbf{Z}$, so $H^1(\Gamma, \mathbf{Q}/\mathbf{Z}) = \mathrm{Hom}(\Gamma, \mathbf{Q}/\mathbf{Z}) = (0)$ since Γ is perfect. Thus $\pi_1(S) \cong \Gamma \times \mathbf{Z}$, whence $H_1(S, \mathbf{Z}) \ne (0)$, a contradiction. Therefore N is a finite cyclic group (being a homomorphic image of $\pi_1(F) \cong \mathbf{Z}$).

This completes the proof of Lemma (3.2).

Lemma 3.3. *Let* Γ *be a group as in* Lemma (3.2). *Assume* $m_i > 1$ *for* $i = 1, \ldots, n$. *Then*
(1) Γ *is a perfect group implies* $(m_1, \ldots, m_n) = 1$.
(2) *If* $n = 2$, $\Gamma = (1)$ *iff* Γ *is perfect.*
(3) *If* $n = 3$, Γ *is a perfect group iff* $\{m_1, m_2, m_3\}$ *are coprime.*
(4) *Suppose* $n \ge 3$ *and* Γ *is perfect. Then* Γ *is finite iff* $\{m_1, m_2, m_3\} = \{2, 3, 5\}$.

Proof. $\Gamma/[\Gamma, \Gamma]$ is defined by the generators $\xi_1, \ldots, \xi_n$ with relations

$$m_1 \xi_1 = \cdots = m_n \xi_n = \xi_1 + \cdots + \xi_n = 0.$$

From this (1), (2) and (3) are easy to deduce.

For (4), we know that $m_1, \ldots, m_n$ are pairwise coprime by virtue of (1). Suppose $m_1 < m_2 < \cdots < m_n$. Then specializing $\gamma_4, \ldots, \gamma_n$ to be trivial, we may assume $n = 3$ (if $n \ge 4$, Γ is infinite by well-known results in group theory). Then Γ is finite iff $\{m_1, m_2, m_3\} = \{2, 3, 5\}$.   *q.e.d.*

If $p: V \to B$ has a singular fiber of the form

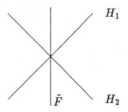

then, by blowing up the point $H_1 \cap H_2$, we can replace $\tilde{F}$ by a reducible singular fiber whose dual graph is

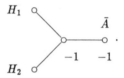

So, we assume hereafter that $H_1 \cap H_2 = \emptyset$. This simplifies some of the later arguments. Note however, that such a normal pair (V, D) is not minimal along fibers.

We will assume that the base of the fibration $\pi: S \to C$ is $\mathbf{P}^1$.

Let $\tilde{F}$ be a singular fiber of $p: V \to B$. By Lemma (3.2), we can contract first the component $\bar{A}$, then the component adjacent to $\bar{A}$ which becomes a (-1) curve after the contraction of $\bar{A}$ and so on till $\tilde{F}$ has been blown down to a smooth rational curve and (H_1^2) does not increase, i.e., the last (-1) curve to be contracted in the above process does not contract to a point on H_1 (in Lemma (3.2), (V, D) is assumed to be minimal along fibers, but even with the new convention above, the argument is valid).

Applying the above operation to all singular fibers of p, we obtain a minimal ruled surface Σ_a $(a \geq 0)$, on which the images $\bar{H}_1$, $\bar{H}_2$ of H_1, H_2 are disjoint from each other. Hence $(\bar{H}_1^2) \leq 0$ or $(\bar{H}_2^2) \leq 0$. We may assume $(\bar{H}_1^2) \leq 0$, a priori, by interchanging the roles of H_1 and H_2 if necessary. Then $\bar{H}_1$ is a minimal section, $\bar{H}_1^2 = -a$ and $\bar{H}_2 \sim \bar{H}_1 + a\ell$, where ℓ is a fiber of Σ_a. Conversely, V is obtained from Σ_a by starting the blowing-ups with centers at points on $\bar{H}_2$ or fibers of Σ_a, while no points of $\bar{H}_1$ are blown up. Then $(H_1^2) = (\bar{H}_1^2) = -a$. Let ρ denote the birational morphism $V \to \Sigma_a$.

As before, let $\tilde{F}_1, \ldots, \tilde{F}_n$ exhaust all singular fibers of $p: V \to B$, where, for each i, $\tilde{F}_i - m_i \bar{A}_i$ is supported by the components of D. Let $L(D)$ be the subgroup of $\text{Pic}(V)$ generated by the components of D. Let $a = \sum_{i=1}^{n} \alpha_i$ be a fixed partition of a $(\alpha_i \geq 0)$. Then we have

$$\rho^* H_2 - \sum_{i=1}^{n} \delta_i \bar{A}_i \in L(D)$$

and

$$\sum_{i=1}^{n} (\alpha_i m_i - \delta_i) \bar{A}_i \in L(D), \quad \text{where } 0 \leq \delta_i < m_i.$$

δ_i is the coefficient of $\bar{A}_i$ in the total transform $\rho^* \bar{H}_2$. Now $\mathrm{Pic}(S)$ is defined by generators $\xi_1, \ldots, \xi_n$ and relations;

$$\sum_{i=1}^{n} (\alpha_i m_i - \delta_i)\xi_i = 0, \qquad m_1 \xi_1 = \cdots = m_n \xi_n.$$

Therefore $\mathrm{Pic}(S) = (0)$ iff the following $n \times n$ matrix is unimodular.

$$\begin{pmatrix} \alpha_1 m_1 - \delta_1 & \alpha_2 m_2 - \delta_2 & \cdots & \cdots & \alpha_n m_n - \delta_n \\ m_1 & -m_2 & 0 & \cdots & 0 \\ \vdots & \vdots & \vdots & \ddots & \vdots \\ m_1 & 0 & 0 & \cdots & -m_n \end{pmatrix}$$

This gives

$$m_1 m_2 \ldots m_n a - \sum_{i=1}^{n} m_1 \ldots \hat{m}_i \ldots m_n \delta_i = \pm 1.$$

Now we prove the

Lemma 3.4. *With the above notations, the following hold true.*
(1) $m_i = 1$ *if* $\delta_i = 0$ *and* $(m_i, \delta_i) = 1$ *if* $\delta_i \neq 0$.
(2)   *If* $\delta_i = 0$, *the dual graph of* $\tilde{F}_i$ *is:*

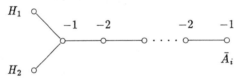

(3)   *If* $\delta_i \neq 0$, *the dual graph of* $\tilde{F}_i$ *is:*

$$G$$

```
  o——————o · · · · o——————o——————o · · · · o——————o
  H₁                                              H₂
                           o  -2
                           ·
                           ·
                           o  -2

                   Āᵢ      o  -1
```

$$\begin{array}{cccc} -2 & -2 & -2 & -1 \end{array}$$

where there might be no chain $\underset{\bar{A}_i}{\circ\!\!-\!\!-\!\!-\!\!-\!\!\circ \cdots\cdots \circ\!\!-\!\!-\!\!-\!\!-\!\!\circ}$ *and* $G = \bar{A}_i$.

Proof.   (1) follows from the unimodularity of the above matrix.

(2) is clear because $m_i = 1$.

(3)   By Lemma (3.2), we have only to consider the case $A_i = \bar{A}_i \cap S \cong \mathbf{A}^1$. Then the dual graph has to be as stated above (otherwise, we would have $(m_i, \delta_i) > 1$). q.e.d.

We say that a singular fiber $\tilde{F}$ is a *singular fiber of type* $(1,0)$ (resp. (m, δ)) if the dual graph of $\tilde{F}$ is as given in (2) (resp. (3)) of Lemma (3.4), where the (-1) curve $\bar{A}$ has coefficients m and δ in $\tilde{F}$ and the total transform $\rho^* \bar{H}_2$, respectively. In case of a singular fiber of type (m, δ) with $\delta > 0$, we distinguish two cases.

(i) $A_i \cong \mathbf{C}^*$.   In this case $\tilde{F}_i$ is called a singular fiber of type $(m, \delta)^*$.

(ii) $A_i \cong \mathbf{A}^1$. $\tilde{F}_i$ is called a singular fiber of type $(m, \delta)^+$.

Lemma 3.5. $0 < \delta < m$ *be integers with* $(m, \delta) = 1$. *Then blowing up a point on* $\bar{H}_2$ *and its infinitely near points, we can produce a singular fiber* $\tilde{F}$ *of type* (m, δ). *If* $\tilde{F}$ *is of type* $(m, \delta)^*$, *the above sequence of blowing-ups is uniquely determined by* m, δ *and a point on* $\bar{H}_2$ *to be blown up first.*

Proof.   It suffices to verify the assertion for a singular fiber of type $(m, \delta)^*$.

In this case, $\tilde{F}$ together with H_1, H_2 form a linear chain. Then the choice of successive blow-up centers is determined uniquely by the continued fraction expansion

$$\frac{m}{\delta} = m_1 + \cfrac{1}{m_2 + \cfrac{\ddots}{\quad + \cfrac{1}{m_\alpha}}}$$

For details, see [3; pp. 78-80].

Let $p: V \to \mathbf{P}^1$ be a $\mathbf{P}^1$-fibration and D a reduced effective divisor on V with simple normal crossings. Let $S = V - D$. Suppose D contains two horizontal components, disjoint from each other. Let $\tilde{F}_1, \ldots, \tilde{F}_n$ be all the singular fibers of p. Suppose that, for each i, $\tilde{F}_i$ is a singular fiber of type $(1,0)$ or (m_i, δ_i). We assume that $\tilde{F}_i$ is of type $(1,0)$ for $1 \le i \le r$, of type $(m_i, \delta_i)^+$ with $\delta_i > 0$ for $r + 1 \le i \le r + s$ and of type $(m_i, \delta_i)^*$ with $\delta_i > 0$ for $r + s + 1 \le i \le n$. Moreover, we assume that V is obtained from Σ_a $(a \ge 0)$ by blowing-up points on $\bar{H}_2$ or on fibers and their infinitely near points, but no points of $\bar{H}_1$, where $\bar{H}_1$ is the minimal section of Σ_a and $\bar{H}_2$ is a section with $\bar{H}_1 \cap \bar{H}_2 = \emptyset$.

Lemma 3.6.   *With the above notations and assumptions, the following hold true.*

(1)   *Assume* $\Gamma(S)^* = \mathbf{C}^*$. *Then* S *is affine iff* $r + s > 0$.

(2) $\text{Pic}(S) = (0)$ *iff we have*

$$m_1 \ldots m_n a - \sum_{i=1}^{n} m_1 \ldots \hat{m}_i \ldots m_n \delta_i = \pm 1.$$

(3) $\text{rank}\, H^2(S, \mathbf{Z}) = r + s - 1$, *provided* $\text{Pic}\, S$ *is finite and* S *is affine.*

Proof.   (1) If S is affine then D is connected and so (1) is clear.

Assume $r + s > 0$. Computing the Picard number of V and using $\Gamma(S)^* = \mathbf{C}^*$, we see that the natural map

$$H_2(D, \mathbf{Q}) \longrightarrow H_2(V, \mathbf{Q})$$

is an isomorphism. On the other hand, from the long exact sequence

$$0 \to H^0(V, \mathbf{Z}) \to H^0(D, \mathbf{Z}) \to H^1(V, D; \mathbf{Z}) \to H^1(V, \mathbf{Z}) \to$$

we see that $H^1(V, D; \mathbf{Z}) \cong H_3(S, \mathbf{Z})$ is trivial provided D is connected. Hence for $r + s > 0$, $H_3(S, \mathbf{Q}) = (0)$. Now we invoke Theorem 1 in [1] to show that S is affine.

(2)   This was already observed in the arguments preceding Lemma (3.4).

(3)   We know that V is rational since $\bar{\kappa}(S) \leq 1$, S is affine and $\text{Pic}(S)$ is finite (cf. arguments in Lemma (2.2)). Hence $H_2(D, \mathbf{Q}) \to H_2(V, \mathbf{Q})$ is a surjection. This implies that $H^2(V, \mathbf{Q}) \to H^2(D, \mathbf{Q})$ is an injection. Now from the exact sequence

$$\to H^1(V, \mathbf{Q}) \to H^1(D, \mathbf{Q}) \to H^2(V, D; \mathbf{Q}) \to H^2(V, \mathbf{Q}) \to H^2(D, \mathbf{Q}),$$

we know that $H^1(D, \mathbf{Q}) \xrightarrow{\sim} H^2(V, D; \mathbf{Q})$. It is easy to see that $\text{rank}\, H^1(D, \mathbf{Q}) = r + s - 1$. Hence

$$\text{rank}\, H_2(S, \mathbf{Q}) = \text{rank}\, H^2(V, D; \mathbf{Q}) = r + s - 1.$$

q.e.d.

Remark.   The condition $\Gamma(S)^* = \mathbf{C}^*$ is guaranteed by the condition

$$m_1 m_2 \ldots m_n a - \sum_{i=1}^{n} m_1 \ldots \hat{m}_i \ldots m_n \delta_i \neq 0.$$

For, from the arguments preceding Lemma (3.4), the latter condition implies that $\text{Pic}(S)$ is finite. On the other hand, $\text{rank}\, \text{Pic}(V)$ can be easily seen to be equal to the number of irreducible components of D. So the components of D are linearly independent in $\text{Pic}(V)$ if

$$m_1 m_2 \ldots m_n a - \sum_{i=1}^{n} m_1 \ldots \hat{m}_i \ldots m_n \delta_i \neq 0.$$

Lemma 3.7. *Let the notations be as in the discussion before* Lemma (3.6). *Suppose there exist smooth, rational curves C_1 and C_2 on V such that $C_1 \not\subset D$ and $C_2 \subset D$, $C_1^2 = -1$, $(C_1 \cdot C_2) = (C_1 \cdot D) = 1$ and C_1, C_2 are contained in the same fiber F of p. Let $\tau\colon V \to W$ be the contraction of C_1, $D' = \tau_*(D - C_2)$ and $S' = W - D'$. Then the following hold true:*

(1) $\mathrm{Pic}(S) = (0)$ *and* $\Gamma(S)^* = \mathbf{C}^*$ *iff* $\mathrm{Pic}(S') = (0)$ *and* $\Gamma(S')^* = \mathbf{C}^*$.

(2)   *If* $\tilde{F} - (C_1 \cup C_2)$ *is connected and non-empty, then* $\bar{\kappa}(S') = \bar{\kappa}(S)$. *Otherwise* $\bar{\kappa}(S') \leq \bar{\kappa}(S)$.

Proof. (1) Assume $\mathrm{Pic}(S) = (0)$ and $\Gamma(S)^* = \mathbf{C}^*$. Write $D - C_2 = \sum_i G_i$ and $G_i' = \tau(G_i)$. Since $\mathrm{Pic}(S) = (0)$, we have a relation $C_1 \sim \alpha C_2 + J$, J supported on irreducible components of $D - C_2$. From $(C_1^2) = -1$ and $(C_1 \cdot C_2) = (C_1 \cdot D) = 1$, we have $-1 = (C_1^2) = \alpha$. Hence $C_1 + C_2 \sim J$. This implies $\tau_*(C_2) \sim \tau_*(J)$. This implies $\mathrm{Pic}(S') = (0)$. Suppose there is a relation $\sum \beta_i \tau(G_i) \sim 0$ on W. Then $\sum \beta_i G_i \sim 0$ on V because G_i's are disjoint from C_1. Hence $\beta_i = 0$ for all i. So $\Gamma(S')^* = \mathbf{C}^*$.

Converse is proved similarly.

(2)   Assume that $\tilde{F} - (C_1 \cup C_2)$ is connected and non-empty. $\tilde{F}$ can be contracted to a nonsingular rational curve. We see easily from the process of contraction, that $\tau(C_2)$ is a (-1) curve meeting D' in an irreducible component C_3' such that $\tau(C_2) \cdot C_3' = \tau(C_2) \cdot D' = 1$. Then $\kappa(D + K_V, V) = \kappa(D' + \tau(C_2) + K_W, W) = \kappa(D' + K_W, W)$. This means $\bar{\kappa}(S) = \bar{\kappa}(S')$. If $\tilde{F} - (C_1 \cup C_2)$ is not connected, we can only assert that

$$\kappa(D + K_V, V) = \kappa(D' + \tau(C_2) + K_W, W) \geq \kappa(D' + K_W, W)$$

i.e., $\bar{\kappa}(S) \geq \bar{\kappa}(S')$.   q.e.d.

Lemma 3.8. *With the notations and assumptions as in the discussion before* Lemma (3.6)

$$\bar{\kappa}(S) = 1 \quad \textit{iff} \quad (n-2) - \sum_{i=r+s+1}^{n} \frac{1}{m_i} > 0.$$

Proof. By virtue of Lemma (3.7), part (2), we can replace any singular fiber of type $(1,0)$ or $(m, \delta)^+$ by a smooth rational curve which is to be included in the boundary divisor (note that $\bar{\kappa}(S) \leq 1$ always, since S has a $\mathbf{C}^*$-fibration). On Σ_a, we have $\bar{H}_2 \sim \bar{H}_1 + a\ell$, where ℓ is a fiber of the $\mathbf{P}^1$-fibration $\Sigma_a \to \mathbf{P}^1$. Also $K_{\Sigma_a} \sim -(\bar{H}_1 + \bar{H}_2) - 2\ell$, ℓ a general fiber. Thus we can write

$$K_V \sim -H_1 - \rho^*(\bar{H}_2) - 2\ell + \sum_{i,j} \lambda_{ij} C_{ij},$$

where $\sum_{i,j} \lambda_{ij} C_{ij}$ is an effective divisor supported on the exceptional locus of the morphism $\rho\colon V \to \Sigma_a$. We write $\rho^*(\bar{H}_2) = H_2 + \sum_{i,j} \mu_{ij} C_{ij}$, where $\sum_{i,j} \mu_{ij} C_{ij}$

is also an effective divisor supported on the exceptional locus of ρ. We have $r + s + 1 \leq i \leq n$ and C_{ij} are the irreducible components of the singular fiber $\tilde{F}_i$.

The boundary D consists of H_1, H_2, $(r + s)$ smooth fibers of the fibration p and all the irreducible components of $\tilde{F}_i$ except for $\bar{A}_i$, $r + s + 1 \leq i \leq n$.

We have therefore

$$D + K_V \sim (r + s)\ell + {\sum_{i,j}}' C_{ij} - \sum_{i,j} \mu_{ij} C_{ij} - 2\ell + \sum_{i,j} \lambda_{ij} C_{ij},$$

where ${\sum_{i,j}}' C_{ij}$ means the components $\bar{A}_i$ are omitted for $r + s + 1 \leq i \leq n$.

For each i, $\tilde{F}_i = \sum_j \alpha_{ij} C_{ij} \sim \ell$ and the coefficient of the component $\bar{A}_i$ is $m_i > 0$. We have

$$\bar{A}_i + {\sum_{i,j}}' \frac{\alpha_{ij}}{m_i} C_{ij} \sim \frac{\ell}{m_i} \quad \text{as } \mathbf{Q}\text{-divisors.}$$

We write

$$D + K_V \sim (r + s - 2)\ell + \sum_{i,j} C_{ij} + \sum_{i,j}(\lambda_{ij} - \mu_{ij}) C_{ij} - \sum_i \bar{A}_i.$$

It is easy to see that $\lambda_{ij} \geq \mu_{ij}$ for all i, j. Actually, one can see by keeping track of the coefficients λ_{ij}, μ_{ij}, that

$$\sum_j C_{ij} + \sum_j (\lambda_{ij} - \mu_{ij}) C_{ij} = \tilde{F}_i.$$

Thus we get

$$D + K_V \sim (r + s + t - 2)\ell - \sum_i \bar{A}_i, \quad \text{where } r + s + t = n.$$

This gives further,

$$D + K_V \sim (n - 2)\ell - (\sum_i \frac{1}{m_i})\ell + {\sum}' \frac{\alpha_{ij}}{m_i} C_{ij}.$$

The term ${\sum_{i,j}}' \frac{\alpha_{ij}}{m_i} C_{ij}$ is an effective divisor and the intersection matrix for its components is negative definite. If $(n - 2) - \sum_i \frac{1}{m_i} > 0$, then for multiples N of $m_1 \ldots m_n$, $|N(D + K_V)|$ has dimension ≥ 1, showing that $\bar{\kappa}(S) = 1$.

If $(n - 2) - \sum_i \frac{1}{m_i} \leq 0$, then by a similar argument, $|N(D + K_V)|$ has dimension < 1, hence $\bar{\kappa}(S) \leq 0$.

The proof of the Lemma (3.8) is now complete.

We consider the special cases $s + t = 2$ and 3.

Lemma 3.9. *Suppose $s + t = 2$, i.e., there are exactly two singular fibers of types (m_1, δ_1) and (m_2, δ_2). Assume $\mathrm{Pic}(S) = (0)$ and $\Gamma(S)^* = \mathbf{C}^*$. Then we have*

(1) $a = 1$ and $(m_1 - \delta_1)(m_2 - \delta_2) - \delta_1 \delta_2 = \pm 1$.

(2)   *If $n = 3$, $s = 0$ and $(m_1, m_2) = 1$, then S is an affine contractible surface with $\bar{\kappa}(S) = 1$.*

Proof. (1)   The condition in Lemma (3.6), (2) reads as $m_1 m_2 a - m_1 \delta_2 - m_2 \delta_1 = \pm 1$. Since $\delta_i > 0$, $a \neq 0$. Using $m_i > \delta_i$, we see easily that $a = 1$.

(2)   We have $r = 1$, $s = 0$ and $t = 2$. Then S is affine by Lemma (3.6). Also $(n - 2) - (\dfrac{1}{m_1} + \dfrac{1}{m_2}) > 0$ since $(m_1, m_2) = 1$. Hence $\bar{\kappa}(S) = 1$ by Lemma (3.8). By Lemmas (3.2) and (3.3), $\pi_1(S) = (1)$ and by Lemma (3.6), $H_2(S, \mathbf{Z}) = (0)$. Hence S is contractible.

Remarks. (1)   An example of $\{(m_1, \delta_1), (m_2, \delta_2)\}$ satisfying the condition (1) of Lemma (3.9) is $(m_1, \delta_1) = (2, 1)$ and $(m_2, \delta_2) = (3, 1)$. Actually, the equation (1) has infinitely many solutions. By looking at the boundary D, we see that there are infinitely many distinct contractible surfaces S with $\bar{\kappa}(S) = 1$.

(2)   With the assumptions of Lemma (3.9), if $r = 0$, $s = 1$, $n = 2$ and $(m_1, m_2) = 1$, then $\bar{\kappa}(S) = -\infty$ and hence $S \cong \mathbf{A}^2$. (For $(n-2) - (\dfrac{1}{m_1} + \dfrac{1}{m_2}) < 0$ and this implies $\bar{\kappa}(S) = -\infty$ by the proof of Lemma (3.8).)

Lemma 3.10. *Suppose that $s + t \geq 3$, $\mathrm{Pic}(S) = (0)$ and $\Gamma(S)^* = \mathbf{C}^*$. Let the singular fibers be of types (m_i, δ_i), $1 \leq i \leq s + t$ (in addition to possible singular fibers of type $(1, 0)$). Then we have the following:*

(1)   *We can take $a \leq \left\lceil \dfrac{s + t}{2} \right\rceil$.*

(2)   *Suppose S is affine, i.e., either $n > s + t$ or $n = s + t$ and $s > 0$. Then S is not simply-connected and $\bar{\kappa}(S) = 1$ unless $n = 3$, $\{m_1, m_2, m_3\} = \{2, 3, 5\}$.*

(3)   *Suppose $s + t = 3$, $a = 1$ and $m_1 \leq m_2 \leq m_3$. Then $\pi_1(S)$ is finite if and only if $(m_1, \delta_1) = (2, 1)$, $(m_2, \delta_2) = (3, 1)$, $(m_3, \delta_3) = (5, 1)$.*

Proof. (1)   We have the relation

$$m_1 \ldots m_p a - \sum_{i=1}^{p} m_1 \ldots \hat{m}_i \ldots m_p \delta_i = \pm 1, \quad \text{where } p = s + t.$$

If $a \geq p$, then we have

$$(a - p)m_1 \ldots m_p + \sum_{i=1}^{p} m_1 \ldots \hat{m}_i \ldots m_p(m_i - \delta_i) = \pm 1,$$

which is impossible. Hence $a < p$. If $p - a < a$, then we can contract all singular fibers $\tilde{F}_1, \ldots, \tilde{F}_p$ of type (m, δ) down to the cross-section H_1 instead of H_2. Hence we can take $a \leq \left\lceil \dfrac{p}{2} \right\rceil$.

(2)   We know that $\bar{\kappa}(S) = 1$ if

$$(n - 2) - \sum_{i=1}^{p} \frac{1}{m_i} > 0.$$

If $n > p$, then

$$(n - 2) - \sum_{i=1}^{p} \frac{1}{m_i} \geq (p - 1) - \sum_{i} \frac{1}{m_i} > p - 1 - \frac{p}{2} = \frac{p - 2}{2} > 0$$

since $\{m_1, \ldots, m_p\}$ are mutually coprime. Hence $\bar{\kappa}(S) = 1$ in this case. Suppose $n = p$ and $s > 0$. Again,

$$(p - 2) - \sum_{i=1}^{p} \frac{1}{m_i} > (p - 2) - \frac{p}{2}$$

since $m_1, \ldots, m_p$ are relatively prime. So, for $p \geq 4$, $\bar{\kappa}(S) = 1$.

When $n = p = 3$, $(p-2) - \sum_{i=1}^{p} \frac{1}{m_i} \leq 0$ only when $p = 3$ and $\{m_1, m_2, m_3\} = \{2, 3, 5\}$ (again because m_1, m_2, m_3 are relatively prime).

By Lemmas (3.2) and (3.3), $\pi_1(S)$ is non-trivial.

(3)   By Lemmas (3.2) and (3.3), $\pi_1(S)$ is finite and non-trivial iff $\{m_1, m_2, m_3\} = \{2, 3, 5\}$. We may assume $a = 1$ by part (1). Then $m_1 m_2 m_3 - m_1 m_2 \delta_3 - m_1 m_3 \delta_2 - m_2 m_3 \delta_1 = \pm 1$ iff $\delta_i = 1$ for each i.

Summarizing Lemmas (3.2)–(3.10), we can state the

Theorem 3.   *Let S be a nonsingular, affine surface such that $\bar{\kappa}(S) = 1$,* $\text{Pic}(S) = (0)$ *and* $\Gamma(S)^* = \mathbf{C}^*$. *Suppose S has a $\mathbf{C}^*$-fibration $\pi : S \to C \cong \mathbf{P}^1$. Then we have the following assertions.*

(1)   *$\pi : S \to C$ is an untwisted $\mathbf{C}^*$-fibration, and all the fibers are irreducible. The fibration $\pi : S \to C$ extends to a $\mathbf{P}^1$-fibration $p : V \to C$ on a normal completion (V, D) of S so that*

   (i)   *D has two horizontal components H_1, H_2 which are disjoint from each other.*

   (ii)   *D is minimal along the fibers of p, i.e., there are no exceptional curves of the first kind in D which are contained in fibers of p and can be contracted without losing the condition (i) above.*

   (iii)   *Set $a = -(H_1^2)$. Then we can assume $a \geq 0$ and V is obtained from a Hirzebruch surface Σ_a by blowing-up points on fibers and their infinitely near points as in the discussion preceding Lemma (3.6).*

   (iv)   *Any singular fiber of p is of type $(1, 0)$, $(m, \delta)^+$ or $(m, \delta)^*$ (cf. Lemmas (3.2) and (3.4)).*

(2)   *S is simply-connected iff the following two conditions are satisfied.*

(i)   *All singular fibers of p are of type $(1,0)$ except for at most two singular fibers of type (m, δ) with $\delta > 0$.*

(ii)   *If there is only one singular fiber of type (m, δ) with $\delta > 0$, then either $a = 1$ and $m = \delta + 1$ or $a = 0$ and $\delta = 1$. Moreover, π has more than two singular fibers.*

(iii)   *If there are exactly two singular fibers of type (m_i, δ_i), $\delta_i > 0$, then $a = 1$ and $m_1 m_2 - m_1 \delta_2 - m_2 \delta_1 = \pm 1$; moreover π has more than two singular fibers.*

(3)   *S is topologically contractible iff the following conditions are satisfied.*

(i)   *$a = 1$.*

(ii)   *The fibration p has one singular fiber of type $(1,0)$ and two singular fibers of types $(m_1, \delta_1)^*$ and $(m_2, \delta_2)^*$, where $m_i > \delta_i > 0$ and $m_1 m_2 - m_1 \delta_2 - m_2 \delta_1 = \pm 1$.*

(4)   *$\pi_1(S)$ is a nontrivial finite group iff the following conditions are satisfied.*

(i)   *$a = 1$ or 2.*

(ii)   *Singular fibers of p are all of type $(1,0)$ except for three singular fibers of types*

$$(2,1), \ (3,1), \ (5,1) \quad if \quad a = 1.$$
$$(2,1), \ (3,2), \ (5,4) \quad if \quad a = 2.$$

Indeed, these two cases are the same, and the configuration of D in the second case is obtained from the first case by making it upside-down.

(iii)   *There is at least one singular fiber of type $(1,0)$ or $(m, \delta)^+$.*

Proof.   (1)   This follows from Lemmas (3.2) and (3.4).

(2)   Suppose S is simply-connected. Then by Lemmas (3.2) and (3.3), the fibration $\pi: S \to C$ has at most two singular fibers of multiplicity > 1. Suppose that there is only one singular fiber of multiplicity > 1 in π. Then $p: V \to C$ has a unique singular fiber of type (m, δ) with $m > \delta > 0$, where $am - \delta = \pm 1$, by Lemma (3.6). Hence $a = 0$ or 1. If $a = 0$, then $\delta = 1$. If $a = 1$, $m = \delta + 1$. Moreover, by Lemma (3.8), we have $n \geq 3$. Suppose that $p: V \to C$ has two singular fibers of types (m_i, δ_i), $i = 1, 2$. Then by Lemma (3.9), we have $a = 1$ and $m_1 m_2 - m_1 \delta_2 - m_2 \delta_1 = \pm 1$.

"If" part.   By Lemmas (3.2) and (3.3), we know that $\pi_1(S)$ is abelian. Hence $\pi_1(S) \cong H_1(S, \mathbf{Z})$. We have $\mathrm{Pic}(S) = (0)$ by assumption, hence $H_1(S, \mathbf{Z}) = (0)$ by Lemma (2.2).

(3)   Suppose S is topologically contractible. Then $H_2(S, \mathbf{Z}) = (0)$. By Lemma (3.6), $r + s = 1$. In view of (2) above, this implies the conditions (i) and (ii).

"If" part.   Under the hypothesis, we know that $\pi_1(S) = (1)$ and D consists of a tree of non-singular rational curves. Hence $H_2(S, \mathbf{Z}) \cong H^1(D, \mathbf{Z}) = (0)$ by

Lemma (2.2). Now $H_i(S, \mathbf{Z}) = (0)$ for all $i \geq 1$ and hence S is contractible by a theorem of J. H. C. Whitehead.

(4) Assume $\pi_1(S)$ is a non-trivial finite group. By Lemmas (3.2) and (3.3), there are exactly three singular fibers of type (m_i, δ_i) with $\delta_i > 0$, where $\{m_1, m_2, m_3\} = \{2, 3, 5\}$. As in the proof of Lemma (3.10), the conditions (i) and (ii) can be verified. Since S is affine, we have the condition (iii), by Lemma (3.6).

"If" part follows from Lemmas (3.2), (3.3), (3.6) and (3.8).   *q.e.d.*

We shall next prove

Theorem 4. *Let S be a nonsingular, affine surface such that* $\mathrm{Pic}(S) = (0)$, $\Gamma(S)^* = \mathbf{C}^*$. *Suppose S has a $\mathbf{C}^*$-fibration* $\pi: S \to C \cong \mathbf{A}^1$. *Then we have the following.*

(1) *π has exactly one reducible fiber with two irreducible components, all the other singular fibers are reduced and irreducible.*

(2) *If $H_2(S, \mathbf{Z}) = (0)$, then $S \cong \mathbf{A}^2$. Hence, if $\bar{\kappa}(S) \geq 0$, $H_2(S, \mathbf{Z}) \neq (0)$ and S is not contractible.*

Proof. (1) has already been proved in Lemma (3.2), part (1).

(2) Suppose $H_2(S, \mathbf{Z}) = (0)$. Then D does not contain a loop by Lemma (2.2). Since $H_1 \cup \ell_\infty \cup H_2$ is connected (recall, ℓ_∞ is the complete fiber of p contained in D), $\mathrm{Supp}(\tilde{F}_1 \cap D)$ is disconnected. Since S is affine, $\tilde{F}_1 \cap D$ has two connected components U and L, where $U \cap H_1 \neq \emptyset$ and $L \cap H_2 \neq \emptyset$. Let $A_1 = C_1 \cap S$ and $A_1' = C_1' \cap S$ (for the notation, see the proof of Lemma (3.2), part (1)). We claim the following:

(i) Either $A_1 \cong \mathbf{A}^1$ or $A_1' \cong \mathbf{A}^1$.

(ii) There is no (-1) curve in $\tilde{F} - C_1 \cup C_1'$. Hence, either C_1 or C_1' must be a (-1) curve.

(iii) One of the following two possibilities does occur.

(a) $A_1 \cong \mathbf{C}^*$, $A_1' \cong \mathbf{A}^1$, $U \cap C_1 \neq \emptyset$ and $L \cap C_1 \neq \emptyset$.

(b) $A_1 \cong A_1' \cong \mathbf{A}^1$, $C_1 \cap C_1' \neq \emptyset$, $C_1 \cap U \neq \emptyset$, $C_1 \cap L = \emptyset$, $C_1' \cap U = \emptyset$ and $C_1' \cap L \neq \emptyset$.

Indeed, (i) follows from Lemma (2.4). As for (ii), let E be an exceptional component of $\tilde{F}_1$ with $E \neq C_1, C_1'$. Then E cannot meet both H_1 and H_2, for otherwise D contains a loop. This contradicts the minimality of D along the fibers of p.

For (iii), note that the two connected components U, L of $\tilde{F}_1 - C_1 \cup C_1'$ must be connected by one (or both) of C_1 and C_1'. Then the assertion is easy to verify.

In the case (a) above, C_1' must be connected to U and L, by a chain of irreducible curves in $D \cap \tilde{F}_1$. Suppose C_1' is connected to L. Then we claim:

(iv) The dual graph of $\ell_\infty + H_1 + U + C_1$ is linear.

For, there are no (-1) curves in U. Assume the assertion to be false. Then U contains a portion as indicated below:

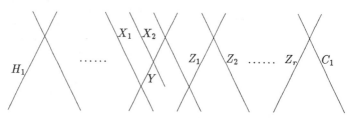

where there are no other components of U meeting the chain $Z_1 + \cdots + Z_r$. Note that $\tilde{F}_1$ is contracted to the component which meets H_1, by a succession of blowing-downs. Hence the component Y must be contracted at some stage, i.e., Y becomes a (-1) curve after a succession of blowing-downs. After contracting Y, the images of X_1 and X_2 must meet the image of H_2, which is a cross-section. This is a contradiction. So the assertion is verified.

Now $(\ell_\infty^2) = 0$ and $\ell_\infty + H_1 + U$ is a linear chain. Then it is easy to show that $\bar{\kappa}(S) = -\infty$, (cf. the proof of Cor. 2.4.3 in [3; p. 16]). Now $S \cong \mathbf{A}^2$ by [3; Theorem (4.1), p. 47].

§4. Examples $\bar{\kappa}(S) = 2$.

In this section we give some examples of affine, nonsingular surfaces with $\mathrm{Pic}(S) = (0)$, $\Gamma(S)^* = \mathbf{C}^*$ and $\bar{\kappa}(S) = 2$. In the example (1) below, $\pi_1(S) = (1)$, so that the answer to the question $(*)$ in the Introduction is affirmative in this case. The surface in the example (2) is contractible and non-isomorphic to the example of C. P. Ramanujam in [6].

Example 1. Let $V \subset \mathbf{P}^3$ be a general nonsingular surface. It is known that $\mathrm{Pic}(V) \cong \mathbf{Z}$ if degree $V \geq 4$ and any hyperplane section $H \cap V$ generates $\mathrm{Pic}(V)$. Let $S = V - V \cap H$, where H is any hyperplane. It follows that $\mathrm{Pic}(S) = (0)$ and $\Gamma(S)^* = \mathbf{C}^*$. Since $\kappa(V) \geq 0$ if degree $V \geq 4$, $\bar{\kappa}(S) = 2$ if degree $V \geq 4$.

By a result of M. Nori, S is simply-connected (even if $H \cap V$ has bad singularities! See [5]).

Example 2. We construct a contractible surface S with $\bar{\kappa}(S) = 2$. It is easy to prove that for a nonsingular contractible surface S, the following holds: (i) S is affine, (ii) $\mathrm{Pic}(S) = (0)$ and (iii) $\Gamma(S)^* = \mathbf{C}^*$ (cf. Fujita [1; §1]).

Construction of S. Let $V = \Sigma_2$ be the Hirzebruch surface with the minimal section M and a fiber ℓ. Let C be a cuspidal rational curve (with exactly one singular point which is an ordinary cusp of multiplicity 2), $C \sim 2M + 4\ell$

and D a nonsingular rational curve, $D \sim M + 3\ell$, such that $C \cdot D = 5Q + P$ as in the figure below.

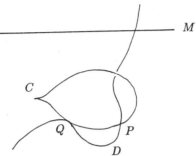

Let $\sigma : V' \to V$ be the blowing-up at P and C', D', M' the proper transforms of C, D, M respectively. Let $S = V' - (C' \cup D' \cup M')$. Then an easy calculation shows that $[C']$, $[D']$, $[M']$ generate $\mathrm{Pic}(V')$ freely. This implies $\mathrm{Pic}(S) = (0)$ and $\Gamma(S)^* = \mathbf{C}^*$.

First we show the existence of such curves C, D.

Existence of C. Starting from a nonsingular conic in $\mathbf{P}^2$ and a line tangent to it, we blow up at a point on the tangent line, different from the point of tangency. We get the following picture.

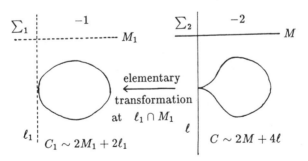

Existence of D. By construction, C is a 2-section of the $\mathbf{P}^1$-fibration $\Sigma_2 \to \mathbf{P}^1$. Hence there exists a unique fiber ℓ_0 on Σ_2, $\ell_0 \neq \ell$ such that $C \cdot \ell_0 = 2P_0$. Then $C \cdot (M + 3\ell_0) = 6P_0$. We use the exact sequence

$$0 \to H^0(\Sigma_2, \mathcal{O}(-M - \ell)) \to H^0(\Sigma_2, \mathcal{O}(M + 3\ell)) \to H^0(C, \mathcal{O}_C(M + 3\ell)) \to 0,$$

where $H^1(\Sigma_2, \mathcal{O}(-M - \ell)) = (0)$ because $M + 3\ell$ is ample.

Since $\mathrm{Pic}^0(C) \cong C - \{\mathrm{cusp}\} \cong G_a$, we can find points P, Q so that $5Q + P \sim 6P_0$ in $\mathrm{Pic}^0(C)$. Hence there exists a curve $D \sim M + 3\ell$ and that $C \cdot D = 5Q + P$.

It is easy to show that D is an irreducible and nonsingular rational curve.

This demonstrates the existence of C and D. After resolving the singularity of C and blowing up successively, we make the set-theoretic total transform of $M' \cup C' \cup D'$ a divisor with normal crossings.

The dual graph of this total transform has the following picture.

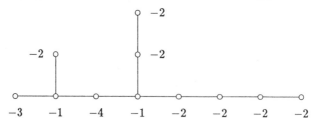

This gives a normal completion (W, D) of S. First we prove that S is contractible.

Since the dual graph of D is a tree of rational curves, it follows from Lemma (2.2) that $H_2(S, \mathbf{Z}) = (0)$ and $H_1(S, \mathbf{Z}) = (0)$. It suffices to show that $\pi_1(S) = (1)$.

On V' we have the following picture.

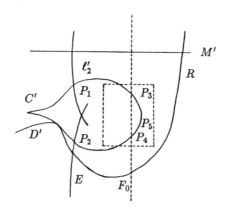

Here E is the exceptional curve obtained after blowing up P, ℓ_2' is the proper transform of the fiber ℓ_2 of $\Sigma_2 \to \mathbf{P}^1$ passing through P, $\{P_1, P_2\} = C' \cap (\ell_2' \cup E')$ and $R = D' \cap M'$. Then $\ell_2' \cup E$ is the inverse image of ℓ_2 on V'. F_0 is a general fiber which is close to the fiber touching the curve C' at the point P_5, and $\{P_3, P_4\} = C' \cap F_0$. The dotterd square (which contains P_3, P_4) is a neighbourhood of P_5.

Since $S = V' - (C' \cup D' \cup M')$, we get a fibration $S \to \mathbf{P}^1$ with a general fiber isomorphic to $\mathbf{P}^1 - \{4 \text{ points}\}$. Each fiber is clearly reduced, so by Lemma (1.5) in [5], we have an exact sequence

$$\pi_1(F) \longrightarrow \pi_1(S) \longrightarrow \pi_1(\mathbf{P}^1) \longrightarrow (1),$$

where F is a general fiber of $S \to \mathbf{P}^1$. $\ell_2' \cup E - \{4 \text{ points}\}$ is a special fiber of $S \to \mathbf{P}^1$ and it is easy to see that it is a strong deformation retract of a suitable neighbourhood in S. Such a neighbourhood contains a general fiber F of $S \to \mathbf{P}^1$, so we get a surjection

$$\pi_1(\ell_2' \cup E - \{4 \text{ points}\}) \longrightarrow \pi_1(S) \longrightarrow (1).$$

We shall show that $\pi_1(S)$ is abelian. We know from the above argument that $\pi_1(S)$ is generated by a loop in ℓ_2' near P_1 and a loop in E near P_2. But these loops can be seen to be homotopic respectively to the loops in F_0 around P_3 and P_4. Now the dotted square minus the curve C' has the fundamental group $\mathbf{Z}$ (because by change of variables, we can assume C' is given by $z_1 = 0$ near P_5). Thus, in $\pi_1(S)$, the above two loops commute. Hence $\pi_1(S)$ is abelian. But $H_1(S, \mathbf{Z}) = (0)$. Hence $\pi_1(S) = (1)$. Now S is contractible.

Now the dual graph of D is minimal (every (-1) curve meets at least three other components). Since the dual graph is not linear, $\bar{\kappa}(S) \neq -\infty$ (otherwise, S would be isomorphic to $\mathbf{A}^2$ and then the dual graph of D would be linear by [6]).

By the results in §2, $\bar{\kappa}(S) \neq 0$. From Theorem 3, part (3), we know that if $\bar{\kappa}(S) = 1$, then the dual graph of a suitable normal completion (W', D') will look like

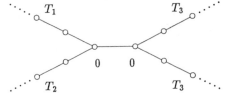

Here T_i are negative definite, linear branches with all the weights ≤ -2. But then D' can be obtained from D by blowing-ups and blowing-downs of exceptional curves. This can be easily seen to be impossible.

Finally, we must have $\bar{\kappa}(S) = 2$. (This can also be proved directly by elementary calculations.)

The configuration of the divisor at infinity for the example of C. P. Ramanujam is as follows.

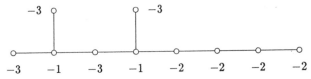

By the same kind of argument used for proving $\bar{\kappa}(S) \neq 1$ above, we see that S is not isomorphic to the example of C. P. Ramanujam in [6].

In view of the results of this paper, we ask the following

Question 1. Are there infinitely many non-isomorphic contractible smooth surfaces with logarithmic Kodaira dimension equal to 2?

Question 2. Is every smooth, contractible surface rational?

References

[1] T. Fujita, On the topology of non-complete algebraic surfaces, J. Fac. Sci. Univ. of Tokyo, 29 (1982), 503–566.

[2] R. V. Gurjar and A. R. Shastri, The fundamental group at infinity of affine surfaces, Comment. Math. Helv., 59 (1984), 459–484.

[3] M. Miyanishi, *Non-complete algebraic surfaces*, Lecture Notes in Mathematics, No. 857, Berlin-Heidelberg-New York: Springer, 1981.

[4] M. Miyanishi, Étale endomorphisms of algebraic varieties, Osaka J. Math., 22 (1985), 345–364.

[5] M. Nori, Zariski conjecture and related problems, Ann. Scient. Éc. Norm. Sup., 4eme série, 16 (1983), 305–344.

[6] C. P. Ramanujam, A topological characterization of the affine plane as an algebraic variety, Ann. of Math., 94 (1971), 69–88.

[7] A. R. Shastri, Divisors with finite local fundamental group on a surface, to appear in the proceedings of the 1985 AMS Summer Institute on Algebraic Geometry, Bowdoin College, Maine.

[8] E. Weiss, *Cohomology of groups*, Academic Press, Inc., 1969.

R. V. GURJAR

Tata Institute of Fundamental Research
Homi Bhabha Road
Bombay, 400005
India

M. MIYANISHI

Department of Mathematics
Osaka University
Toyonaka, Osaka, 560
Japan

Algebraic Geometry and Commutative Algebra
in Honor of Masayoshi NAGATA
pp. 125–140 (1987)

On the Convolution Algebra of Distributions on Totally Disconnected Locally Compact Groups

Hiroaki HIJIKATA

§0. Introduction.

The purpose of this paper is to supply more intrinsic proofs and improvements in conclusion on certain general results, concerning representations of some Hecke algebras, used in my recent paper [3].

In more detail, we consider a triple (G, Z, ω) consisting of a T. D. L. C.(= totally disconnected locally compact) group G, its closed normal subgroup Z, and a locally constant homomorphism $\omega: Z \to \mathbf{C}^\times$, nomalized by G, i.e., $\omega(gzg^{-1}) = \omega(z)$ for any $g \in G$ and $z \in Z$. Let $S(G, \omega)$ denote the vector space of all locally constant complex valued functions f of which supports are compact mod Z, and which are ω-semiinvariant, i.e., $f(zg) = \omega(z)f(g)$ for any $z \in Z$. A linear map $T: S(G, \omega) \to \mathbf{C}$ will be called an ω-distribution on G. Let $\mathcal{D}(G)$ denote the vector space of all ω-distributions of which supports are compact mod Z. By the convolution product, $\mathcal{D}(G)$ is an associative algebra over $\mathbf{C}$.

Let (π, V) be a smooth representation of G on the complex vector space V. If (π, V) is an ω-representation, i.e., if Z acts on V as ω, it is extended to a representation of the algebra $\mathcal{D}(G)$. Let H be a closed subgroup of G containing Z, and having a compact quotient H/Z. Let $\epsilon: H \to \mathbf{C}^\times$ be a locally constant homomorphism which coincides with ω on Z. Let $\mathcal{E} = \mathcal{E}(G, H, \epsilon)$ denote the subalgebra of $\mathcal{D}(G)$ consisting of all locally constant ω-distributions T which are ϵ-semiinvariant under right and left shift by H, i.e., $R(h)T = L(h)T = \epsilon(h)T$ for $h \in H$. Let $V(H, \epsilon)$ denote the ϵ-eigenspace of V under H, then $V(H, \epsilon)$ is an $\mathcal{E}$-subspace. The main results of this paper will be:

Theorem 1. *The functor $V \mapsto V(H, \epsilon)$ defines an equivalence of the category of 'irreducibre smooth ω-representations of G having non-zero ϵ-eigenspace' with the category of 'non-zero irreducible $\mathcal{E}$-modules'.*

Theorem 2. *Assume G has a topological antiautomorphism τ satisfying:*

(0) $\tau(Z) = Z$, $\omega \circ \tau = \omega$,

(1) $\tau(H) = H$, $\epsilon \circ \tau = \epsilon$,

Received April 6, 1987.

(2)   *the automorphism $g \mapsto \tau(g^{-1})$ is of finite order,*

(3)   *each double coset HgH contains a τ-fixed element.*

Then the algebra $\underline{\mathcal{E}}$ is commutative.

If Z is the trivial subgroup $\{1\}$, hence ω is also trivial, and moreover if H is an open subgroup of G, the content of section 2.10 of Bernshtein-Zelevinskii [1] is exactly our Theorem 1. A part of Theorem 1 (for non-trivial ω but H still open), which we used in [3], is given in Casselman [2]. We are going to modify the method of [1] to cover our case.

§1. Finite ω-distribution.

Let (G, Z, ω) be a triple as in §0, and let $p \colon g \mapsto \bar{g}$ denote the canonical projection from G to $\bar{G} = G/Z$. A subset X of G will be called a Z-subset iff $p^{-1}(\bar{X}) = X$, i.e., $X = ZX = XZ$. In the following, X *always denote a locally closed Z-subset of G*. For a complex vector space V, let $C(X, \omega; V)$ denote the vector space of all locally constant V-valued functions on X satisfying:

$$F(zx) = \omega(z)F(x) \quad \text{for } z \in Z, \ x \in X.$$

Set:

$$S(X, \omega; V) := \{ F \in C(X, \omega; V) \mid p(\operatorname{supp} f) \text{ is compact} \},$$
$$S(X, \omega) := S(X, \omega; \mathbf{C}),$$
$$S^*(X, \omega) := \operatorname{Hom}_{\mathbf{C}}(S(X, \omega), \mathbf{C}).$$

An element T of $S^*(X, \omega)$ will be called an ω-*distribution* on X, and its value at $f \in S(X, \omega)$ is denoted by $\langle T, f \rangle$.

There are associated several natural maps which will be given in this and the next section, with indication of some abbreviations to simplify the notation.

If a Z-subset X' is an open subset of X, and F' is an element of $C(X', \omega; V)$, we denote by iF' the function on X, which coincides F' on X' and zero outside. We often identify iF' with F'. Since $\operatorname{supp} F' = \operatorname{supp} iF'$, the linear map $i = i(X', X) \colon F' \mapsto iF'$ induces:

$$i \colon S(X', \omega; V) \rightarrowtail S(X, \omega; V),$$

$$i^* \colon S^*(X, \omega) \twoheadrightarrow S^*(X', \omega), \qquad \langle i^*T, f' \rangle := \langle T, if' \rangle.$$

If Y is a locally closed Z-subset of X, we have the restriction map $r = r(X, Y) \colon F \mapsto rF = F|_Y$:

$$r \colon C(X, \omega; V) \longrightarrow C(Y, \omega; V).$$

If a Z-subset Y is a closed subset of X, then the compactness of $p(\operatorname{supp} F)$ implies the compactness of $p(\operatorname{supp} rF)$, and we have the following exact sequences:

(1) $$S(X - Y, \omega; V) \xrightarrow{i} S(X, \omega; V) \xrightarrow{r} S(Y, \omega; V),$$

(1*) $$S^*(Y, \omega) \xrightarrow{r^*} S^*(X, \omega) \xrightarrow{i^*} S(X - Y, \omega),$$

$$\langle r^* T', f \rangle := \langle T', rf \rangle \quad \text{for } T' \in S^*(Y, \omega), \ f \in S(X, \omega).$$

Define the support $\operatorname{supp} T$ of an ω-distribution T on X by the intersection $\cap Y$ of all Z-subsets Y closed in X and satisfying $T \in \operatorname{Im} r(X, Y)^* = \operatorname{Ker} i(X - Y, X)^*$. In other words, $\operatorname{supp} T$ consists of the points x satisfying:

(2) If W is a Z-subset of X, open in X, containing x, then $\langle T, f \rangle \neq 0$ for some $f \in S(W, \omega) \subset S(X, \omega)$.

An ω-distribution T on X is called *finite* iff $p(\operatorname{supp} T)$ is compact. Let $\underline{\mathcal{D}}(X)$ denote the vector space of all finite ω-distributions on X.

An ω-distribution T on X is finite iff there is a Z-subset K of X satisfying:

(3) K is open in X, $K \supset \operatorname{supp} T$, and $p(K)$ is compact.

If K is any such subset, then K is closed in X and we have

(3') $$\langle T, f \rangle = \langle T, f|_K \rangle \quad \text{for any } f \in S(X, \omega).$$

Indeed, if $i = i(K, X)$, $r = r(X, K)$, $T' \in \underline{\mathcal{D}}(K)$ and $T = r^* T'$, by our convention $f|_K$ in fact means irf, hence $\langle T, irf \rangle = \langle r^* T', irf \rangle = \langle T', rirf \rangle = \langle T', rf \rangle = \langle r^* T', f \rangle = \langle T, f \rangle$.

In general if Y is open and closed in X, $i(Y, X)$ is a cross section of $r(X, Y)$, hence r in (1) is surjective and the sequence splits. Thus if X is a disjoint union of finite number of open and closed subsets X_i, we can identify as:

(4) $$S(X, \omega; V) = \oplus S(X_i, \omega; V),$$

(4*) $$S^*(X, \omega) = \oplus S^*(X_i, \omega).$$

§2. Action of homeomorphisms and multiplication by functions.

Let $\operatorname{Homeo}(G, Z, \omega)$ denote the subgroup of the homeomorphism group of G, consisting of homeomorphisms γ satisfying:

(1) γ induces a homeomorphism $\bar{\gamma}$ of $\bar{G}$, hence in particular $\gamma(zg) = z'\gamma(g)$, $\gamma^{-1}(zg) = z''\gamma^{-1}(g)$ with some z', $z'' \in Z$,

(2) $\omega(z) = \omega(z') = \omega(z'')$.

For a function F in $C(X, \omega; V)$ and γ in $\mathrm{Homeo}(G, Z, \omega)$, we set $\gamma F :=$ $F \circ \gamma^{-1}$. We have $\gamma F \in C(\gamma X, \omega; V)$ and $\mathrm{supp}(\gamma F) = \gamma(\mathrm{supp}\, F)$, and a linear isomorphism $\gamma \colon F \to \gamma F$;

$$(3) \qquad\qquad \gamma \colon S(X, \omega; V) \longrightarrow S(\gamma X, \omega; V).$$

For an ω-distribution T, we set $\gamma T := (f' \mapsto \langle T, f' \circ \gamma \rangle)$ for $f' \in S(\gamma X, \omega)$. We have $\mathrm{supp}(\gamma T) = \gamma(\mathrm{supp}\, T)$, and linear isomorphisms:

$$(3^*) \qquad\qquad \gamma \colon S^*(X, \omega) \longrightarrow S^*(\gamma X, \omega),$$

$$(3') \qquad\qquad \gamma \colon \underline{\mathcal{D}}(X) \longrightarrow \underline{\mathcal{D}}(\gamma X)$$

For example, the right (resp. left) shift $R(g_0)$ (resp. $L(g_0)$) by an element g_0 of G belongs to $\mathrm{Homeo}(G, Z, \omega)$ for any ω normalized by G:

$$(4) \qquad\qquad R(g_0) \colon g \longmapsto g g_0, \quad z' = z'' = z,$$

$$(4') \qquad\qquad L(g_0) \colon g \longmapsto g_0 g, \quad z' = g_0 z g_0^{-1}, \quad z'' = g_0^{-1} z g_0.$$

If τ is a topological antiautomorphism of G satisfying the condition (0) of Theorem 2, then τ belongs to $\mathrm{Homeo}(G, Z, \omega)$ with $z' = \gamma(g)\gamma(z)\gamma(g)^{-1}$.

If $\omega' \colon Z \to \mathbf{C}^\times$ is another locally constant homomorphism normalized by G, there is the bilinear map $C(X, \omega') \times C(X, \omega; V) \to C(X, \omega'\omega; V)$ defined by $(f, F) \mapsto f \cdot F = (x \mapsto f(x)F(x))$. We have $\mathrm{supp}(f \cdot F) = \mathrm{supp}\, f \cap \mathrm{supp}\, F$. If $\delta \in C(X, \omega')$ is never zero on X, the map $\delta \cdot \colon F \mapsto \delta \cdot F$ has the inverse, hence we have the linear isomorphism:

$$(5) \qquad\qquad \delta \cdot \colon S(X, \omega; V) \longrightarrow S(X, \omega'\omega; V).$$

The dual $(\delta \cdot)^*$ for $V = \mathbf{C}$ is defined by $\langle (\delta \cdot)^* T', f \rangle := \langle T', \delta \cdot f \rangle$,

$$(5^*) \qquad\qquad (\delta \cdot)^* \colon S^*(X, \omega) \longrightarrow S^*(X, \omega'\omega).$$

Since $\mathrm{supp}((\delta \cdot)^* T') = \mathrm{supp}\, T'$, $(\delta \cdot)^*$ maps a finite distribution to a finite distribution. In particular, if $\omega' = 1$, we have a linear automorphism:

$$(5') \qquad\qquad (\delta \cdot)^* \colon \underline{\mathcal{D}}(X) \longrightarrow \underline{\mathcal{D}}(X).$$

If $\gamma \in \mathrm{Homeo}(G, Z, \omega) \cap \mathrm{Homeo}(G, Z, \omega')$, we have the following commutative diagrams:

$$(6) \qquad
\begin{array}{ccc}
S(X, \omega; V) & \xrightarrow{\;\delta \cdot\;} & S(X, \omega'\omega; V) \\
{\scriptstyle \gamma} \downarrow & & \downarrow {\scriptstyle \gamma} \\
S(\gamma X, \omega; V) & \xrightarrow{\;\gamma \delta \cdot\;} & S(\gamma X, \omega'\omega; V),
\end{array}$$

$$(6^*) \qquad
\begin{array}{ccc}
S^*(X, \omega) & \xleftarrow{\;(\delta \cdot)^*\;} & S^*(X, \omega'\omega) \\
{\scriptstyle \gamma} \downarrow & & \downarrow {\scriptstyle \gamma} \\
S^*(\gamma X, \omega) & \xleftarrow{\;(\gamma \delta \cdot)^*\;} & S^*(\gamma X, \omega'\omega).
\end{array}$$

Note that we have taken γ for $S^*(X, \omega)$, not the dual of γ for $S(X, \omega)$ but the dual of γ^{-1} in (3^*).

§3. Generators of $S(X, \omega; V)$.

Recall that G is T. D. L. C. group iff it has a fundamental system of neighbourhoods $\mathcal{U}$ of 1, consisting of open and compact subgroups U. Since ω is locally constant, we may and shall assume that $\mathcal{U}$ consists of U satisfying:

(0) $U \cap Z \subset \operatorname{Ker} \omega$.

Hence there is a unique homomorphism $\nu \colon ZU \to \mathbf{C}^{\times}$ satisfying:

(1) $\nu = \omega$ on Z, $\nu = 1$ on U.

Consider the composite of isomorphisms $\nu \cdot$ of (5) and $L(x)$ of $(4')$ §2,

(2) $$S(ZU, 1; V) \xrightarrow{\ \nu\cdot\ } S(ZU, \omega; V) \xrightarrow{\ L(x)\ } S(xZU, \omega; V).$$

If $u \in U$, since $R(u)\nu = \nu$, $R(u)$ commutes with $\nu\cdot$ and $L(x)$. By definition, any element f of $S(ZU, 1; V)$ has the form $f = \bar{f} \circ p$ by the unique element $\bar{f}$ in $S(\bar{U}, 1; V)$, and the isomorphism $f \mapsto \bar{f}$ is equivariant for U-action, $\overline{R(u)f} = R(\bar{u})\bar{f}$, $\overline{L(u)f} = L(\bar{u})\bar{f}$. Thus we may identify $S(ZU, 1; V)$ with $S(\bar{U}, 1; V)$ not only as a vector space but also as a U-bi-module.

Since $S(\bar{U}, 1; V)$ contains a unique function $\bar{f}$ characterized by $\bar{f}(1) = 1$ and $R(\bar{u})\bar{f} = \bar{f}$ for $u \in U$, i.e., the constant function 1, we can conclude that:

(3) $S(xZU, \omega; V)$ contains a unique function f characteized by $f(x) = 1$ and $R(u)f = f$ for $u \in U$, i.e., the function $L(x)\nu$.

Since $S^*(\bar{U}, 1)$ contains a unique distribution $\bar{T}$ characterized by $\langle \bar{T}, 1 \rangle = 1$ and $R(\bar{u})\bar{T} = \bar{T}$ for $u \in U$, i.e., the right (hence left) invariant Haar measure $\bar{\mu}$ of $\bar{U}$ normalized as total volume 1, $\langle \bar{T}, \bar{f} \rangle = \langle \bar{\mu}, \bar{f} \rangle$, and it also satisfies $L(\bar{u})\bar{T} = \bar{T}$ for $u \in U$, we can conclude that:

(3*) $S^*(xZU, \omega)$ contains a unique ω-distribution T_x characterized by $\langle T_x, L(x)\nu \rangle = 1$ and $R(u)T_x = T_x$ for $u \in U$, i.e., $\langle T_x, f \rangle = \langle \bar{\mu}, \nu^{-1} \cdot (L(x)^{-1}f) \rangle$, and it also satisfies $L(xux^{-1})T_x = T_x$ for $u \in U$.

Let X be a locally closed Z-subset of G. If F is in $C(X, \omega; V)$, $y \in X$ and F is constant on $yU \cap X$, then

(4) $F(x) = (L(y)\nu)(x)F(y)$ for $x \in ZyU \cap X$.

Consequently, F is constant on $xU \cap X$ for any $x \in ZyU \cap X$. Indeed, if $x = zyu$, $z \in Z$ and $u \in U$, then $yu \in yU \cap X$, and $F(x) = \omega(z)F(y)$, while $(L(y)\nu)(x) = \nu(y^{-1}zyu) = \omega(z)$.

Lemma 1. Let $F \in S(X, \omega; V)$. Let K be a Z-subset satisfying:

(5) K is open in X, $\operatorname{supp} F \subset K$, and $p(K)$ is compact.

*Then, there is $U \in \underline{\mathcal{U}}$, such that $xU \cap K$ is compact and F is constant on $xU \cap X$
for any $x \in K$.*

*Let Y be a complete system of representatives of the double cosets $Z \backslash KU/U$,
taken from K, thus K is the disjoint union of the finite number of open and
closed subsets $ZyU \cap X$ ($y \in Y$). Then $F = \sum F_y$ with*

$$F_y = (x \longmapsto (L(y)\nu)(x)F(y)) \in S(ZyU \cap X, \omega; V).$$

Proof. For each $x \in K$, choose $U \in \underline{\mathcal{U}}$ such that $xU \cap X = xU \cap K$
is closed in xU (hence compact), and F is constant on $xU \cap X$. Then K is
covered by $xU \cap X$, and since $p(K)$ is compact, K is covered by finite number
of $Z(x_iU_i \cap X) = Zx_iU_i \cap X$. Take any $U \in \underline{\mathcal{U}}$ contained in the intersection
$\cap U_i$. For any $x \in K$, it is contained in some $Zx_iU_i \cap X$, by (1), F is constant
on $xU_i \cap X$, hence constant on $xU \cap X$. This proves our first claim. Since f lies
in $S(K, \omega; V)$, our second claim is a consequence of (4) §1 and the above (4).

§4. Action of T on vector valued functions.

By Lemma 1, the map r (resp. r^*) of (1) (resp. (1*)) §1 is surjective (resp. injective). It then follows that $\operatorname{supp} r^* T' = \operatorname{supp} T'$, hence we have the linear
injection

(1) $r^* : \underline{\mathcal{D}}(Y) \rightarrowtail \underline{\mathcal{D}}(X)$.

The bilinear map $(f, v) \mapsto (x \mapsto f(x)v)$ induces the linear injection

(2) $S(X, \omega) \otimes V \rightarrowtail S(X, \omega; V)$.

By Lemma 1, it is surjective, hence a linear isomorphism. We denote the
image of $f \otimes v$ by the same letter, thus,

$$f \otimes v = (x \longmapsto f(x)v).$$

By the above isomorphism, for $T \in S(X, \omega)$, there is a unique linear map
$T \otimes \operatorname{id} : S(X, \omega; V) \rightarrow V$, characterized by $\langle T \otimes \operatorname{id}, f \otimes v \rangle = \langle T, f \rangle v$. Now we
define the bilinear map $S^*(X, \omega) \times S(X, \omega; V) \rightarrow V$ by

$$(T, F) \longmapsto \langle T, F \rangle := \langle T \otimes \operatorname{id}, F \rangle.$$

Then, for $T \in \underline{\mathcal{D}}(X)$, taking K as in (3) §1, define the bilinear map $\underline{\mathcal{D}}(X) \times
C(X, \omega; V) \rightarrow V$ by

$$(T, F) \longmapsto \langle T, F \rangle := \langle T, F|_K \rangle.$$

Since if $K \subset K'$ and $F|_{K'} = \sum f \otimes v$, then $F|_K = \sum f|_K \otimes v$, the above
map is well defined and bilinear.

Lemma 2. *If $A : V \rightarrow V'$ is a linear map to another vector space V', then,
$A \langle T, F \rangle = \langle T, A \circ F \rangle$, where $A \circ F := (x \mapsto A(F(x))) \in C(X, \omega; V')$.*

Proof. Since $(A \circ F)|_K = A \circ (F|_K)$, it suffices to check when $F = f \otimes v$.
Then the claim is obvious.

§5. Tensor product of distributions.

Let (G_i, Z_i, ω_i), $i = 1, 2$ be triples as in §0. Then $(G_1 \times G_2, Z_1 \times Z_2, \omega_1 \times \omega_2)$ is also such a triple. Let X_i be a locally closed Z_i-subset of G_i. The bilinear map $(f_1, f_2) \mapsto ((x_1, x_2) \mapsto f(x_1)f(x_2))$ induces the linear injection:

(1)
$$S(X_1, \omega_1) \otimes S(X_2, \omega_2) \longrightarrow S(X_1 \times X_2, \omega_1 \otimes \omega_2)$$

$$f_1 \otimes f_2 \longmapsto f_1 \otimes f_2 := ((x_1, x_2) \longmapsto f(x_1)f(x_2)).$$

Again by Lemma 1, it is a linear isomorphism. Hence, for $T_i \in S^*(X_i, \omega_i)$, there is a unique element $T_1 \otimes T_2$ in $S^*(X_1 \times X_2, \omega_1 \times \omega_2)$, characterized by $\langle T_1 \otimes T_2, f_1 \otimes f_2 \rangle = \langle T_1, f_1 \rangle \langle T_2, f_2 \rangle$. From the definition (2) §1 of $\operatorname{supp} T$, it is clear that

(2)
$$\operatorname{supp}(T_1 \otimes T_2) = \operatorname{supp} T_1 \times \operatorname{supp} T_2.$$

Hence if $T_i \in \underline{\mathcal{D}}(X_i)$, then $T_1 \otimes T_2 \in \underline{\mathcal{D}}(X_1 \times X_2)$, and the following equalities (3)–(7) are immediately checked by computing the value at $f_1 \otimes f_2$ of both sides of each equality.

If a Z_i-subset Y_i is closed in X_i,

(3)
$$r(X_1, Y_1)^* T_1 \otimes r(X_2, Y_2)^* T_2 = r(X_1 \times X_2, Y_1 \times Y_2)^* (T_1 \times T_2).$$

If $\gamma_i \in \operatorname{Homeo}(G_i, Z_i, \omega_i)$, then $\gamma_1 \times \gamma_2 \in \operatorname{Homeo}(G_1 \times G_2, Z_1 \times Z_2, \omega_1 \times \omega_2)$, and

(4)
$$\gamma_1 T_1 \otimes \gamma_2 T_2 = (\gamma_1 \times \gamma_2)(T_1 \times T_2).$$

If $\delta_i \in C(X_i, \omega_i)$, $\delta_i \neq 0$ on X_i, then $\delta_1 \otimes \delta_2 \in C(X_1 \times X_2, \omega_1 \times \omega_2)$, $\delta_1 \otimes \delta_2 \neq 0$ on $X_1 \times X_2$, and

(5)
$$(\delta_1 \cdot)^* T_1 \otimes (\delta_2 \cdot)^* T_2 = (\delta_1 \otimes \delta_2 \cdot)^* (T_1 \otimes T_2).$$

If $x_i \in X_i$, and $T(x_i) := (f \mapsto f(x_i))$ is the Dirac distribution, then

(6)
$$T(x_1) \otimes T(x_2) = T((x_1, x_2)).$$

If $G_i = G$, $i = 1, 2$, and $\sigma := ((g_1, g_2) \mapsto (g_2, g_1))$, then $\sigma \in \operatorname{Homeo}(G \times G, Z \times Z, \omega \otimes \omega)$, and

(7)
$$\sigma(T_1 \times T_2) = T_2 \times T_1.$$

Lemma 3. (*Fubini theorem.*) *If $T_i \in \underline{\mathcal{D}}(X_i)$ and $F \in C(X_1 \times X_2, \omega_1 \otimes \omega_2; V)$, then*

(i)
$$F(x_1, \ ) := (x_2 \longmapsto F(x_1, x_2)) \in C(X_2, \omega_2; V),$$
$$F(\ , x_2) := (x_1 \longmapsto F(x_1, x_2)) \in C(X_1, \omega_1; V).$$

(ii)
$$F_1 := (x_1 \longmapsto \langle T_2, F(x_1, \ ) \rangle) \in C(X_1, \omega_1; V),$$
$$F_2 := (x_2 \longmapsto \langle T_1, F(\ , x_2) \rangle) \in C(X_2, \omega_2; V).$$

(iii)
$$\langle T_1 \otimes T_2, F \rangle = \langle T_i, F_i \rangle, \qquad i = 1, 2.$$

Proof. (i) and (ii) are obvious. (iii). If F is in $S(X_1 \times X_2, \omega_1 \times \omega_2; V)$, by the above (1) and (2) §4, it suffices to check for $F = f_1 \otimes f_2 \otimes v$, then the claim is obvious. In general, for each T_i, take K_i as in (3) §1, then $\langle T_1 \otimes T_2, F \rangle = \langle T_1 \otimes T_2, F|_{K_1 \times K_2} \rangle = \langle T_1, (F|_{K_1 \times K_2})_1 \rangle$, while $\langle T_1, F_1 \rangle = \langle T_1, (F_1)|_{K_1} \rangle$. Hence it suffices to see $(F|_{K_1 \times K_2})_1 = (F_1)|_{K_1}$, it amounts to see

$$F(x_1, \ )|_{K_2} = F|_{K_1 \times K_2}(x_1, \ )$$

for $x_1 \in K_1$, and the last equality is obvious.

§6. Convolution.

Let $m: G \times G \to G$ denote the product map $(x, y) \mapsto xy$. Define the bilinear map $\mathcal{D}(G) \times \mathcal{D}(G) \to S^*(G, \omega)$, $(T_1, T_2) \mapsto T_1 * T_2$, by

$$\langle T_1 * T_2, f \rangle := \langle T_1 \otimes T_2, \ f \circ m \rangle, \qquad f \in S(G, \omega).$$

If $\operatorname{supp} T_i = K_i$, then $p(K_1 K_2)$ is compact, $G - K_1 K_2$ is open and $\langle T_1 * T_2, f \rangle = 0$ for $f \in S(G - K_1 K_2, \omega)$, hence $\operatorname{supp} T_1 * T_2 \subset K_1 K_2$. Thus $(T_1, T_2) \mapsto T_1 * T_2$ defines an algebra structure on $\mathcal{D}(G)$. The product $T_1 * T_2$ is called the *convolution*, it is in fact associative as will be seen in §8.

Lemma 4. (i) *If G' is a closed subgroup of G containing Z, then the linear injection*

$$r^*: \mathcal{D}(G') \longrightarrow \mathcal{D}(G)$$

defined in (1) §4, is an algebra embedding.

(ii) *If $\delta \in C(G, 1)$ is a homomorphism, then the linear automorphism $(\delta \cdot)^*$ of $\mathcal{D}(G)$ defined in (5') §2 is an algebra automorphism of $\mathcal{D}(G)$.*

(iii) *If τ is a topological antiautomorphism of G with the property (0) §0, the linear automorphism τ of $\mathcal{D}(G)$ defined in (3') §2, is an algebra antiautomorphism of $\mathcal{D}(G)$.*

(iv) *The map $T(\ ): G \to \mathcal{D}(G)$, which maps $g \in G$ to the Dirac distribution $T(g)$, is a semigroup homomorphism.*

(v) *If $F \in C(G, \omega; V)$, then $\langle T_1 * T_2, F \rangle = \langle T_1 \otimes T_2, F \circ m \rangle = \langle T_i, F_i \rangle$, $i = 1, 2$ with $F_1 = (g \mapsto \langle T_2, L(g)^{-1} F \rangle)$, $F_2 = (g \mapsto \langle T_1, R(g)^{-1} F \rangle)$.*

Proof. (i) Let $m': G' \times G' \to G'$ denote the product map. The claim is a consequence of (3) §5 and the obvious relation, $(r(G, G')f) \circ m' = r(G \times G, G' \times G')(f \circ m)$.

(ii) The claim is a consequence of (5) §5, and the obvious relation, $(\delta \otimes \delta) \cdot (f \circ m) = (\delta \cdot f) \circ m$.

(iii) The claim is a consequence of (4) and (7) §5, and the obvious relation, $m \circ (\tau \times \tau) = \tau \circ m \circ \sigma$.

(iv) The claim is a consequence of (6) §5.

(v) For each T_i, take K_i as in (3) §2. If $K \supset K_1 K_2$, $(F \circ m)|_{K_1 \times K_2} = (F|_K \circ m)|_{K_1 \times K_2}$, and $\langle T_1 \otimes T_2, F \circ m \rangle = \langle T_1 \otimes T_2, F|_K \circ m \rangle$. Thus to see the first equality, it suffices to check for $F = f \otimes v$, which is obvious. The second equality is direct from Lemma 3.

§7. Representation of G.

Let (π, V) be a left (resp. right) G-module over $\mathbf{C}$, i.e., G acts on V through the homomorphism (resp. antihomomorphism) $\pi \colon G \to \mathrm{Aut}_{\mathbf{C}}(V)$. For any closed subgroup G' of G and a homomorphism $\chi \colon G' \to \mathbf{C}^{\times}$, set:

$$V(\pi, G', \chi) = V(G', \chi)$$
$$:= \{\, v \in V \mid \pi(g')v = \chi(g')v \text{ for any } g' \in G' \,\}.$$

A vector $v \in V$ is called *smooth* if it lies in $V(U, 1)$ for some $U \in \underline{U}$. Let $(\pi, V)_0 = V_0$ denote the set of all smooth vectors, which is a G-subspace of V. A left (or right) module (π, V) is called *smooth* iff $V = V_0$. A left (resp. right) module (π, V) is called an ω-representation (resp. ω-antirepresentation) iff $V = V(Z, \omega)$.

Let (π, V) be a smooth ω-representation (resp. ω-antirepresentation). For $v \in V$, set

$$\tilde{v} := (g \longmapsto \pi(g)v) \in C(G, \omega; V).$$

For $T \in \underline{\mathcal{D}}(G)$, set

$$\pi(T)v := \langle T, \tilde{v} \rangle$$

Since $v \mapsto \tilde{v}$ is linear, $\pi(T)$ is an endomorphism of V, and the linear map $\pi \colon \underline{\mathcal{D}}(G) \to \mathrm{End}_{\mathbf{C}}(V)$ is defined.

Lemma 5. (i) $\pi \colon \underline{\mathcal{D}}(G) \to \mathrm{End}_{\mathbf{C}}(V)$ *is an algebra homomorphism (resp. antihomomorphism).*

(ii) *If $g \in G$, $\pi(T(g)) = \pi(g)$.*

Proof. (i) Let (π, V) be an ω-representation of G. By Lemma 4 (v),

$$\pi(T_1 * T_2)v = \langle T_1 * T_2, \tilde{v} \rangle = \langle T_1, F_1 \rangle$$

with $F_1 = (g_1 \mapsto \langle T_2, L(g_1)^{-1}\tilde{v} \rangle)$. Since $L(g_1)^{-1}\tilde{v} = (g \mapsto \pi(g_1 g)v) = \pi(g_1) \circ \tilde{v}$, by lemma 2, $\langle T_2, L(g_1)^{-1}\tilde{v} \rangle = \pi(g_1)\langle T_2, \tilde{v} \rangle = \pi(g_1)\pi(T_2)v$, hence $F_1 = \widetilde{\pi(T_2)}v$. Thus we have $\pi(T_1 * T_2)v = \langle T_1, F_1 \rangle = \pi(T_1)(\pi(T_2)v)$. The case of ω-antirepresentation is entirely similar.

(ii) $\pi(T(g)v) = \langle T(g), \tilde{v} \rangle = \tilde{v}(g) = \pi(g)v$.

§8. Regular representation.

By the left (resp. right) shift $L(g)$ (resp. $R(g)$), $\underline{\mathcal{D}}(G)$ is a left (resp. right) G-module. Since $R(z)T = L(z)T = \omega(z)T$ for $z \in Z$, $(L, \underline{\mathcal{D}}(G))$ (resp. $(R, \underline{\mathcal{D}}(G))$) is an ω-representation (resp. ω-antirepresentation) of G.

Lemma 6. (i) $(L, \underline{D}(G))_0 = (R, \underline{D}(G))_0$.

(ii) Set $\underline{S}(G) := (L, \underline{D}(G))_0$. As a **C**-vectorspace, $\underline{S}(G)$ is generated by T_x of (3^*) §5 for various $x \in G$, and $U \in \underline{U}$.

(iii) The pairing $\underline{S}(G) \times S(G, \omega) \to \mathbf{C}$ is non-degenerate.

(iv) If $T_0 \in \underline{S}(G)$ and $T \in \underline{D}(G)$, then

$$T * T_0 = L(T)T_0 \in \underline{S}(G),$$
$$T_0 * T = R(T)T_0 \in \underline{S}(G).$$

Consequently, $\underline{S}(G)$ is a two sided ideal of $\underline{D}(G)$.

(v) The algebra homomorphism $L: \underline{D}(G) \to \mathrm{End}_{\mathbf{C}}\, \underline{S}(G)$ is injective. In particular, $\underline{D}(G)$ is an associative algebra.

Proof. (i),(ii). If $T \in \underline{D}(G)$, $\operatorname{supp} T = K$, and $R(u)T = T$ for any $u \in U$, then $K = ZKU$, and $T \in r^*S^*(K, \omega) \simeq S^*(K, \omega) = \oplus S^*(ZxU, \omega)$, where x runs over a complete system of representatives of the double cosets $Z\backslash K/U$. Hence by (3^*) §3, $T = \sum c_x T_x$ with $c_x \in \mathbf{C}$, and $L(u)T = T$ for $u \in \cap x U x^{-1}$.

(iii) If f is a non-zero element of $S(G, \omega)$, there is some $x \in G$, and $U \in \underline{U}$, such that $f(x) \neq 0$, $f(xu) = f(x)$ for any $u \in U$, hence $\langle T_x, f \rangle \neq 0$.

(iv) By Lemma 4 (v), $\langle T * T_0, f \rangle = \langle T, F_1 \rangle$ with $F_1 = (g \mapsto \langle T_0, L(g)^{-1}f \rangle = \langle L(g)T_0, f \rangle)$. Set $A: \underline{S}(G) \to \mathbf{C}$, $T_0 \mapsto \langle T_0, f \rangle$, then $F_1 = A \circ \tilde{T}_0$, where $\tilde{T}_0 = (g \mapsto L(g)T_0)$. By Lemma 2, $\langle T, F_1 \rangle = A \langle T, \tilde{T}_0 \rangle = AL(T)T_0 = \langle L(T)T_0, f \rangle$. The other case is similar.

(v) By (iv), $L(T) = 0$ implies $T * T_0 = 0$ for any $T_0 \in \underline{S}(G)$. Then, for any $f \in S(G, \omega)$, by Lemma 4 (v), $0 = \langle T * T_0, f \rangle = \langle T_0, f_2 \rangle$ with $f_2 = (g \mapsto \langle T, R(g)f \rangle)$. Hence, by (iii), $f_2 = 0$, and $f_2(1) = \langle T, f \rangle = 0$, i.e., $T = 0$.

§9. Projection operator.

Let H be a closed subgroup of G containing Z, and its projection $p(H)$ be compact. Let $\epsilon: H \to \mathbf{C}^\times$ be a locally constant homomorphism which coincides with ω on Z. Consider the sequence:

$$S^*(\bar{H}, 1) \xrightarrow[\simeq]{} S^*(H, 1) \xrightarrow[\simeq]{(\epsilon^{-1}\cdot)^*} S^*(H, \omega) \xrightarrow{r^*} S^*(G, \omega).$$

Let $\mu_H \in S^*(\bar{H}, 1)$ denote the left (hence right) invariant Haar measure of $\bar{H}$ normalized as total volume 1, and let $\dot{\epsilon}$ denote its image in $S^*(G, \omega)$ by the above map, i.e.,

$$\langle \dot{\epsilon}, F \rangle = \langle \mu_H, \epsilon^{-1} \cdot (F_{|H}) \rangle \quad \text{for } F \in C(G, \omega; V).$$

Since $L(\bar{h})\mu_H = \mu_H$, $L(h)\epsilon^{-1} = \epsilon(h)\epsilon^{-1}$, by (6^*) §2, $L(h)\dot{\epsilon} = \epsilon(h)\dot{\epsilon}$.

Lemma 7. Let (π, V) be a smooth ω-representation or ω-antirepresentation of G.

(i) $\operatorname{Im} \pi(\acute{\epsilon}) \subset V(\pi, H, \epsilon)$.

(ii) $\pi(\acute{\epsilon})$ is the identity on $V(\pi, H, \epsilon)$.

Consequently, $\pi(\acute{\epsilon})$ is the projection operator to $V(\pi, H, \epsilon)$.

Proof. (i) By Lemma 6 (iv), $\pi(h)\pi(\acute{\epsilon})v = \pi(T(h) * \acute{\epsilon})v = \pi(L(h) * \acute{\epsilon})v = \epsilon(h)\pi(\acute{\epsilon})v$, hence $\pi(\acute{\epsilon})v \in V(H, \epsilon)$.

(ii) If $v \in V(H, \epsilon)$, then $\tilde{v}|_H = (h \mapsto \pi(h)v = \epsilon(h)v)$, and $\epsilon^{-1} \cdot (\tilde{v}|_H) = (\bar{h} \mapsto v) = 1 \otimes v \in S(\tilde{H}, 1, V)$. Hence $\pi(\acute{\epsilon})v = \langle \acute{\epsilon}, \tilde{v} \rangle = \langle \mu_H, \epsilon^{-1} \cdot (\tilde{v}|_H) \rangle = \langle \mu_H, 1 \otimes v \rangle = v$.

Now we take the smooth ω-representation $(L, \underline{S}(G))$ and ω-antirepresentation $(R, \underline{S}(G))$. By Lemma 6, 7, we have:

(1) $$\underline{S}(G)(L, H, \epsilon) = \acute{\epsilon} * \underline{S}(G).$$

(1') $$\underline{S}(G)(R, H, \epsilon) = \underline{S}(G) * \acute{\epsilon}.$$

Set $\underline{\mathcal{E}} = \underline{\mathcal{E}}(G, H, \epsilon) := \underline{S}(G)(L, H, \epsilon) \cap \underline{S}(G)(R, H, \epsilon)$. Since $L(g)$ and $R(g')$ commutes for g, $g' \in G$, we have

(2) $$\underline{\mathcal{E}}(G, H, \epsilon) = \acute{\epsilon} * \underline{S}(G) * \acute{\epsilon}.$$

If H' is a closed subgroup of G containing H with compact projection $p(H')$, and $\epsilon': H' \to \mathbf{C}^\times$ is a locally constant homomorphism coincides with ϵ on H, $V(\pi, H', \epsilon')$ is a subspace of $V(\pi, H, \epsilon)$. In particular, $\acute{\epsilon} * \acute{\epsilon}' * T_0 = \acute{\epsilon}' * T_0$ for any $\underline{S}(G)$, and, by Lemma 6 (v), we can conclude:

(3) $$\acute{\epsilon} * \acute{\epsilon}' = \acute{\epsilon}' * \acute{\epsilon} = \acute{\epsilon}'.$$

Taking (ZU, ν) of §3 as (H, ϵ), we have seen that $\{\acute{\nu}\}$ is a system of idempotents in $\underline{S}(G)$, indexed by the directe set $\underline{\mathcal{U}}$. Also note that from our definition we have:

(4) $$T_x \text{ in §3 } = L(x)\acute{\nu}.$$

§10. $\underline{\mathcal{D}}$-modules and $\underline{S}$-modules.

Since in this and the next section, we shall be concerned with purely ring theoretic feature of the theory where no group element will appear, to avoid the repeated use of dot notation, set $u := \acute{\nu}$. Thus we have an associative algebra $\underline{\mathcal{D}} = \underline{\mathcal{D}}(G)$, its two sided ideal $\underline{S} = \underline{S}(G)$, a system of idempotents $u = u(U)$ indexed by the directed set $\underline{\mathcal{U}}$, satisfying:

(1) $$U \subset U' \Longrightarrow u * u' = u' * u = u' \quad \text{(by (3) §9),}$$

(2) $$\underline{S} = \cup u * \underline{S} * u \quad \text{(by Lemma 6 (i) and (2) §9).}$$

For a left $\underline{\mathcal{D}}$-module (resp. $\underline{S}$-module) (π, V) and a subset Y of V, let $\pi(\underline{\mathcal{D}})Y$ (resp. $\pi(\underline{S})Y$) denote the set of all finite linear combinations $\sum \pi(T_i)v_i, v_i \in Y$, $T_i \in \underline{\mathcal{D}}$ (resp. $\underline{S}$).

If the $\mathcal{D}$-module (π, V) is obtained from a smooth ω-representation (π, V) of G as in §7, $V(\pi, U, 1) = V(\pi, ZU, \nu)$ equals to $\pi(u)V$ by Lemma 7, and we have

(3) $$V = \cup \pi(u)V,$$

(3') $$V = \pi(\underline{S})V.$$

If a $\mathcal{D}$ (or $\underline{S}$)-module (π, V) satisfies one of the above two mutually equivalent conditions (3) and (3'), by abuse of language, we call (π, V) to be a *smooth* $\mathcal{D}$ (or $\underline{S}$)-module. The category of smooth $\mathcal{D}$ (or $\underline{S}$)-module is a full subcategory of $\mathcal{D}$ (or $\underline{S}$)-modules.

Lemma 8. *The functor* $(\pi, V) \mapsto (\pi|_{\underline{S}}, V)$ *defines an isomorphism of the category of smooth $\mathcal{D}$-modules with the category of smooth S-modules. Namely:*

(i) *If* (π_i, V_i), $i = 1, 2$ *is smooth $\mathcal{D}$-module, then*

$$\mathrm{Hom}_{\underline{\mathcal{D}}}(V_1, V_2) = \mathrm{Hom}_{\underline{S}}(V_1, V_2).$$

(ii) *If* (π, V) *is a smooth $\underline{S}$-module, then there is a unique $\mathcal{D}$-module structure* (π', V) *on V such that* $\pi'|_{\underline{S}} = \pi$.

Proof. (i) If $A: V_1 \to V_2$ is a $\underline{S}$-morphism, $v \in V_1$, $T \in \mathcal{D}$, take u such that $\pi_1(u)v = v$, $\pi_1(u)Av = Av$, then $A\pi_1(T)v = A\pi_1(T)\pi_1(u)v = A\pi_1(T * u)v = \pi_2(T * u)Av = \pi_2(T)Av$, i.e., A is a $\mathcal{D}$-morphism.

(ii) For $v \in V$, $T \in \mathcal{D}$, take u such that $\pi(u)v = v$, and set $\pi'(T)v := \pi(T * u)v$. Since $\{u\}$ is directed, $\pi'(T)v$ is well defined and gives an action of $\mathcal{D}$ on V, obviously $\pi'|_{\underline{S}} = \pi$.

§11. $\mathcal{D}$-modules and $\underline{\mathcal{E}}$-modules.

Let h be an idempotent in $\mathcal{D}$, and set $\underline{\mathcal{E}} := h * \underline{S} * h$. (If we take the $\dot{\epsilon}$ of §9 as h, $\underline{\mathcal{E}} = \underline{\mathcal{E}}(G, H, \epsilon)$.) For any $\mathcal{D}$-module (π, V), set $V^h := \{ v \in V \mid \pi(h)v = v \}$, then we have:

(1) $$V^h = \pi(h)V.$$

Hence V^h is an $\underline{\mathcal{E}}$-module, and for any sub $\mathcal{D}$-module V',

(2) $$(V/V')^h = V^h/V'^h.$$

Let (π, V) be a smooth $\mathcal{D}$-module and W be a C-subspace:

(3) If $V^h \supset W \supset \pi(\underline{\mathcal{E}})W$, then $(\pi(\underline{S})W)^h = W$.

Indeed,

$$W = W^h \subset (\pi(\underline{S})W)^h = \pi(h)\pi(\underline{S})W = \pi(h)\pi(\underline{S})\pi(h)W = \pi(\underline{\mathcal{E}})W \subset W.$$

Lemma 9. *The functor* $V \mapsto V^h$ *defines an equivalence of the category of 'irreducible smooth $\underline{D}$-module V having non-zero V^h' with the category of 'non-zero irreducible $\underline{E}$-modules'. Namely:*

(i) *If* (π_i, V_i) *is an irreducible smooth $\underline{D}$-module and* $W_i = V_i^h \neq 0$ ($i = 1, 2$), *then the map* $A \mapsto A|_{W_1}$,

$$\mathrm{Hom}_{\underline{D}}(V_1, V_2) \longrightarrow \mathrm{Hom}_{\underline{E}}(W_1, W_2)$$

is a bijection.

(ii) *If* (ρ, W) *is a non-zero irreducible $\underline{E}$-module, then there is an irreducible smooth $\underline{D}$-module* (π, V) *such that* V^h *is isomorphic to* (ρ, W) *as $\underline{E}$-module.*

Proof. (i) If $A \neq 0$, then A is an isomorphism, hence its restriction to W_1 is not zero. To see the surjectivity, suppose a non-zero $\underline{E}$-morphism $B: W_1 \to W_2$ be given. Set:

$$(\pi, V) := (\pi_1 \oplus \pi_2, V_1 \oplus V_2), \quad p_i: V \longrightarrow V_i, \quad (v_1, v_2) \longmapsto v_i,$$

$$W := \{ (w, Bw) \mid w \in W_1 \} \subsetneq W_1 \oplus W_2 = V^h \subset V,$$

$$W' := \pi(\underline{S})W \quad \text{and} \quad p_i' = p_i|_{W'}: W' \longrightarrow V_i,$$

hence by (3),

(4) $$(W')^h = W \subsetneq W_1 \oplus W_2.$$

Since p_i' is a non-zero $\underline{D}$-morphism, it is surjective. If $\mathrm{Ker}\, p_i' = W' \cap \mathrm{Ker}\, p_i$ is not zero, we have $W' = V$, and $(W')^h = W_1 \oplus W_2$, contradicting (4). Hence p_i' is a $\underline{D}$-isomorphism, and $A = p_2' \circ (p_1'^{-1})$ is a $\underline{D}$-isomorphism of V_1 to V_2 and its restriction to W_1 coincides with B.

(ii) Take a non-zero vector w in W, and let M be its anihilator ideal in $\underline{E}$, $M = \{ T \in \underline{E} \mid \rho(T)w = 0 \}$. In the smooth $\underline{D}$-module $(L, \underline{S})$, set $V_1 = L(\underline{S})\underline{E}$, $V_2 = L(\underline{S})M$. Since $\underline{S}^h = h * \underline{S} \supset \underline{E}$, by (3)

(5) $$(V_1)^h = \underline{E}, \qquad (V_2)^h = M.$$

The quotient $V_3 = V_1/V_2$ is a smooth $\underline{D}$-module by the induced action $\bar{L}$ of L, and by (2), we have

(6) $$(V_3)^h = (V_1)^h/(V_2)^h = \underline{E}/M \simeq W.$$

The irreducibility of $(V_3)^h$ and definitions of V_i imply:

(7) If $(V_3)^h \ni \bar{T} \neq 0$, then $\bar{L}(\underline{S})\bar{T} = V_3$.

If V_4 is a $\underline{D}$-subspace of V_3, $(V_4)^h$ is either 0 or $(V_3)^h$. In the latter case, by (7), we have $V_4 = V_3$. Thus, if $V_4 \neq V_3$, we always have $(V_3/V_4)^h \simeq W$. Again by (7), V_3 is a finitely generated $\underline{D}$-module. Hence we can find some V_4 with the irreducible quotient $V = V_3/V_4$.

§12. Proof of Theorem 1.

Let $T(\ ):G \to \underline{D} = \underline{D}(G)$ denote the semi-group homomorphism $g \mapsto T(g) = (f \mapsto f(g))$.

Lemma 10. *The functor* $\pi \mapsto \pi \circ T(\ )$ *defines an isomorphism of the category of smooth* $\underline{D}(G)$-*modules with the category of smooth* ω-*representations of* G. *Namely:*

(i) *If* (π_i, V_i), $i = 1, 2$ *are smooth* $\underline{D}$-*modules, and a linear map* $A:V_1 \to V_2$ *satisfies* $\pi_2(T(g)) \circ A = A \circ \pi_1(T(g))$ *for any* $g \in G$, *then*

(1) $\pi_2(T) \circ A = A \circ \pi_1(T)$ *for any* $T \in \underline{D}$.

(ii) *If* (ρ, V) *is a smooth* ω-*representation of* G, *then there is a unique* $\underline{D}$-*module structure* (π, V) *on* V *such that* $\rho(g) = \pi(T(g))$ *for any* $g \in G$.

Proof. (i) By Lemma 8, it suffices to check (1) for $T \in \underline{S} = \underline{S}(G)$. For given $T \in \underline{S}$, and $v \in V_1$, if $U \in \underline{U}$ is small enough, $v \in V_1(ZU, \nu)$, $Av \in V_2(ZU, \nu)$, and, by Lemma 6 (ii), T is a linear combination of $T_x = L(x)u = L(T(x))u = T(x) * u$ with $u = \dot{\nu} \in \underline{S}$. Now $\pi_2(T_x)Av = \pi_2(T_x)\pi_2(u)Av = \pi_2(T(x))Av = A\pi_1(T(x))v = A\pi_1(T(x))\pi_1(u)v = A\pi_1(T_x)v$.

(ii) The existence of π is in Lemma 5. The uniqueness is obvious since $\pi(T)v$ for $T \in \underline{S}$ is determined by $\pi(T(x))v$ as in (i).

Since $V(H, \epsilon) = V^h$ by $h = \dot{\epsilon}$, Theorem 1 clearly follows from lemma 8, 9 and 10.

§13. Proof of Theorem 2.

Let $\mu \in S^*(\bar{G}, 1)$ be a left invariant positive Haar measure of $\bar{G}$, $\bar{\Delta}: G \to \mathbf{R}_+^\times$ be the modulus function, $L(\bar{g})\mu = \mu$, $R(\bar{g})\mu = \bar{\Delta}(g)^{-1}\mu$. Let τ be a topological antiautomorphism of G with the property (0) §0, $\iota: g \mapsto g^{-1}$ be the inverse map, and $\bar{\tau}$, $\bar{\iota}$ denote the antiautomorphism of $\bar{G}$ induced by τ, ι. As is well known and easily seen we have

(0) $\bar{\iota}\mu = (\bar{\Delta}^{-1} \cdot)^* \mu$.

Since $\bar{\tau}\bar{\iota} = \bar{\iota}\bar{\tau}$ is an automorphism of $\bar{G}$ we have

(1) $\bar{\tau}\bar{\iota}\mu = c\mu$ with some positive constant c.

By the non-degenerated bilinear map

$$C(G, \omega^{-1}) \times S(G, \omega) \longrightarrow S(G, 1) \longrightarrow S(\bar{G}, 1)$$

$$(\varphi, f) \longmapsto \varphi \circ f \longmapsto \overline{\varphi \circ f} = (\bar{g} \mapsto \varphi(g)f(g)),$$

each element φ of $C(G, \omega^{-1})$ defines an ω-distribution $T(\varphi) = (f \mapsto \langle \mu, \overline{\varphi \circ f} \rangle)$, then a linear injection,

(2) $T(\ ): C(G, \omega^{-1}) \longrightarrow S^*(G, \omega)$.

Set $\Delta = \bar{\Delta} \circ p \colon G \to \mathbf{R}_+^\times$. The following relations are obvious.

$$\text{(3)} \qquad T(L(g)\varphi) = L(g)T(\varphi),$$

$$\text{(3')} \qquad T(R(g)\varphi) = \Delta(g)R(g)T(\varphi),$$

$$\text{(4)} \qquad T(\nu^{-1}) = \mu(\overline{ZU})\check{\nu},$$

where ν is that of §5, and $\mu(\overline{ZU})$ is the volume of $\overline{ZU}$ by μ.

By Lemma 1, $S(G, \omega^{-1})$ is generated by $L(x)\nu^{-1}$, while by Lemma 6 (ii), $\underline{S}(G)$ is generated by $T_x = L(x)\check{\nu}$, hence by the above (3), (4), we conclude that $T(\ )$ induces the linear isomorphism

$$\text{(5)} \qquad T(\ ) \colon S(G, \omega^{-1}) \longrightarrow \underline{S}(G).$$

Lemma 11. *The locally constant homomorphism*

$$\tau\Delta^{-1} = (g \longmapsto \Delta^{-1}(\tau^{-1}(g)))$$

factors through $p \colon G \to \bar{G}$, hence by Lemma 4 (ii), defines an algebra automorphism $(\tau\Delta^{-1}\cdot)^*$ of $\underline{\mathcal{D}}(G)$. Set $\tau' = (\tau\Delta^{-1}\cdot)^* \circ \tau$, by Lemma 4 (iii), it is an antiautomorphism of $\underline{\mathcal{D}}(G)$. We have

$$\text{(6)} \qquad T(\tau\varphi) = c^{-1}\tau'T(\varphi) \quad \text{for } \varphi \in S(G, \omega^{-1}).$$

Proof.

$$I = \langle T(\tau\varphi), f \rangle = \left\langle \mu, \overline{(\tau\varphi) \cdot f} \right\rangle = \left\langle \mu, \check{\tau}\left(\overline{\varphi \cdot \tau^{-1}f}\right) \right\rangle = \left\langle \check{\tau}^{-1}\mu, \overline{\varphi \cdot \tau^{-1}f} \right\rangle.$$

Since, by (0), (1), $\check{\tau}^{-1}\mu = c^{-1}(\bar{\Delta}^{-1}\cdot)^*\mu$, $I = c^{-1}\left\langle \mu, \bar{\Delta}^{-1}\cdot\overline{\varphi \cdot \tau^{-1}f} \right\rangle$. Since $\bar{\Delta}^{-1}\cdot\left(\overline{\varphi \cdot \tau^{-1}f}\right) = \overline{\varphi \cdot \tau^{-1}(\tau\Delta^{-1}\cdot f)}$,

$$I = c^{-1}\left\langle T(\varphi), \tau^{-1}(\tau\Delta^{-1}\cdot f) \right\rangle = c^{-1}\left\langle \tau T(\varphi), \tau\Delta^{-1}\cdot f \right\rangle$$
$$= \left\langle c^{-1}((\tau\Delta^{-1}\cdot)^* \circ \tau)T(\varphi), f \right\rangle = \left\langle c^{-1}\tau'T(\varphi), f \right\rangle$$

as wanted.

We now prove Theorem 2. Let (H, ϵ, τ) be as in §0. Set $S(G, H, \epsilon^{-1}) := \{\varphi \in S(G, \omega^{-1}) \mid L(h)\varphi = R(h)\varphi = \epsilon(h)\varphi \text{ for } h \in H\}$. Since $\Delta = 1$ on H, by (3), (3'), the assumption (1) §0 implies that the map $T(\ )$ of (5) induces the bijection

$$T(\ ) \colon S(G, H, \epsilon^{-1}) \longrightarrow \underline{\mathcal{E}}(G, H, \epsilon).$$

The assumption (2) §0 implies $c = 1$. The assumption (3) §0 implies $\tau\varphi = \varphi$ for $\varphi \in S(G, H, \epsilon^{-1})$. Thus τ is the identity map on $S(G, H, \epsilon^{-1})$, hence by (6), the antiautomorphism τ' is the identity map on $\underline{\mathcal{E}}$, so $\underline{\mathcal{E}}$ is commutative.

References

[1] I. N. Bernshtein-A. V. Zelevinskii, Representations of the group GL(n, F) where F is a non-archimedean local field, Russian Math. Surveys **31:3** (1976) 1–68.

[2] W. Casselmann, Introduction to the theory of admissible representations of p-adic reductive groups, preprint.

[3] H. Hijikata, Any irreducible smooth GL$_2$-module is multiplicity free for any anisotropic torus, preprint.

Hiroaki HIJIKATA

Kyoto University
Kyoto 606
Japan

Algebraic Geometry and Commutative Algebra
in Honor of Masayoshi NAGATA
pp. 141–153 (1987)

The Local Cohomology Groups
of an Affine Semigroup Ring

Masa-Nori ISHIDA

Introduction.

A commutative semigroup ring $k[S]$ over a field k is said to be an *affine semigroup ring* if $k[S]$ is an integral domain of finite type over k. This is equivalent to the condition that S is finitely generated and is contained in a free $\mathbf{Z}$-module M of finite rank. An affine semigroup ring $k[S]$ has a natural structure of an M-graded ring with respect to the free $\mathbf{Z}$-module M.

In this paper, we define the complex $I^\bullet(S)$ which represents, in the derived category, the dualizing complex of the affine semigroup ring $k[S]$. This complex $I^\bullet(S)$ consists of M-graded $k[S]$-modules $I^p(S)$ for $-\dim k[S] \le p \le 0$ with M-homogeneous coboundary homomorphisms of degree zero. Furthermore, each $I^p(S)$ behaves as an injective $k[S]$-module for M-homogeneous homomorphisms of M-graded $k[S]$-modules. This complex is described in terms of the cone π associated to S in the dual vector space $N_{\mathbf{R}}$ of $M_{\mathbf{R}} = M \otimes_{\mathbf{Z}} \mathbf{R}$ similarly as in [I1].

When k^* is the set of units in $k[S]$, the affine semigroup ring has a unique M-homogeneous maximal ideal $P(\pi)$. Then the local cohomology groups $H^p_{P(\pi)}(k[S])$ are equal to the cohomology groups of the complex $L^\bullet(S)$ defined as the "k-dual" of $I^\bullet(S)$. When $k[S]$ satisfies Serre's condition (S_2), this description of the local cohomology groups is equivalent, by the Alexander duality, to the topological description of them due to Trung and Hoa [TH].

§1. Affine semigroup rings and the associated cones.

Let M be a free $\mathbf{Z}$-module of rank $r \ge 0$. For a field k of an arbitrary characteristic, we denote by $k[M]$ the k-vector space

$$\oplus_{m \in M} k e(m)$$

with the free basis $\{\, e(m) \,;\, m \in M \,\}$. $k[M]$ has the structure of a commutative k-algebra if we define $e(0) = 1$ and $e(m)e(m') = e(m + m')$ for $m,\, m' \in M$.

Received January 6, 1987.

If we fix a basis $\{m_1, \ldots, m_r\}$ of M, then

$$k[M] = k[X_1, \ldots, X_r, X_1^{-1}, \ldots, X_r^{-1}]$$

for $X_1 = e(m_1)$, $\ldots$, $X_r = e(m_r)$. In particular, $k[M]$ is an integral domain of dimension r.

Let S be a finitely generated subsemigroup of M with $0 \in S$. Then

$$k[S] := \oplus_{m \in S} k e(m)$$

is a k-subalgebra of $k[M]$. Let $\{m_1', \ldots, m_s'\} \subset S$ be a set of generators of the semigroup S, i.e., any element m of S is written as $m = a_1 m_1' + \cdots + a_s m_s'$ for non-negative integers $a_1, \ldots, a_s$. Then the ring $k[S]$ is generated by $e(m_1')$, $\ldots$, $e(m_s')$ over k.

It is clear that the quotient field of $k[S]$ is equal to that of $k[M]$ if and only if S generates M as a group. In this paper, we assume that S satisfies this condition. This is not an essential restriction for the study of the semigroup ring $k[S]$, since we can replace M by its subgroup generated by S.

By the above assumption, the closed convex cone $\mathbf{R}_0 S$ spanned by S in $M_{\mathbf{R}} = M \otimes_{\mathbf{Z}} \mathbf{R}$ is of dimension r. Let N be the dual $\mathbf{Z}$-module of M. The natural pairing $\langle\ ,\ \rangle: M \times N \to \mathbf{Z}$ is extended to a nondegenerate bilinear map $\langle\ ,\ \rangle: M_{\mathbf{R}} \times N_{\mathbf{R}} \to \mathbf{R}$ for $N_{\mathbf{R}} = N \otimes_{\mathbf{Z}} \mathbf{R}$. We denote by π the dual cone $\{\, a \in N_{\mathbf{R}} \ ; \ \langle x, a \rangle \geq 0$ for every $x \in \mathbf{R}_0 S \,\}$. For the set $\{m_1', \ldots, m_s'\}$ of generators of S, π is equal to $\{\, a \in N_{\mathbf{R}} \ ; \ \langle m_1', a\rangle, \ldots, \langle m_s', a\rangle \geq 0 \,\}$. Since $\mathbf{R}_0 S$ is of maximal dimension by assumption, π is a strongly convex rational polyhedral cone. $\mathbf{R}_0 S$ is equal to the dual cone $\pi^{\vee}$ of π.

Proposition 1.1. *The semigroup ring $k[M \cap \pi^{\vee}]$ is the normalization of $k[S]$ in its quotient field.*

Proof. By our assumption, $k[M \cap \pi^{\vee}]$ is contained in the quotient field of $k[S]$. It is known that $k[M \cap \pi^{\vee}]$ is normal [TE, Chap. 1, §1, Lemma 1]. Hence it is sufficient to show that $k[M \cap \pi^{\vee}]$ is integral over $k[S]$. Let m be an element of $M \cap \pi^{\vee}$. By Carathéodory's theorem [Gr], m is written as $a_1 m_1 + \cdots + a_r m_r$ for some linearly independent elements $m_1, \ldots, m_r$ of S and non-negative real numbers $a_1, \ldots, a_r$. Since m is in M, $a_1, \ldots, a_r$ are rational numbers. Hence there exists a positive integer d with $dm \in S$. This implies $e(m)^d = e(dm) \in k[S]$.   q.e.d.

We denote $\bar{S} = M \cap \pi^{\vee}$. The proof of the above proposition says that $\bar{S} = \{\, m \in M \ ; \ dm \in S$ for a positive integer $d \,\}$. Since $\bar{S}$ is also finitely generated [TE, Chap. 1, §1, Lemma 2], there exists a positive integer ℓ such that $\ell \bar{S} \subset S$.

A subset E of M is said to be *S-closed* if $E + S = E$. If E is S-closed, then

$$k[E] := \oplus_{m \in E} k e(m)$$

is a $k[S]$-submodule of $k[M]$. A subset $E \subset M$ is said to be *weakly S-closed* if the set-theoretic difference $(E + S) \setminus E$ is S-closed.

Lemma 1.2. *A subset E of M is weakly S-closed if and only if $-E$ is weakly S-closed.*

Proof. By definition, E is not weakly S-closed if and only if there exist $m_0 \in E$ and $m_1, m_2 \in S$ such that $m_0 + m_1 + m_2 \in E$ and $m_0 + m_1 \notin E$. If we set $m_0' = -(m_0 + m_1 + m_2) \in -E$, then the last condition implies that $-E$ is not weakly S-closed. Actually, $m_0' + m_2 = -(m_0 + m_1) \notin -E$ and $m_0' + m_2 + m_1 = -m_0 \in -E$. The converse is proved similarly. q.e.d.

Assume that E is weakly S-closed. Since $E+S$ is obviously S-closed, $k[E+S]$ and $k[(E+S) \setminus E]$ are $k[S]$-modules. We denote by $k[E]$ the quotient $k[S]$-module

$$k[E + S]/k[(E + S) \setminus E].$$

For $m \in E$, we also denote by $e(m)$ its image in $k[E]$. Hence we write also

$$k[E] = \oplus_{m \in E} ke(m).$$

However this is not a $k[S]$-submodule of $k[M]$ in general.

We denote by $\Gamma(\pi)$, or simply Γ, the set of faces of π. The zero cone $\mathbf{0} = \{0\}$ and π are elements of Γ. For each $\rho \in \Gamma$, $\pi^\vee \cap \rho^\perp$ is a face of $\pi^\vee$. Actually, $\{\pi^\vee \cap \rho^\perp \; ; \; \rho \in \Gamma\}$ is the set of faces of $\pi^\vee$, and $\dim \rho + \dim(\pi^\vee \cap \rho^\perp) = r$ for every $\rho \in \Gamma$ [MO, Proposition 3.1]. For a face σ of π or $\pi^\vee$, we denote by rel. int σ the *relative interior of* σ, i.e., the interior of σ in the linear subspace of $N_\mathbf{R}$ or $M_\mathbf{R}$ generated by σ. We write int $\pi^\vee := $ rel. int $\pi^\vee$ since $\dim \pi^\vee = r$.

For each $\rho \in \Gamma$, we set $S_\rho = S \cap \rho^\perp$. Then $S \setminus S_\rho$ is S-closed and S_ρ is weakly S-closed. We denote by $P(\rho)$ the ideal $k[S \setminus S_\rho]$ of $k[S]$. We also set $\bar{S}_\rho = \bar{S} \cap \rho^\perp$ and $\bar{P}(\rho) = k[\bar{S} \setminus \bar{S}_\rho]$.

Proposition 1.3. $\{P(\rho) \; ; \; \rho \in \Gamma\}$ *is the set of M-homogeneous prime ideals of $k[S]$.*

Proof. It is known that this is true when $S = \bar{S}$, i.e., in the case of normal semigroup rings [MO, Remark in (5.3)]. In particular, $P(\rho) = \bar{P}(\rho) \cap k[S]$ is a prime ideal of $k[S]$ for every ρ. Conversely, let E be an S-closed subset of S such that $P = k[E]$ is a prime ideal. Then we have $P \cap k[\ell\bar{S}] = k[\ell(\bar{S} \setminus \bar{S}_\rho)]$ for an element $\rho \in \Gamma$, since $k[\ell\bar{S}]$ is a subring of $k[S]$ and is normal. This means $E \cap \ell\bar{S} = \ell\bar{S} \setminus \ell\bar{S}_\rho$. If $m \in S_\rho$, then m is not in E, since $m + (\ell - 1)m = \ell m$ is outside E and E is S-closed. If $m \in S \setminus S_\rho$, then $m \in E$, since $\ell m \in E$ and $k[E]$ is a prime ideal. Hence $E = S \setminus S_\rho$. q.e.d.

For a nonempty subset E of S, there exists a unique maximal element $\rho \in \Gamma$ such that E is contained in the face $\pi^\vee \cap \rho^\perp$ of $\pi^\vee$, i.e., $\pi^\vee \cap \rho^\perp$ is the minimal face of $\pi^\vee$ which contains E. If, furthermore, E is a subsemigroup of S, then we see easily that E contains an element in rel. $\operatorname{int}(\pi^\vee \cap \rho^\perp)$.

For a subsemigroup U of S, we denote by $[U]$ the subset $\{\, e(m) \; ; \; m \in U \,\}$ of $k[S]$. Clearly, $[U]$ is a multiplicatively closed subset of $k[S]$. It is easy to see that the localization $[U]^{-1}k[S]$ is equal to the semigroup ring $k[S-U]$, where $S-U = \{\, m_1 - m_2 \; ; \; m_1 \in S, m_2 \in U \,\}$.

Lemma 1.4. *Let U be a nonempty subsemigroup of S. If ρ is the maximal element of Γ with $U \subset \pi^\vee \cap \rho^\perp$, then $S - U$ is equal to $S - S_\rho$.*

Proof. Since $U \subset S_\rho$, $S - U$ is contained in $S - S_\rho$. Hence it is sufficient to show that $-S_\rho \subset S - U$. Let m be an element of S_ρ. By the maximality of ρ, U contains an element m_0 in rel. $\operatorname{int}(\pi^\vee \cap \rho^\perp)$. Hence, for a sufficiently large integer d, we have $dm_0 \in m + \pi^\vee \cap \rho^\perp$. Hence $\ell(dm_0 - m) \in \ell M \cap (\pi^\vee \cap \rho^\perp) = \ell \bar{S}_\rho \subset S_\rho$. If we set $m_1 = \ell(dm_0 - m) \in S$, then we have $-m = (m_1 + (\ell-1)m) - d\ell m_0$, which is an element of $S - U$. q.e.d.

For an ideal A of $k[S]$, we denote

$$T(A) := \{\, m \in S \; ; \; e(m) \in A \,\}.$$

The following lemma is essentially a special case of [GW, Prop.1.2.2].

Lemma 1.5. *Let P be a prime ideal of $k[S]$. Then $k[T(P)]$ is the largest M-homogeneous prime ideal of $k[S]$ which is contained in P.*

Proof. Let $U = S \setminus T(P)$. Since P is a prime ideal, U is a subsemigroup of S. By Lemma 1.4, $S - U = S - S_\rho$ for the maximal $\rho \in \Gamma$ with $U \subset \rho^\perp$. Since $S_\rho - S_\rho \subset S - S_\rho$, every $e(m) \in [S_\rho]$ is a unit of $[U]^{-1}k[S]$. Hence $T(P) \cap S_\rho = \emptyset$ and $T(P) = S \setminus S_\rho$. $k[T(P)] = P(\rho)$ is the largest, since $[S_\rho]$ is outside P.

q.e.d.

For an M-graded $k[S]$-modules $L = \oplus_{m \in M} L_m$ and K, we denote

$$\operatorname{GHom}_{k[S]}(L, K) = \oplus_{m \in M} \operatorname{GHom}_{k[S]}(L, K)_m$$

and

$$\operatorname{GHom}_k(L, k) = \oplus_{m \in M} \operatorname{Hom}_k(L_m, k),$$

where $\operatorname{GHom}_{k[S]}(L, K)_m$ is the k-module of M-homogeneous homomorphisms of degree m. These are $k[S]$-modules and $\operatorname{GHom}_{k[S]}(L, K)$ is isomorphic to the usual $\operatorname{Hom}_{k[S]}(L, K)$ if L is finitely generated as a $k[S]$-module (cf. [GW, §1]).

$\operatorname{GHom}_k(L, k)$ is called the graded k-dual of L which was introduced and studied in more general situation in [GW, Chap. 2]. For a weakly S-closed subset E of M, we see easily that $\operatorname{GHom}_k(k[E], k)$ is equal to the $k[S]$-module $k[-E]$.

Proposition 1.6. *Let L be a finitely generated M-graded $k[S]$-module, and let ρ be an element of Γ. Then we have*

$$\mathrm{Ext}^p_{k[S]}(L, k[-(S - S_\rho)]) = 0 \quad \text{for } p > 0.$$

Proof. Let

$$\cdots \longrightarrow F^{-2} \longrightarrow F^{-1} \longrightarrow F^0 \longrightarrow L \longrightarrow 0$$

be a resolution of L by M-graded free $k[S]$-modules of finite ranks with M-homogeneous coboundary homomoprhisms. Then we have

$$
\begin{aligned}
\mathrm{Hom}_{k[S]}&(F^\bullet, k[-(S - S_\rho)]) \\
&= \mathrm{GHom}_{k[S]}(F^\bullet, k[-(S - S_\rho)]) \\
&= \mathrm{GHom}_{k[S]}(F^\bullet, \mathrm{GHom}_k(k[S - S_\rho], k)) \\
&= \mathrm{GHom}_k(F^\bullet \otimes_{k[S]} [S_\rho]^{-1}k[S], k).
\end{aligned}
$$

We are done, since the functors $\otimes_{k[S]}[S_\rho]^{-1}k[S]$ and $\mathrm{GHom}_k(\ , k)$ are exact.
$$\text{q.e.d.}$$

By replacing S and S_ρ by $\ell\bar{S}$ and $\ell\bar{S}_\rho$, respectively, in the above proof, we have the following:

Lemma 1.7. *Let L be a finitely generated M-graded $k[\ell\bar{S}]$-module. Then, for each $\rho \in \Gamma$, we have*

$$\mathrm{Ext}^p_{k[\ell\bar{S}]}(L, k[-\ell(\bar{S} - \bar{S}_\rho)]) = 0$$

for $p > 0$.

§2. Complexes associated to an affine semigroup ring.

A subset Φ of $\Gamma = \Gamma(\pi)$ is said to be (1) *star closed* if $\sigma \in \Phi$, $\tau \in \Gamma$ and $\sigma \prec \tau$ imply $\tau \in \Phi$, (2) *star open* if $\tau \in \Phi$, $\sigma \in \Gamma$ and $\sigma \prec \tau$ imply $\sigma \in \Phi$ and (3) *locally star closed* if $\sigma, \rho \in \Phi$, $\tau \in \Gamma$ and $\sigma \prec \tau \prec \rho$ imply $\tau \in \Phi$. By a *subcomplex of* Γ, we mean a star open subset of Γ.

We recall the definitions in [I2, §1] of the complexes $C^\bullet(\Phi, k_{1,0})$ and $C^\bullet(\Phi, k_{0,1})$ defined for a locally star closed subset Φ of Γ.

For each integer p, we set $\Gamma(p) = \{\, \rho \in \Gamma \,;\, \mathrm{codim}\,\rho = p \,\}$, where $\mathrm{codim}\,\rho = r - \dim \rho$. $\Gamma(p)$ is empty for $p > r$ or $p < \mathrm{codim}\,\pi$. For each $\rho \in \Gamma(p)$, we set $M[\rho] = M \cap \rho^\perp$ and $\mathbf{Z}(\rho) = \Lambda^p M[\rho]$. Since $M[\rho]$ is a free $\mathbf{Z}$-module of rank p, $\mathbf{Z}(\rho)$ is isomorphic to $\mathbf{Z}$.

For $\sigma \in \Gamma(p)$ and $\tau \in \Gamma(p - 1)$ with $\sigma \prec \tau$, the isomorphism $q_{\sigma/\tau} \colon \mathbf{Z}(\sigma) \to \mathbf{Z}(\tau)$ is defined as follows. We choose an element $n_1 \in N$ so that $\langle m, n_1 \rangle = 0$ for every $m \in M[\tau]$ and $\langle m_0, n_1 \rangle = 1$ for some $m_0 \in M[\sigma] \cap \tau^\vee$. We define

$q_{\sigma/\tau}(m_1 \wedge \cdots \wedge m_p) = \langle m_1, n_1 \rangle m_2 \wedge \cdots \wedge m_p$ if $m_1 \in M[\sigma]$ and $m_2, \ldots, m_p \in M[\tau]$.

For each σ in Γ, we denote $\mathbf{Z}(\sigma)^* := \mathrm{Hom}_{\mathbf{Z}}(\mathbf{Z}(\sigma), \mathbf{Z})$. For $\sigma \in \Gamma(p)$ and $\tau \in \Gamma(p-1)$ with $\sigma \prec \tau$, we set $q_{\tau/\sigma}^* = (-1)^p (q_{\sigma/\tau})^* : \mathbf{Z}(\tau)^* \to \mathbf{Z}(\sigma)^*$.

Let Φ be a locally star closed subset of Γ. We set $\Phi(p) = \Phi \cap \Gamma(p)$ for each integer p. The complex $C^\bullet(\Phi, k_{1,0})$ is defined by

$$C^p(\Phi, k_{1,0}) = \oplus_{\rho \in \Phi(-p)} k(\rho),$$

where $k(\rho) = k \otimes_{\mathbf{Z}} \mathbf{Z}(\rho)$. The coboundary homomorphism

$$d^p : C^p(\Phi, k_{1,0}) \longrightarrow C^{p+1}(\Phi, k_{1,0})$$

consists of $1_k \otimes q_{\sigma/\tau}$'s for all pairs of $\sigma \in \Phi(-p)$ and $\tau \in \Phi(-p-1)$ with $\sigma \prec \tau$. Similarly, the complex $C^\bullet(\Phi, k_{0,1})$ is defined by

$$C^p(\Phi, k_{0,1}) = \oplus_{\rho \in \Phi(p)} k(\rho)^*,$$

where $k(\rho)^* = k \otimes_{\mathbf{Z}} \mathbf{Z}(\rho)^*$. The coboundary homomorphisms are defined similarly in terms of $1_k \otimes q_{\tau/\sigma}^*$'s.

Now assume Φ is a nonempty subcomplex of Γ. In particular, Φ contains the zero cone $\mathbf{0}$. Let H be the hyperplane $\{ a \in N_{\mathbf{R}} ; \langle x_0, a \rangle = 1 \}$ for a fixed element x_0 in int $\pi^\vee$. Let $d = \dim \pi$. Then $P_\pi = \pi \cap H$ is a $(d-1)$-dimensional bounded convex polytope and $\{ P_\rho = \rho \cap H ; \rho \in \Gamma \setminus \{\mathbf{0}\} \}$ is the set of faces of P_π. We define the *geometric realization* of Φ as the base space $|R(\Phi)|$ of the polyhedral complex

$$R(\Phi) = \{ P_\rho ; \rho \in \Phi \setminus \{\mathbf{0}\} \}.$$

Proposition 2.1. *Let Φ be a nonempty subcomplex of Γ and let $|R(\Phi)|$ be its geometric realization. For each integer p, we have isomorphisms*

$$\mathrm{H}^p(C^\bullet(\Phi, k_{1,0})) \simeq \tilde{\mathrm{H}}^{r+p-1}(|R(\Phi)|, k)$$

and

$$\mathrm{H}^p(C^\bullet(\Phi, k_{0,1})) \simeq \tilde{\mathrm{H}}_{r-p-1}(|R(\Phi)|, k),$$

where the right hand sides are the reduced cohomology groups and homology groups of the topological space $|R(\Phi)|$.

Proof. Let $C^\bullet(\Phi, k_{1,0})[-r+1]$ and $C^\bullet(\Phi, k_{0,1})[r-1]$ be the complexes $C^\bullet(\Phi, k_{1,0})$ and $C^\bullet(\Phi, k_{0,1})$ with the degree shifted to the right and left, respectively, by $r-1$. Since $\dim P_\rho = r - 1 - \mathrm{codim}\,\rho$ for every $\rho \in \Gamma \setminus \{\mathbf{0}\}$, we see that the cochain and chain complexes which give the reduced cohomology and homology groups of the polyhedral complex $R(\Phi)$ are equal to $C^\bullet(\Phi, k_{1,0})[-r+1]$ and $C_\bullet(\Phi, k_{0,1})[r-1]$, respectively, where the chain complex $C_\bullet(\Phi, k_{0,1})[r-1]$ is defined by $C_i(\Phi, k_{0,1})[r-1] = C^{-i}(\Phi, k_{0,1})[r-1]$ with the boundary maps equal to the corresponding coboundary maps of $C^\bullet(\Phi, k_{0,1})[r-1]$. We get the proposition by taking the cohomology and the homology groups of these complexes.   q.e.d.

For the semigroup S, we define the complexes $I^\bullet(S)$ and $L^\bullet(S)$ of M-graded $k[S]$-modules as follows:
For each integer p, we set

$$I^p(S) = \oplus_{\rho \in \Gamma(-p)} k[-(S - S_\rho)] \otimes_k k(\rho)$$

and

$$L^p(S) = \oplus_{\rho \in \Gamma(p)} k[S - S_\rho] \otimes_k k(\rho)^*.$$

Here note that $S - S_\rho$ is S-closed and $-(S - S_\rho)$ is weakly S-closed in M. If $\sigma, \tau \in \Gamma$ and $\sigma \prec \tau$, then $S - S_\tau \subset S - S_\sigma$ since $S_\tau \subset S_\sigma$. Since $k[S - S_\tau] = [S_\tau]^{-1} k[S]$ and $k[S - S_\sigma] = [S_\sigma]^{-1} k[S]$, there exists a natural inclusion $k[S - S_\tau] \to k[S - S_\sigma]$ which we denote by $i^*_{\tau/\sigma}$. We denote by $i_{\sigma/\tau}: k[-(S - S_\sigma)] \to k[-(S - S_\tau)]$ the dual surjection with respect to the functor $\mathrm{GHom}_k(\ , k)$. The component of the coboundary homomorphism $d^p: I^p(S) \to I^{p+1}(S)$ with respect to the direct summands $k[-(S - S_\sigma)] \otimes_k k(\sigma)$ and $k[-(S - S_\tau)] \otimes_k k(\tau)$ for $\sigma \in \Gamma(-p)$ and $\tau \in \Gamma(-p - 1)$ is the zero map if σ is not a face of τ and is $i_{\sigma/\tau} \otimes q_{\sigma/\tau}$ if $\sigma \prec \tau$. Similarly, the coboundary homomorphism $d^p: L^p(S) \to L^{p+1}(S)$ consists of $i^*_{\tau/\sigma} \otimes q^*_{\tau/\sigma}$'s for $\sigma \in \Gamma(p)$ and $\tau \in \Gamma(p-1)$ with $\sigma \prec \tau$.

§3. The dualizing complex and the local cohomology groups.

Since $k[\bar{S}]$ is a normal semigroup ring, $\mathrm{Spec}\, k[\bar{S}]$ is an affine torus embedding. It is known (cf. [Ho]) that $k[\bar{S}]$ is a Cohen-Macaulay ring and the canonical module $\omega_{k[\bar{S}]}$ of $k[\bar{S}]$ is equal to $k[\bar{S} \cap (\mathrm{int}\, \pi^\vee)]\omega_0$, where ω_0 is the r-form

$$\frac{de(m_1)}{e(m_1)} \wedge \cdots \wedge \frac{de(m_r)}{e(m_r)}$$

for a basis $\{m_1, \ldots, m_r\}$ of M (cf. [TE, Chap. 1, Th. 14], [I2, §2]). We regard $\omega_{k[\bar{S}]}$ as an M-graded $k[\bar{S}]$-module with $\deg(\omega_0) = 0$.
We define an injective $k[\bar{S}]$-homomorphism

$$\lambda: \omega_{k[\bar{S}]} \longrightarrow I^{-r}(\bar{S}) = k[M] \otimes_{\mathbf{Z}} \Lambda^r M$$

by $\lambda(f\omega_0) = f \otimes (m_1 \wedge \cdots \wedge m_r)$. It is easy to see that λ does not depend on the choice of the basis $\{m_1, \ldots, m_r\}$. The canonical module $\omega_{k[\bar{S}]}[r]$, with the degree shift to the left by r as a complex, represents the normalized dualizing complex of $k[\bar{S}]$ in the derived category $D_c^+(k[\bar{S}])$, where we say a dualizing complex is *normalized* if its localization at every maximal ideal is the normalized one in the sense of [RD, Chap. 5, §6].

Proposition 3.1. *The homomorphism of complexes*

$$\omega_{k[\bar{S}]}[r] \longrightarrow I^\bullet(\bar{S})$$

induced by λ *is a quasi-isomorphism. In particular,* $I^\bullet(\bar{S})$ *also represents the normalized dualizing complex of* $k[\bar{S}]$.

Proof. It is sufficient to show that $\mathrm{H}^i(I^\bullet(\bar{S})) = 0$ for $i \neq -r$ and that λ induces an isomorphism $\omega_k[\bar{S}] \simeq \mathrm{H}^{-r}(I^\bullet(\bar{S}))$.

Since each $I^i(\bar{S})$ and $\omega_k[\bar{S}]$ are M-graded modules and since the coboundary maps in $I^\bullet(\bar{S})$ and the map λ are M-homogeneous homomorphisms of degree zero, we can decompose the complex and λ into the direct sum of M-homogeneous components associated to each degree in M. For each $m \in M$, $(\bar{S} - \bar{S}_\rho)$ contains m if and only if $m \in \rho^\vee$, since $\bar{S} - \bar{S}_\rho = M \cap \rho^\vee$ by [O, Prop. 1.3]. Hence the component $I^\bullet(\bar{S})_m$ is equal to $C^\bullet(\Gamma_{-m}, k_{1,0})$ where $\Gamma_x = \{ \rho \in \Gamma ; \langle x, a \rangle \geq 0 \text{ for every } a \in \rho \}$ for $x \in M_\mathbf{R}$.

Assume $m \notin M \cap (\mathrm{int}\, \pi^\vee)$. Then by [I1, Prop. 2.3], Γ_{-m} is homologically trivial. Hence $\mathrm{H}^i(I^\bullet(\bar{S}))_m = 0$ for every i. Since $\omega_{k[\bar{S}]} = k[M \cap \mathrm{int}\, \pi^\vee]\omega_0$, the m-component of $\omega_k[\bar{S}]$ is also zero for $m \notin M \cap \mathrm{int}\, \pi^\vee$. Hence the assertion is true in this case.

When $m \in M \cap (\mathrm{int}\, \pi^\vee)$, we have $\Gamma_{-m} = \{0\}$ since $\langle m, a \rangle > 0$ for $a \in \pi \setminus \{0\}$. Hence $\mathrm{H}^i(I^\bullet(\bar{S}))_m = 0$ for $i \neq -r$ is obvious in this case. On the other hand, the m-component of $\omega_k[\bar{S}]$ is $ke(m)\omega_0$. Since λ is injective, we see the homomorphism $(\omega_k[\bar{S}])_m \to \mathrm{H}^{-r}(I^\bullet(\bar{S}))_m$ is isomorphic.   q.e.d.

Theorem 3.2. *The complex* $I^\bullet(S)$ *represents the dualizing complex of the affine semigroup ring* $k[S]$.

Proof. As in Section 1, we take a positive integer ℓ such that $\ell\bar{S}$ is contained in S. By Proposition 3.1, $I^\bullet(\ell\bar{S})$ represents the dualizing complex of the normal semigroup ring $k[\ell\bar{S}]$. Since $k[\bar{S}]$ is finite over $k[\ell\bar{S}]$, so is $k[S]$ over $k[\ell\bar{S}]$. Hence by [RD, Chap. 3, §6], the dualizing complex of $k[S]$ is equal to

$$\mathbf{R}\,\mathrm{Hom}_{k[\ell\bar{S}]}(k[S], I^\bullet(\ell\bar{S})).$$

Since $k[S]$ is a finitely generated M-graded $k[\ell\bar{S}]$-module, we have

$$\mathrm{Ext}^p_{k[\ell\bar{S}]}(k[S], I^i(\ell\bar{S})) = \{0\} \quad \text{for } p > 0$$

and $i \in \mathbf{Z}$ by Lemma 1.7. Hence the dualizing complex is represented by the complex of $k[S]$-modules

$$\mathrm{Hom}_{k[\ell\bar{S}]}(k[S], I^\bullet(\ell\bar{S})).$$

For each integer i, we have

$$\mathrm{Hom}_{k[\ell\bar{S}]}(k[S], I^i(\ell\bar{S}))$$
$$= \mathrm{Hom}_{k[\ell\bar{S}]}(k[S], \oplus_{\rho\in\Gamma(-i)}k[-\ell(\bar{S}-\bar{S}_\rho)])$$
$$= \oplus_{\rho\in\Gamma(-i)}\mathrm{GHom}_k(k[S]\otimes_{k[\ell\bar{S}]}[\ell\bar{S}_\rho]^{-1}k[\ell\bar{S}], k)$$
$$= \oplus_{\rho\in\Gamma(-i)}\mathrm{GHom}_k([\ell\bar{S}_\rho]^{-1}k[S], k)$$
$$= \oplus_{\rho\in\Gamma(-i)}\mathrm{GHom}_k(k[S-S_\rho], k) \qquad \text{(by Lemma 1.4)}$$
$$= \oplus_{\rho\in\Gamma(-i)}k[-(S-S_\rho)]$$
$$= I^i(S).$$

We see easily that the induced coboundary maps are equal to those of $I^\bullet(S)$.

q.e.d.

If $\dim \pi = r$, then $S_\pi = \{0\}$ and $P(\pi) = k[S\setminus\{0\}]$ is a maximal ideal of $k[S]$.
In the rest of this section, we assume $\dim \pi = r$. In this case, $k[-S]$ is the injective hull of the $k[S]_{P(\pi)}$-module $k[S]/P(\pi)$ by [LC, Prop. 4.12 and Example 1 to it].

Theorem 3.3. *The local cohomology group* $\mathrm{H}^p_{P(\pi)}(k[S])$ *is isomorphic to* $\mathrm{H}^p(L^\bullet(S))$ *for every* $p \in \mathbf{Z}$.

Proof. By the local duality [RD, Chap. 5, Cor. 6.3], we have

$$\mathrm{H}^p_{P(\pi)}(k[S]) = \mathrm{Hom}_{k[S]}(\mathrm{H}^{-p}(I^\bullet(S)), k[-S]),$$

since $I^\bullet(S)$ is a dualizing complex of $k[S]$. Since $\mathrm{H}^{-p}(I^\bullet(S))$ is a $k[S]$-module of finite type, this is isomorphic to $\mathrm{GHom}_{k[S]}(\mathrm{H}^{-p}(I^\bullet(S)), k[-S])$. Since $k[-S] = \mathrm{GHom}_k(k[S], k)$ and since the functor $\mathrm{GHom}_k(\ ,k)$ is exact, we have

$$\mathrm{GHom}_{k[S]}(\mathrm{H}^{-p}(I^\bullet(S)), k[-S])$$
$$\simeq \mathrm{GHom}_k(\mathrm{H}^{-p}(I^\bullet(S)), k)$$
$$\simeq \mathrm{H}^p(\mathrm{GHom}_k(I^\bullet(S), k))$$
$$\simeq \mathrm{H}^p(L^\bullet(S)).$$

q.e.d.

Since $L^\bullet(S)$ is a complex of M-graded modules and since the coboundary maps are M-homogeneous of degree zero, we can consider the M-homogeneous component $L^\bullet(S)_m$ of $L^\bullet(S)$ for each $m \in M$.

For each $m \in M$, we set

$$\Gamma(S,m) = \{\, \rho \in \Gamma \,;\, S \cap (m+S_\rho) \neq \emptyset \,\}.$$

Then the zero cone $\mathbf{0}$ is in $\Gamma(S,m)$ for every $m \in M$, since $S-S = M$. Since $\sigma \prec \tau$ implies $S_\tau \subset S_\sigma$, $\Gamma(S,m)$ is a subcomplex of Γ. Since $m \in S - S_\rho$ if and only if $\rho \in \Gamma(S, m)$, the M-homogeneous component $L^\bullet(S)_m$ of $L^\bullet(S)$ is isomorphic to $C^\bullet(\Gamma(S, m), k_{0,1})$.

By Proposition 2.1 and Theorem 3.3, the m-components of the local cohomology groups $H^p_{P(\pi)}(k[S])$ are described by the reduced homology groups of the geometric realization $|R(\Gamma(S, m))|$ as follows.

Theorem 3.4. $H^p_{P(\pi)}(k[S])_m = \tilde{H}_{r-p-1}(|R(\Gamma(S, m))|, k)$ *for every* $m \in M$.

§4. Serre's condition (S_2).

Let $\{\gamma_1, \ldots, \gamma_t\}$ be the set of one-dimensional faces of π. We denote $S_i = S_{\gamma_i}$ and $P_i = P(\gamma_i)$ for each $i = 1, \ldots, t$.

Lemma 4.1. *The equality*

$$S = \bigcap_{i=1}^{t}(S - S_i)$$

holds if and only if the ring $k[S]$ satisfies Serre's condition (S_2)[EGA4, 5.7].

Proof. Since $k[S]$ is an integral domain, $k[S]$ satisfies (S_2) if and only if

$$k[S] = \bigcap_{\text{ht } P=1} k[S]_P$$

by [EGA4, Prop. 5.10.14], where P runs over all prime ideals of height one. Hence, it is sufficient to show that

$$(1) \qquad \bigcap_{\text{ht } P=1} k[S]_P = \bigcap_{i=1}^{t}[S_i]^{-1}k[S].$$

Let $P \subset k[S]$ be a prime ideal of height one. By Lemma 1.5, $k[T(P)]$ is a prime ideal contained in P. Hence $T(P) = \emptyset$ and $k[M] \subset k[S]_P$ if P is not equal to one of P_i, since $P_1, \ldots, P_t$ are prime ideals of height one. Thus we see that the right hand side of (1) is contained in the left hand side.

Since $k[\bar{S}]$ is normal and finite over $k[S]$, the left hand side of (1) is contained in $k[\bar{S}]$. The opposite inclusion holds since

$$k[S]_{P_i} \cap k[\bar{S}] \subset k[S]_{P_i} \cap k[M] = [S_i]^{-1}k[S]$$

for every i.   *q.e.d.*

For each element ρ of Γ, we define

$$J(\rho) := \{\, i \,;\, \gamma_i \prec \rho \,\}.$$

Lemma 4.2. *If $k[S]$ satisfies* (S_2), *then*

$$S - S_\rho = \bigcap_{i \in J(\rho)} (S - S_i)$$

for every $\rho \in \Gamma$.

Proof. Since $k[S]$ satisfies (S_2), so does the localization $[S_\rho]^{-1}k[S]$. We have $S - S_i = (S - S_\rho) - (S_i - S_\rho)$ for every $i \in J(\rho)$, since $S_\rho \subset S_i$. Since ρ is the associated cone of the semigroup $S - S_\rho$, we have the equality in the lemma by Lemma 4.1 applied to the semigroup ring $k[S - S_\rho]$. *q.e.d.*

We also define

$$I(m) := \{\, i \; ; \; m \in S - S_i \,\}$$

for each $m \in M$. By Lemma 4.2, we have the following:

Proposition 4.3. *Assume $k[S]$ satisfies* (S_2). *Let m be an element of M. Then we have*

$$\Gamma(S, m) = \{\, \rho \in \Gamma \; ; \; J(\rho) \subset I(m) \,\}.$$

Let P_π be the $(r - 1)$-dimensional convex polytope which appeared in Section 2. We denote by v_i the intersection point of P_π and γ_i. Then $\{v_1, \ldots, v_t\}$ is the set of vertices of P_π. For a subset I of $\{1, \ldots, t\}$, we denote $V(I) := \{\, v_i \; ; \; i \in I \,\}$. For each element m of M, Proposition 4.3 implies that the polyhedral complex $R(\Gamma(S, m))$ is equal to the set of the faces of P_π with their vertices in $V(I(m))$.

Goto-Watanabe [GW, Cor. 3.3.7] and Trung-Hoa [TH, Th. 3.3] defined a complex which gives the local cohomology groups $\mathrm{H}^p_{P(\pi)}(k[S])$ when $k[S]$ satisfies the condition (S_2). By using this complex, Trung and Hoa proved the following:

Theorem 4.4. ([TH, Th. 3.4].) *Let $I(m)^c = \{1, \ldots, t\} \setminus I(m)$ and $\Delta(m)$ the simplicial complex*

$$\{\, I \subset I(m)^c \; ; \; I \neq \emptyset \text{ and } V(I) \subset P_\rho \text{ for some } \rho \neq \pi \,\}$$

for each m in M. If $k[S]$ satisfies (S_2), *then*

$$\mathrm{H}^p_{P(\pi)}(k[S])_m = \tilde{\mathrm{H}}^{p-2}(\Delta(m), k).$$

For each proper face P_ρ of P_π, we denote by $\mathrm{Star}(P_\rho)$ the union of the relative interiors of the proper faces of P_π which contains P_ρ. Clearly, $\mathrm{Star}(P_\rho)$ is a contractible open subset of the boundary ∂P_π if $\rho \neq \pi$. When $P_\rho = \{v_i\}$, we denote $U_i := \mathrm{Star}(P_\rho)$. For a subset I of $\{1, \ldots, t\}$, we have $\bigcap_{i \in I} U_i = \mathrm{Star}(P_\rho)$, where P_ρ is the minimal face of P_π which contains $V(I)$.

Let

$$U(m) := \bigcup_{i \in I(m)^c} U_i.$$

Then we see easily that

$$U(m) = \partial P_\pi \setminus |R(\Gamma(S, m))|.$$

The complex $C^\bullet(\{\, U_i \; ; \; i \in I(m)^c \,\}, k)$ associated to the covering of $U(m)$ is equal to the cochain complex $C^\bullet(\Delta(m), k)$, since $\cap_{i \in I} U_i$ for $I \subset I(m)^c$ is not empty if and only if $I \in \Delta(m)$. By the contractibility of $\mathrm{Star}(P_\rho)$, we have $H^p(\mathrm{Star}(P_\rho), k) = 0$ for $p > 0$. Hence $H^p(\Delta(m), k)$ is isomorphic to the cohomology group $H^p(U(m), k)$ of the open subset $U(m) \subset \partial P_\pi$ by [Go, Th. 5.4.1]. It follows that the reduced cohomology groups $\tilde{H}^p(\Delta(m), k)$ and $\tilde{H}^p(U(m), k)$ are also isomorphic for every $p \in \mathbf{Z}$. Hence Theorem 4.4 is equivalent to

$$H^p_{P(\pi)}(k[S])_m \simeq \tilde{H}^{p-2}(U(m), k).$$

Since ∂P_π is homeomorphic to the $(r - 2)$-sphere and $U(m) = \partial P_\pi \setminus |R(\Gamma(S, m))|$, this theorem is equivalent to our Theorem 3.4 by the Alexander duality Theorem.

References

[EGA4]   A. Grothendieck and J. Dieudonné, Élément de géométrie algébrique, Inst. Hautes Études Sci. Publ. Math. 20, 24, 28, 32 (1964–1967).

[Go]   R. Godement, Topologie Algébrique et Théorie des Faisceaux, Hermann, Paris (1958).

[Gr]   B. Grünbaum, Convex Polytopes, Interscience, New York, 1967.

[GW]   S. Goto and K. Watanabe, On graded rings, II, Tokyo J. Math. 1 (1978), 237–261.

[Ho]   M. Hochster, Rings of invariants of tori, Cohen-Macaulay rings generated by monomials, and polytopes, Ann. of Math. 96 (1972), 318–337.

[I1]   M.-N. Ishida, Torus embeddings and dualizing complexes, Tohoku Math. J. 32 (1980), 111–146.

[I2]   M.-N. Ishida, Torus embeddings and de Rham complexes, to appear in Commutative Algebra and Combinatorics, (M. Nagata ed.), Adv. studies in Pure Math. 11, Kinokuniya, Tokyo and North-Holland, Amsterdam, New York, Oxford, 1987.

[LC]   A. Grothendieck, Local cohomology, Lecture Notes in Math. 41, Springer-Verlag, Berlin, Heidelberg, New York, 1967.

[MO]   T. Oda, Lectures on torus embeddings and applications (Based on joint work with K. Miyake), Tata Inst. Fund. Res. 58, Bombay, Springer-Verlag, Berlin, Heidelberg, New York, 1978.

[O]   T. Oda, Convex Bodies and Algebraic Geometry, Springer-Verlag, Berlin, Heidelberg, New York, Tokyo, 1987.

[RD]   R. Hartshorne, Residues and duality, Lecture Notes in Math. 20, Springer-Verlag, Berlin, Heidelberg, New York, 1966.

[TE]   G. Kempf, F. Knudsen, D. Mumford and B. Saint-Donat, Toroidal embeddings I, Lecture Notes in Math. 339, Springer-Verlag, Berlin, Heidelberg, New York, 1973.

[TH]   N. V. Trung and L. T. Hoa, Affine semigroups and Cohen-Macaulay rings generated by monomials, Trans. AMS, 298(1986), 145-167.

Masa-Nori ISHIDA

Mathematical Institute
Tohoku University
Sendai, 980
Japan

Algebraic Geometry and Commutative Algebra
in Honor of Masayoshi NAGATA
pp. 155–182 (1987)

Quaternion Extensions

Christian U. JENSEN* and Noriko YUI†

Abstract.

Quaternion extensions over fields of characteristic different from 2 are investigated. We have a constructive version of Witt's theorem on embeddability of biquadratic extensions into a quaternion extension. We determine the number of quaternion extensions for some special classes of fields.

Definitions, Notations and Some Necessary Facts.

1. By the *quaternion* group, denoted by Q_8, we mean the group $\{\pm 1, \pm i,$ $\pm j, \pm k\}$ with the relations $i^2 = j^2 = -1$ and $ij = -ji = k$. As a permutation group of degree 8, Q_8 is a subgroup of the alternating group A_8.

We denote by Z_4, $Z_{2^n} (n \in \mathbf{N})$, V_4 and D_4, respectively, the cyclic group of order 4, the cyclic group of order 2^n, the Klein four group $Z_2 \times Z_2$ and the dihedral group of order 8.

2. Let G be a group (finite in this paper). Let K be a field. A field N over K is called a G-*extension* over K if N is a Galois extension over K with the Galois group $\mathrm{Gal}(N/K) \cong G$. In particular, if $G \cong Q_8$, N is called a *quaternion extension* over K.

3. A field K is called a *Hilbertian* field, if the Hilbert irreducibility theorem holds true for K. (Cf. Kuyk and Lenstra [5] and Lang [6].)

4. The *Pythagoras number* of a field K, denoted by $d(K)$, is the smallest number $n \in \mathbf{N}$ such that any sum of squares in K can be expressed as a sum of n squares in K.

5. The *level* of a field K, denoted by $\mathrm{level}(K)$, is the smallest number $n \in \mathbf{N}$ such that -1 is expressed as a sum of n squares in K. We say that $\mathrm{level}(K) = \infty$ if -1 is not expressible as a sum of squares in K. In particular, $\mathrm{level}(K) = 1$ if and only if -1 is a square in K. In general, $\mathrm{level}(K) = \infty$ or 2^t $(t \in \mathbf{N})$.

*The first author was a visiting professor at the University of Toronto in the fall of 1985, partially supported by the Department of Mathematics, University of Toronto, and by the Danish Natural Science Research Council.

†The second author was partially supported by the Natural Sciences and Engineering Research Council of Canada (NSERC).

Received January 21, 1987.

6. Let K be a field. Let a, $b \in K$. The symbol $\left(\dfrac{a, b}{K} \right)$ stands for the quaternion algebra $\{ \alpha 1 + \beta i + \gamma j + \delta k \mid \alpha, \beta, \gamma, \delta \in K \}$ subject to the relations $i^2 = a$, $j^2 = b$ and $ij = -ji$.

7. We denote by $\mathrm{Br}(K)$ the Brauer group of a field K, which is the group of similarity classes of central division algebras over K under the tensor operation $\otimes$. $\mathrm{Br}_2(K)$ denotes the 2-torsion subgroup of $\mathrm{Br}(K)$.

8. Let K be a field. Let $K^* = K \setminus \{0\}$, and let K^2 (resp. K^{*2}) denote the set $\{ \alpha^2 \mid \alpha \in K \text{ (resp. } K^*) \}$. K^*/K^{*2} is the subgroup of square classes of K, and its order is equal to 2^m for some $m \in \mathbf{N}$.

9. Let K be a field. For elements a, $b \in K^*$, we write $a \underset{(2)}{=} b$ if $a/b \in K^{*2}$.

10. For a pair $(G; K)$ where G is a group and K a field, $\nu(G; K)$ denotes the number of K-isomorphism classes of Galois extensions over K with Galois group $\cong G$.

11. For a set S (resp. a group G), $|S|$ (resp. $|G|$) denotes its cardinality (resp. order).

12. If $f(X) \in K[X]$ is an irreducible polynomial over a field K, $\mathrm{Gal}(f/K)$ stands for the Galois group of the splitting field of $f(X)$ over K. If N is a Galois extension over a field K, $\mathrm{Gal}(N/K)$ denotes the Galois group of N over K.

Introduction.

The purpose of this paper is to give a unified treatment of quaternion extensions over an arbitrary ground field K of characteristic different from 2.

The first example of a quaternion extension over $\mathbf{Q}$ was given by Dedekind [2] in 1886. Later contributions to the theory of quaternion fields were given by Bucht [1], Mertens [9], Reichardt [10], and Witt [12], among others.

Apparently, Bucht's paper has been overlooked by many later authors. In this paper, we, in particular, point out that Bucht's theorem (Theorem (I.2.1)), proved in 1910, is actually equivalent to Witt's theorem (Theorem (I.1.1)), proved in 1936, in the case that the level of K is greater than 2 (Theorem (I.2.2)). We consider separately the case that the level of K is less than or equal to 2, which Bucht and Mertens did not take into account (Theorem (I.3.1) and Theorem (I.3.3)).

We generalize the result of Reichardt (in the case that the ground field K is $\mathbf{Q}$) to an arbitrary Hilbertian field K concerning the embeddability of quadratic extensions over K into quaternion extensions over K (Theorem (II.1.1)). Characterization theorems for local ground fields K that admit quaternion extensions are proved (Theorem (III.3.1) and Theorem (II.3.2)), and the number $\nu(Q_8; K)$ is determined explicitly (Theorem (II.3.5) and Corollary (II.3.6), and Theorem (II.3.7)).

Subsequently, we give results connecting the realizability of Q_8 as a Galois group to other 2-groups, namely, V_4, Z_4, Z_{2^n} $(n \in \mathbf{N})$ and D_4. We compute the number $\nu(Q_8; K)$ for some special fields, and, in particular, we show that for any field K with level$(K) = 1$, $\nu(D_4; K) = 3\nu(Q_8; K)$ (Theorem (III.3.4)), and that for any field K with level$(K) = 2$, $\nu(Q_8; K) \leq \nu(D_4; K) \leq 3\nu(Q_8; K)$ (Theorem (III.3.5)). The range for $\nu(Q_8; K)$ is also discussed (Remark (III.3.8)).

We finish the paper by presenting "versal" families of generic polynomials with Galois group Q_8 ((IV.1), (IV.2) and Theorem (IV.3)).

The paper is organized as follows.

I. Quaternion extensions and quadratic forms.

Let K be a field of characteristic different from 2. Throughout this section this assumption remains in force.

I.1. Witt's Theorem.

Witt [12] has given a characterization of quaternion extensions over K in terms of quadratic forms of rank 3. Witt's theorem is formulated as follows.

(I.1.1) **Theorem.** *Let K be a field and let u, $v \in K \setminus K^2$ be such that $uv \notin K^2$. Then the following statements are equivalent:*
  (i) *The biquadratic extension $L = K(\sqrt{u}, \sqrt{v})$ of K admits a quaternion extension over K.*

(ii)   *The quadratic form* $Q(X_1, X_2, X_3) = uX_1^2 + vX_2^2 + \dfrac{1}{uv}X_3^2$ *is equivalent to the diagonal quadratic form* $Y_1^2 + Y_2^2 + Y_3^2$, *with an equivalence given by*

$$X_i = \sum_{i,j=1}^{3} p_{ij}Y_j, \qquad P = (p_{ij}) \quad \text{with } \det(P) = 1.$$

When this is the case, the quaternion extensions over K that contain $K(\sqrt{u}, \sqrt{v})$ *are given by*

$$K\left(\sqrt{r(1 + p_{11}\sqrt{u} + p_{22}\sqrt{v} + p_{33}(1/\sqrt{uv}))}\right) \quad \text{with } r \in K^*.$$

(There is a misprint in Witt's paper [12] where $1 +$ is missing in the field expression.)

Witt's original proof is very condensed. We include a proof of Witt's theorem here, as we believe that our proof is more accessible to the general reader.

We first need the following elementary fact, whose proof is left to the reader as an exercise.

(I.1.2) Lemma. *Let a_1, a_2, $a_3 \in K \setminus K^2$ be such that $a_1 a_2 a_3 = 1$. Let $L = K(\sqrt{a_1}, \sqrt{a_2}, \sqrt{a_3})$ be a V_4-extension over K. Let $\alpha \in L \setminus L^2$. Then $L(\sqrt{\alpha})$ is a quaternion extension over K if and only if*

$$\begin{aligned}
\alpha\sigma(\alpha) &= \alpha_1^2 a_2 && \text{with some } \alpha_1 \in K(\sqrt{a_1}), \\
\alpha\tau(\alpha) &= \alpha_2^2 a_3 && \text{with some } \alpha_2 \in K(\sqrt{a_2}), \\
\alpha\sigma\tau(\alpha) &= \alpha_3^2 a_1 && \text{with some } \alpha_3 \in K(\sqrt{a_3}),
\end{aligned}$$

where σ and τ are generators of the Galois group $\mathrm{Gal}(L/K) \cong V_4$ which are defined by

$$\begin{aligned}
\sigma(\sqrt{a_1}) &= \sqrt{a_1}, & \sigma(\sqrt{a_2}) &= -\sqrt{a_2}, & \sigma(\sqrt{a_3}) &= -\sqrt{a_3}, \\
\tau(\sqrt{a_1}) &= -\sqrt{a_1}, & \tau(\sqrt{a_2}) &= \sqrt{a_2}, & \tau(\sqrt{a_3}) &= -\sqrt{a_3}.
\end{aligned}$$

Reichardt [10] has obtained a very simple criterion for embeddability of V_4-extensions over K into quaternion extensions over K, which becomes quite useful in the subsequent discussions, and is formulated as follows.

(I.1.3) Proposition. *Let a_1, a_2, $a_3 \in K \setminus K^2$ be such that $a_1 a_2 a_3 = 1$. Let $L = K(\sqrt{a_1}, \sqrt{a_2}, \sqrt{a_3})$ be a V_4-extension over K. Then L is embeddable into a quaternion extension over K if and only if there exists $b \in K^*$ such that the system of equations*

$$\begin{cases}
b = a_2 x^2 - a_3 y^2 \\
b = a_3 z^2 - a_1 w^2 \\
b = a_1 r^2 - a_2 s^2
\end{cases}$$

has a solution $(x, y, z, w, r, s) \in K \times K \times K \times K \times K \times K$.

Proof. This can be deduced easily from Lemma (I.1.2).

(I.1.4) Proof of the implication (ii) $\Rightarrow$ (i) of Theorem (I.1.1).

From the hypothesis, there exists a unimodular matrix $P = (p_{ij})$ with $1 \leq i, j \leq 3$ such that $P^*DP = E$ where E denotes the identity matrix of rank 3,

$$D = \begin{pmatrix} u & 0 & 0 \\ 0 & v & 0 \\ 0 & 0 & \dfrac{1}{uv} \end{pmatrix}$$ and P^* the transpose of P. P^* is of the form $P^{-1}D^{-1}$.

This yields the system of identities

(I.1.5)
$$\begin{cases} up_{11}^2 + vp_{21}^2 + \dfrac{1}{uv}p_{31}^2 = 1 \\[2mm] up_{12}^2 + vp_{22}^2 + \dfrac{1}{uv}p_{32}^2 = 1 \\[2mm] up_{13}^2 + vp_{23}^2 + \dfrac{1}{uv}p_{33}^2 = 1 \\[2mm] p_{11}^2 + p_{12}^2 + p_{13}^2 = \dfrac{1}{u} \\[2mm] p_{21}^2 + p_{22}^2 + p_{23}^2 = \dfrac{1}{v} \\[2mm] p_{31}^2 + p_{32}^2 + p_{33}^2 = uv \\[2mm] up_{11} = p_{22}p_{33} - p_{32}p_{23} \\[1mm] vp_{22} = p_{33}p_{11} - p_{31}p_{13} \\[1mm] \dfrac{1}{uv}p_{33} = p_{11}p_{22} - p_{21}p_{12}. \end{cases}$$

Now take $\alpha = 1 + p_{11}\sqrt{u} + p_{22}\sqrt{v} + p_{33}(1/\sqrt{uv})$. Then $\alpha \in L \setminus L^2$. We claim that $L(\sqrt{\alpha})$ is a quaternion extension over K that contains $L = K(\sqrt{u}, \sqrt{v})$. For this, we show that α satisfies the conditions of Lemma (I.1.2) with $a_1 = u$, $a_2 = v$ and $a_3 = \dfrac{1}{uv}$. We give a proof only of the first identity in Lemma (I.1.2); the remaining cases can be proved similarly. By using the identities in (I.1.5), we get

$$\alpha\sigma(\alpha) = (1 + p_{11}\sqrt{u})^2 - (p_{22}\sqrt{v} + p_{33}(\frac{1}{\sqrt{uv}}))^2$$

$$= 1 + p_{11}^2 u - p_{22}^2 v - p_{33}^2 \frac{1}{uv} + 2(p_{11}\sqrt{u} - p_{22}p_{33}\sqrt{v}\frac{1}{\sqrt{uv}})$$

$$= (p_{23}^2 v + p_{32}^2 \frac{1}{uv}) - 2p_{23}p_{32}\sqrt{v}\frac{1}{\sqrt{uv}}$$

$$= (p_{23}\sqrt{v} - p_{32}\frac{1}{\sqrt{uv}})^2$$

$$= v(p_{23} - p_{32}\frac{1}{v\sqrt{u}})^2 \quad \text{with } p_{23} - p_{32}\frac{1}{v\sqrt{u}} \in K(\sqrt{u}).$$

Therefore, $L = K(\sqrt{u}, \sqrt{v}) = K(\sqrt{u}, \sqrt{v}, 1/\sqrt{uv})$ admits a quaternion extension over K.

In order to prove the implication (i) $\Rightarrow$ (ii) of Theorem (I.1.1), we need the following facts on quaternion algebras over K.

Let a, $b \in K^*$. By the symbol (a, b) we mean the element of order 2 in the Brauer group $\mathrm{Br}(K)$ of K defined by the quaternion algebra $\left(\dfrac{a, b}{K}\right)$. Then the following assertions hold true. Proofs can be found in T. Y. Lam [7], in particular, in Chapters III and IV.

(I.1.6) Proposition. (a) *For a, b, a', $b' \in K^*$, we have $(a, b) = (a', b')$ if and only if the two quadratic forms $-aX_1^2 - bX_2^2 + abX_3^2$ and $-a'X_1^2 - b'X_2^2 + a'b'X_3^2$ are equivalent.*

(b) *For a, b, $c \in K^*$, we have*

$$(a, b)(a, c) = (a, bc), \quad (a, a) = (-1, a) \quad \text{and} \quad (a, -a) = 1.$$

(c) *For a, $b \in K^*$, the following statements are all equivalent:*

(i) $(a, b) = 1$,

(ii) $Z^2 - aX^2 - bY^2 = 0$ *has a non-trivial solution* $(X, Y, Z) \in K \times K \times K$,

(iii) *a is a norm from $K(\sqrt{b})$ to K, that is, $a = x^2 - by^2$ for suitable elements x, $y \in K^*$.*

(I.1.7) Proof of the implication (i) $\Rightarrow$ (ii) of Theorem (I.1.1).

From the assumption, the biquadratic field $L = K(\sqrt{u}, \sqrt{v}, 1/\sqrt{uv})$ is embeddable into a quaternion extension over K. So the system of equations of Proposition (I.1.3) is solvable. This gives rise to the identities

$$bu = u^2 r^2 - \frac{1}{uv}(suv)^2 \in \mathrm{Norm}_{K(1/\sqrt{uv})/K}(K(1/\sqrt{uv})^*)$$

$$bv = v^2 x^2 - u(y/u)^2 \in \mathrm{Norm}_{K(\sqrt{u})/K}(K(\sqrt{u})^*)$$

$$b\frac{1}{uv} = (\frac{1}{uv})^2 z^2 - v(w/v)^2 \in \mathrm{Norm}_{K(\sqrt{v})/K}(K(\sqrt{v})^*)$$

which imply by Proposition (I.1.6)(c) that in $\mathrm{Br}(K)$,

$$\left(bu, \frac{1}{uv}\right) = 1, \quad (bv, u) = 1 \quad \text{and} \quad \left(b\frac{1}{uv}, v\right) = 1.$$

With further computations using the rules of Proposition (I.1.6)(b), we can conclude that

$$(-u, -v) = (-1, -1) \quad \text{in } \mathrm{Br}(K).$$

Therefore, by Proposition (I.1.6)(a), we get the required equivalence:

$$uX_1^2 + vX_2^2 + \frac{1}{uv}X_3^2 \sim Y_1^2 + Y_2^2 + Y_3^2.$$

Next we recall that Witt's construction of quaternion extensions over K is "versal" in the following sense.

(I.1.8) Proposition (Witt [12] and Reichardt [10]). *Let K be a field and let a_1, a_2, $a_3 \in K \setminus K^2$ be such that $a_1 a_2 a_3 = 1$. Let $\theta = c + \sqrt{a_1} + \sqrt{a_2} + \sqrt{a_3}$ with $c \in K$. Suppose that $K(\sqrt{\theta})$ is a quaternion extension over K. Then any quaternion extension over K that contains a_1 and a_2 is of the form $K(\sqrt{r\theta})$ with some $r \in K^*$.*

Furthermore, for θ given as above, $K(\sqrt{r_1\theta})$ and $K(\sqrt{r_2\theta})$, $r_1, r_2 \in K^*$ define the same field over K if and only if $r_1/r_2 \in L^{*2}$ where $L = K(\sqrt{a_1}, \sqrt{a_2}, \sqrt{a_3})$. Consequently, quaternion extensions over K of this form are parameterized by $K^*/(K^* \cap L^{*2})$.

I.2. Bucht's parametrization (level$(K) > 2$).

Bucht [1] and Mertens [9] (with less encompassing results) gave constructions and characterizations of quaternion extensions over K. Their argument assumed that the ground field K has level$(K) > 2$, although this was not mentioned in the articles. In the section I.3 below, we consider separately the case that level$(K) \leq 2$.

We state Bucht's theorem in the following form.

(I.2.1) Theorem (Bucht). *Let k be a field with level$(K) > 2$. Let u, v be quadratically independent elements of K. Then the biquadratic extension $L = K(\sqrt{u}, \sqrt{v})$ can be embedded into a quaternion extension over K if and only if there exist parameters λ, η, ρ in K such that u and v have one of the following parametrical representations:*

(a) *Generic case:*

$$u \underset{(2)}{=} \frac{1 + \lambda^2 + \lambda^2\eta^2}{1 + \rho^2 + \lambda^2\rho^2} \underset{(2)}{=} (1 + \rho^2 + \lambda^2\rho^2)(1 + \lambda^2 + \lambda^2\eta^2)$$

$$v \underset{(2)}{=} \frac{1 + \lambda^2 + \lambda^2\eta^2}{1 + \eta^2 + \eta^2\rho^2} \underset{(2)}{=} (1 + \eta^2 + \eta^2\rho^2)(1 + \lambda^2 + \lambda^2\eta^2)$$

$$uv \underset{(2)}{=} \frac{(1 + \lambda^2 + \lambda^2\eta^2)^2}{(1 + \rho^2 + \lambda^2\rho^2)(1 + \eta^2 + \eta^2\rho^2)}$$

$$\underset{(2)}{=} (1 + \rho^2 + \lambda^2\rho^2)(1 + \eta^2 + \eta^2\rho^2).$$

(b)   *Degenerate case*:

$$u \underset{(2)}{=} \frac{1+\lambda^2+\lambda^2\eta^2}{\rho^2+\lambda^2\rho^2} \underset{(2)}{=} (1+\lambda^2+\lambda^2\eta^2)(1+\lambda^2)$$

$$v \underset{(2)}{=} \frac{1+\lambda^2+\lambda^2\eta^2}{\eta^2\rho^2} \underset{(2)}{=} 1+\lambda^2+\lambda^2\eta^2 \underset{(2)}{=} \frac{1}{1+\lambda^2+\lambda^2\eta^2}$$

$$uv \underset{(2)}{=} \frac{(1+\lambda^2+\lambda^2\eta^2)^2}{\eta^2\rho^4(1+\lambda^2)} \underset{(2)}{=} 1+\lambda^2 \underset{(2)}{=} \frac{1}{1+\lambda^2}$$

(I.2.2) Theorem. *Let K be a field with* level(K) > 2. *Then Bucht's theorem* (I.2.1) *is equivalent to Witt's theorem* (I.1.1).

The following proof of Bucht's theorem based on Witt's theorem can immediately be inverted, thereby showing (I.2.2).

(I.2.3) Proof of the "if" part of Bucht's theorem.

For this purpose, we exhibit a unimodular matrix $P = (p_{ij})$ with $1 \le i, j \le 3$, giving the equivalence of the quadratic form $uX_1^2 + vX_2^2 + \frac{1}{uv}X_3^2$ to the diagonal quafratic form $Y_1^2 + Y_2^2 + Y_3^2$.

(a) Generic case: $P = (p_{ij})$ where p_{ij} are explicitly given as follows.

$$p_{11} = \frac{1-\lambda\eta\rho}{1+\lambda^2+\lambda^2\eta^2}$$

$$p_{12} = \frac{\lambda(1-\lambda\eta\rho)}{1+\lambda^2+\lambda^2\eta^2}$$

$$p_{13} = \frac{\lambda\eta+\rho(1+\lambda^2)}{1+\lambda^2+\lambda^2\eta^2}$$

$$p_{21} = -\frac{\eta\rho+\lambda(1+\eta^2)}{1+\lambda^2+\lambda^2\eta^2}$$

$$p_{22} = \frac{1-\lambda\eta\rho}{1+\lambda^2+\lambda^2\eta^2}$$

$$p_{23} = \frac{\eta(1-\lambda\eta\rho)}{1+\lambda^2+\lambda^2\eta^2}$$

$$p_{31} = -\rho\frac{(1-\lambda\eta\rho)(1+\lambda^2+\lambda^2\eta^2)}{(1+\rho^2+\lambda^2\rho^2)(1+\eta^2+\eta^2\rho^2)}$$

$$p_{32} = -\frac{\{\lambda\rho+\eta(1+\rho^2)\}(1+\lambda^2+\lambda^2\eta^2)}{(1+\rho^2+\lambda^2\rho^2)(1+\eta^2+\eta^2\rho^2)}$$

$$p_{33} = \frac{(1-\lambda\eta\rho)(1+\lambda^2+\lambda^2\eta^2)}{(1+\rho^2+\lambda^2\rho^2)(1+\eta^2+\eta^2\rho^2)}.$$

(b) Degenerate case: $P = (p_{ij})$ where p_{ij} are explicitly given as follows.

$$p_{11} = \frac{\lambda^2\eta}{(1+\lambda^2)(1+\lambda^2+\lambda^2\eta^2)}$$

$$p_{12} = \frac{\lambda\eta}{(1+\lambda^2)(1+\lambda^2+\lambda^2\eta^2)}$$
$$p_{13} = \frac{1}{1+\lambda^2+\lambda^2\eta^2}$$
$$p_{21} = -\lambda$$
$$p_{22} = -1$$
$$p_{23} = \lambda\eta$$
$$p_{31} = 1$$
$$p_{32} = -\lambda$$
$$p_{33} = 0.$$

(I.2.4) Proof of the "only if" part of Bucht's theorem, i.e., Witt's theorem $\Longrightarrow$ Bucht's theorem.

By Witt's theorem (I.1.1), we know that there exists a unimodular matrix P sending the quadratic form $uX_1^2 + vX_2^2 + \frac{1}{uv}X_3^2$ to the diagonal quadratic form $Y_1^2 + Y_2^2 + Y_3^2$. It is slightly more convenient to use the inverse matrix $Q = (q_{ij})$ with $1 \le i, j \le 3$ of P sending $Y_1^2 + Y_2^2 + Y_3^2$ to $uX_1^2 + vX_2^2 + \frac{1}{uv}X_3^2$. We first note that at most one of the elements q_{11}, q_{22} and q_{33} can be equal to zero. (In fact, for instance, if $q_{11} = q_{22} = 0$, then the relation $\sum_{i=1}^{3} q_{i1}q_{i2} = 0$ implies that $q_{31}q_{32} = 0$, and hence either $u = q_{11}^2 + q_{21}^2 + q_{31}^2$ or $v = q_{11}^2 + q_{22}^2 + q_{32}^2$ would be a square, contradicting the assumption that u, v are not squares in K.)

Now we first consider the generic case. In this case, none of q_{11}, q_{22} or q_{33} is equal to zero. We can then write the matrix Q as

$$Q = \begin{pmatrix} q_{11} & -fq_{22} & -\rho q_{33} \\ \lambda q_{11} & q_{22} & -gq_{33} \\ eq_{11} & \eta q_{22} & q_{33} \end{pmatrix}$$

for suitable elements λ, η, ρ and e, f, g in K. Then from the orthogonality relations

$$-fq_{11}q_{22} + \lambda q_{11}q_{22} + \eta eq_{11}q_{22} = 0$$
$$-\rho q_{11}q_{33} - \lambda gq_{11}q_{22} + eq_{11}q_{33} = 0$$
$$\rho fq_{22}q_{33} - gq_{22}q_{33} + \eta q_{22}q_{33} = 0$$

we deduce

$$e(1 - \lambda\eta\rho) = \rho + \lambda\eta + \rho\lambda^2$$
$$f(1 - \lambda\eta\rho) = \lambda + \eta\rho + \lambda\eta^2$$
$$g(1 - \lambda\eta\rho) = \eta + \lambda\eta + \eta\rho^2.$$

Here $1 - \lambda\eta\rho \neq 0$, since otherwise $0 = e(1 - \lambda\eta\rho)\rho = \rho^2 + 1 + \lambda^2\rho^2$, contradicting the assumption that $\text{level}(K) > 2$. Hence we get

$$u \underset{(2)}{=} 1 + \lambda^2 + \frac{(\rho + \lambda\eta + \rho\lambda^2)^2}{(1 - \lambda\eta\rho)^2} \underset{(2)}{=} (1 + \rho^2 + \rho^2\lambda^2)(1 + \lambda^2 + \lambda^2\eta^2),$$

$$v \underset{(2)}{=} 1 + \eta^2 + \frac{(\lambda + \eta\rho + \lambda\eta^2)^2}{(1 - \lambda\eta\rho)^2} \underset{(2)}{=} (1 + \eta^2 + \eta^2\rho^2)(1 + \lambda^2 + \lambda^2\eta^2),$$

and

$$uv \underset{(2)}{=} 1 + \rho^2 + \frac{(\eta + \lambda\rho + \eta\rho^2)^2}{(1 - \lambda\eta\rho)^2} \underset{(2)}{=} (1 + \rho^2 + \lambda^2\rho^2)(1 + \eta^2 + \eta^2\rho^2).$$

Now we consider the degenerate case. In this case, one of the entries q_{11}, q_{22} or q_{33} is equal to zero, say, $q_{33} = 0$. Then we can write Q as

$$Q = \begin{pmatrix} q_{11} & -fq_{22} & q_{13} \\ \lambda q_{11} & q_{22} & q_{23} \\ eq_{11} & \eta q_{22} & 0 \end{pmatrix}.$$

From the orthogonality relations as above, we obtain

$$\left. \begin{array}{l} fq_{13} = q_{23} \\ q_{13} = -\lambda q_{23} \end{array} \right\} \quad \text{i.e., } f\lambda = -1,$$

$$-f + \lambda + e\eta = 0.$$

Since λ must be non-zero (otherwise, $\det Q = 0$), we have

$$e = -(1 + \lambda^2)/\lambda\eta \quad \text{and} \quad f = -1/\lambda.$$

Therefore, we get

$$u \underset{(2)}{=} 1 + \lambda^2 + \frac{(1 + \lambda^2)^2}{\lambda^2\eta^2} \underset{(2)}{=} (1 + \lambda^2)(1 + \lambda^2 + \lambda^2\eta^2)$$

$$v \underset{(2)}{=} \frac{1}{\lambda^2} + 1 + \eta^2 \underset{(2)}{=} 1 + \lambda^2 + \lambda^2\eta^2 \underset{(2)}{=} \frac{1}{1 + \lambda^2 + \lambda^2\eta^2}$$

and

$$uv \underset{(2)}{=} 1 + \lambda^2 \underset{(2)}{=} \frac{1}{1 + \lambda^2}.$$

(I.2.5) Remark. The "if" part of Bucht's theorem is valid for any field K. However, the "only if" part of Bucht's theorem requires the condition on K that $\text{level}(K) > 2$.

If K is a field with $\text{level}(K) > 2$, Bucht's parametrization for quaternion extensions over K is "versal" in the sense that any quaternion extension over K can be obtained by Bucht's parametrization (cf. Proposition (I.1.8)).

We shall illustrate one particular parametrization of quaternion extensions over K, which can be derived from Bucht's parametrization by specialization, and becomes quite useful later.

(I.2.6) Proposition. *Let K be a field and let r, $s \in K^*$ be such that*

$$1 + s^2 \notin K^2$$
$$1 + r^2 + r^2 s^2 \notin K^2$$
$$(1 + s^2)(1 + r^2 + r^2 s^2) \notin K^2.$$

Then the biquadratic extension $K(\sqrt{1 + s^2}, \sqrt{1 + r^2 + r^2 s^2})$ over K is embeddable into a quaternion extension over K.

Proof. In Bucht's parametrization (Theorem (I.2.1)), take $\lambda = 0$, $\eta = r$ and $\rho = s$. Then $u \underset{(2)}{=} 1 + s^2$ and $v \underset{(2)}{=} 1 + r^2 + r^2 s^2$, and hence the assertion follows.

(I.2.7) Corollary. *The quaternion group Q_8 is realizable as a Galois group of a regular extension M over $\mathbf{Q}(t)$ (that is, $\mathbf{Q}$ is algebraically closed in M).*

Proof. In Proposition (I.2.6), take $K = \mathbf{Q}(t)$ and $r = s = t$. Let $N = \mathbf{Q}(t)(\sqrt{1 + t^2}, \sqrt{1 + t^2 + t^4})$. Then N is embeddable into a quaternion extension, M, over $\mathbf{Q}(t)$. Moreover, M is a regular extension over $\mathbf{Q}(t)$, that is, $\mathrm{Gal}(M/\mathbf{Q}(t)) \cong Q_8$ and $\mathbf{Q}$ is algebraically closed in M.

As a consequence of Witt's theorem (I.1.1), we also obtain the following result.

(I.2.8) Proposion. *Let K be a field and let $u \in K \setminus K^2$. If $K(\sqrt{u})$ is embeddable into a quaternion extension over K, then u is expressed as a sum of three squares in K.*

Proof. We know from Witt's theorem (I.1.1) that there exists $v \in K \setminus K^2$ such that $uv \notin K^2$ and

$$uX_1^2 + vX_2^2 + \frac{1}{uv}X_3^2 \sim Y_1^2 + Y_2^2 + Y_3^2.$$

Since the quadratic form on the left hand side represents u, so does the quadratic form on the right hand side, in particular, u is a sum of three squares in K.

(I.2.9) Remark. The converse of the assertion of Proposion (I.2.8) is not true for general fields K, as we shall see in Theorem (II.2.1) below.

I.3. Classification of quaternion extensions when level$(K) \leq 2$.

In this section, we focus our discussions on quaternion extensions over fields K with level$(K) \leq 2$. Thus, this section serves as a complement to section I.2.

We first consider the case level$(K) = 1$, i.e., $-1 \in K^2$.

(I.3.1) Theorem. *Let K be a field with* level$(K) = 1$, *i.e.,* $-1 \in K^2$. *Let u, v be quadratically independent elements of K. Then the biquadratic extension $L = K(\sqrt{u}, \sqrt{v})$ can be embedded into a quaternion extension over K if and only if there exist elements r, s in K^* such that*

$$u \underset{(2)}{=} 1 + s^2 \quad and \quad v \underset{(2)}{=} 1 + r^2 + r^2 s^2.$$

Proof. It follows from Witt's theorem (I.1.1) that $K(\sqrt{u}, \sqrt{v})$ is embeddable into a quaternion extension over K if and only if $(-u, -v) = (-1, -1)$ in Br(K). Since $-1 \in K^2$, this boils down to the condition $(-u, v) = 1$. Further, by Proposition (I.1.6)(c), this is equivalent to the fact that the equation $x^2 + uy^2 = v$ has a non-trivial solution $(x, y) \in K \times K$. Since $-1 \in K^2$, we can write $u \underset{(2)}{=} 1 + s^2$ for some $s \in K^*$. Hence $v = x^2 + (1 + s^2)y^2 \underset{(2)}{=} 1 + (1 + s^2)r^2$ for some $r \in K^*$.

(I.3.2) Remarks. Let K be a field with level$(K) = 1$.

(a)   The parametrization of quaternion extensions over K given in Theorem (I.3.1) is a specialization of Bucht's parametrization (cf. the proof of Proposition (I.2.6)). Thus, this parametrization is also "versal" in the sense that this family parametrizes quaternion extensions over K when level$(K) = 1$.

(b)   We have a constructive version of Witt's theorem (I.1.1). In fact, a unimodular matrix $P = (p_{ij})$, $1 \leq i, j \leq 3$, giving the equivalence of $uX_1^2 + vX_2^2 + \dfrac{1}{uv}X_3^2$ to $Y_1^2 + Y_2^2 + Y_3^2$ is given explicitly as follows.

$$\begin{pmatrix} 1 & 0 & s \\ -rs & 1 & r \\ \dfrac{-s}{(1+s^2)(1+r^2+r^2s^2)} & \dfrac{-r}{1+r^2+r^2s^2} & \dfrac{1}{(1+s^2)(1+r^2+r^2s^2)} \end{pmatrix}$$

Now we consider the case that level$(K) = 2$.

(I.3.3) Theorem. *Let K be a field with* level$(K) = 2$. *Let u, v be quadratically independent elements in K. Then $K(\sqrt{u}, \sqrt{v})$ is embeddable into a quaternion extension over K if and only if there exist elements r, s in K^* such that*

$$u \underset{(2)}{=} r \quad and \quad v \underset{(2)}{=} s$$

and $r + s = -1$ in K.

Proof. When level$(K) = 2$, we have $(-1, -1) = 1$, and hence $K(\sqrt{u}, \sqrt{v})$ is embeddable into a quaternion extension over K if and only if $(-u, -v) = 1$. Further, by Proposition (I.1.6)(c), this is equivalent to saying that the equation $uX^2 + vY^2 + Z^2 = 0$ has a non-trivial solution $(X, Y, Z) \in K \times K \times K$. In return,

$K(\sqrt{u}, \sqrt{v})$ admits a quaternion extension if and only if there exist elements r, $s \in K^*$ such that $u \underset{(2)}{=} r$ and $v \underset{(2)}{=} s$ with $r + s = -1$.

We have a constructive version of Witt's theorem (I.1.1) also in the case that level$(K) = 2$. That is, we can determine a unimodular matrix $P = (p_{ij})$, $1 \le i, j \le 3$, giving the equivalence of $uX_1^2 + vX_2^2 + \dfrac{1}{uv}X_3^2$ to $Y_1^2 + Y_2^2 + Y_3^2$.

(I.3.4) Theorem. *Let K be a field with* level$(K) = 2$. *Let $-1 = \alpha^2 + \beta^2$ for some α, $\beta \in K$. Let u and v be elements in K given as in Theorem (I.3.3), that is, $u \underset{(2)}{=} r$ and $v \underset{(2)}{=} s$ with r, $s \in K^*$ such that $r + s = -1$. Then a unimodular matrix $P = (p_{ij})$, $1 \le i, j \le 3$, giving the equivalence of the quadratic form*

$$uX_1^2 + vX_2^2 + \frac{1}{uv}X_3^2 = rX_1^2 - (r+1)X_2^2 - \frac{1}{r(r+1)}X_3^2$$

with the diagonal quadratic form $Y_1^2 + Y_2^2 + Y_3^2$ is explicitly determined as follows. Write $r = t^2 - w^2$ with $t = \dfrac{r+1}{2}$ and $w = \dfrac{1-r}{2}$ in K. Then

$$P = \begin{pmatrix} -\dfrac{t}{r} & -\dfrac{w\beta}{r} & -\dfrac{w\alpha}{r} \\[2mm] -\dfrac{w}{r+1} & \dfrac{\alpha - \beta t}{r+1} & -\dfrac{\alpha t + \beta}{r+1} \\[2mm] w & r\alpha + t\beta & \alpha t - \beta r \end{pmatrix}.$$

Proof.   Just carry out the transformation

$$\begin{pmatrix} X_1 \\ X_2 \\ X_3 \end{pmatrix} = P \begin{pmatrix} Y_1 \\ Y_2 \\ Y_3 \end{pmatrix}.$$

II. Fields that admit quaternion extensions.

The problems that we are concerned with here are presented in the following. Let K be a field of characteristic different from 2.

(A)   Characterize fields K that admit quaternion extensions.

(B)   For fields K that admit quaternion extensions, determine the number $\nu(Q_8; K)$.

II.1. Arbitrary ground fields.

We first consider the case that K is an arbitrary field. Our results are rather rudimentary: We have only partial answers to the problems (A) and (B) in this situation.

As an immediate consequence of Witt's theorem (I.1.1), we can derive the following assertion.

(II.1.1) Proposition. *Let K be a field such that $\mathrm{Br}_2(K) = \{1\}$. Then any biquadratic extension over K can be embedded into a quaternion extension over K.*

Proof. Let u, $v \in K \setminus K^2$ be such that $uv \notin K^2$. If $\mathrm{Br}_2(K) = \{1\}$, then we have, in $\mathrm{Br}(K)$, $(-u, -v) = (-1, -1)$. Therefore, by Proposition (I.1.6), the biquadratic extension $K(\sqrt{u}, \sqrt{v})$ is embeddable into a quaternion extension over K.

(II.1.2) Proposition. *Let K be a field with $\mathrm{level}(K) = 2$. Then any biquadratic extension over K containing $\sqrt{-1}$ can be embedded into a quaternion extension over K.*

Proof. Let $u \in K \setminus K^2$. Then $K(\sqrt{u}, \sqrt{-1})$ is embeddable into a quaternion extension over K as $(-u, 1) = (-1, -1)$ in $\mathrm{Br}(K)$.

From Proposition (I.2.6), we easily obtain the following result.

(II.1.3) Theorem. *Let K be a field with Pythagoras number $d(K) \geq 3$. Then K admits a quaternion extension.*

Proof. By assumption, there exists an element $v \in K$ which is a sum of three squares, but not a sum of two squares in K. Then v has a representation of the form

$$v \underset{(2)}{=} 1 + r^2 + r^2 s^2 \quad \text{for some } r,\, s \in K^*.$$

Since v is not a sum of two squares in K, neither $1 + s^2$ nor $(1 + s^2)(1 + r^2 + r^2 s^2)$ can be squares in K. Therefore, by virtue of Proposition (1.2.6), K admits a quaternion extension.

(II.1.4) Remark. If K is a field with $\mathrm{level}(K) = 2$, then $d(K) \leq 3$. (In general, we have $d(K) \leq 1 + \mathrm{level}(K)$.) Thus, the assertion of Proposition (II.1.2) and that of Theorem (II.1.3) are complementary.

(II.1.5) Examples. Here are some fields that do not admit any quaternion extension.
 (1)   The field of real numbers $\mathbf{R}$, and the field of complex numbers $\mathbf{C}$.
 (2)   Any finite field $\mathbf{F}_q$.
 (3)   The fields $\mathbf{R}((t))$, $\mathbf{C}((t))$, or more generally, $\mathbf{R}(((t_1)), ((t_2)), \ldots, ((t_n)))$ and $\mathbf{C}(((t_1)), ((t_2)), \ldots, ((t_n)))$.
 (4)   Any field K with Pythagoras number $d(K) = 1$.

II.2. Hilbertian ground fields.

For Hilbertian fields K, we have complete answers to the problems (A) and (B).

(II.2.1) Theorem. *Let K be a Hilbertian field, and let $u \in K \setminus K^2$. Then $K(\sqrt{u})$ is embeddable into a quaternion extension over K if and only if u is expressed as a sum of three squares in K.*

Proof. In view of Proposition (I.2.8), it suffices to prove the "if" part. By Proposition (II.1.3), we may assume that u is a sum of two squares, say, $u = 1 + s^2$ with some $s \in K^*$. By virtue of Proposition (I.2.6), we just have to find an element $r \in K^*$ such that $1 + r^2 + r^2 s^2 \notin K^2$ and $(1 + s^2)(1 + r^2 + r^2 s^2) \notin K^2$. For this purpose, we notice that the polynomials

$$X^2 - (1 + Y^2 + s^2 Y^2) \quad \text{and} \quad X^2 - (1 + s^2)(1 + Y^2 + s^2 Y^2)$$

are irreducible in $K[X, Y]$, since $1 + Y^2 + s^2 Y^2$ and $(1 + s^2)(1 + Y^2 + s^2 Y^2)$ are not squares in $K(Y)$. Because K is Hilbertian, there exist infinitely many elements $r \in K^*$ such that the specializations

$$X^2 - (1 + r^2 + s^2 r^2) \quad \text{and} \quad X^2 - (1 + s^2)(1 + r^2 + r^2 s^2)$$

are irreducible in $K[X]$. Hence $1 + r^2 + r^2 s^2$ and $(1 + s^2)(1 + r^2 + r^2 s^2)$ are not squares in K, and the proof is completed by Proposition (I.2.6).

(II.2.2) Remark. For $K = \mathbf{Q}$, Reichardt [10] proved the assertion of Theorem (II.2.1) using reciprocity laws.

For $\nu(Q_8; K)$, we have the following result.

(II.2.3) Proposition. *Let K be a Hilbertian field. Then $\nu(Q_8; K) = \infty$.*

Proof. If K is Hilbertian, then K^*/K^{*2} has infinitely many elements, that is, K contains infinitely many square classes. Therefore, $\nu(Q_8; K)$ is infinite by Theorem (II.2.1) and Proposition (I.1.8).

(II.2.4) Examples.
(a) Here are some examples of Hilbertian fields:

(1) $\mathbf{Q}$, and any algebraic number field of finite degree.
(2) The maximal abelian extension $\mathbf{Q}^{ab}$ of $\mathbf{Q}$.
(3) Any pure transcendental extension of any field.
(4) The quotient field of any unique factorization domain of Krull dimension > 1.

(b) Here are some examples of fields which are not Hilbertian.

(1) Any algebraically closed field, e.g., $\mathbf{C}$.
(2) The field of p-adic rational numbers $\mathbf{Q}_p$, and any extension of $\mathbf{Q}_p$.
(3) Any field K admitting a henselian valuation, for instance, $K((t))$.

II.3. Local ground fields.

In the case where K is a local field, we have satisfactory answers to the problems (A) and (B).

We first fix the notations. If K is a field with discrete rank 1 valuation ν, we denote by R, $\mathfrak{M}$ and k, respectively, the valuation ring of K, the maximal ideal of R and the residue field $R/\mathfrak{M}$ of K.

We assume throughout this section that the characteristic of K is different from 2, and that the characteristic of the residue field k is also different from 2 with the exception of Theorem (II.3.1), Theorem (II.3.7) and Example (II.3.8).

(II.3.1) Theorem. *Let K be a local field equipped with a discrete rank 1 valuation ν. If K is not formally real and* level$(K) \neq 1$, *then K admits a quaternion extension.*

Proof. Let π be a uniformizing element of R. Since π and -1 are quadratically independent elements in K, the assertion follows from Proposition (II.1.2) when level$(K) = 2$. If level$(K) = 2^n$ with $2 \leq n < \infty$, the Pythagoras number $d(K)$ satisfies $d(K) \geq 2^n > 3$, so that the proof is completed by Theorem (II.1.3).

(II.3.2) Proposition. *Let K be a local field with a discrete rank 1 henselian valuation ν. Let R be the corresponding valuation ring, π a uniformizing element of R, and $k = R/\mathfrak{M}$ with $\mathfrak{M} = \pi R$ the residue field of K.*

(a) *If $u, v \in R \setminus \mathfrak{M}$, then u, v, $uv \notin K^2$ if and only if $\bar{u}$, $\bar{v}$, $\bar{u}\bar{v} \notin k^2$ where $\bar{a} := a \bmod \mathfrak{M}$ for $a \in R$.*

(b) *With u, v as in (a), $K(\sqrt{u}, \sqrt{v})$ is embeddable into a quaternion extension over K if and only if $k(\sqrt{\bar{u}}, \sqrt{\bar{v}})$ is embeddable into a quaternion extension over k.*

Proof. (a) The assertion holds since ν is henselian and the characteristic of k is different from 2.

(b) By Proposition (I.1.3), $K(\sqrt{u}, \sqrt{v})$ is embeddable into a quaternion extension over K if and only if the system of equations (with $a_1 = uv$, $a_2 = u$ and $a_3 = v$)

$$(*) \qquad \begin{cases} b = ux^2 - vy^2 \\ b = vz^2 - uvw^2 \\ b = uvr^2 - us^2 \end{cases}$$

has a non-trivial solution $(b, x, y, z, w, r, s) \in K \times K \times K \times K \times K \times K \times K$, and $k(\sqrt{\bar{u}}, \sqrt{\bar{v}})$ is embeddable into a quaternion extention over k if and only if the system of equations

$$(**) \qquad \begin{cases} \bar{b} = \bar{u}\bar{x}^2 - \bar{v}\bar{y}^2 \\ \bar{b} = \bar{v}\bar{z}^2 - \bar{u}\bar{v}\bar{w}^2 \\ \bar{b} = \bar{u}\bar{v}\bar{r}^2 - \bar{u}\bar{s}^2 \end{cases}$$

has a non-trivial solution $(\bar{b}, \bar{x}, \bar{y}, \bar{z}, \bar{w}, \bar{r}, \bar{s}) \in k \times k \times k \times k \times k \times k \times k$. If $(*)$ has a non-trivial solution in $K \times K \times K \times K \times K \times K \times K$, it has a non-trivial solution in $R \times R \times R \times R \times R \times R \times R$. By dividing out by a suitable power of π, we obtain a solution in $R \times R \times R \times R \times R \times R \times R$ such that not all components are divisible by π. Hence we get a non-trivial solution for the system of equations $(**)$ by reducing it modulo π.

Conversely, a non-trivial solution for the system of equations $(**)$ can be lifted to a non-trivial solution for the system of equations $(*)$, since ν is henselian and the characteristic of k is different from 2.

(II.3.3) Theorem. *Let K be a local field with a discrete rank 1 henselian valuation ν. Let R be the corresponding valuation ring, and let π be a uniformizing element of R. Then the following assertions hold true.*

(a) *If* level$(K) \neq 2$, *then any biquadratic extension over K which is embeddable into a quaternion extension over K is of the form*

$$K(\sqrt{u}, \sqrt{v}) \quad \text{with } u, v \in R \setminus \mathfrak{M} \text{ and } u, v, uv \notin K^2.$$

(b) *If* level$(K) = 2$, *then any biquadratic extension over K which is embeddable into a quaternion extension over K has one of the following forms:*

$$K(\sqrt{u}, \sqrt{v}) \quad \text{with } u, v \in R \setminus \mathfrak{M} \text{ and } u, v, uv \notin K^2$$

or

$$K(\sqrt{u\pi}, \sqrt{-1}) \quad \text{with } u \in R \setminus \mathfrak{M}.$$

Proof. Note that level$(K) = $ level(k). We first consider the case that level$(K) \geq 4$. In this case, $K(\sqrt{u\pi})$ with $u \in R \setminus \mathfrak{M}$ is never embeddable into a quaternion extension over K, since $u\pi$ is not a sum of three squares in K. In fact, assume that

$$u\pi = X^2 + Y^2 + Z^2 \quad \text{in } K.$$

Then by multiplying with a suitable power of π (if necessary), we can derive the identity

$$u\pi^{n+1} = x^2 + y^2 + z^2 \quad \text{with } x, y, z \in R \text{ and } x \notin M.$$

Passing to the residue field k, we get

$$0 = \bar{x}^2 + \bar{y}^2 + \bar{z}^2 \quad \text{in } k,$$

where $\bar{a}$ denotes $a \pmod{\pi}$ for $a \in R$. This implies that

$$-1 = (\bar{y}/\bar{x})^2 + (\bar{z}/\bar{x})^2 \quad \text{in } k,$$

contradicting the hypothesis that level$(k) = $ level$(K) \geq 4$. Therefore, any biquadratic extension over K which is embeddable into a quaternion extension over K must be of the form $K(\sqrt{u}, \sqrt{v})$ with $u, v \in R \setminus \mathfrak{M}$ and $u, v, uv \notin K^2$. (Cf. Theorem (I.2.1).)

Since $\mathrm{level}(k) = \mathrm{level}(K) = 2^n (n \in \mathbf{N})$ or ∞, it remains to consider the cases that $\mathrm{level}(K) = 1$ or 2.

Suppose that $K(\sqrt{u\pi}, \sqrt{v})$, $u, v \in R \setminus \mathfrak{M}$ and $v \notin K^2$ is embeddable into a quaternion extension over K. Then by Proposition (I.1.3), there exists $b \in K^*$ such that the system of equations

$$(*) \qquad \begin{cases} b = vx^2 - u\pi y^2 \\ b = u\pi z^2 - u\pi v w^2 \\ b = u\pi v r^2 - v s^2 \end{cases}$$

has a solution $(x, y, z, w, r, s) \in K \times K \times K \times K \times K \times K$. We may assume without loss of generality that all elements x, y, z, w, r, s are in R and that at least one of them is not in $\mathfrak{M}$. We can deduce from the above identities that

$$b = \pi(uz^2 - uvw^2) \quad \text{and} \quad x, s \in \mathfrak{M}.$$

So we have one of the following possibilities:

(a) $y \notin \mathfrak{M}$, (b) $r \notin \mathfrak{M}$, (c) $z \notin \mathfrak{M}$ or (d) $w \notin \mathfrak{M}$.

In every case, we have $b/\pi \notin \mathfrak{M}$. (In the cases (a) and (b), the assertion is derived from the first, respectively, the third identity in $(*)$. In the case (c), the second identity in $(*)$ is read as $b/\pi = uz^2 - uvw^2$. If $b/\pi \in \mathfrak{M}$, then by reducing modulo $\mathfrak{M}$, we get $\bar{z}^2 = \bar{v}\bar{w}^2$ in k. But as $\bar{z} \neq 0$ in k, we must have $\bar{v} \in k^{*2}$. Since R is henselian, this implies that there is a lifting, v, of $\bar{v}$ to K having the property that $v \in K^2$, contradicting the choice of v. Therefore, $b/\pi \notin \mathfrak{M}$. In the case (d), the same vein of argument as for (c) can be applied to yield $b/\pi \notin \mathfrak{M}$.)

Now, reducing the first and the third identities in $(*)$ modulo $\mathfrak{M}$, we get

$$\overline{b/\pi} = -\bar{u}\bar{y}^2 \quad \text{and} \quad \overline{b/\pi} = \bar{u}\bar{v}\bar{r}^2 \quad \text{in } k.$$

This yields the identity

$$\bar{v} = (-1)(\bar{y}/\bar{r})^2 \quad \text{in } k.$$

Since R is henselian, this implies further that there is a lifting, v of $\bar{v}$ to K with the property that

$$v = (-1)(\text{a square}) \quad \text{in } K.$$

Therefore, if $K(\sqrt{u\pi}, \sqrt{v})$ were to be embeddable into a quaternion extension over K, it must be of the form $K(\sqrt{u\pi}, \sqrt{-1})$ with $u \in R \setminus \mathfrak{M}$.

This completes the proof.

(II.3.4) Examples. Here we list some local fields that admit quaternion extensions.

(1) A local field K for which $-1 \notin K^2$, e.g., the field of p-adic rational numbers $\mathbf{Q}_p$ with $p \equiv 3 \pmod 4$ and the field of 2-adic rational numbers $\mathbf{Q}_2$.

(2) A finite extension K of $\mathbf{Q}_p$ with $p \equiv 3 \pmod 4$ or $p = 2$ such that $-1 \notin K^2$.

(3) $\mathbf{F}_p((t))$ with $p > 2$ such that $p \equiv 3 \pmod 4$.

We now determine the number $\nu(Q_8; K)$ for certain local fields K, thereby giving answers to the problem (B).

(II.3.5) Theorem. *Let K be a local field with a discrete rank 1 henselian valuation ν. Let R be the valuation ring of K with a uniformizing element π and let k be the residue field of K. Then the following assertions hold true.*

(a) *If* level$(K) \neq 2$, $\nu(Q_8; K) = 2\nu(Q_8; k)$.

(b) *If* level$(K) = 2$, $\nu(Q_8; K) = \nu(Q_8; k) + 2^{2n-2}$ *where n is defined by* $2^n = |k^*/k^{*2}|$.

In particular, $\nu(Q_8; K) = 0$ *if* level$(K) =$ level$(k) = 1$, *and* $|k^*/k^{*2}| = 2$.

Proof. If x_i, $1 \leq i \leq 2^n$ are elements of R such that $\bar{x}_i$, $1 \leq i \leq 2^n$ are complete set of representatives of square classes of k^*, then $\{(x_i, \pi x_i) \mid 1 \leq i \leq 2^n\}$ forms a complete set of representatives of square classes of K^*. Hence $|K^*/K^{*2}| = 2^{n+1}$.

(a) This follows from Proposition (I.1.8), Proposition (II.3.2) and Theorem (II.3.3).

(b) This follows from Proposition (I.1.8), Proposition (I.1.2) and Theorem (II.3.3).

For the last assertion, we have only to observe that $\nu(Q_8; k) = 0$ since there is no biquadratic extension over k.

(II.3.6) Corollary. *Let K be a $\mathfrak{p}$-adic number field with the residue field k, where $\mathfrak{p}$ is a prime in K over a rational prime $p \neq 2$, i.e., $\mathfrak{p} \nmid 2$. Then we have the following assertions.*

(a) level$(K) =$ level$(k) \leq 2$, *and it is equal to 1 if and only if -1 is a square in k.*

(b) $\nu(Q_8; K) = \begin{cases} 1 & \text{if } \mathrm{level}(K) \neq 1 \\ 0 & \text{if } \mathrm{level}(K) = 1. \end{cases}$

In particular, $K = \mathbf{Q}_p$ admits a (unique) quaternion extension if and only if $p \equiv 3 \pmod 4$.

Proof. (a) We just notice that -1 can be represented as a sum of two squares in the finite field k, and this representation can be lifted to K.

(b) This follows from Theorem (II.3.5), since $|k^*/k^{*2}| = 2$, i.e., $n = 1$.

Now we shall discuss the complement of the above situation, namely, the case that K is the field $\mathbf{Q}_2$ of 2-adic rationals, or any finite extension of $\mathbf{Q}_2$. In this case, the residue field k is of characteristic 2.

(II.3.7) Theorem. *Let K be a finite extension of $\mathbf{Q}_2$ with $[K : \mathbf{Q}_2] = n$. Then the following assertions hold true.*

(a)   *Let ν_0 denote the number of biquadratic extensions $K(\sqrt{u}, \sqrt{v})$ that are embeddable into a quaternion extension over K. Then ν_0 is given as follows.*

(1)   *If $(-1,-1) = -1$, $\nu_0 = \dfrac{(2^{n+1} - 1)^2}{3}$ (n is necessarily odd).*

(2)   *If $-1 \in K^{*2}$, $\nu_0 = \dfrac{(2^{n+2} - 1)(2^n - 1)}{3}$ (n is necessarily even).*

(3)   *If $-1 \notin K^{*2}$ but $(-1,-1) = 1$, $\nu_0 = \dfrac{2^{2n+2} - 2^{n+2} + 2^{n+1} + 1}{3}$ (n is necessarily even).*

(b) $\nu(Q_8; K) = 2^n \nu_0$.

Proof. (a) We use Witt's theorem (I.1.1). What we are looking for is the number of quadratically independent pairs $\{u, v\}$, u, $v \in K$ such that the quadratic form $uX_1^2 + vX_2^2 + \dfrac{1}{uv}X_3^2$ has the Hasse-Witt invariant $(u, v)(u, uv) \times (v, uv) = (-1, u)(-u, v) = 1$.

We consider this case by case, noting the following fact. One knows (e.g., Hasse [3, p. 222]) that if $|K^*/K^{*2}| = 2^s$, then $s = 2 + n$. If $u \notin K^2$, then $\left| \{ v \in K^*/K^{*2} \mid (u, v) = 1 \} \right| = \left| \{ v \in K^*/K^{*2} \mid (u, v) = -1 \} \right| = 2^{s-1}$. We show how the number ν_0 is obtained in the case (1) and leave the cases (2) and (3) as an exercise to the reader.

Case (i). If $u = -1 \pmod{K^{*2}}$, there are no v's for which $(-1, u)(-u, v) = 1$.

Case (ii). If $(u, -1) = -1$ and $u \neq -1 \pmod{K^{*2}}$, there are 2^{s-1} square classes v such that $(-u, v) = -1$. Thus, we get $2^{s-1} - 1$ u's for which $(u, -1) = -1$ and such that for each u there are 2^{s-1} v's having $(-1, u)(-u, v) = 1$. Altogether there are $2^{s-1}(2^{s-1} - 1)$ pairs $\{u, v\}$ that meet the requirement.

Case (iii). If $(u, -1) = 1$, there are $2^{s-1} - 2$ square classes v, $v \neq u$, $v \neq 1$ such that $(-u, v) = 1$. The number of square classes u, $u \neq 1$ for which $(u, -1) = 1$ is $(2^{s-1} - 1)$. Therefore, altogether there are $(2^{s-1} - 2)((2^{s-1} - 1)$ pairs $\{u, v\}$ that we are after.

Putting all three cases together, the number of ordered pairs $\{u, v\}$ for which $(-1, u)(-u, v) = 1$ is $(2^{s-1} - 1)(2^s - 2)$. Since the six pairs $\{u, v\}$, $\{v, u\}$, $\{u, uv\}$, $\{v, uv\}$, $\{uv, u\}$ and $\{uv, v\}$ all give rise to the same biquadratic extension over K, ν_0 is then equal to $(2^{s-1} - 1)^2/3$.

(b)   For each quadratic form $uX_1^2 + vX_2^2 + \dfrac{1}{uv}X_3^2$ with the Hasse-Witt invariant 1, there are 2^n non-isomorphic quaternion extensions over K containing the biquadratic field $K(\sqrt{u}, \sqrt{v})$ by Proposition (I.1.8). Therefore, $\nu(Q_8; K) = 2^n \nu_0$.

(II.3.8) Example. Let $K = \mathbf{Q}_2$. Then by Theorem (II.3.7), we obtain $\nu(Q_8; \mathbf{Q}_2) = 2(2^2 - 1)^2/3 = 6$. In this case, we can determine the associated quadratic forms explicitly. Note first that there are seven biquadratic extensions over $\mathbf{Q}_2$, and among these, only three fields, namely, $\mathbf{Q}_2(\sqrt{2}, \sqrt{-5})$, $\mathbf{Q}_2(\sqrt{3}, \sqrt{10})$ and $\mathbf{Q}_2(\sqrt{5}, \sqrt{6})$, are embeddable into quaternion extensions over

$\mathbf{Q}_2$, since the quadatic form $uX_1^2 + vX_2^2 + \dfrac{1}{uv}X_3^2$ has the Hasse-Witt invariant 1 only for these cases. For each of these biquadratic extensions over $\mathbf{Q}_2$, there are 2 quaternion extensions over $\mathbf{Q}_2$ that contain the biquadratic extension in question. This yields $\nu(Q_8; \mathbf{Q}_2) = 6$.

(II.3.9) Remark. Massy and Nguyen-Quang-Do [8] have obtained a formula for the number $\nu(Q_8; K)$ for a finite extension K over $\mathbf{Q}_2$ with $[K : \mathbf{Q}_2] = n$. Their formula is read as $\nu(Q_8; K) = (2^n - 1)\nu_0$, while our formula is $\nu(Q_8; K) = 2^n \nu_0$.

III. Automatic realizations.

Let K be a field of characteristic different from 2.
In this section, we consider the following problems.

(A) If Q_8 is realizable as a Galois group over K, are the groups Z_4, V_4 and D_4 automatically realizable as Galois group over K? (Automatic realizations of Z_4, V_4 and D_4.) Is the converse also true?

(B) Determine the numbers $\nu(G; K)$ for $G = Z_4$, V_4, D_4 and Q_8. Give relations among the numbers $\nu(D_4; K)$ and $\nu(Q_8; K)$, etc.

(C) Determine the range of the "function" $\nu(G; K)$ for $G = Z_4$, V_4, D_4 and Q_8, respectively.

III.1. Folklore Theorems.

We first recall some "folklore" facts needed for the subsequent discussion.

(III.1.1) Lemma. *Let K be a field. Let $f(X) = X^4 + bX^2 + d \in K[X]$ be an irreducible polynomial over K. Then*

$$\mathrm{Gal}(g/K) \cong \begin{cases} V_4 & \text{if and only if } d \in K^{*2} \\ Z_4 & \text{if and only if } d \notin K^{*2} \text{ but } d(b^2 - 4d) \in K^{*2} \\ D_4 & \text{if and only if } d \notin K^{*2} \text{ and } d(b^2 - 4d) \notin K^{*2}. \end{cases}$$

(III.1.2) Proposition. *Let K be a field. Let $u \in K \setminus K^{*2}$. Then $K(\sqrt{u})$ is embeddable into a Z_4-extension over K if and only if u is expressed as a sum of two squares in K.*

III.2. V_4- and Z_4-extensions.

We first consider V_4-extensions over K. Since V_4 is a homomorphic image of Q_8, we immediately obtain the following result.

(III.2.1) Proposition. *Let K be a field. If K admits a quaternion extension, then K automatically admits a V_4-extension.*

It is immediate to derive the following result.

(III.2.2) Proposition. *Let K be a field for which $|K^*/K^{*2}| = 2^m$. Then $\nu(V_4; K) = (2^m - 1)(2^{m-1} - 1)/3$. In particular, no even numbers can occur for $\nu(V_4; K)$.*

Now we turn our discussion to Z_4-extensions, and more generally, Z_{2^m}-extensions over K.

(III.2.3) Theorem. *Let K be a field. If K admits a quaternion extension, then K automatically admits a Z_4-extension, and more generally, a Z_{2^m}-extension over K for every $m \in \mathbf{N}$.*

Proof. Let $K(\sqrt{u})$ be a quadratic subfield of a quaternion extension over K. Then $u \notin K^2$ and u is a sum of three squares in K, say $u = x^2 + y^2 + z^2$. If u is not a sum of two squares, then $w := x^2 + y^2$ is not a square in K, and hence by Proposition (III.1.2), $K(\sqrt{w})$ can be embedded into a Z_4-extension over K. If u is a sum of two squares in K, $K(\sqrt{u})$ can be embedded into a Z_4-extension by Proposition (III.1.2).

The assertion on Z_{2^m}-extensions over K follows from a result of Whaples [11], and Kuyk and Lenstra [5] which proves that K admits a Z_4-extension if and only if K admits a Z_{2^m}-extension for every $m \in \mathbf{N}$.

(III.2.4) Remark. The converse of the assertion in Theorem (III.2.3) is not valid. For instance, $K = \mathbf{Q}_5$ admits a Z_4-extension (by Proposition (III.1.2)), but not a quaternion extension (by Corollary (II.3.6)(b)).

(III.2.5) Proposition. *If K is a field for which $|K^*/K^{*2}| = 2^m$, then $\nu(Z_4; K) = 2^{m-1}(2^r - 1)$ for some $r \leq m$.*

Conversely, for any r, m with $r \leq m$, there exists a field K such that $|K^*/K^{*2}| = 2^m$ and that $\nu(Z_4; K) = 2^{m-1}(2^r - 1)$.

Proof. We first derive the formula for $\nu(Z_4; K)$. A quadratic extension $K(\sqrt{u})$, $u \notin K^{*2}$, is embeddable into a Z_4-extesnion over K if and only if u is a sum of two squares by Proposition (III.1.2). The elements in K^* which are sums of two squares form a subgroup, H, of K^* containing K^{*2}. Then H/K^{*2} has order 2^r for some $r \leq m$. Therefore, there are $(2^r - 1)$ quadratic extensions $K(\sqrt{u})$, $u \in H$, $u \notin K^{*2}$ which are embeddable into Z_4-extensions over K. If $K(\sqrt{u})(\sqrt{\theta})$, $\theta \in K(\sqrt{u})$ is a Z_4-extension over K, then $K(\sqrt{u}, \sqrt{c\theta})$, $c \in K^*$ yield all Z_4-extensions over K containing $K(\sqrt{u})$. This gives the formula for $\nu(Z_4; K)$.

Now we prove the converse. Let m, r be integers ≥ 0 such that $m \geq r$. If $m = r$, the field $\mathbf{C}(((t_1)), \ldots, ((t_m)))$ has exactly 2^m square classes, each of which is a sum of two squares. If $m > r$, let $p_1, \ldots, p_r$ be prime numbers such that $p_i \equiv 1 \pmod 4$ for every i. By Zorn's Lemma, there exists a real algebraic extension, N, of $\mathbf{Q}$ which is maximal with respect to the property that $N \cap \mathbf{Q}(\sqrt{p_1}, \ldots, \sqrt{p_r}) = \mathbf{Q}$. Then the numbers $-1, p_1, \ldots, p_r$ form a basis for the square classes of N, and the numbers $p_1, \ldots, p_r$ form a basis for the square classes which are sums of two squares. Now let $K = N(((t_1)), \ldots, ((t_{m-r-1})))$. Then K has 2^m square classes 2^r of which are sums of two squares. Hence $\nu(Z_4; K) = 2^{m-1}(2^r - 1)$.

III.3. D_4-extensions.

Now we shall discuss D_4-extensions over K.

(III.3.1) Theorem. *Let K be a field with* level$(K) > 1$, *i.e.,* $-1 \notin K^2$. *If K admits a quaternion extension, then K automatically admits a D_4-extension.*

Proof. Since Q_8 is realizable as a Galois group over K, by Proposition (III.2.1), V_4 is also realizable as a Galois group over K, and hence there exists an element $d \in K^*$ such that -1 and d are quadratically independent, i.e., $d, -d \notin K^{*2}$. Then the polynomial $g(X) = X^4 + d \in K[X]$ is irreducible over K and $\mathrm{Gal}(g/K) \cong D_4$ by Lemma (III.1.1).

(III.3.2) Remark. The converse of Theorem (III.3.1) is not true. Indeed there exists a field K for which D_4 is realizable, but Q_8 is not. For instance, take $K = \mathbf{R}((t))$. Then level$(K) > 1$, and D_4 is realizable, since the polynomial $g(X) = X^4 - t \in K[X]$ is irreducible over K and $\mathrm{Gal}(g/K) \cong D_4$ by Lemma (III.1.1). However, K does not admit any quaternion extension by Example (II.1.5).

Now we consider the case where level$(K) = 1$, that is, $-1 \in K^2$. For this, we insert some remarks about D_4-extensions over arbitrary fields K (i.e., with level(K) not necessarily equal to 1).

Let K be an arbitrary field, and let u, v be quadratically independent elements in K. Let $L = K(\sqrt{u}, \sqrt{v})$. Let M be a D_4-extension over K containing L. Then M is a Z_4-extension over exactly one of the following quadratic subfields of L: $K(\sqrt{u})$, $K(\sqrt{v})$ or $K(\sqrt{uv})$. Assume that M is a Z_4-extension over $K(\sqrt{uv})$. Then there exist two quadratic extensions $K(\sqrt{a + b\sqrt{u}})$ and $K(\sqrt{a - b\sqrt{u}})$, $a, b \in K$ over $K(\sqrt{u})$ contained in M such that M is the normal closure of each of these fields over K and such that $a^2 - b^2 u = v$, i.e., $(u, v) = 1$. Conversely, if there are elements $a, b \in K$ satisfying the equation $a^2 - b^2 u = v$, that is, $(u, v) = 1$, then there exists a D_4-extension over K containing L which is a Z_4-extension over $K(\sqrt{uv})$. Similarly, M is a D_4-extension over K containing L which is a Z_4-extension over $K(\sqrt{u})$ (resp. $K(\sqrt{v})$) if and only if $(v, uv) = 1$ (resp. $(u, uv) = 1)$) in $\mathrm{Br}(K)$.

This yields the following result.

(III.3.3) Proposition. *Let K be an arbitrary field. Let u, v be quadratically independent elements in K, and let $L = K(\sqrt{u}, \sqrt{v})$ be a biquadratic extension over K. Then L is embeddable into a D_4-extension over K which is a Z_4-extension over $K(\sqrt{uv})$ (resp. $K(\sqrt{u})$; resp. $K(\sqrt{v})$) if and only if $(u, v) = 1$ (resp. $(v, uv) = 1$; resp. $(u, uv) = 1$) in $\mathrm{Br}(K)$.*

Now it is straightforward to check that any D_4-extension over K containing L, which is a Z_4-extension over $K(\sqrt{uv})$, is the normal closure over K of an extension $K(\sqrt{(a + b\sqrt{u})\cdot c})$ with some $c \in K^*$. Moreover, $K(\sqrt{(a + b\sqrt{u})\cdot c'})$ and $K(\sqrt{(a + b\sqrt{u})\cdot c''})$ with c', $c'' \in K^*$ have the same normal closure over K if and only if $c' \cdot c''$ is a square in L. Let A be the subgroup of K^*/K^{*2} generated by the square classes of u and v, and let $K^*/K^{*2} = A \times B$, then there is a 1-1 correspondence between the elements in B and the D_4-extensions over K containing L which are Z_4-extensions over $K(\sqrt{uv})$.

Now suppose that $\mathrm{level}(K) = 1$, i.e., $-1 \in K^2$. Then we have furthermore, $(u, v) = (v, uv) = (u, uv)$ and $(-u, -v)(-1, -1) = (u, v)$.

Therefore, the above discussion yields the following result.

(III.3.4) Theorem. *Let K be a field with $\mathrm{level}(K) = 1$, i.e., $-1 \in K^2$. Then $\nu(D_4; K) = 3\nu(Q_8; K)$.*

If K is a field with $\mathrm{level}(K) = 2$, the situation is slightly more complicated; a similar counting argument as above gives the following result.

(III.3.5) Theorem. *Let K be a field with $\mathrm{level}(K) = 2$, and $|K^*/K^{*2}| = 2^m$ $(m \in \mathbf{N})$. Further, let $|H/K^{*2}| = 2^r$ $(r \leq m)$, where H is the subgroup of K^* consisting of the elements which are sums of two squares in K. Then*

$$\nu(D_4; K) = 3\nu(Q_8; K) - (2^{2m-2} - 2^{m+r-2}).$$

(An alternative formulation is

$$\nu(D_4; K) = 3\nu(Q_8; K) - \frac{1}{4}(\nu(Z_2; K)^2 + \nu(Z_2; K) - 2\nu(Z_4; K)).)$$

In particular, we have

$$\nu(Q_8; K) \leq \nu(D_4; K) \leq 3\nu(Q_8; K).$$

Proof. Use a similar counting argument as for the proof of Theorem (III.3.4). (Note that $\nu(Z_2; K) = 2^m - 1$.)

Finally, we obtain the following result.

(III.3.6) Theorem. *Let K be an arbitrary field. If K admits a quaternion extension, then K automatically admits a D_4-extension.*

Proof. If $-1 \notin K^2$, this is Theorem (III.3.1). If $-1 \in K^2$, this follows from Theorem (III.3.4) (cf. also Massy and Nguyen-Quang-Do [8]).

(III.3.7) Remark. The converse of Theorem (III.3.6) is not true. However, one can show that if K is an arbitraty field which D_4 is realizable, then either Q_8 or D_8 (the dihedral group of order 16) is realizable over K.

(III.3.8) **Remark.** A few words are in order on the range of $\nu(Q_8; K)$ for an arbitrary field K. From Witt's theorem (I.1.1), it can be proved that if $\nu(Q_8; K)$ is finite, then it is either 1 or an even number. It is not known which even numbers can occur for the range of $\nu(Q_8; K)$. By modification of constructions in Jensen [4], one can obtain a field K containing $\sqrt{-1}$ for which $\nu(Q_8; K) = 1$. From Theorem (II.3.5), it follows $\nu(Q_8; K(((x_1)), \ldots, ((x_n)))) = 2^n$. Consequently, all powers of 2 can occur in the range of $\nu(Q_8; K)$.

By Example (II.3.8), $\nu(Q_8; K) = 6$. A. Prestel has pointed out that an intersection of $\mathbf{Q}_3$ and $\bar{\mathbf{Q}}((x))$ (embedded in a suitable way into $\mathbf{C}$) where $\bar{\mathbf{Q}}$ denotes the algebraic closure of $\mathbf{Q}$, has exactly 10 quaternion extensions.

Since level$(\mathbf{Q}_2) = 4$, Theorem (II.3.5) and Example (II.3.8) imply that $\nu(Q_8; \mathbf{Q}_2((x))) = 12$. Again by modification of constructions in Jensen [4], one can obtain a field K for which $\nu(Q_8; K) = 14$. Consequently, all even numbers ≤ 16 occur in the range of $\nu(Q_8; K)$. However, from Witt's theorem (I.1.1), it is not difficult to see 18 is not in the range for $\nu(Q_8; K)$.

(III.3.9) **Remark.** Related results hold for the range of $\nu(D_4; K)$. If $\nu(D_4; K)$ is finite, then $\nu(D_4; K) = 1$, 3 or an even number. There exists a field K for which $\nu(D_4; K) = 2$, while no field has exactly 4 D_4-extensions. (Further investigation on the range of $\nu(D_4; K)$ is in progress.)

IV. Polynomials with Galois group Q_8.

Let K be a field of characteristic different from 2.

Let K be a field with level$(K) > 2$. Then from Bucht's theorem (I.2.1) we can derive "versal" families of generic polynomials having Q_8 as Galois group.

We first consider the generic case. Let $K(t_1, t_2, t_3, t_4)$ denote the function field over K with indeterminates t_1, t_2, t_3 and t_4. Consider the octic polynomial in one variable $F_t(X) = F(t_1, t_2, t_3, t_4; X)$ defined as follows:

$$
\begin{aligned}
(\text{IV.1}) \quad F_t(X) &= F(t_1, t_2, t_3, t_4; X) \\
&= X^8 - 4ct_4 X^6 + t_4^2(6c^2 - 2u - 2v - 2uv)X^4 \\
&\quad - 4t_4^3(c^3 - cu - cv - cuv + 2uv)X^2 \\
&\quad + 2t_4^4(c^4 + u^2 + v^2 + u^2v^2 - 2c^2u - 2c^2v \\
&\quad - 2c^2uv - 2u^2v - 2uv^2 - 2uv + 8cuv)
\end{aligned}
$$

where

$$
u = \frac{1 + t_1^2 + t_1^2 t_2^2}{1 + t_3^2 + t_1^2 t_3^2}
$$

$$
v = \frac{1 + t_1^2 + t_1^2 t_2^2}{1 + t_2^2 + t_2^2 t_3^2}
$$

and

$$c = \frac{1 + t_1^2 + t_1^2 t_2^2}{\pm t_1 t_2 t_3 + 1}.$$

(Note that there is a misprint in Bucht's paper [1], where $+1$ in the denominater was incorrectly printed as -1.) It is easily seen that $F_t(X)$ has cofficients in $K(t_1, t_2, t_3, t_4)$. Moreover, $F_t(X)$ is irreducible over $K(t_1, t_2, t_3, t_4)$, as it is the minimal polynomial of the element $\sqrt{t_4(c + \sqrt{u} + \sqrt{v} + \sqrt{uv})}$ over $K(t_1, t_2, t_3, t_4)$. By Bucht's theorem (I.2.1), F_t has the Galois group

$$\mathrm{Gal}(F_t / K(t_1, t_2, t_3, t_4)) \cong Q_8.$$

Furthermore, for $\mathbf{a} = (a_1, a_2, a_3, a_4) \in K \times K \times K \times K^*$, if the specialized polynomial $F_\mathbf{a}(X)$ is irreducible over K, then $\mathrm{Gal}(F_\mathbf{a}/K) \cong Q_8$.

Now consider the degenerate case. Let $K(t_1, t_2, t_3)$ denote the function field over K with indeterminates t_1, t_2 and t_3. Consider the octic polynomial in one variable $G_t(X) = F(t_1, t_2, t_3; X)$ ($\mathbf{t} = (t_1, t_2, t_3)$) which is defined as follows:

$$\text{(IV.2)} \quad \begin{aligned} G_t(X) &= G(t_1, t_2, t_3; X) \\ &= X^8 - 4cX^6 + (6c^2 - 2u - 2v)X^4 \\ &\quad - 4(c^3 - cu - cv)X^2 \\ &\quad + (c^4 + u^2 + v^2 - 2c^2 u - 2c^2 v - 2uv) \end{aligned}$$

where

$$u = \frac{1 + t_1^2 + t_1^2 t_2^2}{(1 + t_1^2) t_3^2}$$

$$v = \frac{1 + t_1^2 + t_1^2 t_2^2}{t_2^2 t_3^2}$$

and

$$c = \frac{1 + t_1^2 + t_1^2 t_2^2}{\pm t_1 t_2 t_3}.$$

It is easy to see that $G_t(X)$ has coefficients in $K(t_1, t_2, t_3)$. Moreover, $G_t(X)$ is irreducible over $K(t_1, t_2, t_3)$, as it is the minimal polynomial of the element $\sqrt{c + \sqrt{u} + \sqrt{v}}$ over $K(t_1, t_2, t_3)$. Again by Bucht's theorem (I.2.1),

$$\mathrm{Gal}(G_t / K(t_1, t_2, t_3)) \cong Q_8.$$

If the specialized polynomial $G_\mathbf{a}(X)$ with $\mathbf{a} = (a_1, a_2, a_3) \in K \times K \times K$ is irreducible over K, then $\mathrm{Gal}(G_\mathbf{a}/K) \cong Q_8$.

By Proposition (I.1.8) and Remark (II.1.5), the polynomials $F_t(X)$ of (IV.1) and $G_t(X)$ of (IV.2) are "versal" in the following sense.

(IV.3) Theorem. *Let K be a field with* level$(K) > 2$. *Then any octic polynomial over K with Galois group Q_8 is Tschirnhausen equivalent either to a specialization of the polynomial $F_t(X)$ of* (IV.1) *or to a specialization of the polynomial $G_t(X)$ of* (IV.2).

By a similar argument, one can get "versal" families of generic polynomials when level$(K) \leq 2$, using Theorem (I.3.1) and Theorem (I.3.3). The details are left to the reader.

(IV.4) Example. In (IV.1), take $K = \mathbf{Q}$, $t_1 = t_3 = 1$ and $t_2 = 0$. Then $u = \frac{2}{3}$, $v = 2$ and $c = 2$. Then $F_t(X) = F(1, 0, 1, t_4; X)$ is the minimal polynomial of the element $\sqrt{t_4(2 + \sqrt{\frac{2}{3}} + \sqrt{2} + \sqrt{\frac{4}{3}})}$ over K. For instance, if $t_4 = 3$, we get $F(1, 0, 1, 3; X) = X^8 - 24X^6 + 144X^4 - 288X^2 + 144$ whose discriminant is $2^{52}3^{14}$, and if $t_4 = 2$, we get the example of Dedekind [2]: $F(1, 0, 1, 2; X) = X^8 - 24X^6 + 108X^4 - 144X^2 + 36$ whose discriminant is $2^{50}3^{14}$. (The quaternion extension over $\mathbf{Q}$ of Dedekind is $\mathbf{Q}(\sqrt{(2 + \sqrt{2})(3 + \sqrt{6})})$.) It is easy to see that $\frac{2}{3}(2+\sqrt{2})(3+\sqrt{3})$ differs by the square factor $(1+\sqrt{2}+\sqrt{3})^2(2+\sqrt{2})^2 \in \mathbf{Q}(\sqrt{2}, \sqrt{3})$ from $(2 + \sqrt{2})(3 + \sqrt{3})$. Thus, the minimal polynomial of $\sqrt{(2 + \sqrt{2})(3 + \sqrt{6})}$ over $\mathbf{Q}$ is equivalent to $F(1, 0, 1, 2; X)$ under Tschirnhausen transformation.

Acknowledgement.

The authors are grateful to Anders Thorup for having checked some of our calculations on a computer.

References

[1] Bucht, G., Über einige algebraische Körper achten Grades, Arkiv för Matematik, Astronomi och Fysik, Bd. **6**, No. 30 (1910), pp. 1–36.

[2] Dedekind, R., Konstruktion von Quaternionkörpern, Gesammelte mathematische Werke Bd. **2**, Vieweg Braunschweig, 1931, pp. 376–384.

[3] Hasse, H., Number Theory, Grundlehren der math. Wissenschaften, 229, Springer-Verlag, Berlin/Heidelberg/New York, 1980. [Translated from German.]

[4] Jensen, C. U., On the general inverse problem of Galois Theory, C. R. Math. Rep. Acad. Sci. Canada **8** No. 2 (1986), pp. 145–149.

[5] Kuyk, W., and Lenstra, H. W. Jr., Abelian extensions of arbitrary fields, Math. Ann. **216** (1975), pp. 99–104.

[6] Lang, S., Diophantine Geometry, Interscience Tracts 11, New York, London, 1962.

[7] Lam, T.Y., The Algebraic Theory of Quadratic Forms, Mathematics Lecture Note Series, W.A. Benjamin, Inc., Reading, Massachusetts, 1973.

[8] Massy, R., and Nguyen-Quang-Do, T., Plongement d'une extension de degré p^2 dans une surextension non abélienne de degré p^3, J. Reine Angew. Math. **291** (1977), pp. 149–161.

[9] Mertens, F., Gleichungen, deren Gruppe eine Qauternionen-gruppe ist, Akad. der Wissenschaften Wien, Sitzungsberichte Mathem-Naturwiss. Klasse **130**, Ab. 2a (1921), pp. 69–90.

[10] Reichardt, H., Über Normalkörper mit Quaternionengruppe, Math. Zeitschrift **41** (1936), pp. 218–221.

[11] Whaples, G., Algebraic extensions of arbitrary fields, Duke J. Math. **24** (1957), pp. 201–204.

[12] Witt, E., Konstruktion von galoisschen Körpern der Charakteristik p zu vorgegebener Gruppe der Ordung p^f, J. Reine Angew. Math. **174** (1936), pp. 237–245.

Christian U. Jensen

Matematisk Insitut
Københavns Universitet
Universitetsparken 5, DK-2100 København Ø
Denmark

Noriko Yui

Department of Mathematics
University of Toronto
Toronto, Ontario, M5S 1A1
Canada

Current Address of Noriko Yui
Department of Mathematics and Statistics
Queen's University
Kingston, Ontario, K7L 3N6
Canada

Algebraic Geometry and Commutative Algebra
in Honor of Masayoshi NAGATA
pp. 183–201 (1987)

On the Discriminants of the Intersection Form
on Néron-Severi Groups

Toshiyuki KATSURA

§0. Introduction.

Let S be a non-singular complete algebraic surface defined over an algebraically closed field k of characteristic $p \geq 0$. In case $p > 0$, the surface S is said to be supersingular if the Picard number $\rho(S)$ is equal to the second Betti number $B_2(S)$. For a supersingular K3 surface S , the Néron-Severi group $NS(S)$ was investigated by Artin [A]. He showed that the discriminant of the intersection form on $NS(S)$ is equal to $-p^{2\sigma_0}$ for an integer σ_0, $1 \leq \sigma_0 \leq 10$, and he used the result to examine the structure of the moduli of supersingular K3 surfaces. The integer σ_0 is called the Artin invariant. Recently, T.Ekedahl found an algorithm based on the results of N.Nygaard and B.Mazur for calculating σ_0 when S has a suitable group action. Inspired by his method, T.Shioda constructed some explicit examples of K3 surfaces with big Artin invariant (cf. [S4]). In this paper we propose another method to calculate the discriminant disc $NS(S)$ of the intersection form on the Néron-Severi group. We can use our method to calculate explicitly the discriminants for some elliptic surfaces (cf. Section 4), in particular, for some supersingular K3 surfaces (cf. Section 5).

We give a brief outline of this paper. In Section 1 we summarize the known facts on abelian varieties. In Section 2, for polarized abelian varieties (A, D) and (A', D'), we introduce a bilinear form on $\text{Hom}(A, A')$, and we examine the properties. In Section 3 we give a formula of disc $NS(C \times E)$ of a surface $C \times E$, where C is a curve of genus $n \geq 2$ and E is an elliptic curve. In Section 4 we give examples to which we can apply our method. We really calculate disc $NS(S)$ of certain elliptic surfaces. In Section 5, as an application to our method, we show that a supersingular K3 surface which we introduced in Katsura [K2] is a Kummer surface.

Related topics are found in Tate [T1],[T2], Milne [M1],[M2], Shioda [S1],[S3], Shioda and Inose [SI].

Notations and conventions.

Throughout this paper, we denote by k an algebraically closed field of char-

Received April 17, 1987.

acteristic $p \geq 0$, unless otherwise mentioned. Let V be a non-singular complete algebraic variety of dimension n over k. We use the following notations:

$\mathbf{Z}$: the ring of rational integers,

$\mathbf{Q}$: the field of rational numbers,

$\mathbf{F}_{p^a}$: the finite field with p^a elements where p is a prime number,

$\mathrm{ord}_\ell(a)$: the ℓ-adic order of the integer a for a prime number ℓ,

id : the identity mapping,

O_V : the structure sheaf of V,

K_V : the canonical divisor of V,

$\mathrm{Pic}^0(V)$: the Picard variety of V,

$\mathrm{NS}(V)$: the Néron-Severi group of V,

$H^i_{et}(V, \mathbf{Q}_\ell)$: the i-th ℓ-adic étale cohomology group,

$c_2(V)$: the second Chern number of V,

$B_2(V)$: the second Betti number of V,

$\rho(V)$: the Picard number of V.

For a coherent sheaf F on V,

$H^i(F) := H^i(V, F)$

$\chi(F) := \sum_{i=0}^{n} (-1)^i \dim H^i(V, F)$.

For divisors D_1 and D_2 on V,

$D_1 \sim D_2$: the linear equivalence,

$D_1 \equiv D_2$: the algebraic equivalence.

For abelian varieties $\tilde{A}$, A and a homomorphism $f : \tilde{A} \to A$, $\hat{f} : \mathrm{Pic}^0(A) \to \mathrm{Pic}^0(\tilde{A})$ denotes the dual homomorphism of f. For an integer n, $[n]_A$ denotes the multiplication by n on A. We denote by α_p the local-local group scheme of rank p over k. For a free $\mathbf{Z}$-module U of finite rank with bilinear form, we denote by $\mathrm{disc}\, U$ the discriminant of the bilinear form. For a group G and an element σ of G, we denote by $< \sigma >$ the subgroup generated by σ. We denote by $|\, G\,|$ the order of G. In case G acts on U, U^G denotes the $\mathbf{Z}$-submodule of G-invariants in U with the bilinear form restricted to U^G. Sometimes, a Cartier divisor, the associated invertible sheaf and the associated line bundle are identified. The curve (resp. the surface) defined by an equation means the non-singular complete model of the curve (resp. a minimal non-singular complete model of the surface) defined by the equation, unless otherwise mentioned.

§1. Preliminaries.

In this section, we recall some known facts on abelian varieties. Let k be an algebraically closed field of characteristic $p \geq 0$, and let A be an abelian variety defined over k. For a point x of A, we denote by T_x the translation by x. For a divisor D on A, we have the homomorphism φ_D defined by

$$\varphi_D \ : \quad A \quad \longrightarrow \quad \mathrm{Pic}^0(A).$$
$$\qquad \qquad \cup \qquad \qquad \quad \cup$$
$$\qquad \quad x \quad \longmapsto \quad T_x^* D - D$$

Let E be an elliptic curve defined over k. We denote by o the zero point of E. We set $B = \text{End}^0(E)$ and $\mathcal{O} = \text{End}(E)$. The algebra B is isomorphic to one of the following :

    (i) $\mathbf{Q}$
    (ii)   an imaginary quadratic field,
    (iii)   the quaternion division algebra over $\mathbf{Q}$ with discriminant p.

We note that Case (iii) occurs only in characteristic $p > 0$. The order $\mathcal{O}$ is a free $\mathbf{Z}$-module. We denote by r the rank of $\mathcal{O}$ over $\mathbf{Z}$. Then, r is given respectively by

$$\text{(i) } 1, \text{ (ii) } 2, \text{ (iii) } 4.$$

We denote by $^{-}$ the canonical involution of B (the identity in Case (i)). For an element a of B, we denote by $\text{Tr}(a)$ the reduced trace of a ($\text{Tr}(a) = a$ in Case (i)). Let $\alpha_1, \ldots, \alpha_r$ be a basis of $\mathcal{O}$ over $\mathbf{Z}$. We denote by M the matrix whose ij-component is given by $\text{Tr}\, \bar{\alpha}_i \alpha_j$:

$$\text{(1)} \qquad M = (\text{Tr}\, \bar{\alpha}_i \alpha_j).$$

In Case (i), we have $M = (1)$ and $\det M = 1$. In Case (iii), we have

$$\text{(2)} \qquad \det M = p^2.$$

    We set

$$\text{(3)} \qquad X = E^{n-1} \times \{o\} + E^{n-2} \times \{o\} \times E + \cdots + \{o\} \times E^{n-1}.$$

Then, X is a principal polarization on E^n, and we have an injective homomorphism

$$j \; : \; \text{NS}(E^n) \; \longrightarrow \; \text{End}(E^n) \; \simeq \; M_n(\mathcal{O}).$$

$$D \; \longmapsto \; \varphi_X^{-1} \circ \varphi_D$$

We can easily prove the following lemma by Mumford [14, p.209].

Lemma 1.1. (Ibukiyama, Katsura, Oort [IKO, p.144]) *The image of* $\text{NS}(E^n)$ *by* j *is*

$$\{A \in M_n(\mathcal{O}) : A = {}^t\bar{A}\}$$

and the following holds : for $D \in \text{NS}(E^n)$

$$D^n/n! = \begin{cases} \det(j(D)) & \text{in Case (i) or (ii)} \\ \text{HNm}(j(D)) & \text{in Case (iii),} \end{cases}$$

where HNm *is the Hauptnorm of* $M_n(B)$ *(cf.* Braun and Koecher [BK,Chapter II]). *In particular, if* $n = 2$ *in Case* (iii), *then* $\text{HNm}(j(D)) = \det(j(D))$.

Now, assume $char.k = p > 0$.

Definition 1.2. An abelian variety A of dimension n is said to be *supersingular* (resp. *superspecial*) if it is isogenous (resp. isomorphic) to a product of n supersingular elliptic curves. A non-singular complete curve C is said to be *supersingular* (resp. *superspecial*) if the Jacobian variety $J(C)$ is supersingular (resp. superspecial).

Lemma 1.3. (Deligne, Ogus, Serre) *Superspecial abelian varieties of dimension $n \geq 2$ are isomorphic to each other.*

For the proof, see Shioda [S2,Theorem 3.5] and Ogus [Og].

Lemma 1.4. (Oort [Oo,Corollary 7]) *Let E be a supersingular elliptic curve. Then, any supersingular abelian surface is isomorphic to a quotient surface of $E \times E$ by a subgroup scheme which is isomorphic to α_p.*

§2. Bilinear forms.

Let k be an algebraically closed field of characteristic $p \geq 0$, and let (A, D) (resp.(A', D')) be a polarized abelian variety of dimension n (resp. of dimension m) defined over k. For $f, g \in \mathrm{Hom}(A, A')$, we set

$$(4) \quad (f, g) = (1/(n-1)!)\{((f^*D')\cdot D^{n-1}) \\ + ((g^*D')\cdot D^{n-1}) \quad - (((f+g)^*D')\cdot D^{n-1})\},$$

where f^*D', g^*D' and $(f+g)^*D'$ mean the pull-back as line bundles. (In case $n = m$, see also Lang [L].)

Lemma 2.1. *The pairing defined by (3) is bilinear.*

Proof. Let f, g and h be elements of $\mathrm{Hom}(A, A')$. Then, by Mumford [Mu,p.58, Corollary 2], we have

$$(f + g + h)^*D' \sim (f+g)^*D' + (f+h)^*D' \\ + (g+h)^*D' - f^*D' - g^*D' - h^*D'.$$

The result follows from this formula.   Q.E.D.

Definition 2.2. We call the pairing defined by (4) a bilinear form on $\mathrm{Hom}(A, A')$ associated with D and D'.

Let $Aut(A, D)$ (resp. $Aut(A', D')$) be the subgroup of the automorphism group $Aut(A)$ (resp. $Aut(A')$) of A (resp. A') which preserves the polarization

D (resp. D'). We define the action of an element (σ_1, σ_2) of $Aut(A, D) \times Aut(A', D')$ on $Hom(A, A')$ by

(5)
$$
(\sigma_1, \sigma_2) \quad : \quad Hom(A, A') \quad \longrightarrow \quad Hom(A, A').
$$
$$
f \quad \longmapsto \quad \sigma_2 \circ f \circ \sigma_1
$$

Then, it is easy to see that the action of $Aut(A, D) \times Aut(A', D')$ preserves the bilinear form defined by (4). Now, we consider the case of $m = 1$. Let E be an elliptic curve, and o the zero point of E. Then, the bilinear form associated with D and o defined by (4) becomes as follows.

Lemma 2.3. For $f, g \in Hom(A, E)$,

$$
(f, g) = \quad (1/n!)\{(\mathrm{Ker}(f) + D)^n
$$
$$
+ (\mathrm{Ker}(g) + D)^n - (\mathrm{Ker}(f + g) + D)^n - D^n\}
$$

Proof. For any abelian subvariety A_0 of A of dimension $n - 1$ and any divisor G of A, we have

$$
(A_0^i \cdot G^{n-i}) = 0 \text{ for } i \geq 2.
$$

Therefore, the result follows from the definition of the bilinear form.   Q.E.D.

For the rest of this section, assume

(6)
$$
A = E^n.
$$

We denote by o_i the zero point of the i-th factor of E^n. We have the addition

$$
m : E^n \quad \longrightarrow \quad E
$$
$$
{}^t(x_1, \ldots, x_n) \quad \longmapsto \quad x_1 + \cdots + x_n
$$

We set
(7)
$$
\Delta = \mathrm{Ker}\, m.
$$

Let X be the principal divisor as in (3). Then, we have the following lemma.

Lemma 2.4.

$$
j(\Delta) = \begin{pmatrix} 1 & \cdots & 1 \\ \vdots & \ddots & \vdots \\ 1 & \cdots & 1 \end{pmatrix}
$$

Proof. Let $x = {}^t(x_1, \ldots, x_g)$ be a point of E^n. We have a commutative diagram

$$
\begin{array}{ccccc}
E^n & \xrightarrow{\varphi_X} & \mathrm{Pic}^0(E^n) & \ni & L \\
\varphi_{0_1} \times \cdots \times \varphi_{0_n} & \searrow & \mathrm{res} \downarrow & & \downarrow \\
& & \mathrm{Pic}^0(E)^n & \ni & (L\,|_{L_1}, \cdots, L\,|_{E_n}),
\end{array}
$$

where $L\,|_{E_i}$ is the restriction of L to the i-th factor of E^n. It is well-known that the homomorphism res is an isomorphism. Hence, we have

$$
\begin{aligned}
\varphi_X^{-1} \circ \varphi_\Delta(x) &= {}^t(\varphi_{0_1}^{-1}(T_x^*\Delta - \Delta)\,|_{E_1}, \cdots, \varphi_{0_n}^{-1}(T_x^*\Delta - \Delta)\,|_{E_n}) \\
&= {}^t(x_1 + x_2 + \cdots + x_n, \ldots, x_1 + x_2 + \cdots + x_n).
\end{aligned}
$$

Now, the result follows from the definition of j.   Q.E.D.

We set $\mathcal{O} = \mathrm{End}(E)$. Then, we have the natural isomorphism

$$
(8) \qquad \mathrm{Hom}(E^n, E) \simeq \{(a_1, \ldots, a_n) \mid a_i \in \mathcal{O}(i = 1, \ldots, n)\}.
$$

We identify $\mathrm{Hom}(E^n, E)$ with $\{(a_1, \ldots, a_n) \mid a_i \in \mathcal{O}(i = 1, \ldots, n)\}$ by this isomorphism. For a homomorphism $f = (a_1, \ldots, a_n)$, we set

$$
(9) \qquad \Delta_{a_1, \ldots, a_n} = (a_1 \times \cdots \times a_n)^* \Delta.
$$

As is well-known, we have

$$
(10) \qquad \varphi_X^{-1} \circ (a_1 \times \cdots \times a_n)^{\!\scriptscriptstyle\mathsf{f}} \circ \varphi_X = \bar{a}_1 \times \cdots \times \bar{a}_n.
$$

Therefore, we have a commutative diagram

$$
(11) \qquad
\begin{array}{ccc}
E^n & \xrightarrow{a_1 \times \ldots \times a_n} & E^n \\
\varphi_{\Delta_{a_1, \ldots, a_n}} \downarrow & & \downarrow \varphi_\Delta \\
\mathrm{Pic}^0(E^n) & \xleftarrow{(a_1 \times \ldots \times a_n)^{\mathsf{f}}} & \mathrm{Pic}^0(E^n) \\
\varphi_X \uparrow & & \uparrow \varphi_X \\
E^n & \xleftarrow{\bar{a}_1 \times \ldots \times \bar{a}_n} & E^n
\end{array}
$$

Lemma 2.5.

$$
j(\Delta_{a_1, \ldots, a_n}) =
\begin{pmatrix}
\bar{a}_1 a_1 & \bar{a}_1 a_2 & \cdots & \bar{a}_1 a_n \\
\bar{a}_2 a_1 & \bar{a}_2 a_2 & \cdots & \bar{a}_2 a_n \\
\vdots & & & \vdots \\
\bar{a}_n a_1 & \bar{a}_n a_2 & \cdots & \bar{a}_n a_n
\end{pmatrix}
$$

Proof. Using the diagram (11) and Lemma 2.4, we have

$$j(\Delta_{a_1,\ldots,a_n}) = \varphi_X^{-1} \circ \varphi_{\Delta_{a_1,\ldots,a_n}}$$

$$= \begin{pmatrix} \bar{a}_1 & & 0 \\ & \ddots & \\ 0 & & \bar{a}_n \end{pmatrix} \begin{pmatrix} 1 & \cdots & 1 \\ \vdots & \ddots & \vdots \\ 1 & \cdots & 1 \end{pmatrix} \begin{pmatrix} a_1 & & 0 \\ & \ddots & \\ 0 & & a_n \end{pmatrix}.$$

Q.E.D.

Theorem 2.6. *The bilinear form on* $\mathrm{Hom}(E^n, E)$ *associated with* D *and* o *is given as follows : for elements* $f = (a_1, \ldots, a_n)$ *and* $g = (b_1, \ldots, b_n)$ *of* $\mathrm{End}(E^n, E)$,

$$(f, g) = \begin{cases} \det(j(\Delta_{a_1,\ldots,a_n} + D)) + \det(j(\Delta_{b_1,\ldots,b_n} + D)) \\ \quad - \det(j(\Delta_{a_1+b_1,\ldots,a_n+b_n} + D)) - \det(j(D)) \\ \qquad \text{in Cases (i), (ii),} \\ \mathrm{HNm}(j(\Delta_{a_1,\ldots,a_n} + D)) + \mathrm{HNm}(j(\Delta_{b_1,\ldots,b_n} + D)) \\ \quad - \mathrm{HNm}(j(\Delta_{a_1+b_1,\ldots,a_n+b_n} + D)) - \mathrm{HNm}(j(D)) \\ \qquad \text{in Case (iii).} \end{cases}$$

Proof.   Since $\mathrm{Ker}(f) = \Delta_{a_1,\ldots,a_n}$, $\mathrm{Ker}(g) = \Delta_{b_1,\ldots,b_n}$ and $\mathrm{Ker}(f+g) = \Delta_{a_1+b_1,\ldots,a_n+b_n}$, this theorem follows from Lemmas 1.1 and 2.3.   Q.E.D.

§3. The Discriminant of the intersection form.

Let C be a non-singular complete curve of genus $n \geq 2$, and E an elliptic curve with the zero point o. Let

$$\psi : C \longrightarrow J(C)$$

be a natural morphism. For a point P of C, we may assume that $\psi(P)$ is the zero point of $J(C)$. We denote by Θ the subvariety of $J(C)$ obtained by taking the sum of $\psi(C)$ on $J(C)$ n-1 times. Θ gives a principal polarization on $J(C)$. In particular, we have

(12) $\Theta^n = n!$.

By Matsusaka [Ma], we have

(13) $\Theta^{n-1} \equiv (n-1)! C$

We consider the algebraic surface $C \times E$. We denote by p_1 (resp. p_2) the projection from $C \times E$ to the first factor C (resp. the second factor E). We set

$\mathrm{DC}(C \times E) = \{$line bundle L on $C \times E$ such that both $L\big|_{p_1^{-1}(P)}$ and $L\big|_{p_2^{-1}(o)}$ are trivial$\}$.

As is well-known, we have the natural isomorphism

(14) $\mathrm{NS}(C \times E) \simeq \{\mathbf{Z}C + \mathbf{Z}E\} \bigoplus \mathrm{DC}(C \times E).$

We denote by $\mathcal{P}$ the Poincaré bundle on $\mathrm{Pic}^0(E) \times E$. For the line bundle L on $\mathrm{DC}(C \times E)$, there exists a morphism

$$f_L : C \longrightarrow \mathrm{Pic}^0(E)$$

such that $(f_L \times id)^* \mathcal{P} \simeq L$. By the universality of the Jacobian variety, there exists a homomorphism $\tilde{f}_L : J(C) \longrightarrow \mathrm{Pic}^0(E)$ such that the following diagram commutes :

$$
\begin{array}{ccc}
C & \xrightarrow{f_L} & \mathrm{Pic}^0(E) \\
\psi \downarrow & \nearrow & \tilde{f}_L \\
J(C) & &
\end{array}
$$

we set

$$g_L = \varphi_0^{-1} \circ \tilde{f}_L.$$

Then we have the natural isomorphism

(15)
$$
\begin{array}{ccc}
\mathrm{DC}(C \times E) & \longrightarrow & \mathrm{Hom}(J(C), E). \\
\cup & & \cup \\
L & \longmapsto & g_L
\end{array}
$$

Lemma 3.1. *Under the above notations,*

$$L^2 = -(2/(n-1)!)(\mathrm{Ker}(g_L) \cdot \Theta^{n-1}) \ \text{for } L \in \mathrm{DC}(C \times E).$$

Proof. By the Riemann-Roch theorem, we have

$$L^2 = ((f_L \times id)^* \mathcal{P})^2 = deg(f_L) \cdot \mathcal{P}^2 = 2deg(f_L) \cdot \chi(\mathcal{P}).$$

By Mumford [Mu,Chapter III, Section 13], we have $\chi(\mathcal{P}) = -1$. Hence, this lemma follows from $deg(f_L) = (\mathrm{Ker}(g_L) \cdot C)$ and (13).   Q.E.D.

Corollary 3.2. *The isomorphism* (15) *gives an isometry from* $\mathrm{DC}(C \times E)$ *to* $\mathrm{Hom}(J(C), E)$ *with the bilinear form associated with* Θ *and* o. *In particular,* $disc\,\mathrm{NS}(C \times E) = -disc\,\mathrm{Hom}(J(C), E).$

Proof. This follows from Lemmas 2.3, 3.1 and (14).   Q.E.D.

Let σ_1 (resp. σ_2) be an automorphism of C (resp. E) which fixes the point P (resp. o). The automorphism σ_1 induces an automorphism of $(J(C), \Theta)$. We denote it again by σ_1. For $L \in \mathrm{DC}(C \times E)$, we have

$$f_{(\sigma_1 \times \sigma_2)^* L} = \hat{\sigma}_2 \circ f_L \circ \sigma_1.$$

Therefore, by $\varphi_O^{-1} \circ \hat{\sigma}_2 \circ \varphi_O = \bar{\sigma}_2$, the action of $(\sigma_1 \times \sigma_2)^*$ on $DC(C \times E)$ induces the action on $\mathrm{Hom}(J(C), E)$ which is given by

$$
\begin{array}{ccc}
\mathrm{Hom}(J(C), E) & \longrightarrow & \mathrm{Hom}(J(C), E). \\
\cup & & \cup \\
g_L & \longmapsto & \bar{\sigma}_2 \circ g_L \circ \sigma_1
\end{array}
$$

(16)

This action is nothing but the action of $(\sigma_1, \bar{\sigma}_2)$ which is defined by (5).

Finally, we give two lemmas to calculate the discriminants of the intersection forms on Néron-Severi groups of certain elliptic surfaces.

Lemma 3.3. *Let S be a non-singular complete algebraic surface defined over k. Let σ be an automorphism of S of order two, three, four or six, and let G be a finite group generated by σ. Let T be a non-singular complete model of the quotient surface S/G. Assume the following three conditions :*
  (1) *$|G|$ is prime to p,*
  (2) *every element of G except the identity has only isolated fixed points,*
  (3) *both $\mathrm{NS}(S)$ and $\mathrm{NS}(T)$ are torsion-free.*
Then, for any prime ℓ with $(\ell, |G|) = 1$,

$$\mathrm{ord}_\ell \mathrm{disc}\, \mathrm{NS}(S)^G = \mathrm{ord}_\ell \mathrm{disc}\, \mathrm{NS}(T)$$

Proof. The proof is similar to the one in Katsura [K2,Lemma 5.8] (see also Shioda [S2, Proposition 3.1]). We omit the details. Q.E.D.

The following lemma is essentially given in Shioda [S1, Remark 1.9]. We give here an easy proof.

Lemma 3.4. *Let $f : S \to C$ be a minimal elliptic surface without multiple fibres. Assume $\chi(\mathcal{O}_S) \neq 0$. Then, the Néron-Severi group $\mathrm{NS}(S)$ of S is torsion-free.*

Proof. Let D be a divisor which is numerically equivalent to zero. Then, we have $\chi(\mathcal{O}_S(D)) = \chi(\mathcal{O}_S) > 0$. Therefore, we have $H^0(\mathcal{O}_S(D)) \neq 0$ or $H^2(\mathcal{O}_S(D)) \neq 0$. If $H^0(\mathcal{O}_S(D)) \neq 0$, then D is linearly equivalent to zero. Therefore, D is algebraically equivalent to zero. If $H^2(\mathcal{O}_S(D)) \neq 0$, then we have $H^0(\mathcal{O}_S(K_S - D)) = 0$ by the Serre duality. Therefore, there exists an effective divisor G such that $K_S - D \sim G$. For a general fibre E, we have $G \cdot E = 0$. Therefore, G consists of components of fibres. Since $G^2 = K_S^2 = 0$, G consists of fibres (cf. Bombieri and Mumford [BM, p.28]). Since f has no multiple fibres, we conclude that D is algebraically equivalent to zero. Q.E.D.

§4. Examples.

Let C be a non-singular complete curve of genus two and E an elliptic curve. As before, we set $B = \mathrm{End}^0(E)$ and $\mathcal{O} = \mathrm{End}(E)$. Under the notations in Sections 1, 2 and 3, we have the following theorem.

Theorem 4.1. *Assume that $J(C)$ is isomorphic to $E \times E$. Then,* disc $\mathrm{NS}(C \times E) = -(\det M)^2$. *In particular, in case char.$k = p > 0$, if E is supersingular, then* disc $\mathrm{NS}(C \times E) = -p^4$.

Proof. By Lemma 1.1, the principal polarization $\Theta = \psi(C)$ on $J(C)$ corresponds to a matrix

$$\begin{pmatrix} a & \gamma \\ \bar{\gamma} & b \end{pmatrix} \in M_2(\mathcal{O}),$$

where $a, b \in \mathbf{Z}$ and
(17)
$$ab - \gamma\bar{\gamma} = 1.$$

By Theorem 2.6, the bilinear form of $\mathrm{Hom}(J(C), E)$ associated with Θ and o is given by the matrix

$$\begin{pmatrix} -bM & N \\ {}^tN & -aM \end{pmatrix}$$

with respect to the basis $\{(\alpha_1, 0), \cdots, (\alpha_r, 0), (0, \alpha_1), \cdots, (0, \alpha_r)\}$, where

$$N = \begin{pmatrix} \mathrm{Tr}\,\bar{\alpha}_1\alpha_1\gamma & \cdots & \mathrm{Tr}\,\bar{\alpha}_r\alpha_1\gamma \\ \vdots & & \vdots \\ \mathrm{Tr}\,\bar{\alpha}_1\alpha_r\gamma & \cdots & \mathrm{Tr}\,\bar{\alpha}_r\alpha_r\gamma \end{pmatrix}$$

Since $(\alpha_1, \ldots, \alpha_r)$ is a basis of $\mathcal{O}$ over $\mathbf{Z}$, there exists a matrix $Q \in M_r(\mathbf{Z})$ such that

$$(\alpha_1\gamma, \ldots, \alpha_r\gamma) = (\alpha_1, \ldots, \alpha_r)Q.$$

Therefore, we have

(18)
$$\gamma\bar{\gamma}M = {}^tQMQ \quad \text{and} \quad {}^tN = MQ.$$

Since $\det M \neq 0$, we have
(19)
$$(\det Q)^2 = (\gamma\bar{\gamma})^r.$$

By (17) we have $b \neq 0$. Therefore, we have

$$\begin{aligned} \mathrm{disc}\,\mathrm{Hom}(J(C), E) &= \det \begin{pmatrix} -bM & {}^tQ^tM \\ MQ & -aM \end{pmatrix} \\ &= (\det bM)\det(aM - MQ(bM)^{-1}{}^tQ^tM) \\ &= (\det bM)(\det Q)^{-2}\det(a^tQMQ - ({}^tQMQ)(bM)^{-1}({}^tQMQ)) \\ &= (\det M)(\det Q)^{-2}\det((ab\gamma\bar{\gamma} - (\gamma\bar{\gamma})^2)M) \\ &= (\det M)^2 \end{aligned}$$

by (17) and (19). Hence, by Corollary 3.2, we have disc $\mathrm{NS}(C \times E) = -(\det M)^2$. The latter part follows from (2).   Q.E.D.

For the rest of this section we assume *char.k* $= p > 0$. Let E be a supersingular elliptic curve defined over $\mathbf{F}_p$ such that $\mathrm{End}(E)$ is defined over $\mathbf{F}_{p^2}$ (for the existence of such an elliptic curve, see Waterhouse [W,Theorem 4.1.5]). Assume that $J(C)$ is supersingular and is not superspecial. By Lemma 1.4, there exists an immersion

$$(20) \qquad (a,1) : \alpha_p \longrightarrow \alpha_p \times \alpha_p \subset E \times E \quad \text{with } a \in k$$

such that
$$(21) \qquad J(C) \simeq (E \times E)/(a,1)(\alpha_p).$$

The Lie algebra of α_p is isomorphic to k with trivial Frobenius action. The subgroup schemes of $\alpha_p \times \alpha_p$ bijectively correspond to the Lie subalgebras of the Lie algebra $k \bigoplus k$ with trivial Frobenius action. The immersion (20) corresponds to the homomorphism

$$(22) \qquad \begin{array}{ccc} k & \longrightarrow & k \oplus k \\ \cup\!\!\shortmid & & \cup\!\!\shortmid \\ x & \longmapsto & (ax, x) \end{array}$$

We identify $J(C)$ with $(E \times E)/(a,1)(\alpha_p)$ by the isomorphism in (21). Let

$$\pi : E \times E \longrightarrow (E \times E)/(a,1)(\alpha_p)$$

be the canonical projection. Since $J(C)$ is not isomorphic to a product of two supersingular elliptic curves, we have

$$(23) \qquad a \notin \mathbf{F}_{p^2}$$

(cf. Oort [Oo,Introduction]). We set

$$P(J(C)) = \{D \in \mathrm{NS}(J(C)) \mid D > 0, \, D^2 = 2\},$$

where $D > 0$ means that the divisor class D contains an effective divisor. The set $P(J(C))$ is the set of principal polarizations on $J(C)$. We denote by F the primitive element of $\mathcal{O}$.

Lemma 4.2. (Ibukiyama, Katsura, Oort [IKO, Proposition 2.14])
The set $\pi^{-1}(P(J(C)))$ *bijectively corresponds to*

$$\Lambda = \left\{ \begin{pmatrix} ps & F\gamma \\ F\gamma & pt \end{pmatrix} \mid s,t \in \mathbf{Z}; \; s > 0, t > 0; \; \gamma \in \mathcal{O}; \; pst - \gamma\bar{\gamma} = 1 \right\}.$$

Lemma 4.3. *Under the above notations, the homomorphism*

$$\begin{array}{ccc} \mathrm{Hom}(J(C), E) & \longrightarrow & \{(F\gamma_1, F\gamma_2) \mid \gamma_1, \gamma_2 \in \mathcal{O}\} \subset \mathrm{Hom}(E \times E, E) \\ \cup\!\!\shortmid & & \cup\!\!\shortmid \\ g & \longmapsto & g \circ \pi \end{array}$$

is an isomorphism.

Proof.   We have the natural restriction

$$\text{res} : \text{Hom}(E \times E, E) \longrightarrow \text{Hom}(\alpha_p \times \alpha_p, \alpha_p).$$

Since $\text{End}(\alpha_p) \simeq k$, we have an isomorphism

$$\theta : \text{Hom}(\alpha_p \times \alpha_p, \alpha_p) \simeq k \bigoplus k.$$

By Oort [Oo,Lemma 5], the image of $\theta \circ \text{res}$ is equal to $\mathbf{F}_{p^2} \bigoplus \mathbf{F}_{p^2}$. Let g be an element of $\text{Hom}(J(C), E)$. The element $\theta(\text{res}(g \circ \pi))$ is given by (a_1, a_2) with $a_i \in \mathbf{F}_{p^2}$ $(i = 1, 2)$. Since $g \circ \pi((a, 1)(\alpha_p)) = 0$, we have

$$a_1 a + a_2 = 0.$$

Therefore, by (23) we have $a_1 = a_2 = 0$. Therefore, there exists $\gamma_i \in \mathcal{O}(i = 1, 2)$ such that $g \circ \pi = (F\gamma_1, F\gamma_2)$. Hence, $g \circ \pi$ is contained in $\{(F\gamma_1, F\gamma_2) \mid \gamma_1, \gamma_2 \in \mathcal{O}\}$. The other parts are clear.   Q.E.D.

Theorem 4.4. *Let E be a supersingular elliptic curve as above, and C a supersingular curve of genus two which is not superspecial. Then, $\text{disc} \, \text{NS}(C \times E) = -p^8$.*

Proof.   Let

$$\begin{pmatrix} pa & F\gamma \\ F\gamma & pb \end{pmatrix}$$

be an element of Λ which corresponds to $\pi^*(C)$. Since $(\alpha_1, \alpha_2, \alpha_3, \alpha_4)$ is a basis of $\mathcal{O}$ over $\mathbf{Z}$ under the notations in Section 1, $\{(F\alpha_i, 0), (0, F\alpha_i)\}_{i=1,2,3,4}$ is a basis of $\{(F\gamma_1, F\gamma_2) \mid \gamma_i \in \mathcal{O} \ (i = 1, 2)\}$ over $\mathbf{Z}$. For $L \in \text{DC}(C \times E)$, we have

$$(24) \qquad (C \cdot \text{Ker}(g_L)) = (1/p)(\pi^*(C) \cdot \text{Ker}(g_L \circ \pi)).$$

Therefore, by Theorem 2.6, Corollary 3.2, Lemma 4.3 and (24), the intersection matrix of $\text{NS}(C \times E)$ is given by

$$\begin{pmatrix} 0 & 1 & & 0 \\ 1 & 0 & & \\ & 0 & -pbM & R \\ & & {}^tR & -paM \end{pmatrix}$$

with respect to the basis consisted of $C \times \{o\}$, $\{P\} \times E$ and the elements corresponding to $(F\alpha_i, 0)$ $(i = 1, 2, 3, 4)$ and $(0, F\alpha_i)$ $(i = 1, 2, 3, 4)$. Here, M is the matrix given by (1), and

$$R = \begin{pmatrix} \text{Tr} \, \bar{\alpha}_1 \alpha_1 F\gamma & \cdots & \text{Tr} \, \bar{\alpha}_4 \alpha_1 F\gamma \\ \vdots & & \vdots \\ \text{Tr} \, \bar{\alpha}_1 \alpha_4 F\gamma & \cdots & \text{Tr} \, \bar{\alpha}_4 \alpha_4 F\gamma \end{pmatrix}$$

Since $\{\alpha_1, \cdots, \alpha_4\}$ is a basis of $\mathcal{O}$ over $\mathbf{Z}$, there exists a matrix $Q \in M_4(\mathbf{Z})$ such that

$$(\alpha_1 F\gamma, \cdots, \alpha_4 F\gamma) = (\alpha_1, \cdots, \alpha_4)Q.$$

Then, we have

$$p\gamma\bar{\gamma}M = {}^tQMQ \quad \text{and} \quad {}^tR = MQ.$$

Therefore, by the same method as in Theorem 4.1, we have $\operatorname{disc} \operatorname{NS}(C \times E) = -p^8$. Q.E.D.

Example 4.5. Assume $char.k \geq 5$. Let $h(t)$ be a polynomial of degree six with coefficients in k. Assume that $h(t)$ has no multiple zeros. We consider the minimal non-singular complete model $f : S \to \mathbf{P}^1$ of elliptic surfaces defined by the equation

$$y^2 = 4x^3 - h(t)^2 x - \alpha h(t)^3$$

with $\alpha \in \mathbf{F}_p$ such that $1 - 27\alpha^2 \neq 0$. Here, t is a global coordinate of an affine line $\mathbf{A}^1$ in the projective line $\mathbf{P}^1$. Since $c_2(S) = 6deg(h(t)) = 36$, the Kodaira dimension $\kappa(S)$ of S is equal to one. Let C (resp. E) be the curve of genus two (resp. the elliptic curve) defined by the equation

$$z^2 = f(t) \quad (\text{resp. } Y^2 = 4X^3 - X - \alpha).$$

We have an automorphism σ_1 (resp. σ_2) of C (resp. E) defined by

$$\sigma_1 : t \longmapsto t, \ z \longmapsto -z \quad (\text{resp. } \sigma_2 : X \longmapsto X, \ Y \longmapsto -Y).$$

The elliptic surface S is birationally equivalent to $(C \times E)/ < \sigma_1 \times \sigma_2 >$. Since the action of σ_1 (resp. σ_2) on $H^1_{et}(C, \mathbf{Q}_\ell)$ (resp. $H^1_{et}(E, \mathbf{Q}_\ell)$) is given by the multiplication by -1, we see that the action of $\sigma_1 \times \sigma_2$ on $H^2_{et}(C \times E, \mathbf{Q}_\ell)$ is trivial. Therefore, the action of $\sigma_1 \times \sigma_2$ on $\operatorname{NS}(C \times E)$ is trivial. Now, assume that E is a supersingular elliptic curve such that $\operatorname{End}(E)$ is defined over $\mathbf{F}_{p^2}$. By $\chi(\mathcal{O}_S) = (c_2(S) + K_S^2)/12 = c_2(S)/12$, we see that $B_2(S)$ is even. Therefore, if $\rho(S)$ is equal to $B_2(S)$, then by the Hodge index theorem the discriminant of the intersection form on $\operatorname{NS}(S)$ is negative. Therefore, by Lemmas 3.3, 3.4, Theorems 4.1 and 4.4, we have the following :

1) if C is superspecial, then $\operatorname{disc} \operatorname{NS}(S) = -p^4$,
2) if C is supersingular and is not superspecial, then $\operatorname{disc} \operatorname{NS}(S) = -p^8$.

§5. A K3 surface.

In this section, we assume $char.k = p \geq 5$. In Katsura [K1], we introduced the notion of an irrational unirational elliptic surface of base change type, and under the assumption that it has a section we determined all such surfaces. Among such surfaces, four K3 surfaces appear, that is, the surfaces defined by the following equations :

(I) in case $p \equiv 3(\mathrm{mod}\,4)$ and $p \neq 3$

$$\text{(i)} \quad y^2 = 4x^3 - t^3(t-1)^3 x,$$

(II) in case $p \equiv 5(\mathrm{mod}\,6)$,

$$\text{(i)} \qquad y^2 = 4x^3 - t^4(t-1)^4,$$
$$\text{(ii)} \qquad y^2 = 4x^3 - t^4(t-1)^5,$$
$$\text{(iii)} \qquad y^2 = 4x^3 - t^5(t-1)^5.$$

In Katsura [K1,Proposition 5.1], we showed that the minimal elliptic surfaces in Classes (I) (i), (II) (i), (ii) are generalized Kummer K3 surfaces, and in Katsura [K2,Theorem 5.9], we showed that they are really Kummer surfaces which are obtained from a product of two supersingular elliptic curves. (As for the surface (II)(ii), we didn't prove the latter fact in Katsura [K2], but we can prove it by the same way.) In this section, we prove the following theorem.

Theorem 5.1. *Assume* $p \equiv 5(\mathrm{mod}\,6)$. *The minimal elliptic surface* $f :$ $S \to \mathbf{P}^1$ *defined by the equation* (II) (iii) *is a Kummer surface which is obtained from a product of two supersingular elliptic curves.*

For the rest of this section, we assume $char.k = p \equiv 5(\mathrm{mod}\,6)$. To prove Theorem 5.1, we examine the structure of the minimal elliptic surface $f : S \to$ $\mathbf{P}^1$ defined by the equation (II) (iii). We consider the non-singular complete curve defined by the equation

$$\text{(25)} \qquad z^6 = t(t-1).$$

We set

$$u = (-4)^{1/6} z, \quad v = (-1)^{1/2}(2t - 1).$$

Then the curve C is isomorphic to the curve defined by

$$\text{(26)} \qquad v^2 = u^6 - 1.$$

Let E be an elliptic curve defined by

$$\text{(27)} \qquad Y^2 = X^3 - 1.$$

We set, as before, $B = \mathrm{End}^0(E)$ and $\mathcal{O} = \mathrm{End}(E)$, and we denote by o the zero point of E. By the assumption $p \equiv 5(\mathrm{mod}\,6)$, E is a supersingular elliptic curve and C is a superspecial curve of genus two. The curve C (resp. E) has the automorphisms defined by

$$\text{(28)} \qquad
\begin{array}{rcl}
\iota & : & u \longmapsto u, \ v \longmapsto -v, \\
\sigma & : & u \longmapsto -u, \ v \longmapsto v, \\
\sigma_1 & : & u \longmapsto -\omega u, \ v \longmapsto v, \\
(\text{resp. } \sigma_2 & : & X \longmapsto \omega^2 X, \ Y \longmapsto -Y),
\end{array}$$

where ω is a primitive cube root of unity. We set

$$\tau = \sigma \circ \iota.$$

It is easy to see that the elliptic surface S is birationally equivalent to $(C \times E)/G$ with $G = <\sigma_1 \times \sigma_2>$. We set

$$E_\sigma = C/<\sigma> \quad \text{and} \quad E_\tau = C/<\tau>.$$

Then both E_σ and E_τ are elliptic curves which are isomorphic to the curve defined by (27). We identify E_σ, E_τ with E by an isomorphism. We denote by π_σ (resp. π_τ) the canonical projection from C to E_σ (resp. E_τ). By the universality of the Jacobian variety, we have a diagram :

$$
\begin{array}{ccc}
C & \overset{\psi}{\longrightarrow} & J(C) \\
(\pi_\sigma, \pi_\tau) \quad \searrow & & \downarrow \pi_1 \\
& E_\sigma \times E_\tau &
\end{array}
$$

(cf. Igusa [Ig]). We consider C as a subvariety of $J(C)$ by the injection ψ. Let P be a point of C fixed by σ_1. We choose $\psi(P)$ as the zero point of $J(C)$, and $\pi_1(\psi(P))$ as the zero point of $E_\sigma \times E_\tau$. Then π_1 is a homomorphism. By Igusa [Ig, p.648], we see that Ker π_1 is isomorphic to $\mathbf{Z}/2\mathbf{Z} \times \mathbf{Z}/2\mathbf{Z}$. Therefore, there exists a homomorphism $\pi_2 : E_\sigma \times E_\tau \to J(C)$ of degree four such that

$$\pi_1 \circ \pi_2 = [2]_{E_\sigma \times E_\tau}.$$

By Katsura and Oort [KO,(7.12)], we have

$$(29) \qquad\qquad \pi_2^* C \equiv 2E_\sigma + 2E_\tau.$$

The automorphism σ_1 of C induces the automorphism of E_σ(resp. E_τ) defined by

$$\rho_\sigma : X \longmapsto \omega^2 X, \ Y \longmapsto Y$$

$$(\text{resp. } \rho_\tau : X \longmapsto \omega^2 X, \ Y \longmapsto -Y).$$

We denote again by σ_1 the automorphism of $J(C)$ induced by σ_1. Then we have a commutative diagram

$$
\begin{array}{ccc}
E_\sigma \times E_\tau & \overset{\rho_\sigma \times \rho_\tau}{\longrightarrow} & E_\sigma \times E_\tau \\
\downarrow \pi_2 & & \downarrow \pi_2 \\
J(C) & \overset{\sigma_1}{\longrightarrow} & J(C) \\
\downarrow \pi_1 & & \downarrow \pi_1 \\
E_\sigma \times E_\tau & \overset{\rho_\sigma \times \rho_\tau}{\longrightarrow} & E_\sigma \times E_\tau.
\end{array}
$$

We have a homomorphism

$$
\text{(30)} \quad
\begin{array}{ccc}
\mathrm{Hom}(J(C), E) & \longrightarrow & \mathrm{Hom}(E_\sigma \times E_\tau, E) \\
\cup & & \cup \\
g & \longmapsto & g \circ \pi_2
\end{array}
$$

As in Section 2, we consider the bilinear form on $\mathrm{Hom}(J(C), E)$ (resp. $\mathrm{Hom}(E_\sigma \times E_\tau, E)$) associated with C and o (resp. $\pi_2^*(C)$ and o). For $g \in \mathrm{Hom}(J(C), E)$, we have $4(\mathrm{Ker}(g){\cdot}C) = (\mathrm{Ker}(g \circ \pi_2){\cdot}\pi_2^*(C))$. Therefore, by the definition of the bilinear form, we have

$$
\text{(31)} \qquad (g \circ \pi_2, h \circ \pi_2) = 4(g, h) \quad \text{for } g, h \in \mathrm{Hom}(J(C), E).
$$

As in (8), we identify $\mathrm{Hom}(E_\sigma \times E_\tau, E)$ with $\mathcal{O} \oplus \mathcal{O}$ by the natural isomorphism. As in (3), we set $X = E_\sigma + E_\tau$. Then for $\tilde{g} = (a_1, a_2)$ and $\tilde{h} = (b_1, b_2) \in \mathrm{Hom}(E_\sigma \times E_\tau, E)$ with $a_i, b_i \in \mathcal{O}$ $(i = 1, 2)$, we have by (29) and Theorem 2.6

$$
\text{(32)} \qquad (\tilde{g}, \tilde{h}) = -2(\bar{a}_1 b_1 + \bar{b}_1 a_1 + \bar{a}_2 b_2 + \bar{b}_2 a_2).
$$

The following lemma is well-known, and is easily proved by Ibukiyama [Ib].

Lemma 5.2. *Assume $p \equiv 5 \pmod 6$. Let E be the supersingular elliptic curve defined by (27). Then, $B = \mathrm{End}^0(E)$ and $\mathcal{O} = \mathrm{End}(E)$ is given by*

$$
B = \mathbf{Q} + \mathbf{Q}\alpha + \mathbf{Q}\beta + \mathbf{Q}\alpha\beta,
$$

with $\alpha^2 = -p$, $\beta^2 = -3$, $\alpha\beta = -\beta\alpha$,

$$
\mathcal{O} = \mathbf{Z}\omega_1 + \mathbf{Z}\omega_2 + \mathbf{Z}\omega_3 + \mathbf{Z}\omega_4
$$

with $\omega_1 = 1$, $\omega_2 = \alpha$, $\omega_3 = (-1+\beta)/2$, $\omega_4 = (1+\alpha)(3+\beta)/6$. Moreover, β is given by $2\sigma_2^2 + 1$.

Using the notations in Lemma 5.2, we have

$$
\rho_\sigma \times \rho_\tau = \begin{pmatrix} (-1-\beta)/2 & 0 \\ 0 & (1-\beta)/2 \end{pmatrix} \quad \text{and} \quad \sigma_2 = (1+\beta)/2.
$$

We set $\tilde{\rho} = (\rho_\sigma \times \rho_\tau, \varphi_0^{-1} \circ \hat{\sigma}_2 \circ \varphi_0)$. Then the action of $\tilde{\rho}$ on $\mathrm{Hom}(E_\sigma \times E_\tau, E)$ defined by (5) is explicitly given by

$$
\text{(33)} \quad
\begin{array}{ccc}
\tilde{\rho} : \mathrm{Hom}(E_\sigma \times E_\tau, E) & \longrightarrow & \mathrm{Hom}(E_\sigma \times E_\tau, E) \\
\cup & & \cup \\
(a_1, a_2) & \longmapsto & (\gamma a_1, \gamma a_2) \begin{pmatrix} \delta & 0 \\ 0 & \gamma \end{pmatrix}
\end{array}
$$

where $\gamma = (1-\beta)/2$, $\delta = (-1-\beta)/2$, $a_1, a_2 \in \mathcal{O}$.

Lemma 5.3. *Under the above notations,*

$$
\mathrm{disc}\,\mathrm{Hom}(E_\sigma \times E_\tau, E)^{<\tilde{\rho}>} = 12p^2.
$$

Proof. By (33) and Lemma 5.2, a basis of $\mathrm{Hom}(E_\sigma \times E_\tau, E)^{<\tilde{\rho}>}$ is given by $\{(0,\alpha),(0,\alpha(3+\beta)/2)\}$. Therefore, by (32) the bilinear form on $\mathrm{Hom}(E_\sigma \times E_\tau, E)^{<\tilde{\rho}>}$ with respect to this basis is given by

$$\begin{pmatrix} -4p & -6p \\ -6p & -12p \end{pmatrix}$$

Hence, taking the determinant of this matrix, we get the result. Q.E.D.

Lemma 5.4. *Let S be a K3 surface in Theorem 5.1. Then,*

$$\mathrm{disc}\, \mathrm{NS}(S) = -p^2.$$

Proof. We use the above notations. By Artin [A, p.556] we have

$$(34) \qquad \mathrm{disc}\, \mathrm{NS}(S) = -p^{2\sigma_0} \text{ for an integer } \sigma_0,\ 1 \leq \sigma_0 \leq 10.$$

By (14) and Lemma 3.3, we have

$$(35) \qquad \mathrm{ord}_p \, \mathrm{disc}\, \mathrm{NS}(S) = \mathrm{ord}_p \, \mathrm{disc}\, \mathrm{DC}(C \times E)^G.$$

Therefore, to calculate $\mathrm{disc}\, \mathrm{NS}(S)$, it suffices to calculate the p-adic order of $\mathrm{disc}\, \mathrm{DC}(C \times E)^G$. We consider the automorphism $\sigma_1 \times \sigma_2$ of $C \times E$. Then, as in (16), $\sigma_1 \times \sigma_2$ induces the action on $\mathrm{Hom}(J(C), E)$. We set $\rho = (\sigma_1, \varphi_0^{-1} \circ \hat{\sigma}_2 \circ \varphi_0)$. Then the action of $\sigma_1 \times \sigma_2$ on $\mathrm{Hom}(J(C), E)$ is equal to the action of ρ defined by (5). The homomorphism defined by (15) induces an isometry from $\mathrm{DC}(C \times E)^G$ to $\mathrm{Hom}(J(C), E)^{<\rho>}$. On the other hand, we have the homomorphism induced by (30):

$$\lambda : \mathrm{Hom}(J(C), E)^{<\rho>} \longrightarrow \mathrm{Hom}(E_\sigma \times E_\tau, E)^{<\tilde{\rho}>}.$$

Since $\mathrm{Ker}\,\pi_2 \subset \mathrm{Ker}[2]_{E_\sigma \times E_\tau}$, we see that the image of λ contains $2\,\mathrm{Hom}(E_\sigma \times E_\tau, E)^{<\tilde{\rho}>}$. Therefore, by (31) and Lemma 5.3 there exists an integer m such that

$$4^2 \cdot 12p^2 = m^2 \cdot 4^2 \cdot \mathrm{disc}\, \mathrm{Hom}(J(C), E)^{<\rho>}.$$

Therefore, we have $\mathrm{ord}_p \, \mathrm{disc}\, \mathrm{Hom}(J(C), E)^{<\rho>} = 0$ or 2. Therefore, we have $\mathrm{ord}_p \, \mathrm{disc}\, \mathrm{DC}(C \times E)^G = 0$ or 2. Hence, we conclude $\mathrm{ord}_p \, \mathrm{disc}\, \mathrm{NS}(S) = 2$ by (34) and (35). Q.E.D.

Now, Theorem 5.1 follows from Lemma 5.4 and Ogus [Og,Theorem 7.10].

References

[A]   M. Artin, Supersingular K3 surfaces, Ann. Sci. Ecole Norm. Sup. 4^e serie, 7 (1974), 543-568.

[BM]   E. Bombieri and D. Mumford, Enriques' classification of surfaces in char. p, II, Complex Analysis and Algebraic Geometry, A collection of papers dedicated to K. Kodaira (W. L. Baily, Jr. and T. Shioda, eds.), Iwanami Shoten Publishers, Tokyo, and Princeton Univ. Press, Princeton, 1977, 23-42.

[BK]   H. Braun and M. Koecher, Jordan Algebren, Springer Verlag, 1966.

[Ib]   T. Ibukiyama, A basis and maximal orders in quaternion algebras over the rational number field (in Japanese), Sugaku, 24 (1972),316-318.

[IKO]   T. Ibukiyama, T. Katsura and F. Oort, Supersingular curves of genus two and class numbers, Compositio Math., 57 (1986), 127-152.

[Ig]   J. Igusa, Arithmetic variety of moduli for genus two, Ann. of Math., 72 (1960), 612-649.

[K1]   T. Katsura, Unirational elliptic surfaces in characteristic p, Tôhoku Math. J., 33 (1981), 521-553.

[K2]   T. Katsura, Generalized Kummer surfaces and their unirationality in characteristic p, J. Fac. Sci. Univ. Tokyo Sect.IA Math., 34 (1987), 1-41.

[KO]   T. Katsura and F. Oort, Families of supersingular abelian surfaces, Compositio Math., 62 (1987), 107-167.

[L]   S. Lang, Abelian Varieties, Interscience-Wiley, New York, 1959.

[Ma]   T. Matsusaka, On a characterization of a Jacobian variety, Mem. Coll. Sci., Univ. Kyoto, Series A, 32 (1959), 1-18.

[M1]   J. S. Milne, The Tate-Shaferevich group of a constant abelian variety, Invent. Math., 6 (1968), 91-105.

[M2]   J. S. Milne, On a conjecture of Artin and Tate, Ann. of Math., 102 (1975), 517-533.

[Mu]   D. Mumford, Abelian Varieties, Oxford Univ. Press, 1970.

[Og]   A. Ogus, Supersingular K3 crystals, in Journées de Géométrie Algébrique de Rennes, 1978, Astérisque 64 (1979), 3-86.

[Oo]   F. Oort, Which abelian surfaces are products of elliptic curves ?, Math. Ann., 214 (1975), 35-47.

[S1]   T. Shioda, On elliptic modular surfaces, J. Math. Soc. Japan, 24 (1972), 20-59.

[S2]   T. Shioda, Supersingular K3 surfaces, in Algebraic Geometry, Proc. Copenhagen 1978 (K. Lønsted, ed.), Lecture Notes in Math., 732, Berlin-Heidelberg-New York, Springer-Verlag (1979), 564-591.

[S3]   T. Shioda, Some observations on Jacobi sums, to appear.

[S4]   T. Shioda, Supersingular K3 surfaces with big Artin invariant, to appear.

[SI]   T. Shioda and H. Inose, On singular K3 surfaces, in Complex Analysis and Algebraic Geometry, A collection of papers dedicated to K. Kodaira

(W. L. Baily, Jr. and T. Shioda, eds), Iwanami Shoten Publishers, Tokyo, and Cambridge University Press, Cambridge 1977, 119-136.

[T1] J. Tate, On the conjectures of Birch and Swinnerton-Dyer and a geometric analog, Sém. Bourbaki, 1965/66, $n°$ 306, in Dix Exposés sur la Cohomologie des Schemas, North-Holland, Amsterdam (1968), 189-214.

[T2] J. Tate, Endomorphisms of abelian varieties over finite fields, Invent. Math., 2 (1966), 134-144.

[W] W. C. Waterhouse, Abelian varieties over finite fields, Ann. Sci. Ecole Norm. Sup. 4e serie, 2 (1969), 521-560.

Toshiyuki KATSURA

Department of Mathematics
Yokohama City University
Yokohama, 236
Japan

Current Address

Deparment of Mathematics
Faculty of Scinece
Ochanomizu University
Ôtsuka, Tokyo, 112
Japan

Algebraic Geometry and Commutative Algebra
in Honor of Masayoshi NAGATA
pp. 203–231 (1987)

On Complete Ideals in Regular Local Rings

Joseph LIPMAN*

Introduction
§1. Point bases and completions of ideals in regular local rings.
§2. Simple complete ideals corresponding to infinitely near points.
§3. The length of a complete ideal (dimension 2).
§4. Unique factorization for complete ideals (dimension 2).
References

Introduction.

We present here an approach to unique factorization of complete (=integrally closed) ideals in two-dimensional regular local rings, based on a decomposition theorem ((2.5) below) which is valid in all dimensions.

The theory of complete ideals in two-dimensional regular local rings was founded by Zariski [Z], [ZS₂, appendix 5], and furthur developed in [Ho], [D], [L], and [G]. Recent interest in this subject (and related ones) has been shown in [R], [Sp], [C] and [Hy].

Zariski's work was motivated by the birational theory of linear systems on smooth surfaces [Z′, Chapter 2]. Roughly speaking, the monoid of complete ideals in a normal noetherian ring R (with product $I * J = \{$completion of $IJ\}$) generates the group of locally principal divisors on the Zariski-Riemann space of R; these divisors can also be interpreted more concretely as $\varinjlim$ of divisors on schemes birational over R, and thereby one connects to classical situations involving divisors and linear systems on such schemes (cf. [Z′, appendix to ch.2] for more details).

Zariski first raised the question of higher-dimensional generalizations in [Z, p.152], but not much has happened in this respect during the intervering fifty years. Perhaps the lack of progress is understandable in view of the complexity of birational geometry in dimension ≥ 3. It *is* possible to extend the *definitions* in [Z′, appendix to Ch.2] to higher dimension (cf. [W],[Sn],[ZS₂, pp.356–361], and §1 below) but *not the main results* in [*ibid*, §4]. In itself what this yields is little more than a convenient language for discussing linear systems with base

---

*Partially supported by NSF grant DMS-850994 at Purdue University.
Received February 2, 1987.

conditions, a theory in search of theorems. Zariski himself eventually concluded [ZS$_2$, p.362]: "It is almost certain that the theory ... cannot be generalized to higher dimension without substantial modifications both of statements and proofs."

Indeed, whereas $I * J = IJ$ in dimension two (i.e. the product of complete ideals is still complete), counterexamples to such a statement in three-dimensional regular local rings have recently been given by Huneke[Hu, §3]. And the main theorem in dimension two, on *unique factorization* of complete ideals into simple complete ideals, also breaks down in higher dimension. The first counterexample is due to Cutkosky [C]. Huneke and I subsequently found the following counterexample, in a power series ring $k[[x, y, z]]$, k a field:

$$(0.1) \qquad (x, y, z)(x^3, y^3, z^3, xy, yz, xz) = (x^2, y, z)(x, y^2, z)(x, y, z^2).$$

The ideals appearing here are all complete and *-simple (cf. beginning of §2), and the product ideal on each side is also complete.

In spite of these discouraging developments, the main results in §2 below may offer a scintilla of hope that some substantial generalizations are not entirely out of reach. What we do in §2, after setting up the foundations in §1, is to associate special *-simple complete ideals to certain "infinitely near points"; and then show that any "finitely supported" complete ideal admits a "unique factorization" into special *-simple complete ideals, with possibly *negative* exponents. (Precise statements are given in (2.1) and (2.5).) Thus, in (0.1), the simple complete ideals $(x^2, y, z), (x, y^2, z), (x, y, z^2)$ are all special (associated to the infinitely near points (= quadratic transforms) in the directons $(1 : 0 : 0)$, $(0 : 1 : 0)$, $(0 : 0 : 1)$ respectively), and they all appear in the "factorization" of $(x^3, y^3, z^3, xy, yx, xz)$ with exponent $+1$, while the maximal ideal (x, y, z), which is also special (associated to $k[[x, y, z]]$ itself), may be said to appear with exponent -1.

Unique factorization in dimension two is derived from Theorem (2.5) in §4, via some results in §3. The point is that in dimension two there are no negative exponents.

Acknowledgement. While preparing the paper, I benefited greatly from numerous stimulating conversations with Craig Huneke.

§1. Point bases and completions of ideals in regular local rings.

This section is based on ideas going back to [Z]. The main results are Proposition (1.10) and its elaborations Propositions (1.16) and (1.23).

If R is a noetherian local domain, with maximal ideal $\mathfrak{m}$ and fraction field K, then a *prime divisor* of R is a valuation v of K whose valuation ring R_v dominates R (i.e. $R \subset R_v$ and $\mathfrak{m} \subset \mathfrak{m}_v$, the maximal ideal of R_v), and such that the transcendence degree of the field $R_v/\mathfrak{m}_v$ over $R/\mathfrak{m}$ is as large as possible, viz. $\dim R - 1$; such a v must be a discrete rank one valuation [A, p.330, Thm.1].

Let I be an ideal in any commutative ring R. An element $x \in R$ is *integral over I* if x satisfies a condition of the form

$$x^n + a_1 x^{n-1} + a_2 x^{n-2} + \cdots + a_n = 0 \qquad (a_j \in I^j, 1 \le j \le n),$$

i.e. if for some $n > 0$, $x^n \in I(I + xR)^{n-1}$. The set of all such x, denoted $\bar{I}$, is called the *integral closure*, or *completion*, of I. The completion $\bar{I}$ is itself an ideal, and we have $I \subset \bar{I} = \bar{\bar{I}}$. [1] I is *integrally closed*, or *complete*, if $I = \bar{I}$.

Proposition (1.1). *Let R be an integral domain with fraction field K, and let I be an ideal in R, with completion $\bar{I}$. Then for any valuation v of K whose valuation ring R_v contains R, we have $\bar{I} R_v = I R_v$. Conversely, if R is noetherian, local, and universally catenary [EGA IV, (5.6.2)] (for example R regular [ibid. (5.6.4)]) then for every $x \notin \bar{I}$ there exists a prime divisor v of R such that $x \notin I R_v$.*

Proof: For the first assertion, cf. [ZS$_2$, p.350, proof of Thm.1]. For the second, $\mathfrak{m}$ being the maximal ideal of R, it will suffice to produce a finitely generated R-subalgebra S of K and a height one prime ideal p in S such that $\mathfrak{m} + x^{-1} I \subset p$; for then, since R is universally catenary, the field S_p/pS_p will have transcendence degree $\dim R - 1$ over its subfield $R/\mathfrak{m}$, and so any valuation ring R_v in K dominating S_p (i.e. any localization at a maximal ideal of the integral closure of S_p in K, cf. [B, §2, no.5, Cor.2]) will give a prime divisor v with $x^{-1} I \subset \mathfrak{m}_v$, i.e. $x \notin I R_v$.

Actually we need not assume that the prime ideal p has height one, since this can always be arranged by blowing up (argue as in [ZS$_2$, p.96]). So consider the ring

$$S = R[x^{-1}I] = \bigcup_{n \ge 0} (I + xR)^n / x^n$$

and note that $p = \mathfrak{m} + x^{-1} IS$ is an ideal in S such that S/p is a homomorphic image of $R/\mathfrak{m}$. Moreover $1 \notin p$, since otherwise there would be a relation of the form

$$1 + a \in I(I + xR)^{n-1} / x^n \qquad a \in \mathfrak{m}, \quad n > 0,$$

and multiplying this by $x^n(1 + a)^{-1}$ would yield $x \in \bar{I}$, a contradiction. Thus p is a maximal ideal in S, and we are done.

*   *   *

Next, after a preparatory lemma, we describe the *transform* of an ideal (Definition (1.4), Proposition (1.5)).

---

[1] These assertions follow e.g. from the fact that $x \in \bar{I}$ if and only if, in the polynomial ring $R[T]$, the element xT is integral over the graded subring $R[IT]$.

Lemma (1.2). *Let R be an integral domain with fraction field K, and let p be a prime ideal in R such that the localization R_p is a discrete valuation ring (d.v.r.). Let S be a ring with $R \subset S \subset K$, and set $S_p = S \otimes_R R_p \subset K$ (i.e. S_p is the ring of fractions $S[M^{-1}]$, $M = R - p$). Let*

$$p^S = pS_p \cap S.$$

Then the following are equivalent:

(i) $p^S \neq S$.

(ii) $q = p^S$ *is the unique prime ideal in S whose intersection with R is p; and $S_q = R_p$.*

(iii) *There exists an ideal J in S such that $J \cap R = p$.*

(iii)′ $pS \cap R = p$.

(iv) $S_p \neq K$.

(v) $S_p = R_p$.

(v)′ $S \subset R_p$.

Proof: The implications (ii) $\Rightarrow$ (i) $\Rightarrow$ (iv), (ii) $\Rightarrow$ (iii) $\Leftrightarrow$ (iii)′, and (ii) $\Rightarrow$ (v)′ $\Leftrightarrow$ (v) are all trivial. (iii) $\Rightarrow$ (iv) is immediate (since $J \cap R = p$ implies $0 \neq JS_p \neq S_p$), as is (iv) $\Rightarrow$ (v) (since $R_p \subset S_p \subset K$ and R_p is a d.v.r.).

It remains to prove that (v) $\Rightarrow$ (ii). If $S_p = R_p$, then $q = p^S = pR_p \cap S$ is a prime ideal in S, and $q \cap R = pR_p \cap R = p$; and furthermore $R_p \subset S_q \neq K$, so that R_p being a d.v.r. we must have $R_p = S_q$. Similarly for *any* prime ideal q' in S such that $q' \cap R = p$, we have $S_{q'} = R_p (= S_q)$; and therefore $q' = q$.   q.e.d.

Suppose now that R is a *unique factorization domain* (UFD) with fraction field K. For any non-zero ideal I in R, let x be a greatest common divisor of the elements in I (i.e., among the principal ideals containing I, xR is the smallest) and set

$$I^{-1} = x^{-1}R = \{z \in K \mid zI \subset R\}.$$

Then II^{-1} is the unique ideal J in R such that:

(i) $J^{-1} = R$ (i.e., R itself is the only principal ideal containing J), and

(ii) $I = yJ$ for some y in R.

Thus: *every non-zero ideal I in R is uniquely of the form*

$$(1.3) \qquad I = p_1^{a_1} p_2^{a_2} \cdots p_n^{a_n} J \qquad (a_i > 0)$$

where the p_i are principal prime ideals, the a_i are (strictly) positive integers, and J is an ideal with $J^{-1} = R$.

Definition (1.4). *Let R be a UFD with fraction field K, and let S be a UFD with $R \subset S \subset K$. Let $I \neq (0)$ be an ideal in R, factor I as in (1.3), and* *for each* $i = 1, 2, \ldots, n$, *set* $q_i = p_i^S$ *(cf.* (1.2)*). We define the transform of* I *in* S *to be the ideal*

$$I^S = q_1^{a_1} \cdots q_n^{a_n} (JS)(JS)^{-1}.$$

In particular, if I *is a principal prime ideal then* I^S *is the same here as in* (1.2).

Some basic properties of the "transform" operation are given in the next Proposition.

Proposition (1.5). *Let* $R \subset S \subset K$ *be as in* (1.4)*, let* I, I_1, I_2 *be non-zero ideals in* R *and let* T *be a UFD with* $S \subset T \subset K$.

(i) *If* I *is a principal prime ideal, then either* $IS \cap R = I$, *in which case* I^S *is a principal prime ideal with* $I^S \cap R = I$; *or* $IS \cap R \neq I$, *in which case* $I^S = S$.

(ii) *If* $I^{-1} = R$, *then* $I^S = (IS)(IS)^{-1}$ *(so that* $(I^S)^{-1} = S$*).*

(iii) *(Compatibility of transform with products.)* $(I_1 I_2)^S = I_1^S I_2^S$.

(iv) *(Transitivity of transform.)* $(I^S)^T = I^T$.

(v) *(Localization.)* *If* S *is a ring of fractions of* R, *then* $I^S = IS$.

(vi) *(Compatibility with integral dependence.)* *If* $I_2 \subset I_1 \subset \overline{I_2}$ *(the completion of* I_2*) then* $I_2^S \subset I_1^S \subset \overline{(I_2^S)}$.

Proof: (i)–(v) are left as an exercise. As for (vi), we first check, by applying the first assertion in (1.1) to the discrete valuation rings obtained by localizing R at height one primes p, that $I_1^{-1} = I_2^{-1}$ (note that $I_1^{-1} = J_1^{-1}$, where $J_1 = \bigcap_p I_1 R_p$). Hence for suitable $x \in R$ and with $L_i = I_i I_i^{-1}$ $(i = 1, 2)$ we have

$$I_1 = xL_1, \quad I_2 = xL_2, \quad L_1^{-1} = L_2^{-1} = R, \quad L_2 \subset L_1 \subset \overline{L_2} = x^{-1}\overline{I_2};$$

and if $y \in S$ is such that $(xR)^S = yS$, (cf. (i)) then (by (iii))

$$(1.5.1) \qquad I_1^S = yL_1^S, \qquad I_2^S = yL_2^S.$$

Now since $L_1 \subset \overline{L_2}$, therefore every $z \in L_1$ is integral over $L_2 S$, and hence, since $\overline{L_2^S}$ is an S-ideal, we have

$$L_2 S \subset L_1 S \subset \overline{L_2 S}.$$

As above, then, $(L_1 S)^{-1} = (L_2 S)^{-1}$, and (by (ii))

$$(1.5.2) \qquad L_2^S = (L_2 S)(L_2 S)^{-1} \subset (L_1 S)(L_1 S)^{-1} = L_1^S \subset \overline{L_2^S}.$$

The conclusion follows from (1.5.1) and (1.5.2).

$$* \qquad\qquad * \qquad\qquad *$$

We come now to "infinitely near points" (Definition (1.6 below) and their relation to prime divisors (Proposition (1.7)).

Let K be a field. We denote by Greek letters $\alpha, \beta, \gamma, \ldots$ regular local rings of Krull dimension ≥ 2, with fraction field K; and refer to such objects as "points". For any point α, let m_α be its maximal ideal, and ord_α the corresponding order valuation, i.e., the unique (discrete, rank one) valuation of K such that for $0 \neq x \in \alpha$,

$$\mathrm{ord}_\alpha(x) = \max\{\, n \mid x \in m_\alpha^n \,\}.$$

Recall that a *quadratic transform* of a point α is a local ring of the form $Q = (\alpha[x^{-1}m_\alpha])_p$ where $x \in m_\alpha, x \notin m_\alpha^2$, and p is a prime ideal in the ring $\alpha[x^{-1}m_\alpha]$ such that $m_\alpha \subset p$. Such a Q is necessarily regular (hence a point if its dimension $\dim Q$ is ≥ 2), and the residue field Q/m_Q has transcendence degree $\dim \alpha - \dim Q$ over α/m_α (cf. [A, p.334, Lemma 10], or [EGA IV, (5.6.4)]). There is a unique one-dimensional quadratic transform of α, namely the valuation ring of ord_α.

Definition (1.6). *A point β is* infinitely near *to α, $\beta \succ \alpha$ (or $\alpha \prec \beta$) in symbol, if there exists a sequence*

$$\alpha = \alpha_0 \subset \alpha_1 \subset \cdots \subset \alpha_n = \beta \quad (n \geq 0)$$

such that for each $i = 0, 1, \ldots, n-1$, α_{i+1} is a quadratic transform of α_i. Such a sequence, if it exists, is unique; we call it the quadratic sequence from α to β.

In case $\dim \alpha = 2$, the factorization theorem of Zariski and Abhyankar states that *any* β containing α is infinitely near to α (cf. [A, p.343, Thm.3]. [2] No such statement holds when $\dim \alpha > 2$ (e.g. [Sa]).

If $\alpha \prec \beta$, then ord_β is a prime divisor of α (by the above remarks on quadratic transforms). In fact, *every* prime divisor v of α is of the form ord_β, where β is found as follows: let $\alpha_0 = \alpha$, and having defined α_i for some $i \geq 0$, let α_{i+1} be the unique quadratic transform of α_i dominated by (the valuation ring of) v, unless $v = \mathrm{ord}_{\alpha_i}$ in which case set $\beta = \alpha_i$ and stop; then this process must terminate after a finite number of steps (cf. [A, p.336, Prop.3]); thus β is the largest point infinitely near to α and dominated by v. In summary (with remaining details left to the reader):

Proposition (1.7). *The map of sets*

$$\{\text{points infinitely near to } \alpha\} \longrightarrow \{\text{prime divisors of } \alpha\}$$

which takes β to ord_β is a bijection.

*   *   *

---

[2] A similar result holds if we assume only that α and β are two-dimensional local rings with fraction field K, such that α is rational and β is factorial (cf. [L, p.203, Prop.(3.1)] and [He, Thm.1]).

We are now almost ready to state the first main result (1.10) of this section. Let $I \neq (0)$ be an ideal in a point α. [3] I is finitely generated, so for any valuation v of K whose valuation ring contains α, we can set

$$v(I) = \min\{v(z)|z \in I\}.$$

If $\beta \succ \alpha$ is an infinitely near point, then, since α and β are both regular, hence UFD's, the transform I^β can be defined as in (1.4). In particular, when $\beta = \alpha$, then $I^\beta = I$.

Definition (1.8). *The* point basis *of a non-zero ideal $I \subset \alpha$ is the family of non-negative integers*

$$\mathbf{B}(I) = \{\mathrm{ord}_\beta(I^\beta)\}_{\beta \succ \alpha}.$$

A base point *of I is a point $\beta \succ \alpha$ such that $\mathrm{ord}_\beta(I^\beta) \neq 0$ (i.e., $I^\beta \neq \beta$).*

Remarks (1.9). (a): For any $\beta \succ \alpha$, the point basis $\mathbf{B}(I^\beta)$ is obtained by restricting $\mathbf{B}(I)$ to the set of $\gamma \succ \beta$ (because $(I^\beta)^\gamma = I^\gamma$, by (1.5)(iv)).

(b): For two non-zero ideals I, J in α, (1.5)(iii) gives:

$$\mathbf{B}(IJ) = \mathbf{B}(I) + \mathbf{B}(J).$$

Proposition (1.10). *Two non-zero ideals I, J in α have the same point basis if and only if their integral closures are equal:*

$$\mathbf{B}(I) = \mathbf{B}(J) \Leftrightarrow \overline{I} = \overline{J}.$$

Proof: [4] For any $\beta \succ \alpha$, let R_β be the valuation ring of ord_β. It follows from (1.1) and (1.7) that

$$\{\overline{I} = \overline{J}\} \quad \Leftrightarrow \quad \{IR_\beta = JR_\beta \text{ for all } \beta \succ \alpha\}$$
$$\Leftrightarrow \quad \{\mathrm{ord}_\beta(I) = \mathrm{ord}_\beta(J) \text{ for all } \beta \succ \alpha\}.$$

The question is whether this last condition is equivalent to:

$$\mathrm{ord}_\beta(I^\beta) = \mathrm{ord}_\beta(J^\beta) \text{ for all } \beta \succ \alpha.$$

An affirmative answer can be deduced from the following useful fact:

Lemma (1.11). *Let $\beta \succ \alpha$, and let*

$$\alpha = \alpha_0 \subset \alpha_1 \subset \cdots \subset \alpha_n = \beta$$

---

[3]Note that α, being integrally closed in K, is determined by I: $\alpha = \{z \in K | zI \subset I\}$.

[4]The implication $\overline{I} = \overline{J} \Rightarrow \mathbf{B}(I) = \mathbf{B}(J)$ also follows from (1.5)(vi) and (1.1).

be the quadratic sequence from α to β (cf. (1.6)). Let L be a non-zero ideal in α. *For* $0 \leq j \leq n$ *set*

$$
\begin{aligned}
\mathfrak{m}_j &= \text{maximal ideal of } \alpha_j \\
\text{ord}_j &= \text{ord}_{\alpha_j} \\
L_j &= L^{\alpha_j}.
\end{aligned}
$$

Then for any valuation v whose valuation ring contains β, we have

$$
v(L) = v(L_n) + \sum_{j=0}^{n-1} \text{ord}_j(L_j) v(\mathfrak{m}_j).
$$

Indeed, taking $v = \text{ord}_\beta$, we see from (1.11) that if $\mathbf{B}(I) = \mathbf{B}(J)$ then $\text{ord}_\beta(I) = \text{ord}_\beta(J)$, whence, as above, $\overline{I} = \overline{J}$.

Conversely, if $\overline{I} = \overline{J}$, so that $\text{ord}_j(I) = \text{ord}_j(J)$ for $0 \leq j \leq n$, then we find by induction on n that $\text{ord}_n(I_n) = \text{ord}_n(J_n)$, i.e., $\text{ord}_\beta(I^\beta) = \text{ord}_\beta(J^\beta)$; and thus $\mathbf{B}(I) = \mathbf{B}(J)$.

Proof of (1.11): Proceeding by induction on n, we need only show that

$$
v(L_{n-1}) = v(L_n) + \text{ord}_{n-1}(L_{n-1}) v(\mathfrak{m}_{n-1})
$$

i.e., (since $L_n = (L_{n-1})^{\alpha_n}$, (1.5)(iv)), we need only treat the case $n = 1$. So set $\mathfrak{m} = \mathfrak{m}_0$, and assume that $\beta = \alpha_1$, say β is a localization of $\alpha[x^{-1}\mathfrak{m}], x \in \mathfrak{m}$, so that $\mathfrak{m}\beta = x\beta$ and $v(\mathfrak{m}) = v(x)$. We want to show, with $l = \text{ord}_\alpha(L)$, that

$$
v(L) = v(L_1) + lv(\mathfrak{m}) = v(L_1) + lv(x).
$$

So it will be enough to check that

$$
(1.11.1) \qquad\qquad L_1 = x^{-l}L\beta.
$$

Since "transform" respects products (1.5)(iii), it is in fact enough to check (1.11.1) when

(a): $L^{-1} = \alpha$

and when

(b): $L = p$, a principal prime ideal in α.

For this purpose, note that if q is a prime ideal in β with $\mathfrak{m} \not\subset q$, and if $q' = q \cap \alpha$, then $\alpha_{q'} = \beta_q$ (for, $x \notin q'$, so $\alpha_{q'}$ and β_q are both localizations of $\alpha[x^{-1}\mathfrak{m}]$, and since β_q dominates $\alpha_{q'}$, therefore $\beta_q = \alpha_{q'}$). In particular, every principal prime ideal in β *other than* $x\beta$ intersects α in a principal prime. Since $l = \text{ord}_\alpha(L)$, and since the valuation ring of ord_α is the localization of β at the prime ideal $x\beta$, therefore

$$
x^{-l}L\beta \not\subset x\beta,
$$

and in case (a) it follows that no principal prime contains $x^{-l}L\beta$, i.e., (cf. (1.5)(ii)):

$$x^{-l}L\beta = (L\beta)^{-1}(L\beta) = L^\beta = L_1.$$

As for case (b), it follows that except for $x\beta$ every associated prime ideal of the principal ideal $p\beta$ intersects α in p, whence, (e.g., by[ZS$_1$, p.225, Thm.17]) with notation as in (1.2):

$$x^{-l}p\beta = p\beta_p \cap \beta = p^\beta = L_1.$$

$$q.e.d.$$

Exercise (1.12). Generalize (1.11.1) by showing, for *any* $\beta \succeq \alpha$, that an ideal L' in β equals L^β if and only if:

(i) $L' = y^{-1}L\beta$ for some $y \in \beta$ such that $\mathfrak{m}_\alpha\beta \subset \sqrt{y\beta}$, and
(ii) $(\mathfrak{m}_\alpha\beta) : L' = \mathfrak{m}_\alpha\beta$.
(In other words, $y\beta$ is the smallest principal ideal containing $L\beta$ and all of whose associated primes contain $\mathfrak{m}_\alpha\beta$)

$$* \qquad\qquad * \qquad\qquad *$$

We continue with some simple—but necessary— supplements to the foregoing material.

Definition (1.13). *For any two ideals I, J in a commutative ring R, we set*

$$I * J = \overline{IJ} \quad \text{(the completion of } IJ\text{)}.$$

Lemma (1.14). *Assume that R is a commutative integral domain.*
(i) *For any two ideals I, J in R we have $I * J = \overline{I} * \overline{J}$.*
(ii) *If I is complete and $J \neq (0)$ is finitely generated then*

$$(I * J) : J = I.$$

(iii) *The non-zero complete ideals in R, with the $*$-product, form a commutative monoid $\mathcal{M}_R$ with cancellation (i.e., $I * J = I' * J \Rightarrow I = I'$).*

Proof: (i) is a consequence of the fact that if $x \in \overline{I}$ and $y \in \overline{J}$ then $xy \in \overline{IJ}$, a fact which follows easily from [ZS$_2$, bottom of p.349] (where $N \neq (0)$). Similarly one shows that if J is finitely generated then $xJ \subset I * J$ implies that $x \in \overline{I}$, and (ii) results. As for (iii), associativity of the $*$-product can be shown as follows:

$$(I_1 * I_2) * I_3 = (I_1 I_2) * I_3 \text{ (cf. (i))} = \overline{I_1 I_2 I_3} = I_1 * (I_2 I_3)$$
$$= I_1 * (I_2 * I_3);$$

while commutativity and the existence of an identity (viz. R) are obvious. Cancellation follows from (ii). $q.e.d.$

From (1.10) we have that for complete ideals I, J,

(1.15) $\mathbf{B}(I * J) = \mathbf{B}(IJ) = \mathbf{B}(I) + \mathbf{B}(J)$

and furthermore

$$\{\mathbf{B}(I) = \mathbf{B}(J)\} \Rightarrow \{I = J\}.$$

Thus:

Proposition (1.16).   *By associating to each non-zero complete ideal I in a point α the point basis $\mathbf{B}(I)$, we obtain an injective homomorphism of monoids*

$$\mathcal{M}_\alpha \hookrightarrow \prod_{\beta \succ \alpha} \mathbf{N}_\beta$$

where, for each $\beta \succ \alpha$, $\mathbf{N}_\beta$ is the monoid of non-negative integers (under addition).

*   *   *

Definition (1.17).   *With $R \subset S \subset K$ as in (1.4), and I a non-zero ideal in R, we define the* complete transform $I^{\overline{S}}$ *to be* $\overline{I^S}$, *the completion of I^S.*

Proposition (1.18).   *With notation as in (1.5), we have:*
(i) $(I_1 * I_2)^{\overline{S}} = (I_1 I_2)^{\overline{S}} = I_1^{\overline{S}} * I_2^{\overline{S}}.$
(ii) $(I^{\overline{S}})^{\overline{T}} = (I^S)^{\overline{T}} = I^{\overline{T}}.$

Proof: By (vi) of (1.5) we have

$$(I_1 I_2)^S \subset (I_1 * I_2)^S \subset \overline{(I_1 I_2)^S}$$

and hence

$$(I_1 * I_2)^{\overline{S}} = \overline{(I_1 I_2)^S} = (I_1 I_2)^{\overline{S}}.$$

Furthermore, by (iii) of (1.5) we have

$$(I_1 I_2)^{\overline{S}} = \overline{I_1^S I_2^S} = I_1^S * I_2^S = I_1^{\overline{S}} * I_1^{\overline{S}}$$

(the last equality by (1.14)(i)), proving (i).
    Again by (vi) of (1.5),

$$(I^S)^T \subset (I^{\overline{S}})^T \subset (I^S)^{\overline{T}}$$

and hence

$$(I^{\overline{S}})^{\overline{T}} = (I^S)^{\overline{T}}.$$

And by (iv) of (1.5),

$$(I^S)^{\overline{T}} = \overline{(I^S)^T} = \overline{I^T} = I^{\overline{T}}.$$

Remark (1.19): If S is an integrally closed domain, then every principal ideal in S is complete (for, if $0 \neq y \in S$, then an equation of integral dependence of x over yS yields an equation of integral dependence of x/y over S ...). Moreover, if I is an ideal in S such that $\overline{I}$ is principal, say $\overline{I} = xS$, then $I = \overline{I}$. (This is clear if $x = 0$; and otherwise $x^{-1}I \subset S$ and an equation of integral dependence of x over I yields

$$1 \in x^{-1}I + x^{-2}I^2 + \cdots = x^{-1}I$$

whence $x \in I$.). In particular, with $I \subset R \subset S$ as in (1.17):

$$\{I^S \text{ is principal}\} \Leftrightarrow \{I^{\overline{S}} \text{ is principal}\} \Rightarrow \{I^S = I^{\overline{S}}\};$$

and if $I^{-1} = R$, then (cf. (1.5)(ii)):

$$\{I^S = S\} \Leftrightarrow \{I^{\overline{S}} = S\}$$

*   *   *

Definition (1.20). *An ideal I in α is* finitely supported *if $I \neq (0)$ and I has at most finitely many base points (cf. (1.8)).*

Proposition (1.21). *If I is a finitely supported ideal in α, then for all $\beta \succ \alpha$, the transform I^β is finitely supported, and the ring β/I^β is artinian.*

Proof: The first assertion follows from (1.9) (a). For the second, we may then assume that $\beta = \alpha$. Suppose that I is contained in a non-maximal prime ideal p. Then any $\gamma \succ \alpha$ such that $\gamma \subset \alpha_p$ is a base point of I, because by (1.5)(v) and (1.5)(iv),

$$\alpha_p \neq I\alpha_p = I^{\alpha_p} = (I^\gamma)^{\alpha_p}$$

so that $I^\gamma \neq \gamma$. Thus (1.21) results from the following elementary fact:

Lemma (1.21.1). *Let p be a non-maximal prime ideal in α. Then there exists a quadratic transform α_1 of α with $\alpha_1 \subset \alpha_p$; and hence there is an infinite sequence*

$$\alpha = \alpha_0 \subset \alpha_1 \subset \alpha_2 \subset \cdots \subset \alpha_p$$

where each α_i $(i > 0)$ is a quadratic transform of α_{i-1}.

Proof: The second assertion follows from the first, since $p_1 = p\alpha_p \cap \alpha_1$ is a non-maximal prime (because $p_1 \cap \alpha = p$ whereas α_1 dominates α), and $(\alpha_1)_{p_1} = \alpha_p$, so that we can apply the first assertion to find a quadratic transform α_2 of α_1 with $\alpha_2 \subset \alpha_p$, etc. etc. The first assertion follows from the fact that the map

$X \to \mathrm{Spec}(\alpha)$ obtained by blowing up the maximal ideal $\mathfrak{m}$ of α is surjective and proper. Or, we can argue directly as follows.

The graded ring

$$gr_{\mathfrak{m}}(\alpha/p) = \bigoplus_{n \geq 0} (\mathfrak{m}^n + p)/(\mathfrak{m}^{n+1} + p)$$

is not artinian, hence has non-nilpotent elements of degree one, i.e., there is an $x \in \mathfrak{m}$ such that for all $n > 0$,

(1.21.2) $x^n \notin \mathfrak{m}^{n+1} + p.$

In particular, $x \notin \mathfrak{m}^2 + p$, and so

$$A = \alpha[x^{-1}\mathfrak{m}] = \bigcup_{n > 0} \mathfrak{m}^n/x^n \subset \alpha_p.$$

Set

$$p' = p\alpha_p \cap A = \bigcup_{n > 0} (p \cap \mathfrak{m}^n)/x^n.$$

Then

$$1 \notin \mathfrak{m}A + p' = xA + p'$$

since otherwise, for some $n > 0$, we would have

$$1 \in \mathfrak{m}^{n+1}/x^n + (p \cap \mathfrak{m}^n)/x^n$$

contradicting (1.21.2). So there exists a prime ideal q in A containing $\mathfrak{m}A + p'$, and $\alpha_1 = A_q \subset \alpha_p$ is a quadratic transform as desired.

Corollary (1.22). *If $\beta \succ \alpha$ is a base point of a finitely supported ideal $I \subset \alpha$, then $\dim \beta = \dim \alpha$.*

Proof: Let

$$\alpha = \alpha_0 \subset \alpha_1 \subset \cdots \subset \alpha_n = \beta$$

be the quadratic sequence from α to β (1.6), and argue by induction on n. There being nothing to prove when $n = 0$, assume that $n > 0$ and that $\dim \alpha_{n-1} = \dim \alpha$. For some $x \in \mathfrak{m}$ (the maximal ideal of α_{n-1}), β is of the form $\beta = A_q$, where $A = \alpha_{n-1}[x^{-1}\mathfrak{m}]$ and $q \supset \mathfrak{m}$ is a prime ideal in A. Let $Q \subset A$ be a maximal ideal in A containing q. Then $Q/\mathfrak{m}A$ is a maximal ideal in $A/\mathfrak{m}A$, which is a finitely generated algebra over the field $\alpha_{n-1}/\mathfrak{m}$. It follows that A/Q is a finite field extension of $\alpha_{n-1}/\mathfrak{m}$, and hence that

(1.22.1) $\dim A_Q = \dim \alpha_{n-1} = \dim \alpha$

(cf. remarks preceding (1.6)).

Now if $q \neq \,'Q$, then β is a localization of $\gamma = A_Q$ at a non-maximal prime ideal; but by (1.21) γ / I^γ is artinian, whence, by (1.5)(v) and (1.5)(iv)

$$\beta = (I^\gamma)\beta = (I^\gamma)^\beta = I^\beta,$$

i.e., β is not a base point of I. Thus $q = Q$, $\beta = A_Q$, and by (1.22.1), $\dim \beta = \dim \alpha$. $\hspace{4cm}$ *q.e.d.*

Remark (*not used elsewhere*): If $f : X \to \mathrm{Spec}(\alpha)$ is the map obtained by blowing up I, then I is finitely supported if and only if there exists a sequence

$$\sigma : X_n \xrightarrow[f_n]{} X_{n-1} \to \cdots \to X_1 \xrightarrow[f_1]{} X_0 = \mathrm{Spec}(\alpha)$$

where each f_i $(i > 0)$ is obtained by blowing up a closed point of X_{i-1} (for example a base point of I) and such that X_n dominates X, i.e., there is a map $g : X_n \to X$ such that

$$f \circ g = f_1 \circ f_2 \circ \cdots \circ f_n;$$

in other words, "the indeterminacies of f^{-1} can be eliminated by a finite number of point blow ups". (Indeed, if σ exists, the base points of I must be among those which are blown up in σ.) It follows that *whenever a suitable local version of resolution of singularities is available* (for example if α is excellent and equicharacteristic, with $\alpha / \mathfrak{m}_\alpha$ a perfect field, of characteristic zero if $\dim \alpha > 3$, [Hi, p.142. Thm.II], [A', p.149, (5.2.1)]) *then I is finitely supported if* (and by (1.22) only if) *every base point β of I satisfies* $\dim \beta = \dim \alpha$. (For, the indeterminacies of f^{-1} can then be eliminated by blowing up finitely many base points of I, and I will have no other base points.)

(1.23). For non-zero ideals I, J in α, the complete ideal $I * J$ is finitely supported if and only if both I and J are. (This is because $\mathbf{B}(I * J) = \mathbf{B}(I) + \mathbf{B}(J)$, cf. (1.15).) Thus (and by (1.22)):

Proposition (1.23). *The finitely supported complete ideals in α, together with the $*$-product, form a commutative monoid $\mathcal{M}_\alpha^f$, isomorphic under the injective monoid map (1.16) to a submonoid of the free commutative monoid $\mathcal{F}$ generated by all $\beta \succ \alpha$ with $\dim \beta = \dim \alpha$. ($\mathcal{F}$ is the submonoid $\{\sum n_\beta \beta \mid n_\beta \geq 0 \text{ for all } \beta\}$ of the free abelian group $\mathcal{G}_\alpha^f$ generated by such β).*

(1.24). A basic question now is to understand the structure of the monoid $\mathcal{M}_\alpha^f$. A partial result along these lines is given in Theorem (2.5) below; but it raises more questions than it answers. When $\dim \alpha = 2$, there is a satisfactory result, due to Zariski (Theorem (4.2)): $\mathcal{M}_\alpha^f$ is itself a free commutative monoid.

§2. Simple complete ideals corresponding to infinitely near points.

We say that a complete ideal I in a commutative ring R is $*$-**simple** if $I \neq R$ and if whenever $I = J * L$ with ideals J, L in R (cf. (1.13)) then either $J = R$ or $L = R$.

As in §1, we consider "points" $\alpha, \beta, \gamma, \ldots$ all having the same fraction field K. Recall the definitions of "infinitely near points" (1.6) and of "complete transform" (1.17). The main results in this section are contained in (2.1) and (2.5).

Proposition (2.1). *For each pair of points α, β with $\dim \alpha = \dim \beta$ there exists a unique complete ideal $\mathfrak{p}_{\alpha\beta}$ in α such that for every $\gamma \succ \alpha$: if $\gamma \prec \beta$ then the complete transform $(\mathfrak{p}_{\alpha\beta})^{\overline{\gamma}}$ is $*$-simple, and otherwise $(\mathfrak{p}_{\alpha\beta})^{\overline{\gamma}} = \gamma$.*

Corollary (2.2). (i). *γ is a base point of $\mathfrak{p}_{\alpha\beta}$ if and only if $\alpha \prec \gamma \prec \beta$; and hence $\mathfrak{p}_{\alpha\beta}$ is finitely supported (1.20).*
(ii). *The ring $\alpha/\mathfrak{p}_{\alpha\beta}$ is artinian.*
(iii). *$\mathfrak{p}_{\alpha\alpha}$ is the maximal ideal $\mathfrak{m}_\alpha$ of α.*
(iv). *For all $\gamma \succ \alpha$ with $\dim \gamma = \dim \alpha$ we have*

$$(\mathfrak{p}_{\alpha\beta})^{\overline{\gamma}} = \mathfrak{p}_{\gamma\beta}.$$

Proof of (2.2): (i). γ is a base point of $\mathfrak{p}_{\alpha\beta}$ (1.8) iff $(\mathfrak{p}_{\alpha\beta})^\gamma \neq \gamma$, i.e. (clearly, or by (1.19)) iff $(\mathfrak{p}_{\alpha\beta})^{\overline{\gamma}} \neq \gamma$, i.e. (by (2.1)) iff $\alpha \prec \gamma \prec \beta$. If $\alpha \prec \beta$, there are only finitely many such γ, viz. the members of the quadratic sequence from α to β (1.6); and otherwise there are no such γ. In any case, $\mathfrak{p}_{\alpha\beta}$ has at most finitely many base points.

(ii). This follows from (i) together with (1.21).

(iii). The ideal $(\mathfrak{m}_\alpha)^{\overline{\alpha}} = \mathfrak{m}_\alpha$ is $*$-simple; and for every $\gamma \succ \alpha$ with $\gamma \neq \alpha$ we have $(\mathfrak{m}_\alpha)^{\overline{\gamma}} = \gamma$.

(iv). For any $\delta \succ \gamma$, we have that $((\mathfrak{p}_{\alpha\beta})^{\overline{\gamma}})^{\overline{\delta}} = (\mathfrak{p}_{\alpha\beta})^{\overline{\delta}}$ (cf. (1.18)(ii)), which is $*$-simple if $\delta \prec \beta$ and equal to δ otherwise.

Proof of (2.1): If β is not infinitely near to α, then $\mathfrak{p}_{\alpha\beta} = \alpha$ is an ideal in α having the required properties; and by taking $\gamma = \alpha$ in (2.1) we see that there is no other such ideal. So suppose that $\alpha \prec \beta$, and let

$$\alpha = \alpha_0 \subset \alpha_1 \subset \cdots \subset \alpha_n = \beta$$

be the corresponding quadratic sequence (cf. (1.6)). We proceed by induction on n.

For $n = 0$, i.e. $\beta = \alpha$, we have already noted that $\mathfrak{p}_{\alpha\alpha} = \mathfrak{m}_\alpha$ has the required behavior with respect to $\gamma \succ \alpha$ (proof of (2.2)(iii)). That no other ideal in α has this behavior follows easily from (2.3) below (with $\mathcal{I} = \mathcal{O}_X$).

For $n > 0$, we already have $\mathfrak{p}_{\alpha_1\beta}$ (by the inductive hypothesis). Any $\gamma \succ \alpha$, other than $\gamma = \alpha$, satisfies $\gamma \succ \delta$ for a unique quadratic transform δ of α. As in the proof of (2.2)(iv), we find then that a $*$-simple complete ideal I in α satisfies the defining properties of $\mathfrak{p}_{\alpha\beta}$ if and only if: (i) $I^{\overline{\alpha_1}} = \mathfrak{p}_{\alpha_1\beta}$, and (ii) $I^\delta = \delta$ for every quadratic transform δ of α except $\delta = \alpha_1$.

We consider the map $f : X \to \mathrm{Spec}(\alpha)$ obtained by blowing up $\mathfrak{m}_\alpha$.[5] The quadratic transforms of α are just the local rings $\mathcal{O}_{X,x}$ of points $x \in f^{-1}\{\mathfrak{m}_\alpha\}$ (the closed fibre). Let $y \in f^{-1}\{\mathfrak{m}_\alpha\}$ be such that $\mathcal{O}_{X,y} = \alpha_1$. Since $\alpha_1/\mathfrak{p}_{\alpha_1\beta}$ is artinian (2.2)(ii), there exists a unique coherent $\mathcal{O}_X$-ideal $\mathcal{I}(\alpha,\beta)$ whose stalk at y is $\mathfrak{p}_{\alpha_1\beta}$ and whose stalk at any $x \neq y$ is $\mathcal{O}_{X,x}$. Thus, to complete the proof of (2.1), it suffices to show:

Lemma (2.3). *With $f : X \to \mathrm{Spec}(\alpha)$ as above, and $\mathfrak{m} = \mathfrak{m}_\alpha$, let $\mathcal{I}$ be a coherent $\mathcal{O}_X$-ideal whose stalk $\mathcal{I}_x$ is a complete $\mathcal{O}_{X,x}$-ideal for all $x \in X$, with $\mathcal{I}_x = \mathcal{O}_{X,x}$ if $x \notin f^{-1}\{\mathfrak{m}\}$. Assume also that $\mathcal{I} \not\subseteq \mathfrak{m}\mathcal{O}_X$. Then there exists a unique $\mathfrak{m}$-primary complete ideal I in α such that*
(i): *for every quadratic transform $\gamma = \mathcal{O}_{X,x}$ of α, we have*

$$I^{\overline{\gamma}} = \mathcal{I}_x;$$

and (ii): *any complete ideal $J \subset \mathfrak{m}$ such that $J^{\overline{\gamma}} = I^{\overline{\gamma}} = \mathcal{I}_x$ for all $\gamma = \mathcal{O}_{X,x}$ as in* (i) *must be of the form*

$$J = \mathfrak{m} * \mathfrak{m} * \cdots * \mathfrak{m} * I.$$

Furthermore, if $\mathcal{I}$ is $*$-simple (in the sense that $\mathcal{I} \neq \mathcal{O}_X$ and whenever $\mathcal{J} \neq \mathcal{O}_X$ and $\mathcal{L}$ are $\mathcal{O}_X$-ideals such that for all $x \in X$ we have $\mathcal{I}_x = \mathcal{J}_x * \mathcal{L}_x$, then $\mathcal{L} = \mathcal{O}_X$), then I is $*$-simple.

Proof: The $\mathcal{O}_X$-ideal $\mathfrak{m}\mathcal{O}_X$ is invertible, and hence for every $n \geq 0$ and every $x \in X$, $\mathfrak{m}^n\mathcal{I}_x$ is complete. Also, $\mathfrak{m}\mathcal{O}_X$ is very ample, so for some integer $N > 0$ the $\mathcal{O}_X$-ideal $\mathfrak{m}^N\mathcal{I}$ is generated by its global sections [Ha, p.121, Thm. 5.17]; in other words, if for any $n \geq 0$, I_n is the α-ideal

$$(2.3.1) \qquad I_n = H^0(X, \mathfrak{m}^n\mathcal{I}) = \bigcap_{x \in X} \mathfrak{m}^n\mathcal{I}_x \subseteq \bigcap_{x \in X} \mathcal{O}_{X,x} = \alpha$$

then, for each $x \in X$,

$$\mathfrak{m}^N\mathcal{I}_x = I_N\mathcal{O}_{X,x}.$$

So we can define r to be the *least* among all integers $n > 0$ such that for all $x \in X$, $\mathfrak{m}^n\mathcal{I}_x$ is the *completion* $(I_n\mathcal{O}_{X,x})^-$; and then we set

$$I = I_r.$$

Let us check that this I is as asserted in (2.3).

---

[5]$X = \mathrm{Proj}\, S$, where S is the graded α-algebra $\bigoplus_{n \geq 0} \mathfrak{m}_\alpha^n$. In the language of models, [ZS$_2$, p.116 ff.], [EGA I, §8], X is the projective model determined by any basis of $\mathfrak{m}_\alpha$, and f is the domination mapping to $V(\alpha) = \mathrm{Spec}(\alpha)$.

First of all, for any $n \geq 0$, any $\xi \in \alpha$ which is integral over I_n is integral over $I_n \mathcal{O}_{X,x} \subset \mathfrak{m}^n \mathcal{I}_x$ for every $x \in X$, so that

$$\xi \in \bigcap_{x \in X} \mathfrak{m}^n \mathcal{I}_x = I_n,$$

and thus I_n is *complete*. Also, since $\mathcal{O}_X/\mathcal{I}$ is supported in $f^{-1}\{\mathfrak{m}\}$, there is an integer $N' > 0$ such that $\mathfrak{m}^{N'}\mathcal{O}_X \subset \mathcal{I}$ (this can be checked in each member of a finite affine open covering of X); hence

$$\mathfrak{m}^{n+N'} \subset \bigcap_{x \in X} \mathfrak{m}^{n+N'}\mathcal{O}_{X,x} \subset \bigcap_{x \in X} \mathfrak{m}^n \mathcal{I}_x = I_n,$$

and so I_n is $\mathfrak{m}$-primary, unless $n = 0$ and $\mathcal{I} = \mathcal{O}_X$ in which case $1 \in I_n$, i.e. $I_n = \alpha$. In particular, $I = I_r$ is complete and (since $r > 0$) $\mathfrak{m}$-primary.

Now note that $r = \mathrm{ord}_\alpha(I)$: for if x is the generic point of the closed fibre $f^{-1}(\mathfrak{m})$, so that $\mathcal{O}_{X,x}$ is just the valuation ring of ord_α, then

$$I \subset \alpha \cap \mathfrak{m}^r \mathcal{O}_{X,x} = \{\xi \in \alpha \mid \mathrm{ord}_\alpha(\xi) \geq r\} = \mathfrak{m}^r;$$

and if $I \subset \mathfrak{m}^{r+1}$ then for all $x \in X$:

$$\mathfrak{m}^r \mathcal{I}_x = (I\mathcal{O}_{X,x})^- \subset (\mathfrak{m}^{r+1}\mathcal{O}_{X,x})^- = \mathfrak{m}^{r+1}\mathcal{O}_{X,x},$$

whence $\mathcal{I} \subset \mathfrak{m}\mathcal{O}_X$, contrary to assumption. As in (1.11.1) we see then that for any quadratic transform $\gamma = \mathcal{O}_{X,x}$ ($x \in f^{-1}\{\mathfrak{m}\}$), we have

$$I^\gamma = (\mathfrak{m}\gamma)^{-r}(I\gamma)$$

and so

$$I^{\bar{\gamma}} = (\mathfrak{m}\gamma)^{-r}(I\gamma)^- = (\mathfrak{m}\mathcal{O}_{X,x})^{-r}(\mathfrak{m}^r\mathcal{I}_x) = \mathcal{I}_x,$$

i.e. (i) in (2.3) is satisfied.

Next we prove (2.3) (ii) (from which, in particular, the uniqueness of I follows). Let $s = \mathrm{ord}_\alpha(J)$. Then for any quadratic transform $\gamma = \mathcal{O}_{X,y}$, we have, as above,

$$\mathcal{I}_y = J^{\bar{\gamma}} = (\mathfrak{m}\gamma)^{-s}(J\gamma)^-.$$

Hence

(2.3.2) $$\mathfrak{m}^s \mathcal{I}_y = (J\mathcal{O}_{X,y})^-$$

so that (since $\mathfrak{m}^s \mathcal{I}_x = \mathcal{O}_{X,x}$ for all $x \in X$ such that $x \notin f^{-1}\{\mathfrak{m}\}$)

$$J \subset \bigcap_{x \in X} \mathfrak{m}^s \mathcal{I}_x = I_s$$

(cf. (2.3.1)); and then (since, as above, $\mathfrak{m}^{s+N'} \subset I_s$ so that $I_s\mathcal{O}_{X,x} = \mathcal{O}_{X,x}$ for $x \notin f^{-1}\{\mathfrak{m}\}$):

$$\mathfrak{m}^s \mathcal{I}_x = (I_s\mathcal{O}_{X,x})^-$$

for *all* $x \in X$. By the definition of r, therefore, $s \geq r$, and for all $x \in X$:

$$(2.3.3) \qquad \mathfrak{m}^s \mathcal{I}_x = (\mathfrak{m}^{s-r} I \mathcal{O}_{X,x})^-.$$

Now, I claim, *every element* $\xi \in I_s$ is *integral over* J (whence $I_s = J$ since $J \subset I_s$ and J is complete). Indeed, if v is any valuation of the fraction field K of α whose valuation ring R_v dominates α (i.e. $R_v \supset \alpha$ and $v(\eta) > 0$ for all $\eta \in \mathfrak{m}$) then R_v dominates $\mathcal{O}_{X,y}$ for some $y \in f^{-1}\{\mathfrak{m}\}$ (since f is a proper map, cf. e.g. [ZS$_2$, p.120, (b)]), so by (2.3.2) and (1.1)

$$\xi \in I_s \subset (\mathfrak{m}^s \mathcal{I}_y) R_v = J R_v;$$

and thus (cf. (1.1)) $\xi \in \bar{J}$. Similarly, we see from (2.3.3) that $I_s \, (= J)$ is integral over $\mathfrak{m}^{s-r} I$, i.e. that

$$J = \underbrace{\mathfrak{m} * \mathfrak{m} * \cdots * \mathfrak{m}}_{s-r \ times} * I$$

proving (ii).

Finally, suppose that $\mathcal{I}$ is $*$-simple and that $I = J * L$ where J, L are complete ideals in α; and let us deduce that either $J = \alpha$ or $L = \alpha$ (i.e. I is $*$-simple). Set

$$p = \text{ord}_\alpha(J), \quad q = \text{ord}_\alpha(L).$$

Then, since $I = (JL)^-$, (1.1) gives

$$(2.3.4) \qquad r = \text{ord}_\alpha(I) \quad = \quad \text{ord}_\alpha(J) + \text{ord}_\alpha(L)$$
$$= \quad p + q,$$

and for all $x \in X$, since

$$(2.3.5) \qquad \mathfrak{m}^{p+q} \mathcal{I}_x = \mathfrak{m}^r \mathcal{I}_x \quad = \quad (I \mathcal{O}_{X,x})^-$$
$$= \quad (JL \mathcal{O}_{X,x})^-$$

we see that

$$\mathcal{I}_x = [(\mathfrak{m}\mathcal{O}_{X,x})^{-p} J \mathcal{O}_{X,x}] * [(\mathfrak{m}\mathcal{O}_{X,x})^{-q} L \mathcal{O}_{X,x}].$$

So if $\mathcal{J}, \mathcal{L}$ are the $\mathcal{O}_X$-ideals

$$\mathcal{J} = \mathfrak{m}^{-p} J \mathcal{O}_X, \qquad \mathcal{L} = \mathfrak{m}^{-q} L \mathcal{O}_X,$$

then either $\mathcal{J} = \mathcal{O}_X$ or $\mathcal{L} = \mathcal{O}_X$.

Suppose for example that $\mathcal{L} = \mathcal{O}_X$. Then for all $x \in X$ we have

$$L \mathcal{O}_{X,x} = \mathfrak{m}^q \mathcal{O}_{X,x},$$

and consequently (cf. (2.3.5))

$$I^{\bar{\gamma}} = J^{\bar{\gamma}}, \quad \gamma = \mathcal{O}_{X,x},$$

so that by (ii) above:

$$J = \mathfrak{m} * \mathfrak{m} * \cdots * \mathfrak{m} * I \subset I.$$

It follows at once (from (2.3.4)) that $q = 0$, i.e. $L = \alpha$.   q.e.d.

Remarks (2.4) (*not used elsewhere*):   For *any* $n \geq 0$, the α-ideal I_n defined by (2.3.1) is *contracted*, i.e.

$$(2.4.1) \qquad I_n = H^0(X, I_n \mathcal{O}_X) = \bigcap_{x \in X} I_n \mathcal{O}_{X,x}.$$

Moreover, if $n \geq r$ (r as in the proof of (2.3)), then, for all $x \in X$,

$$\mathfrak{m}^n \mathcal{I}_x = \mathfrak{m}^{n-r}(I_r \mathcal{O}_{X,x})^- = (\mathfrak{m}^{n-r} I_r \mathcal{O}_{X,x})^- \quad \text{(since } \mathfrak{m}^{n-r} \mathcal{O}_{X,x} \text{ is principal)}$$
$$\subset (I_n \mathcal{O}_{X,x})^- \subset \mathfrak{m}^n \mathcal{I}_x{}^6$$

so that $\mathfrak{m}^n \mathcal{I}_x = (I_n \mathcal{O}_{X,x})^-$;[7] and if $\mathcal{I}_x = \mathcal{O}_{X,x}$ then

$$\mathfrak{m}^n \mathcal{O}_{X,x} = \mathfrak{m}^n \mathcal{I}_x = (I_n \mathcal{O}_{X,x})^- = I_n \mathcal{O}_{X,x}$$

(the last equality e.g. by (1.19), since $\mathfrak{m}^n \mathcal{O}_{X,x}$ is a principal $\mathcal{O}_{X,x}$-ideal). Thus if

$$(2.4.2) \qquad J_n = \left(\bigcap_{\substack{x \in X \\ \mathcal{I}_x \neq \mathcal{O}_{X,x}}} I_n \mathcal{O}_{X,x} \right) \cap \alpha$$

then

$$I_n = J_n \cap \left(\bigcap_{x \in X} \mathfrak{m}^n \mathcal{O}_{X,x} \right) = J_n \cap \mathfrak{m}^n.$$

Moreover, if $\mathcal{I}_x \neq \mathcal{O}_{X,x}$ for some $x \in f^{-1}\{\mathfrak{m}\}$ (i.e. $\mathcal{I} \neq \mathcal{O}_X$) then

$$J_n \subset (I_n \mathcal{O}_{X,x}) \cap \alpha \subset \mathfrak{m}^n \mathcal{O}_{X,x} \cap \alpha = \mathfrak{m}^n$$

(for the last equality, just note that $\operatorname{ord}_\alpha(\xi) \geq n$ for any $\xi \in \mathfrak{m}^n \mathcal{O}_{X,x}$). We conclude that

$$(2.4.3) \qquad I_n = J_n \qquad (n \geq r).$$

In particular:

(2.4.4)(cf. [ZS$_2$, top of p.373]).   *Suppose that* $S = \{$ support of $\mathcal{O}_X/\mathcal{I}$ $\}$ *is finite and non-empty*,[8] *and let* $\alpha_\mathcal{I}$ *be the semi-local ring*

$$\alpha_\mathcal{I} = \bigcap_{x \in S} \mathcal{O}_{X,x}.$$

---

[6] For any non-negative integers $p \geq q$, it follows easily from (2.3.1) that $I_q = I_p : \mathfrak{m}^{p-q}$. In particular, $\mathfrak{m}^{n-r} I_r \subset I_n$.

[7] As in the proof of (2.3), it follows that $I_n^{\bar{\gamma}} = \mathcal{I}_x$ for any quadratic transform $\gamma = \mathcal{O}_{X,x}$ of α, whence, by (2.3) (ii), $I_n = \mathfrak{m}^{n-r} * I$.

[8] Equivalently: if $k = \alpha/\mathfrak{m}$ and $L_r(I)$ is the k-vector space of leading forms of elements $\xi \in I$ such that $\operatorname{ord}_\alpha(\xi) = r$, then the set of zeros of $L_r(I)$ in the projective space $\operatorname{Proj}(\bigoplus_{t \geq 0} \mathfrak{m}^t/\mathfrak{m}^{t+1})$ is finite and non-empty.

Then, for $n \geq r$

$$I_n = (I_n \alpha_{\mathcal{I}}) \cap \alpha.$$

*   *   *

The next result is the central one in this paper. It expresses a kind of "unique factorization with possibly negative exponents". In the two-dimensional case, the exponents all turn out to be non-negative (Theorem (4.2)).

Theorem (2.5). *For each finitely supported* (cf. (1.20)) *complete ideal I in a point α there exists a unique family of integers*

$$(n_\beta) = (n_\beta(I))_{\beta \succ \alpha, \dim \beta = \dim \alpha}$$

such that $n_\beta = 0$ for almost all (i.e. all but finitely many) β and such that

$$\left(\prod\nolimits^*_{n_\beta < 0} \mathfrak{p}_{\alpha\beta}^{-n_\beta} \right) * I = \prod\nolimits^*_{n_\gamma > 0} \mathfrak{p}_{\alpha\gamma}^{n_\gamma}$$

where $\mathfrak{p}_{\alpha\beta}$ is as in (2.1), $\prod\limits^*_{n_\beta < 0}$ denotes *-product over all $\beta \succ \alpha$ such that $n_\beta < 0$, and similarly for $\prod\limits^*_{n_\gamma > 0}$.

It is straightforward to see that (2.5) can be restated as follows:

Theorem (2.5)$'$. *For fixed α and variable $\beta \succ \alpha$ with $\dim \beta = \dim \alpha$, the images of the ideals $\mathfrak{p}_{\alpha\beta}$ under the canonical injection described in (1.23) form a basis of the free abelian group $\mathcal{G}_\alpha^f$ generated by all the β.*

Proof of (2.5)$'$: Fixing α, set

$$\Gamma = \{\gamma \mid \gamma \succ \alpha, \quad \dim \gamma = \dim \alpha\}.$$

According to (1.23), the canonical image of $\mathfrak{p}_{\alpha\beta}$ is the family of non-negative integers $(p_{\alpha\beta,\gamma})_{\gamma \in \Gamma}$ where

$$
\begin{aligned}
p_{\alpha\beta,\gamma} &= \operatorname{ord}_\gamma((\mathfrak{p}_{\alpha\beta})^\gamma) &= \operatorname{ord}_\gamma((\mathfrak{p}_{\alpha\beta})^{\overline{\gamma}}) && \text{(cf. (1.1))} \\
&&= \operatorname{ord}_\gamma(\mathfrak{p}_{\gamma\beta}) && \text{(cf. (2.2) (iv)).}
\end{aligned}
$$

By (2.2) (i), then, $p_{\alpha\beta,\gamma} = 0$ unless $\gamma \prec \beta$; and by (2.2) (iii), $p_{\alpha\beta,\beta} = 1$ for all β.

Now (2.5)$'$ asserts that for any family of integers $\mathbf{g} = (g_\gamma)_{\gamma \in \Gamma}$, with $g_\gamma = 0$ for almost all γ, there is a unique family of integers $\mathbf{h} = (h_\beta)_{\beta \in \Gamma}$ with $h_\beta = 0$ for almost all β and such that for all $\gamma \in \Gamma$

$$g_\gamma = \sum_{\beta \in \Gamma} h_\beta p_{\alpha\beta,\gamma}.$$

For the existence of $\mathbf{h}$, argue by induction on the number $\nu_{\mathbf{g}}$ of points γ such that $g_\delta \neq 0$ for some $\delta \succ \gamma$: if $\nu_{\mathbf{g}} > 0$ then choose β such that $g_\beta \neq 0$ and $g_\delta = 0$ for all $\delta \succ \beta$, and set

$$g'_\gamma = g_\gamma - g_\beta p_{\alpha\beta,\gamma} \qquad (\gamma \in \Gamma);$$

then $g'_\gamma = g_\gamma$ unless $\gamma \prec \beta$, and moreover $g'_\beta = 0$, so $\nu_{\mathbf{g}'} < \nu_{\mathbf{g}} \cdots$.

For the uniqueness of $\mathbf{h}$, assume that $g_\gamma = 0$ for all γ, but that $h_\gamma \neq 0$ for some γ. Then for some γ, $h_\gamma \neq 0$ and $h_\beta = 0$ for all $\beta \succ \gamma$, $\beta \neq \gamma$, so $g_\gamma = h_\gamma \neq 0$, contradiction.   q.e.d.

§3. The length of a complete ideal (dimension 2).

In this section we derive a formula of Hoskin and Deligne for the length of an $\mathfrak{m}_\alpha$-primary complete ideal I in a two-dimensional regular local ring α, in terms of the point basis of I,[9] and deduce a number of consequences, some of which will be needed in §4.

We denote the length of an α-module M by $\lambda_\alpha(M)$. If β is a point infinitely near to α (i.e. $\alpha \subset \beta \subset$ fraction field of α, cf. note following (1.6)) then $[\beta : \alpha]$ denotes the (finite) degree of the residue field extension $\beta/\mathfrak{m}_\beta \supset \alpha/\mathfrak{m}_\alpha$.

Theorem (3.1). [Ho, p.85, Thm.(5.2)], [D, p.22, Thm. (2.13)]. *Let α be a two-dimensional regular local ring, with maximal ideal $\mathfrak{m}$, and let I be a complete $\mathfrak{m}$-primary ideal with point basis*

$$\mathbf{B}(I) = \{r_\beta\}_{\beta \succ \alpha} \qquad \text{cf. (1.8).}$$

Then $r_\beta = 0$ for all but finitely many β (i.e. I is finitely supported, cf. (1.20)), and

$$\lambda_\alpha(\alpha/I) = \sum_\beta [\beta : \alpha] r_\beta (r_\beta + 1)/2.$$

Proof: From (1.5) (ii) it follows that at most finitely many quadratic transforms of α, say $\alpha_1, \alpha_2, \ldots, \alpha_n$, are base points of I. Let

$$I_i = I^{\alpha_i} = (I\alpha_i)(I\alpha_i)^{-1} \qquad (1 \leq i \leq n),$$

---

[9]Geometrically speaking, the length of I is the number of conditions imposed on curves of sufficiently high degree by requiring their local equations to lie in I. In other words, if α is the local ring of a point x on a non-singular projectively embedded surface X over an algebraically closed field k, and $\mathcal{I}$ is the $\mathcal{O}_X$-ideal whose stalk at x is I and which coincides with $\mathcal{O}_X$ at all points other than x, then for sufficiently large n there is an exact sequence

$$0 \longrightarrow H^0(X, \mathcal{I}(n)) \longrightarrow H^0(X, \mathcal{O}_X(n)) \longrightarrow H^0(X, \mathcal{O}_X(n)/\mathcal{I}(n)) \longrightarrow 0,$$

and, since the support of $\mathcal{O}_X/\mathcal{I}$ is the single point x, we have, for all n,

$$\dim_k H^0(X, \mathcal{O}_X(n)/\mathcal{I}(n)) = \dim_k(\alpha/I)$$

and set $f_i = [\alpha_i : \alpha]$. By a theorem of Zariski, [ZS$_2$, p.381, Prop.5], [L, p.209, (6.5)], the ideal $I\alpha_i$ is complete, whence so is I_i.

It is clear (cf. (1.9) (a)) that Theorem (3.1) implies:

$$(3.1.1) \qquad \lambda_\alpha(\alpha/I) = \frac{1}{2}r_\alpha(r_\alpha + 1) + \sum_{i=1}^{n} f_i\lambda_{\alpha_i}(\alpha_i/I_i).$$

On the other hand, the validity of (3.1.1) for all α and I implies (by a straightforward induction) the validity of (3.1).

To prove (3.1.1), consider the map $X \to \mathrm{Spec}(\alpha)$ obtained by blowing up $\mathfrak{m}$. Since I is complete, we deduce from (1.1) that

$$(3.1.2) \qquad H^0(X, I\mathcal{O}_X) = \bigcap_{x \in X} I\mathcal{O}_{X,x} = I.$$

(For another proof, cf. [L, p.208, Prop.(6.2)]). The argument which follows applies to any I satisfying (3.1.2), i.e. to any I which is "contracted from X".[10]

Using the affine open covering

$$X = \mathrm{Spec}(\alpha[b/c]) \cup \mathrm{Spec}(\alpha[c/b])$$

where $b, c \in \alpha$ generate $\mathfrak{m}$, one checks that $H^1(X, \mathcal{O}_X) = 0$ (cf. e.g. [L, p.200]), and that for any coherent $\mathcal{O}_X$-module $\mathcal{F}$, $H^2(X, \mathcal{F}) = 0$ (since the Čech complex corresponding to the covering vanishes in dimension $\neq 0, 1$). There exists an exact sequence of the form

$$0 \longrightarrow \mathcal{F} \longrightarrow \mathcal{O}_X^N \longrightarrow I\mathcal{O}_X \longrightarrow 0,$$

whence an exact sequence

$$0 = H^1(X, \mathcal{O}_X^N) \longrightarrow H^1(X, I\mathcal{O}_X) \longrightarrow H^2(X, \mathcal{F}) = 0$$

so that
$$(3.1.3) \qquad H^1(X, I\mathcal{O}_X) = 0.$$

Now with
$$r = r_\alpha = \mathrm{ord}_\alpha(I)$$

we have an exact sequence

$$0 \to I\mathcal{O}_X \to \mathfrak{m}^r\mathcal{O}_X \to \mathfrak{m}^r\mathcal{O}_X/I\mathcal{O}_X \to 0$$

whence an exact sequence

$$(3.1.4) \ 0 \to H^0(X, I\mathcal{O}_X) \to H^0(X, \mathfrak{m}^r\mathcal{O}_X) \to H^0(X, \mathfrak{m}^r\mathcal{O}_X/I\mathcal{O}_X) \to 0$$

---

[10]It also applies, with slight modifications, when α is replaced by any local ring of a two-dimensional (pseudo-) rational singularity.

(where the 0 on the right comes from (3.1.3)). By (3.1.2)

$$H^0(X, I\mathcal{O}_X) = I,$$

and also (as in easily seen)

$$H^0(X, \mathfrak{m}^r \mathcal{O}_X) = \mathfrak{m}^r.$$

Thus

(3.1.5) $\lambda_\alpha(\alpha/I) - (1/2)r(r+1) = \lambda_\alpha(\mathfrak{m}^r/I) = \lambda_\alpha(H^0(X, \mathfrak{m}^r \mathcal{O}_X/I\mathcal{O}_X)).$

Moreover, for each x in the closed fibre $X \otimes_\alpha (\alpha/\mathfrak{m})$, the local ring $\beta = \mathcal{O}_{X,x}$ is a quadratic transform of α, $\mathfrak{m}\beta$ is invertible, and we have, as in (1.11.1),

(3.1.6) $$I^\beta = (\mathfrak{m}\beta)^{-r}(I\beta).$$

Hence $\mathfrak{m}^r \mathcal{O}_X/I\mathcal{O}_X$ is supported in the finite set of closed points $x_1, \ldots x_n \in X$ whose local rings are

$$\mathcal{O}_{X,x_i} = \alpha_i \qquad (1 \le i \le n),$$

and for each i, the stalk $(\mathfrak{m}^r \mathcal{O}_X/I\mathcal{O}_{X,x})_{x_i}$ is isomorphic to α_i/I_i, so that

(3.1.7) $$\lambda_\alpha(H^0(X, \mathfrak{m}^r \mathcal{O}_X/I\mathcal{O}_X)) = \sum_{i=1}^n \lambda_\alpha(\alpha_i/I_i)$$
$$= \sum_{i=1}^n f_i \lambda_{\alpha_i}(\alpha_i/I_i).$$

Together, (3.1.5) and (3.1.7) give (3.1.1).   q.e.d.

Henceforth we write "λ" for "λ_α".

Corollary (3.2). *Let I be an $\mathfrak{m}$-primary ideal satisfying (3.1.2) (e.g. I complete), and set $r = \mathrm{ord}_\alpha(I)$. Then any minimal generating set of I contains $r+1$ elements; in other words:*

$$\lambda(I/\mathfrak{m}I) = (say) \ \mu(I) = r + 1.$$

Proof: The ideal $\mathfrak{m}I$ is also contracted from X (cf. [ZS$_2$, p.376, Cor.1], or [L, p.209, Thm.(7.2)]), so we can replace I by $\mathfrak{m}I$ in (3.1.1) (cf. remarks following (3.1.2)). Since

$$\mathrm{ord}_\alpha(\mathfrak{m}I) = \mathrm{ord}_\alpha(I) + 1 = r + 1,$$

and since for all quadratic transforms β of α, we have

$$(\mathfrak{m}I)^\beta = (\mathfrak{m}\beta)^{-r-1}(\mathfrak{m}I\beta) = (\mathfrak{m}\beta)^{-r}(I\beta) = I^\beta$$

(cf. (3.1.6)), we deduce that

$$\lambda(I/\mathfrak{m}I) = \lambda(\alpha/\mathfrak{m}I) - \lambda(\alpha/I) \;=\; \frac{1}{2}(r+1)(r+2) - \frac{1}{2}r(r+1)$$
$$= \; r+1,$$

proving (3.2).

Remark (3.3): In [Hy,Thm. 2.1], Huneke and Sally prove a converse to (3.2), at least when $\alpha/\mathfrak{m}$ is infinite: *if I is an $\mathfrak{m}$-primary ideal with $\mu(I) =$ $\mathrm{ord}_\alpha(I) + 1$ then I satisfies* (3.1.2). In their proof, they point out that for any $\mathfrak{m}$-primary I,

$$\mathrm{Tor}_2^\alpha(\alpha/I, \alpha/\mathfrak{m}) \cong (I:\mathfrak{m})/I$$

(as can be seen from the Koszul resolution of $\alpha/\mathfrak{m}$), and then, calculating Tor_2 via an exact sequence

$$0 \to \alpha^{\mu-1} \to \alpha^\mu \to \alpha \to \alpha/I \to 0 \quad (\mu = \mu(I)),$$

they conclude that

$$\lambda((I:\mathfrak{m})/I) = \mu(I) - 1.$$

Hence:

Corollary (3.4). *For any I as in* (3.2),

$$\lambda((I:\mathfrak{m})/I) = \mathrm{ord}_\alpha(I).$$

Corollaries (3.2) and (3.4) yield a proof, suggested to me by Craig Huneke, of the following result of Zariski:

Corollary (3.5). (cf. [ZS$_2$, p.368, Prop.3]). *Let I and $r = \mathrm{ord}_\alpha(I)$ be as in* (3.2), *and assume that $I \neq \mathfrak{m}(I:\mathfrak{m})$. Then*

$$\lambda(I/\mathfrak{m}^{r+1} \cap I) = 1.$$

Proof: If I satisfies (3.1.2), then

$$(3.5.1) \qquad x \in I : \mathfrak{m} \iff x\mathfrak{m} \subset I \iff x\mathfrak{m}\mathcal{O}_X \subset I\mathcal{O}_X$$
$$\iff x \in H^0(X, (\mathfrak{m}\mathcal{O}_X)^{-1} I\mathcal{O}_X)$$

and consequently $I : \mathfrak{m}$ also satisfies (3.1.2), so that by (3.4) and (3.2)

$$r = \lambda((I:\mathfrak{m})/I) \;<\; \lambda((I:\mathfrak{m})/\mathfrak{m}(I:\mathfrak{m}))$$
$$= \; \mathrm{ord}_\alpha(I:\mathfrak{m}) + 1 \leq r + 1.$$

Hence
$$(3.5.2) \qquad\qquad \mathrm{ord}_\alpha(I:\mathfrak{m}) = r$$

and

(3.5.3) $\lambda(I/\mathfrak{m}(I : \mathfrak{m})) = \lambda((I : \mathfrak{m})/\mathfrak{m}(I : \mathfrak{m})) - \lambda((I : \mathfrak{m})/I)$
$= 1.$

From (3.5.2) we get

$$\mathfrak{m}(I : \mathfrak{m}) \subset \mathfrak{m}^{r+1} \cap I \subsetneq I,$$

and so (3.5.3) gives the conclusion.

Remark(3.6): Since $I : \mathfrak{m}$ satisfies (3.1.2) whenever I does (proof of (3.5)), (3.4) can be restated as:

$$\lambda(\alpha/I) = \sum_{n=0}^{\infty} \mathrm{ord}_\alpha(I : \mathfrak{m}^n)$$

*   *   *

We conclude with some remarks on "intersection numbers" and Hilbert-Samuel functions of complete finitely supported ideals I, J in α. (cf. (4.1)(C) below).

If I, J have respective point bases

$$\mathbf{B}(I) = \{r_\beta\}_{\beta \succ \alpha} \qquad \mathbf{B}(J) = \{s_\beta\}_{\beta \succ \alpha}$$

then we set

$$(I \cdot J) = \sum_\beta [\beta : \alpha] r_\beta s_\beta.$$

This integer can be interpreted as the "intersection multiplicity at α of generic member of I and a generic member of J" (cf. e.g. [N, p.189, Thm.8]). Its negative is the total intersection number of the curves defined by the ideals $I\mathcal{O}_Y, J\mathcal{O}_Y$, where $g : Y \to \mathrm{Spec}(\alpha)$ is any proper birational map such that $I\mathcal{O}_Y$ and $J\mathcal{O}_Y$ are both invertible (cf. e.g. [D, p.17, Thm.(2.9)]).

The following corollaries of (3.1) may be compared with [L, p.223, (13.1)(c)] and [L, p.253, (23.2)] respectively.

Corollary (3.7). *If I,J are complete finitely supported ideals in α, then*

$$\lambda(\alpha/IJ) = \lambda(\alpha/I) + \lambda(\alpha/J) + (I \cdot J).$$

Proof: The ideal IJ is complete (cf. (4.1)(A) below) and

$$\mathbf{B}(IJ) = \mathbf{B}(I) + \mathbf{B}(J)$$

(cf. (1.9)(b)). So we need only note that

$$(r_\beta + s_\beta)(r_\beta + s_\beta + 1) = r_\beta(r_\beta + 1) + s_\beta(s_{\beta+1}) + 2r_\beta s_\beta,$$

and apply (3.1).

Corollary (3.8). *If I is a complete finitely supported ideal in α, then for every $n \geq 0$ we have*

$$\lambda(I^n/I^{n+1}) = n(I \cdot I) + \lambda(\alpha/I).$$

Proof: The ideal I^n is complete for all $n \geq 0$ (cf. (4.1)(A) below), so (3.7) with $J = I^n$ gives

$$\begin{aligned}
\lambda(I^n/I^{n+1}) = \lambda(\alpha/I^{n+1}) - \lambda(\alpha/I^n) &= \lambda(\alpha/I) + (I \cdot I^n) \\
&= \lambda(\alpha/I) + n(I \cdot I)
\end{aligned}$$

(where the last equality follows from $\mathbf{B}(I^n) = n\mathbf{B}(I)$). q.e.d.

§4. Unique factorization for complete ideals (dimension 2).

We assume that all points $\alpha, \beta, \ldots$ are two-dimensional regular local rings with the same fraction field K. As noted following (1.6), β is then infinitely near to α if and only if $\beta \supseteq \alpha$.

The theory of complete ideals in the two-dimensional case is due to Zariski [Z],[ZS$_2$, Appendix 5.]. (Generalizations to rational singularities can be found in [L, Chapters II, V].) Here we review some of the main results in light of the preceding material in this paper.

First of all, concerning notions introduced in §1 above we have the following simplifications:

(4.1) **(A).** *Any product of complete ideals in α is again complete* (in other words, the *-product of (1.13) is just the usual product), [ZS$_2$, p.385, Thm.2'], [L, (7.1)].

(B). *If I is a complete ideal in α and $\beta \supseteq \alpha$, then $I\beta$ is a complete ideal in β* (and hence $I^\beta = I^{\overline{\beta}}$, cf. (1.5),(1.17)), [ZS$_2$, p.381, Prop.5], [L, (6.5)].

(C). *An ideal I in α is finitely supported if (cf. (3.1) and (1.10)) and only if (cf. (1.21)) α/I is artinian* (i.e. α contains some power $\mathfrak{m}_\alpha^n$ of the maximal ideal $\mathfrak{m}_\alpha$).

We will say that an ideal L in α is **simple** if $L \neq \alpha$ and if whenever $L = IJ$ with ideals I, J in α then $I = \alpha$ or $J = \alpha$.

Since α is a two-dimensional local unique factorization domain, every non-zero ideal in α is uniquely a product IJ, with I principal and J containing some power $\mathfrak{m}_\alpha^n$. Hence a simple ideal must be either principal (and prime) or (by (4.1)(C)) finitely supported.

The main results to be proved here are:

Theorem (4.2). [ZS$_2$, p.386, Thm.3],[L, p.244, Thm.(20.1)].
When dim $\alpha = 2$, all the integers $n_\beta(I)$ in Theorem (2.5) are ≥ 0, and consequently every $\mathfrak{m}_\alpha$-primary complete ideal in α is in a unique way a product of simple complete ideals of the form $\mathfrak{p}_{\alpha\beta}$ ($\beta \supseteq \alpha$).

Corollary (4.3). [ZS$_2$, p.389,(B)]. *Every* $\mathfrak{m}_\alpha$-*primary simple complete ideal is of the form* $\mathfrak{p}_{\alpha\beta}$ *for some* $\beta \supseteq \alpha$.

Corollary (4.4). [ZS$_2$, p.386, Lemma 6],[L, p.247, Prop.(21.5)]. *If* I *is a simple complete ideal in* α, *and* $\beta \supseteq \alpha$, *then the transform* I^β *is a simple complete ideal in* β, *unless* $I^\beta = \beta$.

Proof of (4.4): If I is principal use (1.5)(i). If I is $\mathfrak{m}_\alpha$-primary, use (4.3), (4.1)(B), and (2.2)(iv).

Proof of (4.2): In view of (4.1)(A), it is clear from (2.5) that we can characterize the integers $n_\beta(I)$ as follows:

Let $\mathcal{M}_\alpha^f$ *be the monoid of complete ideals in* α *containing some power* $\mathfrak{m}_\alpha^n$ *(cf.* (1.23) *and* (4.1)(A),(C)) *and let*

$$(4.2.1) \qquad\qquad \nu_\beta : \mathcal{M}_\alpha^f \longrightarrow \mathbf{Z}$$

be a monoid homomorphism such that for $\gamma \supseteq \alpha$

$$(4.2.2) \qquad \nu_\beta(\mathfrak{p}_{\alpha\gamma}) \;=\; 1 \quad if \quad \gamma = \beta$$
$$\qquad\qquad\qquad\qquad\; =\; 0 \quad if \quad \gamma \neq \beta.$$

Then for all $I \in \mathcal{M}_\alpha^f$, *we have*

$$\nu_\beta(I) = n_\beta(I).$$

Thus, to prove (4.2), it will suffice to exhibit (for each β) such a map ν_β which satisfies in addition the property:

$$(4.2.3) \qquad\qquad \nu_\beta(I) \geq 0 \qquad \text{for all} \qquad I \in \mathcal{M}_\alpha^f.$$

For this purpose we use the "characteristic form" $c(I)$ of a non-zero ideal $I \subset \alpha$, defined as follows (cf. [ZS$_2$, p.363]): we fix a basis (x, y) of $\mathfrak{m} = \mathfrak{m}_\alpha$, and correspondingly identify the graded ring $\bigoplus_{n \geq 0} \mathfrak{m}^n/\mathfrak{m}^{n+1}$ with the polynomial ring $k[X,Y]$ ($k = \alpha/\mathfrak{m}$); then, with $r = \mathrm{ord}_\alpha(I)$, we have an identification of the k-vector space

$$(I + \mathfrak{m}^{r+1})/\mathfrak{m}^{r+1} \cong I/\mathfrak{m}^{r+1} \cap I$$

with a k-vector space $L(I)$ consisting of forms of degree r in $k[X,Y]$, and we let $c(I)$ be a greatest common divisor of all the members of $L(I)$. Thus $c(I)$ is a form, uniquely determined up to multiplication by a non-zero element in k; and the *degree* $s(I)$ of $c(I)$ satisfies

$$(4.2.4) \qquad\qquad s(I) \leq \mathrm{ord}_\alpha(I).$$

We define:

$$\nu_\beta(I) = \mathrm{ord}_\beta(I^\beta) - s(I^\beta).$$

Then (4.2.3) is immediate (by (4.2.4)), and the fact that ν_β is a monoid homomorphism (i.e. $\nu_\beta(IJ) = \nu_\beta(I) + \nu_\beta(J)$) follows from the easily proved identities

$$L(IJ) = L(I)L(J)$$
$$c(IJ) = a.c(I)c(J) \qquad 0 \neq a \in k$$

together with the fact that $(IJ)^\beta = I^\beta J^\beta$ (cf. (1.5)(iii)).
It remains then to prove (4.2.2). Since (by (4.1)(B) and (2.2))

$$(\mathfrak{p}_{\alpha\beta})^\beta = \mathfrak{p}_{\beta\beta} = \mathfrak{m}_\beta,$$

it is clear that $\nu_\beta(\mathfrak{p}_{\alpha\beta}) = 1$. If $\alpha \subseteq \gamma$ and $\beta \nsubseteq \gamma$, then, by definition (cf. (2.1)) $(\mathfrak{p}_{\alpha\gamma})^\beta = \beta$, and so $\nu_\beta(\mathfrak{p}_{\alpha\gamma}) = 0$. Suppose then that $\alpha \subseteq \beta \subsetneq \gamma$. Then (cf. (2.2), (4.1)(B)):

$$(\mathfrak{p}_{\alpha\gamma})^\beta = \mathfrak{p}_{\beta\gamma} \neq \mathfrak{m}_\beta,$$

and furthermore $\mathfrak{p}_{\beta\gamma}$ is complete and $*$-simple (cf. (2.1)), hence not divisible by $\mathfrak{m}_\beta$ (i.e. not of the form $\mathfrak{m}_\beta J$ for some β-ideal J). What we have to show then is that

$$\mathrm{ord}_\beta(\mathfrak{p}_{\beta\gamma}) - s(\mathfrak{p}_{\beta\gamma}) = 0$$

or, equivalently, that *the $(\beta/\mathfrak{m}_\beta)$-vector space $L(\mathfrak{p}_{\beta\gamma})$ has dimension one.*
But this is a special case of Corollary (3.5). q.e.d.

References

[A]   S. S. Abhyankar, On the valuations centered in a local domain, Amer. J. Math **78** (1956), 321–348.

[A′]   S. S. Abhyankar, "Resolution of Singularities of Embedded Algebraic Surfaces," Academic Press, New York-London, 1966.

[B]   N. Bourbaki, "Algèbre Commutative, Chapitre 7, Diviseurs," Act. Sci. et Industrielles, no. 1314, Hermann, Paris, 1965.

[C]   S. D. Cutkosky, Factorization of complete ideals, J. Algebra (to appear).

[D]   P. Deligne, Intersections sur les surfaces régulières, in "Groupes de Monodromie ... (SGA 7, II)", Lecture Notes in Math. no. 340, Springer-Verlag, Berlin-Heidelberg-New York, 1973, pp.1–38.

[EGA I]   A. Grothendieck, J. Dieudonné, "Éléments de Géométrie Algébrique, I," Publ. Math. Inst. Hautes Études Sci., no. 4, 1960.

[EGA IV] A. Grothendieck, J. Dieudonné, "Éléments de Géométrie Algébrique, IV," Publ. Math. Inst. Hautes Études Sci., no. 24, 1965.

[G]   H. Göhner, Semifactoriality and Muhly's condition (N) in two dimensional local rings, J. Algebra **34** (1975), 403–429.

[Ha]   R. Hartshorne, "Algebraic Geometry," Springer-Verlag, Berlin-Heidelberg-New York, 1977.

[He]   W. Heinzer, C. Huneke, J. Sally, A criterion for spots, J. Math. Kyoto Univ. **26** (1986), 667–671.

[Hi]   H. Hironaka, Resolution of singularities of an algebraic variety over a field of characteristic zero, Annals of Math. **79** (1964), 109–326.

[Ho]   M. A. Hoskin, Zero-dimensional valuation ideals associated with plane curve branches, Proc. London Math. Soc. (3) **6** (1956), 70–99.

[Hu]   C. Huneke, The primary components of and integral closures of ideals in 3-dimensional regular local rings, Math. Ann. **275** (1986), 617–635.

[Hy]   C. Huneke, J. Sally, Birational extensions in dimension two and integrally closed ideals, J. Algebra (to appear).

[L]   J. Lipman, Rational singularities, with applications to algebraic surfaces and unique factorization, Publ. Math. Inst. Hautes Études Sci. no.36 (1969), 195–279.

[N]   D. G. Northcott, Abstract dilations and infinitely near points, Proc. Cambridge Phil. Soc. **52** (1956), 176–197.

[R]   D. Rees, Hilbert functions and pseudo-rational local rings of dimension two, J. London Math. Soc. (2) **24** (1981), 467–479.

[Sa]   J. Sally, Regular overrings of regular local rings, Trans. Amer. Math. Soc. **171** (1972), 291–300.

[Sn]   E. Snapper, Higher-dimensional field theory, Compositio Math. **13** (1956), 1–46.

[Sp]   M. Spivakovsky, Sandwiched surface singularities and the Nash resolution for surfaces, Thesis, Harvard University, 1985. (cf. also Séminaire Bourbaki, exposé 661, Astérisque **145–146** (1987).)

[W]   B. v. d. Waerden, The invariant theory of linear sets on an algebraic variety, "Proc. International Congress of Math. 1954, vol.III", 542–544, North-Holland, Amsterdam, 1956.

[Z]   O. Zariski, Polynomial ideals defined by infinitely near base points, Amer. J. Math. **60** (1938), 151–204.

[Z']   O. Zariski, "Algebraic Surfaces (second supplemented edition)," Springer-Verlag, Berlin-Heidelberg-New York, 1971.

[ZS₁]   O. Zariski, P. Samuel, "Commutative Algebra," volume I, van Nostrand, Princeton, 1958.

[ZS₂]   O. Zariski, P. Samuel, "Commutative Algebra," volume II, van Nostrand, Princeton, 1960.

Joseph LIPMAN

Department of Mathematics
Purdue University
West Lafayette, Indiana 47907
U.S.A.

Algebraic Geometry and Commutative Algebra
in Honor of Masayoshi NAGATA
pp. 233–260 (1987)

On a Compactification of a Moduli Space of Stable Vector Bundles on a Rational Surface

Masaki MARUYAMA

Introduction. Let $(X, \mathcal{O}_X(1))$ be a couple of a non-singular projective surface X and an ample line bundle $\mathcal{O}_X(1)$ on X and let $M_X(r, c_1, c_2)_0^\mu$ be the moduli space of μ-stable vector bundles of rank r on $(X, \mathcal{O}_X(1))$ with the first Chern class c_1 and the second Chern class c_2. It is known that $M_X(r, c_1, c_2)_0^\mu$ is quasi-projective and it is compactified or projectified by attaching the S-equivalence classes of semi-stable sheaves with the given invariants. We denote the compactified moduli by $\bar{M}_X(r, c_1, c_2)$. In this article we shall construct another compactification dominated by $\bar{M}_X(r, c_1, c_2)$ under the assumption

(0.1) X is rational, $d(c_1, \mathcal{O}_X(1)) = 0$ and $d(K_X, \mathcal{O}_X(1)) < 0$, where K_X is the canonical line bundle of X.

Let us explain our motivations of this new compactification.

(1) An open set of the boundary $\bar{M}_X(2, c_1, c_2) \setminus M_X(2, c_1, c_2)_0^\mu$ is constructed as follows (see [4, Proposition 3.23]): Let E be a member of $M_X(2, c_1, c_2 - 1)_0^\mu$ and x a point of X. If we have a surjective $\theta : E \to k(x)$, then $E(\theta, x) = \ker(\theta)$ is a μ-stable sheaf of rank 2, $c_1(E(\theta, x)) = c_1(E)$ and $c_2(E(\theta, x)) = c_2(E) + 1 = c_2$. When we fix E and x, θ is parametrized by $\mathbf{P}_{k(x)}^1$ and hence when (E, x) varies over $M_X(2, c_1, c_2 - 1)_0^\mu \times X$, the $E(\theta, x)$ runs over a $\mathbf{P}^1$-bundle B on $M_X(2, c_1, c_2 - 1)_0^\mu \times X$. This B is the open part of the boundary. Computing the dimension of B, we see that B is of pure codimension 1 in $\bar{M}_X(2, c_1, c_2)$. Then the following question is naturally raised.

Question 0.2. Can one contract the $\mathbf{P}^1$-fibration of B in $\bar{M}_X(2, c_1, c_2)$?

If one gets an affirmative answer to this question, then it leads him to a result by Hirschowitz and Hulek: $M_{\mathbf{P}^2}(2, c_1, c_2)_0^\mu$ contains a complete curve.

Examining the above question with Hulek and Str∅mme, we found that to get an answer we could exploit the morphism $\psi(c_2)$ of $\bar{M}_{\mathbf{P}^2}(2, 0, c_2)$ to the space P of plane curves of degree c_2 which sent a member E of $\bar{M}_{\mathbf{P}^2}(2, 0, c_2)$ to the curve of jumping lines of E (see [5, Proposition 1.10]). In fact, $\psi(c_2)$ is finite on a non-empty open set of $M_X(2, 0, c_2)_0^\mu$ and maps a fiber of B to a point (see [5,

---

Received March 16, 1987.

Proposition 1.8]). Thus, for a sufficiently large integer n, $\psi(c_2)^*(\mathcal{O}_{\mathbf{P}}(n))$ defines a morphism which contracts the fibres of B. However, the method works only on $\mathbf{P}^2$ and under the assumption that the ground field is of characteristic zero.

(2) Let $M_X(n)$ be the moduli space of ASD $SU(2)$-connections on X with the second Chern class n. Donaldson [1] proved that if one takes a "general" metric on X, then $M_X(n)$ is topologically compactified by attaching a subset of $\coprod_{d=1}^{n} M_X(n-d) \times S^d(X)$, where $S^d(X)$ is the d-th symmetric product of X. As we explained in (1), the boundary of this compactification is smaller than that of $\bar{M}_X(2,0,n)$ and its biggest part is exactly the base space of the $\mathbf{P}^1$-bundle B. Thus the questiuon 0.2 is, roughly speaking, asking whether Donaldson's compactification carries an algebraic structure.

(3) In [6], the author made use extensively an extension of a stable vector bundle on $\mathbf{P}^2$ by a trivial vector bundle, which is called in this article the universal extension (see Definition 2.6). This operation gives rise to an immersion of $M_{\mathbf{P}^2}(r,0,c_2)_0^{\mu}$ to $\bar{M}_{\mathbf{P}^2}(c_2,0,c_2)$. Then, as was mentioned in [6, Question 5.7], we come to the question: What is the closure of the image in $\bar{M}_{\mathbf{P}^2}(c_2,0,c_2)$? Our new compactification is nothing but the closure.

Assuming (0.1), we define the universal extension $U(E)$ of a semi-stable sheaf E on $(X,\mathcal{O}_X(1))$. What the author did not realize in [6] is the fact that $U(E)$ is semi-stable (see Proposition 3.1). This provides us with a morphism $\Phi(r,c_1,c_2)$ of $\bar{M}_X(r,c_1,c_2)$ to $\bar{M}_X(c_2,c_1,c_2)$ which is an immersion on $M_X(r,c_1,c_2)_0^{\mu}$ (see Theorem 3.8). This morphism solves the Question (0.2) (see Proposition 4.1) and in the case where $r = 2$ and $c_1 = 0$, endows the Donaldson's compactification with an algebraic structure if the polarization $\mathcal{O}_X(1)$ is carefully chosen (see Theorem 6.4).

When the rank is greater than 2, a semi-stable sheaf E with $c_1(E) = 0$ may have a filtration $0 = E_0 \subset E_2 \subset \cdots \subset E_\alpha = E$ such that $d(E_i/E_{i-1}, \mathcal{O}_X(1)) = 0$ for all i but $c_1(E_j/E_{j-1}) \neq 0$ for some j. For this reason we could find out only a set which contains $\mathrm{im}(\Phi(r,c_1,c_2))$ (Theorem 4.3) and the question of the surjectivity is left to studies in the future. In §5 we shall work out, however, the study of the set of the points which come from semi-stable sheaves having no such a filtration as above.

Notation and Convension. For a coherent sheaf F on a non-singular projective variety Y, $r(F)$ is the rank of F and $c_i(F)$ denotes the i-th Chern class of F. If Y is a surface, we understand that the first Chern class is defined as a rational equivalence class of divisors and the second is as a numerical equivalence class of 0-dimensional cycles. Thus $c_2(F)$ is regarded as an integer in this case. If $\mathcal{O}_Y(1)$ is an ample line bundle on Y, then $d(F,\mathcal{O}_Y(1))$ is the degree of $c_1(F)$ with respect to $\mathcal{O}_Y(1)$. For a cycle D on Y, $d(D,\mathcal{O}_Y(1))$ means the degree of D with respect to $\mathcal{O}_Y(1)$. Assuming that $r(F) \neq 0$, $P_F(m)$ (or, $\mu(F)$) denotes the polynomial $\chi(F(m))/r(F)$ (or, $d(F, \mathcal{O}_Y(1))/r(F)$, resp.), where $\chi(F(m))$ is the Hilbert polynomial $\sum_{i=0}^{\dim Y} (-1)^i H^i(Y, F(m))$ of F with respect to $\mathcal{O}_Y(1)$.

§1. Some remarks on semi-stable sheaves.

Let us start with a lemma which is quite useful but not necessarily well known.

Lemma 1.1. *Let $(Y, \mathcal{O}_Y(1))$ be a couple of a non-singular projective variety Y and an ample line bundle $\mathcal{O}_Y(1)$ and let E be a torsion free coherent sheaf on $(Y, \mathcal{O}_Y(1))$ with $d(E, \mathcal{O}_Y(1)) = 0$. Then we have the following.*

1) If there is a generically surjective homomorphism $\sigma : \mathcal{O}_Y^{\oplus s} \to E$, then $E \simeq \mathcal{O}_Y^{\oplus r}$, σ is surjective and hence $\ker(\sigma) \simeq \mathcal{O}_X^{\oplus(s-r)}$.

2) If E is μ-semi-stable, then the natural homomorphism $\rho : H^0(Y, E) \otimes_k \mathcal{O}_Y \to E$ is injective and $\mathrm{coker}(\rho)$ is torsion free.

Proof. 1) Let r be the rank of E. Since σ is surjective at the generic point of Y, there are r sections which generate E at the generic point. Thus we may assume that $s = r$ and σ is injective. If $B(E)$ be the set of pinching points of E, then $\mathrm{codim}(B(E)) \geq 2$. For the open set $U = Y - B(E)$, we can define $\det(\sigma|U)$ which is of pure codimension 1 or empty and is exactly the set of points where $\sigma|U$ is not surjective. Since $d(E, \mathcal{O}_Y(1)) = d(\mathcal{O}_Y^{\oplus r}, \mathcal{O}_Y(1)) + d(\det(\sigma|U), \mathcal{O}_Y(1))$, $\det(\sigma|U)$ is of degree 0, and hence $\sigma|U$ is isomorphic. Then, for the injection $i : U \to Y$, the isomorphism $i_* i^*(\sigma) : \mathcal{O}_Y^{\oplus r} \to i_* i^*(E)$ passes through the injection of E to $i_* i^*(E)$. This shows that σ is isomorphic.

2) If one sets $F = \mathrm{im}(\rho)$, then $r(F)\mu(E) \geq d(F, \mathcal{O}_Y(1)) \geq r(F)\mu(\mathcal{O}_Y^{\oplus t})$, where $t = \dim H^0(Y, F)$. Thus $d(F, \mathcal{O}_Y(1)) = 0$. By virtue of 1) this implies that $F \simeq \mathcal{O}_Y^{\oplus u}$ and then $\ker(\rho)$ is isomorphic to $\mathcal{O}_Y^{\oplus(t-u)}$. Since $H^0(\rho)$ must be bijective, we have $u = t$ which shows that ρ is injective. If for the torsion part T of $\mathrm{coker}(\rho)$, $\mathrm{codim}(\mathrm{Supp}(T)) = 1$, the inverse image G of T to E has a positive dgree, which contradicts the μ-semi-stability of E. Then, by the same argument as in the proof of 1), we see that F coincides with G. Q.E.D.

For convenience' sake we shall introduce a convention of stable sheaves and an order among polynomials in $\mathbf{Q}[x]$.

Convention 1.2. Let $(Y, \mathcal{O}_Y(1))$ be as in Lemma 1.1. Every torsion free, coherent sheaf of *rank 1* on $(Y, \mathcal{O}_Y(1))$ is μ-stable.

Definition 1.3. *Let $P_1(x)$ and $P_2(x)$ be polynomials in $\mathbf{Q}[x]$. When $P_1(m) < P_2(m)$ (or, $P_1(m) \leq P_2(m)$) for all sufficiently large integers m, we denote, by abuse of notation, $P_1(m) < P_2(m)$ (or, $P_1(m) \leq P_2(m)$, resp.).*

We shall next give a way to construct a stable reflexive sheaf by using succesive extensions of μ-stable reflexive sheaves.

Proposition 1.4. *Let the couple $(Y, \mathcal{O}_Y(1))$ be as in Lemma 1.1. Assume that a coherent sheaf E has a filtration $0 = E_0 \subset E_1 \subset \cdots \subset E_\alpha = E$ with the following properties:*

(i) $F_i = E_i / E_{i-1}$ *is a μ-stable reflexive sheaf with $\mu(F_i) = \mu(E)$ for all $1 \le i \le \alpha$,*

(ii) $P_{F_{i-1}}(m) \le P_{F_i}(m)$ *for all $2 \le i \le \alpha$,*

(iii) *for every $1 \le i < \alpha$, the extension*

$$0 \longrightarrow E_i / E_{i-1} \longrightarrow E_{i+1} / E_{i-1} \longrightarrow F_{i+1} \longrightarrow 0$$

is non-trivial.

Then we have the following:

(a) *E is a semi-stable reflexive sheaf and moreover, if E' is a coherent subsheaf of E such that $\mu(E') = \mu(E)$ and E/E' is torsion free, then $E' = E_i$ for some i,*

(b) *if there exists a j such that $P_{F_{j-1}}(m) < P_{F_j}(m)$, then E is stable.*

Proof. It is clear that E is a μ-semi-stable reflexive sheaf. To simplify our computation let us prepare inequalities among $P_{E_i}(m)$'s and $P_{F_j}(m)$'s.

Lemma 1.5. *Let $0 = G_0 \subset G_1 \subset \cdots \subset G_\beta = G$ be a filtration of a coherent sheaf G on $(Y, \mathcal{O}_Y(1))$ by coherent subsheaves such that for $H_i = G_i / G_{i-1}$, we have $r(H_i) \ne 0$ and $P_{H_{i-1}}(m) \le P_{H_i}(m)$. Then we get inequalities $P_{G_{i-1}}(m) \le P_{G_i}(m) \le P_{H_i}(m)$ and moreover, the equalities hold if and only if $P_{H_1}(m) = P_{H_2}(m) = \cdots = P_{H_i}(m)$.*

Proof of Lemma 1.5. We shall prove our assertion by induction on i. Since $G_1 = H_1$ and $P_{H_1}(m) \le P_{H_2}(m)$, we have

$$r(G_2)P_{G_2}(m) = r(G_1)P_{G_1}(m) + r(H_2)P_{H_2}(m) \begin{cases} \ge r(G_2)P_{G_1}(m) \\ \le r(G_2)P_{H_2}(m) \end{cases}$$

and the equalities hold if and only if $P_{H_1}(m) = P_{H_2}(m)$. Thus our lemma holds in case $i = 2$. Assume that the inequalities are true up to i. Then our computation proceeds as follows:

$$\begin{aligned} r(G_{i+1})P_{G_{i+1}}(m) &= r(G_i)P_{G_i}(m) + r(H_{i+1})P_{H_{i+1}}(m) \ge r(G_i)P_{G_i}(m) \\ &+ r(H_{i+1})P_{H_i}(m) \ge r(G_i)P_{G_i}(m) + r(H_{i+1})P_{G_i}(m) \\ &= r(G_{i+1})P_{G_i}(m), \end{aligned}$$

where the first (or second) inequality becomes an equality if and only if $P_{H_{i+1}}(m) = P_{H_i}(m)$ (or, $P_{H_1}(m) = P_{H_2}(m) = \cdots = P_{H_i}(m)$, resp.). The proof of another inequality $P_{G_{i+1}}(m) \le P_{H_{i+1}}(m)$ is similar to the above.

Now let us go back to the proof of Proposition 1.4. If $\alpha = 1$, then we have nothing to prove. The filtration $0 = E_0 \subset E_1 \subset \cdots \subset E_{\alpha-1}$ of $E_{\alpha-1}$ has the properties (i), (ii) and (iii). Thus, assuming that (a) and (b) hold for $E = E_{\alpha-1}$, we shall prove (a) and (b) for $E = E_\alpha$. Let E' be a coherent subsheaf of E such that E/E' is torsion free. Since E is μ-semi-stable, we see $\mu(E') \leq \mu(E)$. If $\mu(E') < \mu(E)$, then obviously $P_{E'}(m) < P_E(m)$. Thus it is enough to prove that $P_{E'}(m) \leq P_E(m)$ for (a) and $P_{E'}(m) < P_E(m)$ for (b) under the assumption $\mu(E') = \mu(E)$. Set $j_0 = \min\{j \mid P_{F_{j-1}}(m) \neq P_{F_j}(m)\} \cup \{\alpha + 1\}$. If $E' \subset E_{\alpha-1}$, then $P_{E'}(m) \leq P_{E_{\alpha-1}}(m) \leq P_E(m)$ by virtue of our assumption and Lemma 1.5. If $j_0 \leq \alpha - 1$ (or $j_0 = \alpha$) in addition to the above, then we obtain that $P_{E'}(m) < P_{E_{\alpha-1}}(m) \leq P_E(m)$ (or, $P_{E'}(m) \leq P_{E_{\alpha-1}}(m) < P_E(m)$, resp.) by our assumption and Lemma 1.5 again. Let us consider next the case where $E' \not\subset E_{\alpha-1}$. If we put $F = E' \cap E_{\alpha-1}$, then $E_{\alpha-1}/F$ is a subsheaf of the torsion free E/E' and hence it is torsion free, too. Since $E_{\alpha-1}$ is semi-stable, we have that $\mu(F) \leq \mu(E_{\alpha-1}) = \mu(E)$. On the other hand, since $\mu(E') = \mu(E)$ and E is μ-semi-stable, E/E' is μ-semi-stable and $\mu(E/E') = \mu(E)$. Thus we obtain $\mu(E_{\alpha-1}/F) \leq \mu(E)$. Combining these with the fact that $\mu(E_{\alpha-1}) = \mu(E)$, we see that $\mu(F) = \mu(E) = \mu(E_{\alpha-1})$. By (a) for $E_{\alpha-1}$, we see that for a j with $0 \leq j \leq \alpha - 1$, $F = E_j$. We get therefore the following exact commutative diagram:

$$
\begin{array}{ccccccc}
 & & & & 0 & & \\
 & & & & \uparrow & & \\
 & & & & E/E' & & \\
 & & & & \uparrow & & \\
0 & \to & E_{\alpha-1}/E_j & \to & E/E_j & \to & F_\alpha & \to & 0 \\
 & & \uparrow & & \uparrow & & \\
 & & E'/E_j & = & E'/F & & \\
 & & \uparrow & & \uparrow & & \\
 & & 0 & & 0 & &
\end{array}
$$

Since E/E' is torsion free and E/E_j is reflexive, E'/E_j is reflexive. Thus E'/F is reflexive and $\mu(E'/F) = \mu(E) = \mu(F_\alpha)$. These and the fact that F_α is μ-stable and $E'/F \neq 0$ imply that $E'/F = F_\alpha$. Then the middle row in the above diagram splits or $j = \alpha - 1$. That violates our assumption (iii). In fact, the splitting gives rise to one for the extension

$$0 \longrightarrow E_{\alpha-1}/E_{\alpha-2} \longrightarrow E/E_{\alpha-2} \longrightarrow F_\alpha \longrightarrow 0.$$

Therefore, we have that $E' = E$. $\hspace{4cm}$ Q.E.D.

§2. Semi-stable sheaves on a rational surface.

Let X be a non-singular projective surface over an algebraically closed field k and let $\mathcal{O}_X(1)$ be an ample line bundle on X. We shall assume from now on the following:

$$(2.1) \quad \chi(\mathcal{O}_X) = \sum_{i=0}^{2} (-1)^i \dim H^i(X, \mathcal{O}_X) = 1, \quad e = d(K_X, \mathcal{O}_X(1)) < 0,$$

where K_X is the canonical line bundle of X.

Note that it follows from the above condition that X is rational and hence $H^i(X, \mathcal{O}_X) = 0$ for $i = 1, 2$.

For a coherent sheaf F on X, we have the following by Riemann-Roch Theorem

$$(2.2) \quad \begin{aligned} \chi(F(m)) &= \sum_{i=0}^{2} (-1)^i \dim H^i(X, F(m)) \\ &= r(F) dm^2/2 + \{d(F, \mathcal{O}_X(1)) - r(F)e/2\}m + \\ &\quad c_1(F)\{c_1(F) - K_X\}/2 - c_2(F) + r(F), \end{aligned}$$

where d is the degree of X with respect to $\mathcal{O}_X(1)$ and e is as in (2.1).

Let us introduce an invariant of a coherent sheaf F on X which plays a key role in the sequel.

Definition 2.3. *For a divisor c_1 and a zero-dimensional cycle c_2 on X, we define $f(c_1, c_2)$ to be the integer*

$$\deg\{-c_1(c_1 - K_X)/2 + c_2\}.$$

If F is a coherent sheaf on X, then $f(F)$ is the integer $f(c_1(F), c_2(F))$.

The following is a slight genralization of [6, Lemma 1.3] whose proof is essentially the same as that in [6].

Lemma 2.4. *Let E be a torsion free coherent sheaf on X. If $d(E, \mathcal{O}_X(1)) = 0$, $f(E) \leq 0$ and if E is μ-semi-stable, then E is isomorphic to the trivial bundle $\mathcal{O}_X^{\oplus r}$.*

Proof. As in the proof of [6, Lemma 1.3], we may assume that E is locally free. When $r(E) = 1$, E is a line bundle with $d(E, \mathcal{O}_X(1)) = 0$ and $c_1(E)\{c_1(E) - K_X\} = -2f(E) \geq 0$. This and the Riemann-Roch Theorem provide us with $\chi(E) \geq 1$. Since $d(E^* \otimes K_X, \mathcal{O}_X(1)) < 0$, we have that $\dim H^2(X, E) = \dim H^0(X, E^* \otimes K_X) = 0$. Thus E has a non-zero global section and hence, by using the equality $d(E, \mathcal{O}_X(1)) = 0$ again, we see that E is isomorphic to $\mathcal{O}_X$. Now we can follow completely the same argument as that of [6, Lemma 1.3].

Thanks to the above result, to study the moduli spaces of semi-stable sheaves E on X with $d(E, \mathcal{O}_X(1)) = 0$, we may assume that $f(E) > 0$. Let E be a μ-semi-stable sheaf on $(X, \mathcal{O}_X(1))$ with $r(E) = r$, $d(E, \mathcal{O}_X(1)) = 0$ and $f(E) = n > 0$. Assume that $H^0(X, E) = 0$. This assumption is satisfied if

E is semi-stable and $E \not\simeq \mathcal{O}_X$. In fact, by (2.2) we see that $\chi(\mathcal{O}_X(m)) > P_E(m) = \chi(E(m))/r$. This implies that $H^0(X, E) = 0$ if E is semi-stable. By the second assumption in (2.1) and the Serre duality we see that $H^2(X, E)$ also vanishes. Hence, by (2.2) again, we obtain that $\dim H^1(X, E) = n - r$. On the other hand, the natural isomophism $\mathrm{Ext}^1_{\mathcal{O}_X}(H^1(X, E) \otimes_k \mathcal{O}_X, E) \simeq \mathrm{Hom}_k(H^1(X, E), H^1(X, E))$ provides us with a special element ξ of the first space which corresponds to the identity in the second. The ξ defines an extension

$$(2.5) \qquad 0 \longrightarrow E \longrightarrow U(E) \longrightarrow \mathcal{O}_X^{\oplus(n-r)} \longrightarrow 0.$$

Definition 2.6. *Let E be a μ-semi-stable sheaf on $(X, \mathcal{O}_X(1))$ with $d(E, \mathcal{O}_X(1)) = 0$ and $H^0(X, E) = 0$. The above $U(E)$ is called the universal extension of E.*

The universal extension is a key idea in this article. Let us first show some properties of $U(E)$ which are simply obtained.

Lemma 2.7. *Let E be as in Definition 2.6.*
1) $c_1(U(E)) = c_1(E)$, $c_2(U(E)) = c_2(E)$ and $r(U(E)) = f(E)$.
2) $H^i(X, U(E)) = 0$ for all i and hence $U(E)$ is 1-regular.
3) If a coherent sheaf F fits in an exact sequence

$$0 \longrightarrow E \longrightarrow F \longrightarrow \mathcal{O}_X^{\oplus s} \longrightarrow 0$$

and if $H^i(X, F) = 0$ for $i = 0, 1$, then $s = n - r$ and F is isomorphic to $U(E)$.

Proof. 1) is obvious by the definition of the universal extension. Set $t = n - r$. Assume that we have a non-zero element a in $H^0(X, U(E))$. Since $H^0(X, E) = 0$, a is mapped to a non-zero $v = (a_1, \cdots, a_{n-r})$ in $H^0(X, \mathcal{O}_X^{\oplus t})$, where a_i is an element of k. Let us consider the injection ζ_i of $\mathcal{O}_X$ to the i-th direct factor of $\mathcal{O}_X^{\oplus t}$. We have the following exact commutative diagram:

$$
\begin{array}{ccc}
\mathrm{Hom}_{\mathcal{O}_X}(\mathcal{O}_X^{\oplus t}, \mathcal{O}_X^{\oplus t}) & \xrightarrow{\gamma} & \mathrm{Ext}^1_{\mathcal{O}_X}(\mathcal{O}_X^{\oplus t}, E) \\
\downarrow{\alpha_i} & & \downarrow{\beta_i} \\
\end{array}
$$
$$0 \to \mathrm{Hom}_{\mathcal{O}_X}(\mathcal{O}_X, U(E)) \to \mathrm{Hom}_{\mathcal{O}_X}(\mathcal{O}_X, \mathcal{O}_X^{\oplus t}) \to \mathrm{Ext}^1_{\mathcal{O}_X}(\mathcal{O}_X, E),$$

where α_i and β_i are the maps induced by the ζ_i. The ξ is the image of id by γ. By the definition of ξ, $u_i = \beta_i(\xi)$ are elements which form a basis of $H^1(X, E) \simeq \mathrm{Ext}^1_{\mathcal{O}_X}(\mathcal{O}_X, E)$. Since $\alpha_i(id) = \zeta_i$, the above commutative diagram shows that ζ_i is sent to u_i and hence v is to a non-zero $\sum_{i=1}^t a_i u_i$ in $\mathrm{Ext}^1_{\mathcal{O}_X}(\mathcal{O}_X, E)$. This is a contradiction and hence we have that $H^0(X, U(E)) = 0$. This and (2.5) give rise to an exact sequence of cohomologies:

$$0 \to H^0(X, \mathcal{O}_X^{\oplus t}) \to H^1(X, E) \to H^1(X, U(E)) \to H^1(X, \mathcal{O}_X^{\oplus t})$$
$$\to H^2(X, E) = 0 \to H^2(X, U(E)) \to H^2(X, \mathcal{O}_X^{\oplus t}).$$

Since $\dim H^1(X, E) = t$ and $H^1(X, \mathcal{O}_X) = 0 = H^2(X, \mathcal{O}_X)$, we see that $H^1(X, U(E)) = 0 = H^2(X, U(E))$, which completes the proof of 2). By the exact sequence in 3) and the vanishing of cohomologies of F we have that $s = \dim H^0(X, \mathcal{O}_X^{\oplus s}) = \dim H^1(X, E) = t$. Let us identify $H^0(X, \mathcal{O}_X^{\oplus t})$ with $H^1(X, E)$ through the connecting homomorphism of the cohomology sequence. Then we have the following exact sequence;

$$0 = \mathrm{Hom}_{\mathcal{O}_X}(H^1(X, E) \otimes_k \mathcal{O}_X, F) \to \mathrm{Hom}_k(H^1(X, E), H^1(X, E)) \to$$
$$\mathrm{Ext}^1_{\mathcal{O}_X}(H^1(X, E) \otimes_k \mathcal{O}_X, E) \to \mathrm{Ext}^1_{\mathcal{O}_X}(H^1(X, E) \otimes_k \mathcal{O}_X, F) = 0.$$

The identity in the second term goes to the element τ of the third which defines the extension in 3). On the other hand, the identification between $H^0(X, \mathcal{O}_X^{\oplus t})$ and $H^1(X, E)$ shows that τ is nothing but ξ which defines the universal extension of E.

$$Q.E.D.$$

Remark 2.8. Let $s_1, \cdots, s_t$ be a basis of $H^1(X, E)$. Then $\sigma = (s_1, \cdots, s_t)$ in $H^1(X, E^{\oplus t})$ can be regarded as an element of $\mathrm{Ext}^1_{\mathcal{O}_X}(\mathcal{O}_X^{\oplus t}, E)$ and hence it defines an extension

$$0 \longrightarrow E \longrightarrow F_\sigma \longrightarrow \mathcal{O}_X^{\oplus t} \longrightarrow 0.$$

It is easy to see that the F_σ is isomorphic to the universal extension $U(E)$ of E.

§3. Semi-stability of the universal extension.

We shall study in this section to what extent the universal extension $U(E)$ of E inherits the stability or the semi-stability of E. Throughout this section we work on the non-singular projective surface $(X, \mathcal{O}_X(1))$ which satisfies the condition (2.1).

Proposition 3.1. Let E be a semi-stable sheaf on X with $d(E, \mathcal{O}_X(1)) = 0$. Then the universal extension $U(E)$ of E is semi-stable, too.

Proof. Since E is semi-stable and $d(E, \mathcal{O}_X(1)) = 0$, $U(E)$ is clearly μ-semi-stable and $r(U(E)) = f(U(E)) = f(E)$. Then, by (2.2) we have

$$(3.1.1) \qquad P_{U(E)}(m) = \chi(U(E)(m))/r(U(E)) = dm^2/2 - em/2.$$

Let F be a coherent subsheaf of $U(E)$ such that $r(F) < r(U(E))$ and $U(E)/F$ is torsion free. What we have to show is the inequality $P_F(m) \leq P_{U(E)}(m)$. By the μ-semi-stability of $U(E)$ we have that $d(F, \mathcal{O}_X(1)) \leq 0$. Thus we may assume that $d(F, \mathcal{O}_X(1)) = 0$. Let us first show the following:

$$(3.1.2) \quad \text{We may assume that } H^0(X, U(E)/F) = 0.$$

Suppose that $H^0(X, U(E)/F) \neq 0$. Since $U(E)/F$ is μ-semi-stable and since $d(U(E)/F, \mathcal{O}_X(1)) = 0$, the natural homomorphism $u : H^0(X, U(E)/F) \otimes_k$

$\mathcal{O}_X \to U(E)/F$ is injective by Lemma 1.1, 2). The vanishing of the first co-homology of $\mathcal{O}_X$ implies $H^0(X, \mathrm{coker}(u)) = 0$. Let F' be the inverse image of $H^0(X, U(E)/F) \otimes_k \mathcal{O}_X$ to $U(E)$ by the natural map of $U(E)$ to $U(E)/F$. Then $\chi(F'(m)) = r(F)P_F(m) + s(dm^2/2 - em/2 + 1)$, where $s = \dim H^0(X, U(E)/F)$. If $P_{F'}(m) \leq P_{U(E)}(m)$, then by (3.1.1) we see that $P_F(m) < r(F')P_{F'}(m)/r(F) - sP_{U(E)}(m)/r(F) \leq P_{U(E)}(m)$. Thus, replacing F by F', we may assume that $H^0(X, U(E)/F) = 0$.

The exact sequence

$$0 \longrightarrow F \longrightarrow U(E) \longrightarrow U(E)/F \longrightarrow 0,$$

Lemma 2.7 and (3.1.2) show that $H^0(X, F) = H^1(X, F) = 0$. On the other hand, $H^2(X, F) = 0$ because F is μ-semi-stable, $d(F, \mathcal{O}_X(1)) = 0$ and $d(K_X, \mathcal{O}_X(1)) < 0$. These and the Riemann-Roch Theorem give rise to an equality

$$0 = \chi(F) = -f(F) + r(F).$$

By this and (2.2) we obtain the equality $P_F(m) = dm^2/2 - em/2 = P_{U(E)}(m)$ as required.   Q.E.D.

The following is proved along the same idea as in the above.

Lemma 3.2. Let E be a μ-semi-stable sheaf on $(X, \mathcal{O}_X(1))$ with $d(E, \mathcal{O}_X(1)) = 0$ and $H^0(X, E) = 0$ and F a coherent subsheaf of E such that $d(F, \mathcal{O}_X(1)) = 0$ and E/F is torsion free. Then the universal extension $U(E)$ has a unique subsheaf F' which contains F and is isomorphic to the universal extension $U(F)$. Moreover, if C is the cokernel of the natural homomorphism $H^0(X, E/F) \otimes_k \mathcal{O}_X \to E/F$, then $U(E)/F'$ is isomophic to $U(C)$.

Proof. We first prove the uniqueness of F'. There is an exact commutative diagram because $U(F) \simeq F'$:

$$
\begin{array}{ccccccccc}
 & & & & & & 0 & & \\
 & & & & & & \uparrow & & \\
0 & \to & F' & \to & U(E) & \to & U(E)/F' & \to & 0 \\
 & & \uparrow & & \| & & \uparrow & & \\
0 & \to & F & \to & U(E) & \to & U(E)/F & \to & 0 \\
 & & \uparrow & & & & \uparrow & & \\
 & & 0 & & & & F'/F \simeq \mathcal{O}_X^{\oplus s} & & \\
 & & & & & & \uparrow & & \\
 & & & & & & 0 & &
\end{array}
$$

where $s = f(F) - r(F)$. Using top row of the above and Lemma 2.7 we see that $H^0(X, U(E)/F') = 0$ and hence $H^0(X, F'/F) = H^0(X, U(E)/F)$. This means that $G = H^0(X, U(E)/F) \otimes_k \mathcal{O}_X$ is a subsheaf of $U(E)/F$ and coincides with F'/F. Since F' is the inverse image of G to $U(E)$, we complete the proof of the uniqueness. This proof indicates the way how we can get F'. Since $d(F, \mathcal{O}_X(1)) = 0$, $H = U(E)/F$ is a μ-semi-stable sheaf with $d(H, \mathcal{O}_X(1)) = 0$ and then $I = H^0(X, H) \otimes_k \mathcal{O}_X$ is a subsheaf of H. Let F' be the inverse image of I to $U(E)$. The same argument as in the proof of Proposition 3.1 shows that $H^i(X, F') = 0$ for all i and hence F' is isomorphic to $U(F)$ by virtue of Lemma 2.7, 3). To prove the last assertion let us note first that $H^i(X, U(E)/F') = 0$ for all i because of the vanishing of all the cohomologies of $U(E)$ and F'. Thus we have to show that C is a subsheaf of $D = U(E)/F'$ and $D/C \simeq \mathcal{O}_X^{\oplus \beta}$ for some β. Let us consider the following exact commutative diagram:

$$
\begin{array}{ccccccccc}
& & 0 & & 0 & & 0 & & \\
& & \uparrow & & \uparrow & & \uparrow & & \\
0 & \to & K & \to & C & \xrightarrow{\zeta} & U(E)/F' & \to & L & \to & 0 \\
& & \uparrow & & \uparrow & & \uparrow & & \uparrow & & \\
0 & \to & E/F & \to & U(E)/F & \to & \mathcal{O}_X^{\oplus t} & \to & 0 \\
& & \uparrow & & \uparrow & & \uparrow \theta & & \\
0 & \to & \mathcal{O}_X^{\oplus \alpha} & \to & \mathcal{O}_X^{\oplus s} & \to & \mathcal{O}_X^{\oplus s-\alpha} & \to & 0, \\
& & \uparrow & & \uparrow & & & & \\
& & 0 & & 0 & & & &
\end{array}
$$

where $\mathcal{O}_X^{\oplus \alpha}$ is the $H^0(X, E/F) \otimes_k \mathcal{O}_X$ which is a subsheaf of E/F by virtue of Lemma 1.1 and the μ-semi-stability of E/F and where K and L are the kernel and the cokernel of the natural $\zeta : C \to U(E)/F'$, respectively. Thanks to the snake lemma K is the kernel of θ which is isomorphic to $\mathcal{O}_X^{\oplus \gamma}$ for some γ. By the definition of C and the fact that $H^1(X, \mathcal{O}_X) = 0$ we see that $H^0(X, C) = 0$. Thus we obtain $\gamma = 0$ and hence C is a subsheaf of $U(E)/F'$. Now it is obvious that $L \simeq \mathrm{coker}(\theta) \simeq \mathcal{O}_X^{\oplus(t+\alpha-s)}$.   Q.E.D.

Using the above lemma repeatedly, we see an interesting feature of the universal extension.

Proposition 3.3. Let E be a μ-semi-stable sheaf on $(X, \mathcal{O}_X(1))$ with $d(E, \mathcal{O}_X(1)) = 0$ and $H^0(X, E) = 0$.

1) Assume that E is semi-stable. If we have a filtration $0 = E_0 \subset E_1 \subset \cdots \subset E_\alpha = E$ by coherent subsheaves E_i with $P_E(m) = P_{E_i}(m)$, then there is a filtration $0 = F_0 \subset F_1 \subset \cdots \subset F_\alpha = U(E)$ such that $F_i \simeq U(E_i)$ and $F_i/F_{i-1} \simeq U(E_i/E_{i-1})$.

2) Suppose that there is filtration $0 = E_0' \subset E_1' \subset \cdots \subset E_\beta' = E$ such that (i) E_i'/E_{i-1}' is a μ-stable sheaf with degree 0 for all $1 \leq i \leq \beta$ and (ii) if $E_i'/E_{i-1}' \simeq \mathcal{O}_X$ and $E_{i+1}'/E_i' \not\simeq \mathcal{O}_X$, then $H^0(X, E/E_i') = 0$. Then there exists a filtration $0 = F_0' \subset F_1' \subset \cdots \subset F_\beta' = U(E)$ such that (a) $F_i' \simeq U(E_i')$ (b) $F_i' = F_{i+1}'$ if and only if $E_{i+1}'/E_i' \simeq \mathcal{O}_X$ and (c) if E_{i+1}'/E_i' is not isomorphic to $\mathcal{O}_X$, then $F_{i+1}'/F_i' \simeq U(E_{i+1}'/E_i')$.

Proof. 1) We shall prove our assertion by induction on α. The condition that E is semi-stable and $P_{E_i}(m) = P_E(m)$ implies that both E_i and E/E_i are semi-stable. Applying Lemma 3.2 to the couple (E_1, E), we get a subsheaf F_1 of $U(E)$ such that $E_1 \subset F_1$, $F_1 \simeq U(E_1)$ and $U(E)/F_1 \simeq U(E/E_1)$. Thus our 1) holds in case of $\alpha = 2$. If $\alpha > 2$, then the filtration $0 = E_1/E_1 \subset E_2/E_1 \subset \cdots \subset E_\alpha/E_1 = E/E_1$ satisfies the assumptions in 1). Thus the induction hypothesis provides us with a filtration $0 = G_1 \subset G_2 \subset \cdots \subset G_\alpha = U(E/E_1) = U(E)/F_1$ such that $G_i \simeq U(E_i/E_1)$ and $G_i/G_{i-1} \simeq U(E_i/E_{i-1})$. Let F_i be the inverse image of G_i by the natural projection of $U(E)$ to $U(E)/F_1$ for $i > 0$ and $F_0 = 0$. Then we have an exact commutative diagram:

$$
\begin{array}{ccccccccc}
 & & 0 & & 0 & & 0 & & \\
 & & \uparrow & & \uparrow & & \uparrow & & \\
0 & \to & E_i/E_1 & \to & G_i & \to & \mathcal{O}_X^{\oplus \sigma_i} & \to & 0 \\
 & & \uparrow & & \uparrow & & \uparrow & & \\
0 & \to & E_i & \to & F_i & \to & F_i/E_i & \to & 0 \\
 & & \uparrow & & \uparrow & & \uparrow & & \\
0 & \to & E_1 & \to & F_1 & \to & \mathcal{O}_X^{\oplus \gamma_i} & \to & 0. \\
 & & \uparrow & & \uparrow & & \uparrow & & \\
 & & 0 & & 0 & & 0 & &
\end{array}
$$

Since $H^j(X, F_1) = 0 = H^j(X, G_i)$ for all j, all the cohomologies of F_i vanish. On the other hand, the rightmost column of the above diagram shows that F_i/E_i is isomorphic to $\mathcal{O}_X^{\oplus \gamma_i + \delta_i}$. By virtue of Lemma 2.7 these imply that $F_i \simeq U(E_i)$. Since $F_1/F_0 \simeq F_1$ and $F_i/F_{i-1} \simeq G_i/G_{i-1}$ for all $i > 1$, we have that $F_i/F_{i-1} \simeq U(E_i/E_{i-1})$.

2) Applying Lemma 3.2 to the couple (E, E_1'), we obtain a subsheaf F_1' of $U(E)$ which is isomorphic to $U(E_1')$. If $E_j' \simeq \mathcal{O}_X$ for $1 < j \le k$ and $E_{k+1}' \not\simeq \mathcal{O}_X$, then $F_1' \simeq U(E_j')$ for all $1 \le j \le k$ by the construction of F_1' and $H^0(X, E/E_k') = 0$ by the assumption (ii). Setting $F_j' = F_1'$ for $1 \le j \le k$, we see by Lemma 3.2 that $U(E/E_k') = U(E)/F_k'$. Replacing E by E/E_k' and using induction on β, we get a filteration $0 = G_k' \subset G_{k+1}' \subset \cdots \subset G_{\beta-k}' = U(E/E_k') = U(E)/F_k'$. Let F_j' be the inverse image of G_j' to $U(E)$. Then it is easily seen as in the proof of 1) that the filtration $0 = F_0' \subset F_1' \subset \cdots \subset F_\beta' = U(E)$ meets our requirement.   Q.E.D.

Remark 3.4. If E is a μ-semi-stable sheaf on X with $d(E, \mathcal{O}_X(1)) = 0$ and $H^0(X, E) = 0$, then there is at least one filtration $0 = E_0' \subset E_1' \subset \cdots \subset E_\beta' = E$ which satisfies the assumption of 2) in the above proposition.

As a direct consequence of Proposition 3.3, 1) we have the following.

Corollary 3.5. *If semi-stable sheaves E_1 and E_2 are S-equivalent to each other, then so are $U(E_1)$ and $U(E_2)$ (for the definition of the S-equivalence, see [3, Definition 1.5]).*

Proof. Let $0 = E_{i0} \subset E_{i1} \subset \cdots \subset E_{i\alpha} = E_i$ be a filtration of E_i which gives rise to $gr(E_i)$. From this we get a filtration $0 = F_{i0} \subset F_{i1} \subset \cdots \subset F_{i\alpha} = U(E_i)$ as in Proposition 3.3, 1). Then F_{ij} are semi-stable by virtue of Proposition 3.1 and $P_{U(E_i)}(m) = dm^2/2 - em/2 = P_{F_{ij}}(m)$ for all $i, j \neq 0$. It is easy to see that there is a filtration $0 = G_{i0} \subset G_{i1} \subset \cdots \subset G_{i\beta_i} = U(E_i)$ which passes through the above filtration and gives rise to $gr(U(E_i))$. Now we have isomorphisms $gr(U(E_1)) = \oplus_j G_{1j}/G_{1j-1} \simeq \oplus_k gr(F_{1k}/F_{1k-1}) \simeq \oplus_p gr(F_{2p}/F_{2p-1}) \simeq \oplus_q gr(G_{2q}/G_{2q-1}) = gr(E_2)$ as required.

A μ-stable *vector bundle* goes to a stable vector bundle under the operation of taking the universal extension. Indeed, we have the following whose proof is completely the same as in [6, Lemma 1.4, (2)] after replacing $c_2(E)$ by $f(E)$.

Proposition 3.6. *If E is a μ-stable vector bundle on $(X, \mathcal{O}_X(1))$ with $d(E, \mathcal{O}_X(1)) = 0$ and $E \not\simeq \mathcal{O}_X$, then the universal extension $U(E)$ is a stable vector bundle.*

Note that in the situation of the above proposition $U(E)$ can not be μ-stable unless $f(E) = r(E)$ or $E = U(E)$. Now let us fix our notation of moduli spaces of semi-stable sheaves.

Definition 3.7. *Let $\bar{M}_X(r, c_1, c_2)$ be the moduli space of semi-stable sheaves E on X with $r(E) = r$, $c_1(E) = c_1$ and $c_2(E) = c_2$. $M_X(r, c_1, c_2)$, $M_X(r, c_1, c_2)_0$, $M_X(r, c_1, c_2)^\mu$ and $M_X(r, c_1, c_2)_0^\mu$ are open subscheme of $\bar{M}_X(r, c_1, c_2)$ consisting of points corresponding to stable sheaves, stable vector bundles, μ-stable sheaves and μ-stable vector bundles, respectively.*

Proposition 3.1 and Corollary 3.5 show that by letting the S-equivalence class of E correspond to that of $U(E)$, we obtain a set theoretical map $\phi(r, c_1, c_2)$ of $\bar{M}_X(r, c_1, c_2)$ to $\bar{M}_X(f(c_1, c_2), c_1, c_2)$ if $d(c_1, \mathcal{O}_X(1)) = 0$. From the exact sequence (2.5) we derive the dual exact sequence:

$$0 \longrightarrow \mathcal{O}_X^{\oplus t} \longrightarrow U(E)^* \longrightarrow E^* \longrightarrow 0.$$

Thus if $H^0(X, E^*) = 0$ and E is locally free, then E is uniquely determined by $U(E)$. This obsevation and Proposition 3.6 imply that $\phi(r, c_1, c_2)$ is injective on $M_X(r, c_1, c_2)_0^\mu$. If one notes the fact that $\bar{M}_X(r, c_1, c_2)$ is normal and $M_X(r, c_1, c_2)$ is smooth under the assumption (2.1), then one obtains the following result by the same argument as in [6, Proposition 1.7].

Theorem 3.8. *Let $(X, \mathcal{O}_X(1))$ be a couple of a non-singular projective surface X and an ample line bundle $\mathcal{O}_X(1)$ on X which satisfies the condition (2.1), c_1 be a divisor on X with $d(c_1, \mathcal{O}_X(1)) = 0$ and let r, c_2 be integers with $f(c_1, c_2) \geq r > 0$. There is a morphism $\Phi(r, c_1, c_2)$ of $\bar{M}_X(r, c_1, c_2)$ to $\bar{M}_X(f(c_1, c_2), c_1, c_2)$ which induces $\phi(r, c_1, c_2)$ set-theoretically and whose restriction to $M_X(r, c_1, c_2)_0^\mu$ is an immersion.*

§4. Image of $\Phi(r, c_1, c_2)$.

As was indicated in Proposition 3.3, 2), $\Phi(r, c_1, c_2)$ is in general not injective outside $M_X(r, c_1, c_2)_0^\mu$. In this section we shall study which points are identified by the map $\phi(r, c_1, c_2)$.

Let $(X, \mathcal{O}_X(1))$ be as in (2.1), E a μ-stable sheaf on $(X, \mathcal{O}_X(1))$ with $d(E, \mathcal{O}_X(1)) = 0$, E' the double dual $\underline{\mathrm{Hom}}_{\mathcal{O}_X}(\underline{\mathrm{Hom}}_{\mathcal{O}_X}(E, \mathcal{O}_X), \mathcal{O}_X)$ of E and T the cokernel of the natural injection of E to E'. Pick a point x_1 in $\mathrm{Supp}(T)$ and assume that $\mathrm{length}(T_{x_1}) = a_1$. Then $x_i = x_1$ for $1 \le i \le a_1$ and x_{a_1+1} is a point in $\mathrm{Supp}(T)$ other than x_1 if any. If the length of the stalk of T at x_{a_1+1} is $a_2 - a_1$, then $x_i = x_{a_1+1}$ for $a_1 + 1 \le i \le a_2$. Continuing this process until we exhaust $\mathrm{Supp}(T)$, we obtain an ordered set of points $(x_1, \cdots, x_s)$ such that $s = \dim H^0(X, T)$ and $\mathrm{Supp}(T) = \{x_1, \cdots, x_s\}$ as sets. Under this notation we have one of crucial results.

Proposition 4.1. *Assume that $E' \not\simeq \mathcal{O}_X$. $U(E)$ has a filtration $0 = F_0 \subset F_1 \subset \cdots \subset F_s \subset F_{s+1} = U(E)$ such that $F_{s+1}/F_s \simeq U(E')$ and $F_i/F_{i-1} \simeq m_{x_i}$ for $1 \le i \le s$, where m_{x_i} is the ideal defining the reduced one point scheme x_i. Therefore, $\mathrm{gr}(U(E)) \simeq (\oplus_{i=1}^s m_{x_i}) \oplus U(E')$ and hence the S-equivalence class of $U(E)$ depends only on E' and the positions of $x_1, \cdots, x_s$.*

Proof. Since $H^0(X, E') = 0 = H^1(X, T)$, we have an exact sequence

$$0 \longrightarrow H^0(X, T) \longrightarrow H^1(X, E) \longrightarrow H^1(X, E') \longrightarrow 0.$$

Let $\sigma_1, \cdots, \sigma_t$ be a basis of $H^1(X, E)$ such that $\sigma_1, \cdots, \sigma_s$ span the subspace $H^0(X, T)$. Then $\zeta = (\sigma_1, \cdots, \sigma_t)$ in $\mathrm{Ext}^1_{\mathcal{O}_X}(\mathcal{O}_X^{\oplus t}, E)$ and its image ζ' in $\mathrm{Ext}^1_{\mathcal{O}_X}(\mathcal{O}_X^{\oplus t}, E')$ define extensions F and F' of $\mathcal{O}_X^{\oplus t}$ by E and E', respectively. As was mentioned in Remark 2.8, F is isomorphic to the universal extension $U(E)$ of E. By the construction of F and F' we have the following exact commutative diagram:

$$
\begin{array}{ccccccccc}
& & 0 & & 0 & & & & \\
& & \uparrow & & \uparrow & & & & \\
& & T & \overset{u}{\to} & S & & & & \\
& & \uparrow & & \uparrow & & & & \\
0 & \to & E' & \to & F' & \to & \mathcal{O}_X^{\oplus t} & \to & 0 \\
& & \uparrow & & \uparrow & & \| & & \\
0 & \to & E & \to & F & \to & \mathcal{O}_X^{\oplus t} & \to & 0, \\
& & \uparrow & & \uparrow & & & & \\
& & 0 & & 0 & & & &
\end{array}
$$

where S is the cokernel of the natural homomorphism of F to F'. By the snake lemma u is an isomorphism. Since $\sigma_1, \cdots, \sigma_s$ are mapped to zero in $H^1(X, E')$, the inverse image of a suitable direct factor $\mathcal{O}_X^{\oplus s}$ of $\mathcal{O}_X^{\oplus t}$ to F' splits into a direct sum $E' \oplus \mathcal{O}_X^{\oplus s}$. In other words, F' contains a subsheaf G which is isomorphically mapped to a direct factor of $\mathcal{O}_X^{\oplus t}$. S is generated by the image of $H^0(X, G)$ because $H^0(X, F) = H^1(X, F) = 0$, $\dim H^0(X, S) = s$ and S is generated by its global sections. By the definition of the sequence $(x_1, \cdots, x_s)$ we can easily construct a filtration $0 = S_0 \subset S_1 \subset \cdots \subset S_s = S$ such that $S_i/S_{i-1} \simeq k(x_i)$ $(1 \leq i \leq s)$. Since $\dim H^0(X, S_i) = i$, we have a filtration $0 = G_0 \subset G_1 \subset \cdots \subset G_s = G$ such that $G_i/G_{i-1} \simeq \mathcal{O}_X$ and G_i mapped onto S_i. If one sets $F_i = G_i \cap F$, then one gets an exact sequence

$$0 \longrightarrow F_i/F_{i-1} \longrightarrow G_i/G_{i-1} \longrightarrow S_i/S_{i-1} \longrightarrow 0.$$

Thus we see that $F_i/F_{i-1} \simeq m_{x_i}$ for $1 \leq i \leq s$. From the construction of G we obtain an exact sequence

$$0 \longrightarrow E' \longrightarrow F'/G \longrightarrow \mathcal{O}_X^{\oplus(t-s)} \longrightarrow 0.$$

On the other hand, there an exact commutative diagram

$$
\begin{array}{ccccccccc}
 & & 0 & & 0 & & & & \\
 & & \uparrow & & \uparrow & & & & \\
 & & S & = & S & & & & \\
 & & \uparrow & & \uparrow & & & & \\
0 & \to & G & \to & F' & \to & F'/G & \to & 0 \\
 & & \uparrow & & \uparrow & & \| & & \\
0 & \to & F_s & \to & F & \to & F/F_s & \to & 0. \\
 & & \uparrow & & \uparrow & & & & \\
 & & 0 & & 0 & & & &
\end{array}
$$

The left vertical exact sequence shows that $H^i(X, F_s) = 0$ for all i and then the bottom sequence implies that $H^i(X, F'/G) = H^i(X, F/F_s) = 0$, too. These, above exact sequence and Lemma 2.7 show that F/F_s is isomorphic to $U(E')$.
$$\text{Q.E.D.}$$

We shall next treat the case where $E' \simeq \mathcal{O}_X$.

Proposition 4.2. *Let E, E' and $(x_1, \cdots, x_s)$ be as before Proposition 4.1. If $E' \simeq \mathcal{O}_X$, then there is a filtration $0 = F_0 \subset F_1 \subset \cdots \subset F_s = U(E)$ such that $F_i/F_{i-1} \simeq m_{x_i}$ for all $1 \leq i \leq s$. Therefore, $gr(U(E)) \simeq \oplus_{i=1}^s m_{x_i}$ and hence the S-equivalence class of $U(E)$ is determined by the positions of $x_1, \cdots, x_s$.*

Proof. In this case, E is an ideal sheaf in $\mathcal{O}_X$ such that $T = \mathcal{O}_X/E$ is a torsion sheaf supported by $\{x_1, \cdots, x_s\}$. Thus $\dim H^1(X, E) = \dim H^0(X, T) - \dim H^0(X, \mathcal{O}_X) = s - 1$ and then we have an exact commutative diagram as in the proof of the above proposition:

$$
\begin{array}{ccc}
0 & & 0 \\
\uparrow & & \uparrow \\
T & = & T \\
\uparrow & & \uparrow \\
0 \to \mathcal{O}_X \to \mathcal{O}_X^{\oplus s} & \to & \mathcal{O}_X^{\oplus(s-1)} \to 0 \\
\uparrow \qquad \uparrow & & \parallel \\
0 \to E \to U(E) & \to & \mathcal{O}_X^{\oplus(s-1)} \to 0. \\
\uparrow \qquad \uparrow & & \\
0 \qquad 0 & &
\end{array}
$$

There are filtrations $0 = T_0 \subset T_1 \subset \cdots \subset T_s = T$ and $0 = G_0 \subset G_1 \subset \cdots \subset G_s = \mathcal{O}_X^{\oplus s}$ such that $T_i/T_{i-1} \simeq k(x_i)$, $G_i/G_{i-1} \simeq \mathcal{O}_X$ and that G_i is mapped onto T_i. Setting $F_i = G_i \cap U(E)$, it is easy to see that the filtration $0 = F_0 \subset F_1 \subset \cdots \subset F_s = U(E)$ meets our requirement. $Q.E.D.$

Combining the above results we obtain the following.

Theorem 4.3. *As a set the image of $\phi(r, c_1, c_2)$ is contained in the set*

$$
\coprod \left\{ \prod_{i=1}^{\alpha} M_X(s_i, a_i, b)_0^{\mu} \right\} \times S^d(X),
$$

where the disjoint union $\coprod$ ranges over the set $A = \{ \ (\{(s_1, a_1, b_1), \cdots, (s_\alpha, a_\alpha, b_\alpha)\}, d) \ | \ \alpha > 0, \ 0 < \sum_{i=1}^{\alpha} s_i \leq r, \ f(a_i, b_i) \geq 0, \ d(a_i, \mathcal{O}_X(1)) = 0, \ \sum_{i=1}^{\alpha} a_i = c_1, \ \sum_{i=1}^{\alpha} f(a_i, b_i) + d = f(c_1, c_2), \ and \ d \geq 0 \ \}$ (see Convention 1.2). More precisely, the S-equivalence class of a point of the image of $\Phi(r, c_1, c_2)$ is represented by a $(\oplus_{i=1}^{\alpha} U(E_i)) \oplus (\oplus_{j=1}^{d} m_{x_j})$ with $E_i \in M_X(s_i, a_i, b_i)_0^{\mu}$ and $(x_1, \cdots, x_d) \in S^d(X)$.

Proof. Let E represent a point of $\bar{M}_X(r, c_1, c_2)$. E has a filtration $0 = E_0 \subset E_1 \subset \cdots \subset E_\beta = E$ which satisfies the assumption of Proposition 3.3, 2) (see Remark 3.4). By Proposition 3.3 we obtain a filtration $0 = F_0 \subset F_1 \subset \cdots \subset F_\gamma = U(E)$ such that $\{F_j/F_{j-1} \mid 1 \leq j \leq \gamma\} = \{U(E_i/E_{i-1}) \mid 1 \leq i \leq \beta, \ E_i/E_{i-1} \not\simeq \mathcal{O}_X\}$. Since $gr(U(E)) = \sum_{j=1}^{\gamma} gr(F_j/F_{j-1})$, we may assume, after renumbering suitably, that $F_j/F_{j-1} = U(E_j/E_{j-1})$ $(1 \leq j \leq \gamma)$ and $H_j = ((E_j/E_{j-1})^*)^*$ is isomorphic to $\mathcal{O}_X$ or not according as $1 \leq j \leq \alpha$ or $\alpha + 1 \leq j \leq \gamma$. By virtue of Proposition 4.1 and 4.2 $gr(F_j/F_{j-1}) = (\oplus_\ell m_{x_{j\ell}}) \oplus U(H_j)$ or $gr(F_j/F_{j-1}) = \oplus_k m_{x_{jk}}$ according as $1 \leq j \leq \alpha$ or $\alpha + 1 \leq j \leq \gamma$. Therefore we get that $gr(U(E)) = (\oplus_{i=1}^{\alpha} U(H_i)) \oplus (\oplus_{j=1}^{d} m_{y_j})$ for a suitable set $\{y_1, \cdots, y_d\}$ of points of X. If we set $r(H_j) = s_j$, $c_1(H_j) = a_j$ and $c_2(H_j) = b_j$, then $gr(U(E))$ defines a point of $(\prod_{i=1}^{\alpha} M_X(s_i, a_i, b_i)_0^{\mu}) \times S^d(X)$ because $\phi(s_i, a_i, b_i)$ is injective on $M_X(s_i, a_i, b_i)_0^{\mu}$ by Proposition 3.6 or Theorem 3.8. $Q.E.D.$

Remark 4.4. Obviously we can omit the subset $\{\ (\{\ (s_1,\ a_1,\ b_1),\ \cdots,\ (s_\alpha,\ a_\alpha,\ b_\alpha)\ \},\ d)\ |\ \alpha \geq 2,\ s_1 = \cdots = s_\alpha = 1,\ a_1 = \cdots = a_\alpha = 0\ \}$ from A.

§5. Image of $\Phi(r, 0, c_2)$.

We shall study next which components in the disjoint union of Theorem 4.3 are actually in the image of $\Phi(r, c_1, c_2)$. We shall treat only the case where $c_1 = 0$ and $a_1 = a_2 = \cdots = a_\alpha = 0$. Let A_0 be the subset of A consisting of the members with $a_1 = a_2 = \cdots = a_\alpha = 0$. A necessary and sufficient condition for $M_X(s_i, 0, b_i)_0^\mu$ to be non-empty is either

(5.1.1) $s_i = 1$ and $b_i = 0$, that is, $M_X(s_i, 0, b_i)_0^\mu = \{\mathcal{O}_X\}$

or (5.1.2) $s_i \geq 2$ and $b_i \geq s_i$.

In fact, we have the following which is a generalization of a well-known result on $\mathbf{P}^2$ (see [6, Proposition 5.3] or [2]).

Proposition 5.2. *Let r be an integer with $r \geq 2$. $M_X(r, 0, c)_0^\mu \neq \emptyset$ if and only if $c \geq r$.*

Proof.   First let us recall some of known results which are relevant to our proof.

(5.2.1)   If E is a μ-semi-stable sheaf on $(X, \mathcal{O}_X(1))$ with $d(E, \mathcal{O}_X(1)) = 0$ and if E is simple, then $\mathrm{Ext}^2_{\mathcal{O}_X}(E, E) = 0$ and $\dim \mathrm{Ext}^1_{\mathcal{O}_X}(E, E) = 2rc_2 - r^2 + 1$ (note that (2.1) implies that $\mathrm{Hom}_{\mathcal{O}_X}(E, E \otimes_{\mathcal{O}_X} K_X) = 0$ and hence $H^2(X, \underline{\mathrm{Hom}}_{\mathcal{O}_X}(E, E)) = 0$).

(5.2.2)   The moduli space $M_X(r, 0, c_2)$ is smooth and of pure dimension $2rc_2 - r^2 + 1$ (see the above (5.2.1) and [3, Proposition 6.9]).

Step I.   If E is a stable vector bundle of rank $r \geq 2$ with $c_1(E) = 0$ and $c_2(E) = c_2$, then $H^0(X, E) = 0 = H^2(X, E)$ and hence $\dim H^1(X, E) = c_2 - r$ by virtue of Riemann-Roch Theorem. This shows that the condition $c_2 \geq r$ is necessary.

Step II.   The case of $c_2 = r = 2$:   Let I and J be the ideal sheaves of points x and y, respectively. Locally in a neighborhood of x, we have a standard resolution $0 \to \mathcal{O}_X \to \mathcal{O}_X^{\oplus 2} \to I \to 0$ of I. Thus we see that $\underline{\mathrm{Ext}}^1_{\mathcal{O}_X}(I, J) \simeq k(x)$ or m_x/m_x^2 according as $x \neq y$ or not, where m_x is the maximal ideal of $\mathcal{O}_{X,x}$. On the other hand, it is easy to see that $\underline{\mathrm{Hom}}_{\mathcal{O}_X}(I, J) \simeq J$ or $\mathcal{O}_X$ according as $x \neq y$ or not and hence $H^1(X, \underline{\mathrm{Hom}}_{\mathcal{O}_X}(I, J)) = 0$. These imply that $\dim \mathrm{Ext}^1_{\mathcal{O}_X}(I, J) = 1$ or 2 according as $x \neq y$ or not. Thus the sheaves E which fit in non-trivial extensions

(5.2.3) $0 \longrightarrow J \longrightarrow E \longrightarrow I \longrightarrow 0$

are parametrized by a four dimensional algebraic scheme when x and y vary over X. Taking x and y with $x \neq y$, let us fix a non-trivial extension $0 \longrightarrow J \longrightarrow E \longrightarrow I \longrightarrow 0$. E is locally free at x and obviously E is semi-stable. Since $\mathrm{Hom}_{\mathcal{O}_X}(I, E) = \mathrm{Hom}_{\mathcal{O}_X}(\mathcal{O}_X, E) = H^0(X, E) = 0$, $\mathrm{End}_{\mathcal{O}_X}(E)$ is a subspace of $\mathrm{Hom}_{\mathcal{O}_X}(J, E)$. By using the assumption that $x \neq y$, it is not hard to see that $\mathrm{Hom}_{\mathcal{O}_X}(J, E)$ is one dimensional space generated by the given injection. E is therefore simple. If one picks a sufficiently large integer m, then $E(m)$ is generated by global sections and $H^1(X, E(m)) = 0 = H^2(X, E(m))$. Let us look at the Quot-scheme $Q = \mathrm{Quot}(\mathcal{O}_X^{\oplus N}/X/k)$, where $N = \chi(E(m))$. The choice of m gives rise to an exact sequence

$$(5.2.4) \qquad 0 \longrightarrow F \longrightarrow \mathcal{O}_X^{\oplus N} \xrightarrow{\psi} E(m) \longrightarrow 0$$

such that $H^0(X, \psi)$ is an isomorphism. This sequence provides us with a point z in Q. There is a neighborhood U of z in Q such that U is stable under the natural action of $PGL(N)$ on Q, the restriction of the action to U is free and $\tilde{E}(w) = \tilde{E} \otimes_{\mathcal{O}_U} k(w)$ is semi-stable and simple for all w in $U1$ with $\tilde{E}$ the restriction of the universal quotient sheaf to $X \times U$. The exact sequence (5.6.4) yields an exact sequence

$$\mathrm{End}_{\mathcal{O}_X}(E) \to H^0(X, E(m))^{\oplus N} \to \mathrm{Hom}_{\mathcal{O}_X}(F, E(m)) \to \mathrm{Ext}^1_{\mathcal{O}_X}(E, E)$$
$$\to H^1(X, E(m))^{\oplus N} = 0.$$

and an isomorphism (note that F is locally free)

$$H^1(X, \underline{\mathrm{Hom}}_{\mathcal{O}_X}(F, E(m))) \simeq \mathrm{Ext}^2_{\mathcal{O}_X}(E, E).$$

By (5.2.1) the last spaces are zero and hence we may assume that U is smooth. (5.2.1) and the above exact sequence show that U is of dimension $N^2 - 1 + 5 = \dim PGL(N) + 5$. On the other hand, the set of points w where $\tilde{E}(w)$ fits in the sequence (5.2.3) is of dimension one less. Thus there is a morphism $\mathrm{Spec}(R) \to U$ with R a discrete valuation ring such that the image of the closed point is the z and for $G = (\tilde{E} \otimes p_*(\mathcal{O}_X(m))) \otimes_{\mathcal{O}_U} R$ and the generic point $\bar{z}$, $G(\bar{z})$ does not fit in the extension of the type (5.2.3), where $p : U \times X \to X$ is the projection. Since $G(\bar{z})$ is semi-stable, the degree of a rank 1 subsheaf of $G(\bar{z})$ is not positive. If $\bar{L}$ is a coherent subsheaf of $G(\bar{z}) \otimes \overline{k(\bar{z})}$ of rank 1 with degree 0, then, after changing R suitably, $\bar{L}$ extends to a coherent subsheaf L of G which is flat over R. Since all the coherent subsheaves of E of rank 1 with degree 0 have the first Chern class 0 and since $\mathrm{Pic}(X)$ is a discrete group, we see that $c_1(\bar{L}) = 0$. Then it is easy to see that $G(\bar{z})$ fits in an extension of type (5.2.3), which is a contradiction. Thus $G(\bar{z})$ is μ-stable. If $G(\bar{z})$ is not locally free, its double dual G' is a stable vector bundle with $r(G') = 2$, $c_1(G') = 0$ and $c_2(G') < 2$, which is impossible by virtue of Step I. Now, thanks to the openness of the μ-stability, we have a required vector bundle on X.

Step III.   Let us prove our assertion in the case of rank 2 by induction on
c_2. By virtue of Step II the proposition is true in case $c_2 = 2$ and hence we
assume that $M_X(2, 0, c_2 - 1)_0^\mu$ is not empty. For a member E of $M_X(2, 0, c_2 - 1)_0^\mu$
and a point x of X, we set $E(x, \sigma)$ to be the kernel of a sujective homomorphism $\sigma : E \rightarrow k(x)$. Then the family $\{ E(x, \sigma) \mid E \in M_X(2, 0, c_2 - 1)_0^\mu,$
$x \in X \}$ is parametrized by a $\mathbf{P}^1$-bundle Z over $M_X(2, 0, c_2 - 1)_0^\mu \times X$ which
is of dimension $4c_2 - 4$ by (5.2.2). Since every $E(x, \sigma)$ defines a smooth point
of $M_X(2, 0, c_2)^\mu$ whose dimension is $4c_2 - 3$. Therefore, there are a discrete
valuation ring R and a family of μ-stable sheaves F on $X \times \mathrm{Spec}(R)$ such
that for the closed point z of $\mathrm{Spec}(R)$, $F(z)$ is one of $E(x, \sigma)$'s and for the
generic point $\bar{z}$ of $\mathrm{Spec}(R)$, $F(\bar{z})$ corresponds to a point outside Z. Then it
is not hard to see that $\dim H^0(X, F(\bar{z})'/F(\bar{z})) \leq \dim H^0(X, F(z)'/F(z)) = 1$,
where G' is the double dual of G (use the facts that for a torsion free coher-
ent G, $\dim H^0(X, G'/G) = \dim H^0(X, \underline{\mathrm{Ext}}^1_{\mathcal{O}_X}(G, \mathcal{O}_X))$, $\underline{\mathrm{Ext}}^1_{\mathcal{O}_{X_{\bar{z}}}}(F(\bar{z}), \mathcal{O}_{X_{\bar{z}}}) \simeq$
$\underline{\mathrm{Ext}}^1_{\mathcal{O}_{X_R}}(F, \mathcal{O}_{X_R})(\bar{z})$ and that $\underline{\mathrm{Ext}}^1_{\mathcal{O}_X}(F(z), \mathcal{O}_X) \simeq \underline{\mathrm{Ext}}^1_{\mathcal{O}_{X_R}}(F, \mathcal{O}_{X_R})(z))$. This
means that $F(\bar{z})$ is locally free or corresponds to a point of Z. By the choice of
R and Z, the latter is not the case.

Step IV.   Fixing c_2, we shall prove the proposition by induction on r. The
case of $r = 2$ was proved in Step III. Assume that $c_2 \geq r > 2$ and $M_X(r -$
$1, 0, c_2)_0^\mu \neq \emptyset$. For a member E of $M_X(r - 1, 0, c_2)_0^\mu$, an extension defined by a
non-zero element ξ of $\mathrm{Ext}^1_{\mathcal{O}_X}(\mathcal{O}_X, E) \simeq H^1(X, E)$

$$0 \longrightarrow E \longrightarrow E_\xi \longrightarrow \mathcal{O}_X \longrightarrow 0$$

gives rise to a member E_ξ of $M_X(r, 0, c_2)_0$ by virtue of Proposition 1.4. Since the
family $Y = \{ E_\xi \mid E \in M_X(r - 1, 0, c_2)_0^\mu, \xi \in H^1(X, E) \}$ is of dimension $\{2(r -$
$1)c_2 - r^2 + 2r\} + (c_2 - r) = 2rc_2 - r^2 + 1 - (c_2 - r) - 1 < \dim M_X(r, 0, c_2)_0$, there are
a discrete valuation ring R and a vector bundle F of rank r on $X \times \mathrm{Spec}(R)$ such
that $F(\bar{z})$ on the generic fiber is outside Y and $F(z)$ on the special fiber is in Y.
Suppose that $F(\bar{z})$ is not μ-stable. Then, after extending the ground field, it
contains a coherent subsheaf $\bar{H}$ of rank less than r such that $d(\bar{H}, \mathcal{O}_X(1)) = 0$
and $F(\bar{z})/\bar{H}$ is torsion free. Replacing R by its suitable extension, we may
assume that there is a coherent subsheaf H of F such that $H(\bar{z}) = \bar{H}$ and F/H
is flat over R. $F(z)$ fits in a non-trivial extension

$$0 \longrightarrow E \longrightarrow F(z) \longrightarrow \mathcal{O}_X \longrightarrow 0$$

with E in $M_X(r - 1, 0, c_2)_0^\mu$. Proposition 1.4 shows us that $H(z)$ is a subsheaf
of E, $r(H(z)) = r - 1$ and hence $\dim \mathrm{Supp}(E/H(z)) \leq 0$. Thus we have
$c_2((F/H)(z)) = c_2(\mathcal{O}_X) - \dim H^0(X, E/H(z))$, which is equal to $c_2((F/H)(\bar{z}))$.
On the other hand, since $(F/H)(\bar{z})$ is torsion free, $c_2((F/H)(\bar{z}))$ is non-negative
and 0 if and only if $(F/H)(\bar{z})$ is invertible. Combining all these together, we
see that $H(z) = E$ and $(F/H)(\bar{z})$ is invertible. Then $H(\bar{z})$ is μ-stable because
so is $H(z) = E$ and $(F/H)(\bar{z}) \simeq \mathcal{O}_X$ because $\mathrm{Pic}(X)$ is discrete. This means

that $F(\bar{z})$ is contained in Y, which contradicts the choice of F. Therefore, $M_X(r, 0, c_2)_0^\mu$ is not empty as required.   Q.E.D.

Let us pick a member $(F_1, F_2, \cdots, F_\alpha, (x_1, \cdots, x_d))$ of $(\prod_{i=1}^\alpha M_X(s_i, 0, b_i)_0^\mu) \times S^d(X)$. We arrange the numbering so that $b_1/s_1 \geq b_2/s_2 \geq \cdots \geq b_\alpha/s_\alpha$ and assume that $b_\alpha > 0$ whence $b_i \geq s_i \geq 2$ by virtue of the above proposition. Since $c_2(\underline{\mathrm{Hom}}_{\mathcal{O}_X}(F_i, F_j)) = s_i b_j + s_j b_i \geq 2 s_i s_j$, Riemann-Roch Theorem tells us that $\dim H^1(X, \underline{\mathrm{Hom}}_{\mathcal{O}_X}(F_i, F_j)) \geq s_i b_j + s_j b_i - s_i s_j \geq s_i s_j \geq 4$. Set $E_1' = F_1$. Suppose that for an i, we have a coherent sheaf E_i' and a filtration $0 = E_0' \subset E_1' \subset E_2' \subset \cdots \subset E_{i-1}' \subset E_i'$ by coherent sheaves such that (1) $E_j'/E_{j-1}' \simeq F_j$ for all $1 \leq j \leq i$ and (2) the extension $0 \to E_{j-1}'/E_{j-2}' \to E_j'/E_{j-2}' \to F_j \to 0$ is nontrivial for all $2 \leq j \leq i$. Take a non-zero element $\bar{\eta}$ of $\mathrm{Ext}^1_{\mathcal{O}_X}(F_{i+1}, E_i'/E_{i-1}') \simeq H^1(X, \underline{\mathrm{Hom}}_{\mathcal{O}_X}(F_{i+1}, F_i))$ which was proved to be non-zero in the above. There is a natural exact sequence

$$\mathrm{Ext}^1_{\mathcal{O}_X}(F_{i+1}, E_i') \to \mathrm{Ext}^1_{\mathcal{O}_X}(F_{i+1}, E_i'/E_{i-1}') \to \mathrm{Ext}^2_{\mathcal{O}_X}(F_{i+1}, E_{i-1}')$$

On the other hand, $\mathrm{Ext}^2_{\mathcal{O}_X}(F_{i+1}, E_{i-1}') \simeq H^2(X, \underline{\mathrm{Hom}}_{\mathcal{O}_X}(F_{i+1}, E_i')) = 0$ because $\underline{\mathrm{Hom}}_{\mathcal{O}_X}(F_{i+1}, E_i')$ is a μ-semi-stable vector bundle with degree 0. Hence we can find an element η of $\mathrm{Ext}^1_{\mathcal{O}_X}(F_{i+1}, E_i')$ which is mapped to the $\bar{\eta}$. Let E_{i+1}' be the vector bundle defined by the extension

$$0 \longrightarrow E_i' \longrightarrow E_{i+1}' \longrightarrow F_{i+1} \longrightarrow 0$$

which corresponds to η. By induction on i we can construct E_i''s up to $E' = E_\alpha'$ which have the properties (1) and (2) in the above.

Fix an integer s with $0 \leq s \leq d$ and divide the set $\{x_1, \cdots, x_s\}$ into groups $A_1, \cdots, A_e$ so that $x_i = x_j$ if and only if they are in the same group. For each A_i let us provide an artinian ring $R_i = \mathcal{O}_{X, x_j}/I$ such that $\mathrm{length}(R_i) = \#A_i$, where x_j is an element of A_i. Take a surjection $\zeta : F_1 \to T = \oplus_{i=1}^e R_i$ and set $E_1 = \ker(\zeta)$. Assume that we have a coherent subsheaf E_i of E_i' such that if one sets $E_j = E_j' \cap E_i$, then E_j/E_{j-1} coincides with E_j'/E_{j-1}' by the natural injection for all $2 \leq j \leq i$. Then E_i'/E_i is a torsion sheaf which is isomorphic to T. Since $\mathrm{Ext}^1_{\mathcal{O}_X}(F_{i+1}, T) \simeq H^1(X, \underline{\mathrm{Hom}}_{\mathcal{O}_X}(F_{i+1}, T)) = 0$, the natural map of $\mathrm{Ext}^1_{\mathcal{O}_X}(F_{i+1}, E_i)$ to $\mathrm{Ext}^1_{\mathcal{O}_X}(F_{i+1}, E_i')$ is surjective. By lifting the extension class in $\mathrm{Ext}^1_{\mathcal{O}_X}(F_{i+1}, E_i')$ defining E_{i+1}' to $\mathrm{Ext}^1_{\mathcal{O}_X}(F_{i+1}, E_i)$, we get an exact commutative diagram

$$
\begin{array}{ccccccccc}
0 & \to & E_i' & \to & E_{i+1}' & \to & F_{i+1} & \to & 0 \\
 & & \uparrow & & \uparrow & & \| & & \\
0 & \to & E_i & \to & E_{i+1} & \to & F_{i+1} & \to & 0 \\
 & & \uparrow & & \uparrow & & & & \\
 & & 0 & & 0 & & & &
\end{array}
$$

It is obvious that $E_i' \cap E_{i+1} = E_i$. By induction on i we obtain a coherent subsheaf $E = E_\alpha$ of E' such that for $E_j = E_j' \cap E$, $E_j/E_{j-1} = F_j$ $(2 \leq j \leq \alpha)$ and $E_1' \cap E$ is isomorphic to the given E_1. Let G be a coherent subsheaf of E such that E/G is torsion free and $d(G, \mathcal{O}_X(1)) = 0$. There is a coherent subsheaf G' of E' such that $G' \supset G$, $\dim \mathrm{Supp}(G'/G) \leq 0$ and E'/G' is torsion free. The last two properties of G' and Proposition 1.4 imply that $G' = E_i'$ for some i and hence $G = E_i$. Then, thanks to Lemma 1.5, we see that E is semi-stable.

Bringing the above results together, we have the following:

(5.3.1) E is semi-stable, $c_1(E) = 0$, $c_2(E) = s + \sum_{i=1}^{\alpha} b_i$ and $r(E) = \sum_{i=1}^{\alpha} s_i$.

(5.3.2) E has a filtration $0 = E_0 \subset E_1 \subset \cdots \subset E_{\alpha-1} \subset E_\alpha = E$ such that $E_1 \simeq ker(F_1 \to \oplus_{i=1}^{e} R_i)$ and $E_i/E_{i-1} \simeq F_i$ $(2 \leq i \leq \alpha)$ and hence $E/E_i \simeq E'/E_i'$ if $i \geq 1$.

(5.3.3) If G is a coherent subsheaf such that E/G is torsion free and $d(G, \mathcal{O}_X(1)) = 0$, then $G = E_i$ for some i.

To adjust the rank we need further consideration. First we put an assumption:

Assumption 5.4.1. For all $k > s$, $x_k \neq x_\ell$ if $k \neq \ell$.

Since $\underline{\mathrm{Hom}}_{\mathcal{O}_X}(m_{x_{s+1}}, E_\alpha/E_{\alpha-1}) \simeq E_\alpha/E_{\alpha-1}$ under this assumption, we have the following exact diagram by setting $y = x_{s+1}$ and $F_\alpha' = E_\alpha/E_{\alpha-1}$:

$$
\begin{array}{ccccc}
 & & H^1(X, F_\alpha') & & \\
 & & \downarrow & & \\
\mathrm{Ext}^1_{\mathcal{O}_X}(m_y, E) & \to & \mathrm{Ext}^1_{\mathcal{O}_X}(m_y, F_\alpha') & \to & \mathrm{Ext}^2_{\mathcal{O}_X}(m_y, E_{\alpha-1}) \\
 & & \downarrow & & \\
 & & H^0(X, \underline{\mathrm{Ext}}^1_{\mathcal{O}_X}(m_y, F_\alpha')) & & \\
 & & \downarrow & & \\
 & & H^2(X, F_\alpha') & &
\end{array}
$$

As is easily seen, $\mathrm{Ext}^2_{\mathcal{O}_X}(m_y, E_{\alpha-1}) = 0$ and $H^2(X, F_\alpha') = 0$. Thus we can lift a non-zero element of $H^0(X, \underline{\mathrm{Ext}}^1_{\mathcal{O}_X}(m_y, F_\alpha'))$ to an element ζ of $\mathrm{Ext}^1_{\mathcal{O}_X}(m_y, E)$. Using this and its image to $\mathrm{Ext}^1_{\mathcal{O}_X}(m_y, E')$, we can construct a morphism of non-trivial extensions

$$
\begin{array}{ccccccccc}
0 & \to & E' & \to & G_1' & \to & m_{x_{s+1}} & \to & 0 \\
 & & \uparrow & & \uparrow & & \| & & \\
0 & \to & E & \to & G_1 & \to & m_{x_{s+1}} & \to & 0
\end{array}
$$

Note that both G_1 and G_1' are locally free at x_{s+1}. Similarly we can lift a non-zero element of $H^0(X, \underline{\mathrm{Ext}}^1_{\mathcal{O}_X}(m_{x_{s+2}}, m_{x_{s+1}})) \simeq k$ to $\mathrm{Ext}^1_{\mathcal{O}_X}(m_{x_{s+2}}, G_1)$ and then get an exact commutative diagram

$$
\begin{array}{ccccccccc}
0 & \to & G_1' & \to & G_2' & \to & m_{x_s+2} & \to & 0 \\
  &   & \uparrow & & \uparrow & & \| & & \\
0 & \to & G_1 & \to & G_2 & \to & m_{x_s+2} & \to & 0
\end{array}
$$

Continuing this procedure, we obtain filters of coherent sheaves

$$
\begin{array}{ccccccccc}
E' & = & G_0' & \subset & G_1' & \subset & \cdots & \subset & G_{d-s}' & = & G' \\
   &   & \cup &   & \cup &   &   &   & \cup   &   & \\
E & = & G_0 & \subset & G_1 & \subset & \cdots & \subset & G_{d-s} & = & G
\end{array}
$$

such that $G_i' \cap G_j = G_i$ for all $i \leq j$ and $G_i/G_{i-1} \simeq G_i'/G_{i-1}' \simeq m_{x_s+i}$ for all $1 \leq i \leq d - s$.

Suppose that if H' is a coherent subsheaf of G_i' such that $d(H', \mathcal{O}_X(1)) = 0$ and G'/H_i' is torsion free, then $H' = E_j'$ or G_k'. Let I' be a coherent subsheaf of G_{i+1}' with the same properties as H' in G_i'. Then, for $J' = I' \cap G_i'$, G'/J_i' is a subsheaf of G_{i+1}'/I' and hence it is torsion free. Because of μ-semi-stability of G_i' and G_{i+1}'/I', $d(J', \mathcal{O}_X(1)) = 0$. By these and the assumption we made just before, $J' = E_j'$ or G_k'. Look at the following exact commutative diagram

$$
\begin{array}{ccccccc}
 & & 0 & & 0 & & \\
 & & \uparrow & & \uparrow & & \\
0 \to & G_i'/H \to & G_{i+1}'/I' & \to & C & \to & 0 \\
 & \| & \uparrow & & \uparrow & & \\
0 \to & G_i'/H \to & G_{i+1}'/H & \to & m_{x_s+i+1} & \to & 0, \\
 & & \uparrow & & \uparrow & & \\
 & & I'/H & = & I'/H & & \\
 & & \uparrow & & \uparrow & & \\
 & & 0 & & 0 & &
\end{array}
$$

where $H = E_j'$ or G_k'. Since G'/H_i is locally free and since $d(G'/H_i, \mathcal{O}_X(1)) = d(G_{i+1}'/I', \mathcal{O}_X(1))$, C is torsion free. This means that $C = 0$ or $I'/H = 0$. In the latter case, we have that $I' = H$. In the former case, $H = G_i'$, that is, $I' = G_{i+1}'$ or otherwise the middle row splits, which is not the case because of the construction of G_{i+1}'. Taking Proposition 1.4 into account, we obtain that if H' is a coherent subsheaf of G' such that G'/H' is torsion free and $d(H', \mathcal{O}_X(1)) = 0$, then $H' = E_j'$ or G_ℓ'.

(5.4.2)   If H is a coherent subsheaf of G such that G/H is torsion free and $d(H, \mathcal{O}_X(1)) = 0$, then $H = E_j$ or G_ℓ.

It is easy to prove the above. Indeed, H is contained in such an H' that G'/H' is torsion free and $\dim \operatorname{Supp}(H'/H) = 0$. Then we have proved that $H' = E_k'$ or G_ℓ'. (5.4.2) is deduced from this and the fact that $H' \cap G = H$.

For a coherent sheaf D on X, $C(D)$ denotes the constant term of the polynomial $\chi(D(m))$. Since $C(G_i) = C(E) \leq 0$ for all i, we have the inequalities

$$C(E_j)/r(E_j) \leq C(E)/r(E) \leq C(E)/r(G) = C(G)/r(G),$$
$$C(G_i)/r(G_i) = C(E)/r(G_i) \leq C(E)/r(G) = C(G)/r(G).$$

(5.4.2) and these inequalities prove

(5.4.3) G is semi-stable, $c_1(G) = 0$, $c_2(G) = c_2$ and $r(G) = d - s + \sum_{i=1}^{\alpha} s_i$.

We have to extend G by a trivial bundle. Note that the natural map $H^1(X, E_i) \to H^1(X, F_i)$ is surjective $(1 \leq i \leq \alpha)$ and for every $1 \leq i < \alpha$,

$$\dim H^1(X, E/E_i) = \sum_{j=i+1}^{\alpha} \dim H^1(X, F_j)$$

and $\qquad \dim H^1(X, E) = \dim H^0(X, F_1/E_1) + \sum_{j=1}^{\alpha} \dim H^1(X, F_j).$

Take an integer t such that there is an index j with $\dim H^1(X, E/E_{j-1}) < t \leq \dim H^1(X, E/E_{j-2})$ and that $t \leq \sum_{i=1}^{\alpha} \dim H^1(X, F_i)$. Let $\sigma_{k_i+1}, \cdots, \sigma_{k_{i+1}}$ be the lifting to $H^1(X, E_{\alpha-i})$ of a basis of $H^1(X, F_{\alpha-i})$ $(0 \leq i \leq \alpha - j)$ and let $\sigma_{k_{\alpha-j}+1}, \cdots, \sigma_t$ be one to $H^1(X, E_{j-1})$ of a system of linearly independent elements of $H^1(X, F_{j-1})$. Then $\sigma_1, \cdots, \sigma_t$ can be regarded as linearly independent elements of $H^1(X, G)$. We consider the element $\xi = (\sigma_1, \cdots, \sigma_t)$ in $\mathrm{Ext}^1_{\mathcal{O}_X}(\mathcal{O}_X^{\oplus t}, G) \simeq H^1(X, G)^{\oplus t}$. ξ defines an extension

$$0 \longrightarrow G \longrightarrow F \longrightarrow \mathcal{O}_X^{\oplus t} \longrightarrow 0.$$

Lemma 5.5. If $0 < t < \dim H^1(X, E)$, then F is stable. If $t = \dim H^1(X, E)$, then F is semi-stable.

Proof. The second assertion was proved in Proposition 3.1 because F is nothing but the universal extension of G. To prove the first, pick a coherent sub-sheaf Q of F such that $r(Q) < r(F)$, F/Q is torsion free and $d(Q, \mathcal{O}_X(1)) = 0$. If we set $R = G \cap Q$, we see as before that G/R is torsion free and $d(R, \mathcal{O}_X(1)) = 0$. By the property (5.4.2) of G, $R = E_i$ or G_ℓ. Look at the following exact commutative diagram

$$
\begin{array}{ccccccccc}
 & & & & 0 & & 0 & & \\
 & & & & \uparrow & & \uparrow & & \\
0 & \to & G/R & \to & F/Q & \to & C & \to & 0 \\
 & & \| & & \uparrow & & \uparrow & & \\
0 & \to & G/R & \to & F/R & \xrightarrow{g} & \mathcal{O}_X^{\oplus t} & \to & 0 \\
 & & & & \uparrow & & \uparrow f & & \\
 & & & & Q/R & = & Q/G \cap Q = Q/R & & \\
 & & & & \uparrow & & \uparrow & & \\
 & & & & 0 & & 0 & &
\end{array}
$$

Case I.   Assume that $R = E_i$ with $1 \le i \le \alpha - 1$. Then G/R is locally free and hence C is torsion free. This and the fact that $d(Q/R, \mathcal{O}_X(1)) = 0$ imply that $Q/R \simeq \mathcal{O}_X^{\oplus \beta}$ for some β. Setting $U = g^{-1}(Q/R)$, we have a splitting exact sequence

$$0 \longrightarrow G/R \longrightarrow U \longrightarrow \mathcal{O}_X^{\oplus \beta} \longrightarrow 0.$$

The middle row of the above diagram is defined by the extension class $\bar{\xi}$ in $\mathrm{Ext}^1_{\mathcal{O}_X}(\mathcal{O}_X^{\oplus t}, G/R)$ which is the image of ξ by the natural map $Ext^1_{\mathcal{O}_X}(\mathcal{O}_X^{\oplus t}, G)$ $\to \mathrm{Ext}^1_{\mathcal{O}_X}(\mathcal{O}_X^{\oplus t}, G/R)$. If $\bar{\sigma}_k$ is the image of σ_k in $H^1(X, G/R)$, then $\bar{\xi}$ is is represented by $(\bar{\sigma}_1, \cdots, \bar{\sigma}_t)$. f induces a map $\bar{f} : \mathrm{Ext}^1_{\mathcal{O}_X}(\mathcal{O}_X^{\oplus t}, G/R) \longrightarrow$ $\mathrm{Ext}^1_{\mathcal{O}_X}(\mathcal{O}_X^{\oplus \beta}, G/R)$ which sends $\bar{\xi}$ to the extension class of the above exact sequence, that is, $\bar{f}(\bar{\xi}) = 0$. Let $e_1, \cdots, e_t$ be a standard basis of $\mathcal{O}_X^{\oplus t}$ and $u_1, \cdots, u_\beta$ a basis of $\mathcal{O}_X^{\oplus \beta}$. If we write $u_k = \sum a_{k\ell} e_\ell$, then we come to the equation

$$
\bar{f}(\bar{\xi}) = \begin{pmatrix} a_{11} & \cdots & a_{1t} \\ \vdots & \ddots & \vdots \\ a_{\beta 1} & \cdots & a_{\beta t} \end{pmatrix} \begin{pmatrix} \bar{\sigma}_1 \\ \vdots \\ \bar{\sigma}_\tau \end{pmatrix} = 0.
$$

For $\gamma = \min(t, k_{\alpha-i-1})$, $\bar{\sigma}_1, \cdots, \bar{\sigma}_\gamma$ are linearly independent and $\bar{\sigma}_{\gamma+1} = \cdots = \bar{\sigma}_t = 0$. Therefore, the above equation shows that $a_{km} = 0$ if $m \le \gamma$. On the other hand, the rank of the matrix (a_{km}) must be β. From these we deduce the inequality $t - \gamma \ge \beta$. Thus if $i \le j - 2$, then $\beta = 0$ or $Q = E_i$ and hence $C(Q)/r(Q) = C(E_i)/r(E_i) \le C(G)/r(G) < \{C(G)+t\}/r(G) < \{C(G)+t\}/\{r(G)+t\} = C(F)/r(F)$ because $t < \dim H^1(X, E) = -C(G)$ (for a coherent sheaf D on X, $C(D)$ denotes the constant term of the polynomial $\chi(D(m))$). If $i \ge j - 1$, then $\gamma = \dim H^1(X, E/E_i) = -C(G/E_i)$. We have the following:

$$C(Q) = C(E_i) + \beta,$$
$$C(F) = C(E_i) + C(G/E_i) + t = C(E_i) + t - \gamma \ge C(E_i) + \beta.$$

Since $-C(E_i) = \dim H^1(X, E_i) \ge t - \gamma \ge \beta$, $C(Q) = C(E_i) + \beta$ is not positive. Taking these into account, the following inequalities are obtained:

$$C(Q)/r(Q) < C(Q)/r(F) = \{C(E_i) + \beta\}/r(F) \le C(F)/r(F).$$

We see therefore that $P_Q(m) < P_F(m)$ in both cases.

Case II.   Assume that $R = 0$. Let ξ' be the image of ξ by the natural map $\operatorname{Ext}^1_{\mathcal{O}_X}(\mathcal{O}_X^{\oplus t}, E) \to \operatorname{Ext}^1_{\mathcal{O}_X}(\mathcal{O}_X^{\oplus t}, E')$. ξ' defines an extension F' of $\mathcal{O}_X^{\oplus t}$ by E' and we obtain an exact commutative diagram:

$$
\begin{array}{ccccccccc}
 & & 0 & & 0 & & & & \\
 & & \uparrow & & \uparrow & & & & \\
 & & T & = & T & & & & \\
 & & \uparrow & & \uparrow & & & & \\
0 & \to & G' & \to & F' & \to & \mathcal{O}_X^{\oplus t} & \to & 0 \\
 & & \uparrow & & \uparrow & & \| & & \\
0 & \to & G & \to & F & \to & \mathcal{O}_X^{\oplus t} & \to & 0 \\
 & & \uparrow & & \uparrow & & & & \\
 & & 0 & & 0 & & & &
\end{array}
$$

There is a coherent subsheaf Q' of F' such that $d(Q', \mathcal{O}_X(1)) = 0$, F'/Q' is torsion free, Q' contains Q and $\dim \operatorname{Supp}(Q'/Q) \leq 0$. As is easily seen, $Q' \cap G' = 0$ and then $Q' \simeq \mathcal{O}_X^{\oplus \beta}$ as before. Since $\sigma_1, \cdots, \sigma_t$ are mapped to linearly independent elements in $H^1(X, E')$, β must be 0, in other words, Q' and hence Q is zero.

Case III.   Assume that $R = G_\ell$. Then we have that $C(Q) = C(G_\ell) + C(Q/R) \leq C(G) + t = C(F)$. We see obviously then that $C(Q)/r(Q) < C(F)/r(F)$ or $P_Q(m) < P_F(m)$.

$$Q.E.D.$$

We have proved so far the following.

Proposition 5.6. Let $(F_1, F_2, \cdots, F_\alpha, x_1, \cdots, x_d)$ be a point of $(\prod_{i=1}^\alpha M_X(s_i, 0, b_i)_0^\mu) \times S^d(X)$ with $b_1/s_1 \geq \cdots \geq b_\alpha/s_\alpha$, $b_i \geq s_i \geq 2$ and let r and s be integers such that $\sum_{i=1}^\alpha s_i \leq r \leq \sum_{i=1}^\alpha b_i + d$ and $\max(0, \sum_{i=1}^\alpha s_i + d - r) \leq s \leq \min(d, \sum_{i=1}^\alpha b_i + d - r)$. Assume that $x_i \neq x_j$ if $i > s$ and $i \neq j$. Then there exists a stable sheaf F such that $r(F) = r$, $c_1(F) = 0$, $c_2(F) = \sum_{i=1}^\alpha b_i + d$ and that F has a filtration $0 = E_0 \subset E_1 \subset \cdots \subset E_{\alpha+d-s} \subset E_{\alpha+d-s+1} = F$ with the following properties:

(a) E_1 is a subsheaf of F_1 such that F_1/E_1 is a direct sum of artinian modules R_j supported by a point y_j and that $\sum_j \operatorname{length}(R_j)y_j = \sum_{j=1}^s x_j$ as 0-dimensional cycles.

(b) $E_i/E_{i-1} \simeq F_i$ if $2 \leq i \leq \alpha$.

(c) $E_i/E_{i-1} \simeq m_{x_{s+i-\alpha}}$ if $\alpha < i \leq \alpha + d - s$.

(d) $E/E_{\alpha+d-s} \simeq \mathcal{O}_X^{\oplus t}$ with $t = \sum_{i=1}^\alpha s_i + d - s$.

Proof.   Since $\sum_{i=1}^\alpha s_i + d - s \leq r$, $t = r - \sum_{i=1}^\alpha s_i - d + s$ is a non-negative integer. By using upper bound of s, we see that $t \leq \sum_{i=1}^\alpha b_i - \sum_{i=1}^\alpha s_i = \sum_{i=1}^\alpha \dim H^1(X, F_i)$. Thus the above construction of F can be applied to these s and t.

$$Q.E.D.$$

Combining the above (a), (b), (c) and (d), we see that for F in the above proposition, $gr(U(F)) \simeq (\oplus_{i=1}^{\alpha} U(F_i)) \oplus (\oplus_{j=1}^{d} m_{x_j})$ and hence the class $[F]$ of F in $\bar{M}_X(r, 0, c_2)$ mapped to $((F_1, F_2, \cdots, F_\alpha), (x_1, \cdots, x_d))$ by the map $\Phi(r, 0, c_2)$ (see the interpretation of the image by $\Phi(r, 0, c_2)$ in Theorem 4.3). If r satisfies the first inequalities in the proposition, then clearly there is an s which satisfies the second. Therefore, $\text{im}(\Phi(r, 0, c_2))$ contains $(\prod_{i=1}^{\alpha} M_X(s_i, 0, b_i)_0^{\mu}) \times V$, where $c_2 = d + \sum_{i=1}^{\alpha} b_i$ and V is the open set $\{ (x_1, \cdots, x_d) \in S^d(X) \mid x_i \neq x_j \text{ if } i \neq j \}$. On the other hand, $\text{im}(\Phi(r, 0, c_2))$ is closed in $\bar{M}_X(c_2, 0, c_2)$. What we have obtained is the following.

(5.7)   Let r and c_2 be integers with $c_2 \geq r$ and let $s_1, \cdots, s_\alpha, b_1, \cdots, b_\alpha$ be integers such that $b_i \geq s_i \geq 2$, $c_2 = d + \sum_{i=1}^{\alpha} b_i$ and $b_1/s_1 \geq \cdots \geq b_\alpha/s_\alpha$. Then $(\prod_{i=1}^{\alpha} M_X(s_i, 0, b_i)_0^{\mu}) \times S^d(X)$ is non-empty and in the image of $\Phi(r, 0, c_2)$.

Pick a set of mutually distinct points $x_1, x_2, \cdots, x_{c_2}$ of X. For an integer r with $r \leq c_2$, let E be the ideal $\bigcap_{i=r}^{c_2} m_{x_i}$. Applying the way of the construction of G from E in the above to this E, we get a semi-stable sheaf F with a filtration $0 = F_0 \subset F_1 \subset \cdots \subset F_{r-1} \subset F_r = F$ such that $F_1 = E$ and $F_i/F_{i-1} \simeq m_{x_{i-1}}$ for all $2 \leq i \leq r$. For this F, $gr(U(F))$ is isomorphic to $\oplus_{i=1}^{c_2} m_{x_i}$. By using closedness of the $\text{im}(\Phi(r, 0, c_2))$ again, we have

(5.8) $S^{c_2}(X)$ is contained in $\Phi(r, 0, c_2)$ if $c_2 \geq r$.

Thanks to (5.7) and (5.8), the proof of the following theorem has been completed.

Theorem 5.9. *Let r and c_2 be integers with $c_2 \geq r$ and let $A = \{ (s_1, b_1, s_2, b_2, \cdots, s_\alpha, b_\alpha, d) \mid$ (i) $\alpha > 0$, (ii) $b_i \geq s_i \geq 2$, $b_1/s_1 \geq \cdots \geq b_\alpha/s_\alpha$ and $s_i \geq s_{i+1}$ if $b_i/s_i = b_{i+1}/s_{i+1}$, (iii) $c_2 = d + \sum_{i=1}^{\alpha} b_i$ and (iv) $r \geq \sum_{i=1}^{\alpha} s_i \}$.*
(1) If $a = (s_1, b_1, s_2, b_2, \cdots, s_\alpha, b_\alpha, d)$ is a member of A, then $M(a) = (\prod_{i=1}^{\alpha} M_X(s_i, 0, b_i)_0^{\mu}) \times S^d(X)$ is not empty.
(2) $M(r, c_2) = (\coprod_{a \in A} M(a)) \coprod S^{c_2}(X)$ is contained in the image of $\Phi(r, 0, c_2)$ and moreover $M(r, c_2)$ exhausts all the members H in $\text{im}(\Phi(r, 0, c_2))$ such that $gr(H)$ has only the components with the first Chern class 0.

Corollary 5.9.1. *If $X = \mathbf{P}^2$, then $\text{im}(\Phi(r, 0, c_2))$ is exactly $M(r, c_2)$.*

§6. Good polarizations and the case of rank 2.

We can determine more precisely the image of the map $\Phi(2, 0, c_2)$ if the polarization $\mathcal{O}_X(1)$ is *good*.

Definition 6.1. *Let X be a non-singular, projective, rational surface and m a positive integer. An ample line bundle L is said to be good (with respect to m) if (1) $(K_X, L) < 0$ and (2) for every member M of $L^\perp = \{N \in \text{Pic}(X) \mid (L, N) = 0\}$, we have $-M^2 > m$ or $M = 0$.*

We always have a good polarization.

Lemma 6.2. *Let X and m be as in the above definition. We have a good ample line bundle L with respect to m on X.*

Proof. If $X = \mathbf{P}^2$, then our assertion is trivial. Assume that X is iso-morphic to $F_n = \mathbf{P}(\mathcal{O}_{\mathbf{P}^1} \oplus \mathcal{O}_{\mathbf{P}^1}(n))$ $(n > 0)$. We have a section Γ of F_n with $\Gamma^2 = n$. If F is a fiber of F_n, then $K_X = -2\Gamma + (n-2)F$ is a canonical divisor of F_n. $L = \mathcal{O}_X(\alpha\Gamma + \beta F)$ is ample if and only if both α and β are positive. Since $(L, K_X) = -\alpha n - 2\beta - 2\alpha$, the condition (1) for the good-ness is automatically satisfied if L is ample. For $N = \mathcal{O}_X(\gamma\Gamma + \delta F)$, we have that $(L, N) = \alpha\gamma n + \alpha\delta + \beta\gamma$. Thus N is contained in $L^{\perp}$ if and only if $\delta = -\gamma n - \beta\gamma/\alpha$. Then $N^2 = \gamma^2 n - 2\gamma(\gamma n + \beta\gamma/\alpha) = -\gamma^2(n + \beta/\alpha)$. This shows that for a couple (α, β) with $\beta/\alpha > m - n$, $L = \mathcal{O}_X(\alpha\Gamma + \beta F)$ meets our requirement. When X is not a relatively minimal model, we have a sequence of blowing-ups $X \xrightarrow{f_0} X_1 \xrightarrow{f_1} X_2 \xrightarrow{f_2} \cdots \to F_n$. By induction on the length of the sequence we may assume that we have a good ample line bundle L_1 on X_1. Let E be the exceptional divisor of $f = f_0$ and set $f^*(L_1) = L_0$. If p is a sufficiently large integer, then $L = L_0^{\otimes p} \otimes \mathcal{O}_X(-E)$ is ample. Since $K_X = f^*(K_{X_1}) + E$, we see that $(L, K_X) = p(L_1, K_{X_1}) - E^2 = p(L_1, K_{X_1}) + 1 < 0$ if $p > 1$. Pick a non-zero element M of $L^{\perp}$. Then there are an integer γ and an invertible sheaf N on X_1 such that $M \simeq f^*(N) \otimes \mathcal{O}_X(\gamma E)$. Since L_1 is ample, we can find a positive integer t such that $N^{\otimes t}$ is written in the form $L_1^{\otimes \alpha} \otimes A^{\otimes \beta}$ with α and β integers and with $A \in L_1^{\perp}$. The condition that M is a member of $L^{\perp}$ is equivalent to $(M, L) = p(L_1, N) + \gamma = (p\alpha/t)L_1^2 + \gamma = 0$ or $\gamma = -(p\alpha/t)L_1^2$. The computation of M^2 proceeds as follows:

$$
\begin{aligned}
M^2 = N^2 - \gamma^2 &= (\alpha/t)^2 L_1^2 + (\beta/t)^2 A^2 - \gamma^2 \\
&= (\alpha/t)^2 L_1^2 + (\beta/t)^2 A^2 - (p\alpha/t)^2(L_1^2)^2 \\
&= (\alpha/t)^2 L_1^2 \{1 - p^2 L_1^2\} + (\beta/t)^2 A^2
\end{aligned}
$$

Since $A^2 \leq 0$ thanks to Hodge index theorem, it is enough for our purpose that $L_1^2\{1 - p^2 L_1^2\} < -t^2 m$. If p is sufficiently big, then the last inequality holds because $L_1^2 > 0$.   Q.E.D.

If we fix a good polarization with respect to m and if the second Chern class of a μ-semi-stable sheaf E is relatively small to m, then the filtration of E in Proposition 3.3, 2) has a good property.

Lemma 6.3. *Let E be a coherent sheaf on X with $c_1(E) = 0$ and let $\mathcal{O}_X(1)$ be a good polarization with respect to $c_2(E)$. Assume that there is a filtration $0 = E_0 \subset E_1 \subset \cdots \subset E_\alpha = E$ such that (1) $d(E_i, \mathcal{O}_X(1)) = 0$, (2) $c_2(E_i) \geq 0$ and (3) $c_2(E_i/E_{i-1}) \geq 0$ for all $i > 0$. Then we have that $c_1(E_i) = 0$ for all i.*

Proof. Our proof is by induction on α. If $\alpha = 1$, then our assertion is obvious. Let us look at the exact sequence

$$0 \longrightarrow E_{\alpha-1} \longrightarrow E \longrightarrow F \longrightarrow 0,$$

where $F = E/E_{\alpha-1}$. Since $d(E_{\alpha-1}, \mathcal{O}_X(1)) = d(F, \mathcal{O}_X(1)) = 0$, both $c_1(E_{\alpha-1})$ and $c_1(F)$ are contained in $\mathcal{O}_X(1)^{\perp}$. If one of them is 0, then so is the rest by our assumption. Assume the contrary. Then we see that $c_2(E) < -c_1(F)^2 = c_1(F)c_1(E)_{\alpha-1} = c_2(E) - c_2(F) - c_2(E_{\alpha-1}) \leq c_2(E)$, which is a contradiction. Thus we see that $c_1(F) = c_1(E_{\alpha-1}) = 0$ and then $c_2(E_{\alpha-1}) = c_2(E) - c_2(F) \leq c_2(E)$. Hence $\mathcal{O}_X(1)$ is good with respect to $c_2(E_{\alpha-1})$. Now we can apply our induction hypothesis to $E_{\alpha-1}$ and complete our proof.

Let c_2 be a positive integer and let $\mathcal{O}_X(1)$ be a good polarization on X with respect to c_2. Pick a semi-stable sheaf E of rank 2 on $(X, \mathcal{O}_X(1))$ with $c_1(E) = 0$ and $c_2(E) = c_2$. If E is not μ-stable, then we have an exact sequence

$$0 \longrightarrow L_1 \longrightarrow E \longrightarrow L_2 \longrightarrow 0,$$

where $r(L_i) = 1$ and $d(L_i, \mathcal{O}_X(1)) = 0$. By virtue of Lemma 6.3, $c_1(L_i)$ must be 0. Then $gr(U(E)) = \oplus_{1 \leq j \leq c_2} m_{x_j}$. This and Theorem 5.9 and the proof of Proposition 5.2 yield another theorem.

Theorem 6.4. *Let c_2 be an integer greater than 1 and $\mathcal{O}_X(1)$ a good polarization on X with respect to c_2. Then $\mathrm{im}(\Phi(2, 0, c_2))$ is exactly*

$$\left(\coprod_{d=0}^{c_2-2} M_X(2, 0, c_2 - d)_0^\mu \times S^d(X) \right) \coprod S^{c_2}(X). \text{ Moreover, the closure of}$$

$\Phi(2, 0, c_2)(M_X(2, 0, c_2)_0^\mu)$ *is* $\mathrm{im}(\Phi(2, 0, c_2))$.

This theorem shows that Donaldson's compactification in [1] carries an algebraic structure if the underlying manifolds is a rational algebraic variety.

Question 6.5. Is $\mathrm{im}(\Phi(r, c_1, c_2))$ normal or more weakly, are all the fibers of $\Phi(r, c_1, c_2)$ connected ?

References

[1] S. K. Donaldson, Connections, cohomology and the intersection forms of 4-manifolds, J. of Diff. Geom., 24 (1986), 275-341.

[2] J. M. Drezet and J. Le Potier, Fibrés stables et fibrés exceptionels sur $\mathbf{P}_2$, Annales Scient. Éc. Norm. Sup., 18 (1985), 193-244.

[3] M. Maruyama, Moduli of stable sheaves II, J. Math. Kyoto Univ., 18 (1978), 95-146.

[4] M. Maruyama, Moduli of stable of sheaves - generalities and the curves of jumping lines of vector bundles on $\mathbf{P}^2$, Advanced Studies of Pure Math., I, Alg. Var. and Anal. Var., Kinokuniya and North-Holland (1983), 1-27.

[5] M. Maruyama, Singularities of the curve of jumping lines of a vector bundle of rank 2 on $\mathbf{P}^2$, Alg. Geom., Proc. of Japan-France Conf., Tokyo and Kyoto 1982, Lect. Notes in Math., 1016, Springer-Verlag (1983) 370-411.

[6] M. Maruyama, Vector bundles on $\mathbf{P}^2$ and torsion sheaves on the dual plane, Vector bundles on Algebraic Varieties, Proc. Bombay Colloq. 1984, Oxford Univ. Press (1987) 275-339.

Masaki MARUYAMA

Department of Mathematics
Faculty of Science
Kyoto University
Kyoto 606
Japan

Algebraic Geometry and Commutative Algebra
in Honor of Masayoshi NAGATA
pp. 261–266 (1987)

On the Dimension of Formal Fibres of a Local Ring

Hideyuki MATSUMURA

Introduction.

Let (A, m) be a noetherian local ring and $\hat{A}$ be its m-adic completion. The fibres of the morphism $\mathrm{Spec}(\hat{A}) \to \mathrm{Spec}(A)$ are the formal fibres of A. Since $\hat{A}$ has very good properties, and since $\hat{A}$ is flat over A, A has a good property if the formal fibres have related good properties. This is the philosophy of Grothendieck's theory of excellent rings. But the dimension of formal fibres did not play any significant roles in that theory.

It seems natural to think that as A gets closer to $\hat{A}$ the dimension of the formal fibres becomes smaller. So we have investigated their dimension. The result is somewhat unexpected. The maximum value of the dimension of formal fibres, which we denote by $\alpha(A)$, seems to take only three values, namely $n-1$, $n-2$ and 0 (we do not have any proof of this yet), where $n = \dim A$. But the set $N(A) := \{P \in \mathrm{Spec}(A) | \dim(A/P) \geq 2, \alpha(A/P) = \dim(A/P) - 1\}$ becomes smaller as A gets closer to $\hat{A}$.

§1. Formal fibres.

Let A, m and $\hat{A}$ be as in the introduction and p be a prime ideal of A. The formal fibre of A at p is $\mathrm{Spec}(\hat{A} \otimes_A \kappa(p))$, where $\kappa(p)$ is the residue field A_p/pA_p. We set

$$\alpha(A, p) = \dim \hat{A} \otimes_A \kappa(p), \quad \alpha(A) = \sup\{\alpha(A, p) \mid p \in \mathrm{Spec}(A)\}.$$

It can happen that a non-complete local ring A has $\alpha(A) = 0$. Examples: the local rings of dimension 1; a regular local ring A whose completion $\hat{A}$ is purely inseparable over A, (cf. [LR] p.206, [CA] (34.B)).

Since $\hat{A}$ is flat over A, the going-down theorem holds between A and $\hat{A}$. Moreover, $m\hat{A}$ is the only prime ideal of $\hat{A}$ lying over m. Therefore

$$\alpha(A, p) + \mathrm{ht}\, p \leq \dim A - 1 \quad (p \neq m), \qquad \alpha(A, m) = 0.$$

It follows that
$$\alpha(A) \leq \dim A - 1 \quad \text{if } \dim A \neq 0.$$

Received February 9, 1987.

We will later show that under some conditions this upper bound can be improved by one, but not more than one,

Theorem 1. *If p, $q \in \mathrm{Spec}(A)$ and $q \subset p$, then $\alpha(A, q) \geq \alpha(A, p)$.*

Proof. Since $\hat{A}$ is catenary, this is a special case of the following theorem of the author ([CR] Th. 15.3) : if $A \to B$ is a homomorphism of noetherian rings such that the going-down theorem holds and B is catenary (cf. Appendix), and if p, $q \in \mathrm{Spec}(A)$ with $q \subset p$, then $\dim B \otimes_A \kappa(Q) \geq \dim B \otimes_A \kappa(P)$. (Note that, when the going-up theorem holds, we get the opposite inequality, cf. ibid. Th. 15.2.)

Corollary 1. *If A is a noetherian local domain with quotient field K, then $\alpha(A) = \dim \hat{A} \otimes_A K$.*

The fibre over the "generic point" (0) of $\mathrm{Spec}(A)$ is called the generic fibre.

Corollary 2. *If A is a noetherian local ring and $p \in \mathrm{Spec}(A)$, then $\alpha(A, p) = \alpha(A/p)$.*

The next theorem shows that $\alpha(A) = n - 1$ in the 'normal' case.

Theorem 2. *Let A be a local ring essentially of finite type over a field k. If $\dim A = n > 0$, then $\alpha(A) = n - 1$.*

Proof. First we treat the case $A = k[\underline{X}]_{(\underline{X})}$, where $\underline{X} = (X_1, \cdots, X_n)$. Then $\hat{A} = k[\![\underline{X}]\!]$. Let $k[\![Y]\!]$ be the formal power series ring in one variable Y. Since it has an infinite transcendence degree over k, we can choose power series $u_2(Y), \cdots, u_n(Y) \in k[\![Y]\!]$ without constant terms such that $Y, u_2(Y), \cdots, u_n(Y)$ are algebraically independent over k. Consider the k-algebra homomorphism $\varphi : \hat{A} = k[\![X_1, \cdots, X_n]\!] \to k[\![Y]\!]$ defined by $\varphi(X_1) = Y$, $\varphi(X_i) = u_i(Y)$ ($i = 2, \cdots, n$). Put $\mathrm{Ker}(\varphi) = P$. Then $P \cap A = (0)$. To see this, it suffices to show $P \cap k[\underline{X}] = (0)$. But if $F(\underline{X}) \in k[\underline{X}]$ then $\varphi(F(X)) = F(Y, u_2(Y), \cdots, u_n(Y)) \neq 0$ unless $F(\underline{X}) = 0$. Thus P is on the generic fibre of A, and since $\hat{A}/P \simeq k[\![Y]\!]$ we have $\mathrm{ht}\, P = n - 1$. Therefore $\alpha(A) \geq n - 1$, and since $\alpha(A)$ cannot exceed $n - 1$, the equality holds.

Now we consider the general case. By Theorem 1 and Corollary 2, we can assume that A is an integral domain of dimension n. Then A is of the form $A = C_p$, where $C = k[x_1, \cdots, x_m]$ and p is a prime ideal of C. Thus there is a polynomial ring $R = k[X_1, \cdots, X_m]$ and prime ideals $P \supset Q$ of R such that $A \simeq (R/Q)_P = R_P/QR_P$. If $\mathrm{tr.deg}_k(R/P) = r$ we can renumber the X_i's and assume that the images of $X_1, \cdots, X_r$ in R/P are algebraically independent over k. Then R_P is of the form S_{PS}, where $S = k(X_1, \cdots, X_r)[X_{r+1}, \cdots, X_m]$, and $A \simeq S_{PS}/QS_{PS}$. Therefore, replacing k by $k(X_1, \cdots, X_r)$ we may assume that P is a maximal ideal of R. Then $A/m = R/P$ is finite algebraic over k, and $n = \dim A = \mathrm{tr.deg}_k R/Q = m - \mathrm{ht}(Q)$. Let $y_1, \cdots, y_n$ be a system of parameters of A. The y_i's are analytically independent over k, and $\hat{A}$ is finite over $k[\![y_1, \cdots, y_n]\!]$. Put $B = k[y_1, \cdots, y_n]_{(y_1, \cdots, y_n)}$. Then $\hat{B} = k[\![y_1, \cdots, y_n]\!]$. By what we have already seen, we can find a chain of prime ideals $P_{n-1} \supset \cdots \supset P_0 = (0)$ in $\hat{B}$ such that $P_i \cap B = (0)$ for all i. Since $\hat{A}$ is integral over $\hat{B}$ there exists a chain of prime ideals $P'_{n-1} \supset \cdots \supset P'_0$ in $\hat{A}$ such that $P'_i \cap \hat{B} = P_i$ for all i. We claim that $P'_i \cap A = (0)$ for all i. In fact, if K_A and K_B denote the quotient fields of A and B respectively, we have tr.deg$_k K_A = \dim A = n = \dim B = $ tr.deg$_k K_B$. This means that K_A is finite algebraic over K_B, hence $K_A = K_B[A] = A \otimes_B K_B$. Hence (0) is the only prime ideal of A lying over (0) of B. Since $P'_i \cap A \cap B = P'_i \cap B = P'_i \cap \hat{B} \cap B = P_i \cap B = (0)$, we have $P'_i \cap A = (0)$, as wanted. Therefore $\alpha(A) = n - 1$.   Q.E.D.

Corollary. *Let (A, m) be a local ring of dimension n, essentially of finite type over a field, and P be a prime ideal of A other than m. Then $\alpha(A, P) = n - 1 - \mathrm{ht}(P)$.*

Proof. We have $\alpha(A, P) = \alpha(A/P)$, and A/P is also essentially of finite type over the same field. Moreover, $\dim(A/P) = \dim A - \mathrm{ht}(P)$ as A is catenary.

Definition. *Let A be a noetherian local ring, and put*

$$\mathrm{Spec}_i(A) := \{P \in \mathrm{Spec}(A) \mid \dim(A/P) \geq i\},$$
$$N(A) := \{P \in \mathrm{Spec}_2(A) \mid \alpha(A, P) = \dim(A/P) - 1\}.$$

Open Problem 1. Suppose that $P, Q \in \mathrm{Spec}_2(A)$ and that $P \subset Q$, $P \in N(A)$. Is Q then necessarily in $N(A)$? (Assume, if necessary, that A is catenary.)

§2. Some cases where $\alpha(A)$ is smaller than $\dim A - 1$.

Theorem 3. *Let (A, m) and (A_0, m_0) be noetherian local domains such that*
(1) *A_0 is a subring of A, $m \cap A_0 = m_0$, $A/m = A_0/m_0$, and*
(2) *A_0 is complete and $\dim A > 1$, $\dim A_0 > 0$.*
Then
$$\alpha(A) < \dim A - 1.$$

Proof. Set $\dim A = n$, and suppose $\alpha(A) = n - 1$. Then there exists a prime ideal P of $\hat{A}$ such that $P \cap A = (0)$ and $\dim \hat{A}/P = 1$. Set $B = \hat{A}/P$. The maximal ideal of B is $m\hat{A}/P = mB$, and B contains A and A_0. Since $m_0 B$ is a non-zero ideal in the one-dimensional local domain B it is a primary ideal belonging to mB, so that $m_0 B \supset m^\nu B$ for some ν. Let $mB = \sum u_j B$. Since $B/mB = A/m = A_0/m_0$, we have

$$
\begin{aligned}
B &= A_0 + \sum u_j B = A_0 + \sum u_j A_0 + \sum u_j u_k B = \cdots \\
&= A_0 + \sum u_j A_0 + \sum u_j u_k A_0 + \cdots + \sum u_{j_1} \cdots u_{j_{\nu-1}} A_0 + m_0 B.
\end{aligned}
$$

Thus $B/m_0 B$ is finitely generated as A_0-module, and B is separated in the m_0-adic topology, hence B is finite over A_0 (cf. e.g. [CR] p.58 Th. 8.4). Therefore B is finite over A, hence $\dim A = \dim B = 1$, contradiction.

Corollary. *Under the hypotheses of the theorem, if* $p \in \mathrm{Spec}_2(A)$ *and* $p \not\supseteq m_0 A$ *then* $\alpha(A, p) < \dim(A/p) - 1$, *in other words* $N(A) \subset V(m_0 A) \cap \mathrm{Spec}_2(A)$.

Proof. If $p \not\supseteq m_0 A$ then A/p contains $A_0/(p \cap A_0)$, which is complete and of positive dimension.

When, in particular, $\dim A = 2$ in the theorem, we have $\alpha(A) = 0$. In general, for a noetherian local domain A, the condition $\alpha(A) = 0$ is equivalent to that every height 1 prime ideal P of $\hat{A}$ contracts to a non-zero prime ideal of A. In particular, when $\hat{A}$ is a UFD (and then A is also UFD, cf. [CR] p.169 Ex. 20.4), we have $\alpha(A) = 0$ if and only if every prime element of $\hat{A}$ divides an element of A.

Example 1. Let $A = k[\![X]\!][Y]_{(X,Y)}$, where k is a field. Then $\hat{A} = k[\![X, Y]\!]$, and since A contains $k[\![X]\!]$ we have $\alpha(A) = 0$ by the theorem. This can be proved also by Weierstrass preparation theorem (cf. e.g. [AL] p.215). In fact, every element $f(X, Y)$ of $k[\![X, Y]\!]$ which is not divisible by X is of the form

$$f(X, Y) = u(X, Y)(Y^d + a_1(X)Y^{d-1} + \cdots + a_d(X))$$

where u is a unit and $a_i(X) \in k[\![X]\!]$. Thus f divides the element $Y^d + \cdots + a_d(X)$ of A. Since X is also in A, every prime element of $\hat{A}$ divides an element of A.

Example 2. Let $A = A_{r,n} = k[\![X_1, \cdots, X_r]\!][X_{r+1}, \cdots, X_n]_{(X_1, \cdots, X_n)}$ where $n > r \geq 1$. Then $\hat{A} = k[\![X_1, \cdots, X_n]\!]$ and $\alpha(A) \leq n - 2$ by the theorem. Let $\varphi_2(X_n), \cdots, \varphi_{n-1}(X_n)$ be $n - 2$ power series in $k[\![X_n]\!]$ which are algebraically independent over $k(X_n)$, and consider the homomorphism

$$\Phi : \hat{A} = k[\![X_1, \cdots, X_n]\!] \longrightarrow k[\![X_1, X_n]\!]$$

defined by $\Phi(X_1) = X_1$, $\Phi(X_i) = X_1 \varphi_i(X_n)$ $(1 < i < n)$, $\Phi(X_n) = X_n$. Set $P = \mathrm{Ker}\, \Phi$. Then $\hat{A}/P \simeq k[\![X_1, X_n]\!]$, hence $\mathrm{ht}(P) = n - 2$. We claim $P \cap A = (0)$. Let $F \in P \cap k[\![X_1, \cdots, X_r]\!][X_{r+1}, \cdots, X_n]$. We can write

$$F = \sum_{\nu=0}^{\infty} f_\nu(X_1, \cdots, X_{n-1}, X_n),$$

where f_ν is a homogeneous polynomial of degree ν in $X_1, \cdots, X_{n-1}$ and is a polynomial in X_n. Then

$$\Phi(F) = \sum_{0}^{\infty} X_1^\nu f_\nu(1, \varphi_2(X_n), \cdots, \varphi_{n-1}(X_n), X_n) = 0,$$

hence all f_ν must be identically zero and $F = 0$. Thus $\alpha(A) \geq n - 2$, and so $\alpha(A) = n - 2$.

In this example, if $p \in \mathrm{Spec}_2(A)$ and $p \supset (X_1, \cdots, X_r)$ then A/p is essentially of finite type over k, and so $\alpha(A/p) = \dim(A/p) - 1$. Therefore $N(A_{r,n}) = V((X_1, \cdots, X_r)) \cap \mathrm{Spec}_2(A_{r,n})$, which is closed in $\mathrm{Spec}_2(A)$. Note that, although $\alpha(A_{r,n}) = n - 2$ for $r = 1, 2, \cdots, n - 1$, $N(A_{r,n})$ becomes smaller as r increases.

When A is a noetherian ring and I is an ideal, we will say that A is I-complete if A is complete and separated in the I-adic topology, in other words if $A \simeq \varprojlim A/I^\nu$.

Theorem 4. *Let (A, m) be a noetherian local ring of dimension $n > 1$, and I be a proper ideal of A such that*
(1) *A is I-complete, and*
(2) *$\dim A/I < \dim A$.*
Then
$$\alpha(A) \leq n - 2.$$

Proof. Suppose $\alpha(A) = n - 1$. Then there exists a minimal prime ideal p of A such that $\dim A/p = n$ and $\alpha(A, p) = n - 1$. The local domain A/p is clearly $(I + p)/p$-complete, and $\dim A/I + p \leq \dim A/I < n = \dim A/p$, hence we may replace A by A/p and assume that A is a local domain. Then there exists a prime ideal P of $\hat{A}$ of height $n - 1$ satisfying $P \cap A = (0)$. The ring $B := \hat{A}/P$ is a complete local domain of dimension 1 containing A as a subring. Then IB is a non-zero ideal of B, hence $IB \supset m^\nu B$ for some $\nu > 0$. Since $B = A + mB$ we have $B = A + m^i B$ for all $i > 0$, hence $B = A + IB$. As in the proof of Theorem 3, this implies that $B = A$. But $\dim B = 1 < \dim A$, contradiction.

Let A be a noetherian ring and I, J be ideals. If A is I-complete and J-complete, then A is $(I + J)$-complete ([CR] Ex. 8.1). Therefore there is a unique largest ideal among the ideals I such that A is I-complete. We shall call it the *ideal of completeness* of A.

Let I be the ideal of completeness of our local ring A, and assume that $\dim A = n \geq 2$ and $\dim A/I < n$. Then the following corollary is immediate from the theorem.

Corollary. $N(A) \subset V(I) \cap \mathrm{Spec}_2(A)$.

Example 3. Set $B_{r,n} = k[X_1, \cdots, X_r]_{(X_1, \cdots, X_r)}[X_{r+1}, \cdots, X_n], n > r > 0$. Then a similar argument as in Example 2 shows that $\alpha(B_{r,n}) = n - 2$. Since $B_{r,n}$ contains $k[X_{r+1}, \cdots, X_n]$, this is also a consequence of Theorem 3. Here, $I = (X_{r+1}, \cdots, X_n)$ is the ideal of completeness of $B_{r,n}$, and since $B_{r,n}/I = k[X_1, \cdots X_r]_{(X_1, \cdots, X_r)}$ we have $N(B_{r,n}) = V(I) \cap \mathrm{Spec}_2(B_{r,n})$.

Open Problem 2. Are there local rings A such that $0 < \alpha(A) < \dim A - 2$?

Appendix. (Added in July, 1987)

In the proof of Theorem 1, we quoted Th. 15.3 of [CR], in which the ring B is assumed to be catenary. In fact we can eliminate this assumption from the quoted theorem, as follows.

Theorem. *If $A \to B$ is a homomorphism of noetherian rings such that the going-down theorem holds, and if p, $q \in \mathrm{Spec}(A)$ with $q \subset p$, then*

$$\dim B \otimes_A \kappa(q) \geq \dim B \otimes_A \kappa(p).$$

Proof. We will prove that, if $P_1 \subset P_2 \subset \cdots \subset P_r$ are prime ideals of B lying over p, then there are prime ideals $Q_1 \subset Q_2 \subset \cdots \subset Q_r$ of B lying over q such that $Q_i \subset P_i$ $(1 \leq i \leq r)$. Obviously we may assume that $\mathrm{ht}(p/q) = 1$. We may also assume that $r = 2$ and $\mathrm{ht}(P_2/P_1) = 1$. Since going-down holds we can find $Q_1 \in \mathrm{Spec}(B)$ such that $Q_1 \subset P_1$ and $Q_1 \cap A = q$. Take $x \in p$ such that $x \notin q$, and let $P_1^*, \cdots, P_n^*$ be the prime ideals of B minimal over $Q_1 + xB$. Then $\mathrm{ht}(P_i^*/Q_1) = 1$ by the principal ideal theorem of Krull, and $\mathrm{ht}(P_2/Q_1) \geq 2$, so that $P_2 \not\subseteq P_i^*$ for all i. Therefore by the prime avoidance lemma we can choose an element y of P_2 which is neither in Q_1 nor in any of P_i^*. Let Q_2 be a prime ideal which is contained in P_2 and minimal over $Q_1 + yB$. Then $\mathrm{ht}(Q_2/Q_1) = 1$, so that we have $Q_1 \subset Q_2 \subset P_2$ (strict inclusion). It follows that $q \subseteq Q_2 \cap A \subseteq p$. Since $\mathrm{ht}(p/q) = 1$ the prime ideal $Q_2 \cap A$ must coincide with either p or q . But if $Q_2 \cap A = p$ then Q_2 would contain xB, hence Q_2 must be one of the P_i^*'s. Since $y \in Q_2$ this cannot happen. Therefore $Q_2 \cap A = q$, as wanted.

References

[LR] M. Nagata, Local Rings, John Wiley, 1962.

[CR] H. Matsumura, Commutative Ring Theory, Cambridge Univ. Press 1986.

[CA] H. Matsumura, Commutative Algebra, second edition, Benjamin/Cummings, 1980.

[AL] S. Lang, Algebra, second edition, Addison-Wesley, 1984.

Hideyuki MATSUMURA

Department of Mathematics
Faculty of Sciences
Nagoya University
Nagoya 464
Japan

Algebraic Geometry and Commutative Algebra
in Honor of Masayoshi NAGATA
pp. 267–279 (1987)

On the Classification Problem of Embedded Lines in Characteristic p

Tzuong-Tsieng MOH*

§1. Introduction.

In the year 1970, Professor Nagata came to Purdue University and delivered a lecture on his article entitled "A theorem of Gutwirth" (cf. Nagata [9]). After his speech, he kindly mentioned to me "if $x = f(t)$, $y = g(t) \in \mathbf{C}[t]$ defines a line with $m = \deg f(t) < n = \deg g(t)$, then it is impossible to have $n/2 < m < n$." I thought for a while and remarked "the above inequality is unnatural. A natural statement should be $m|n$". Hence we have the following theorem (cf. Abhyankar-Moh [3]).

Theorem I. *Let k be a field of characteristic zero,*

$$k[f(t), g(t)] = k[t] \quad \text{with } m = \deg f(t), n = \deg g(t).$$

Then either $m|n$ or $n|m$.

The main technique of the proof of the above theorem is the "theory of approximate roots" which has been simplified and generalized (cf. Moh [4], [5]). It is a simple exercise to deduce the following from the above Theorem I.

Theorem II. *Let $x = f(t)$, $y = g(t)$ be a parametrization of an embedded line, i.e., $k[f(t), g(t)] = k[t]$. Then there exists an automorphism $\sigma: k[X, Y] \to k[X, Y]$ such that $\sigma(x) = 0$, $\sigma(y) = t$. In other words, every embedded line of the plane in characteristic zero can be transformed into an axis by a suitable automorphism.*

It is an interesting problem to investigate the corresponding phenomena in characteristic p. We have the following important example due to Professor Nagata (cf. Nagata [8]):

---

*The research of this work is supported in part by a grant from NSF.
Received August 28, 1986.
Revised February 23, 1987.

Nagata's example Let k be a field of characteristic p. Let $p \nmid m$ and $y = t^{p^2}$, $x = t^{mp} + t$. Then $k[x, y] = k[t]$ and there is no automorphism which transforms the embedded line to an axis.

It is easy to explain Nagata's example. The non-existence of the said auto-morphism is due to van der Kulk's theorem (cf. Nagata [8]) on the automorphism group of the affine plane. There are several simple ways to see the parametrized curve is indeed an embedded line. We shall use the following argument which is not the simplest while very useful for our later discussions.

Proposition 1. *Let k be an algebraically closed field of characteristic p. Let $f(t)$, $g(t) \in k[t]$. Then $k[f^p, g] = k[t] \Leftrightarrow k[f, g] = k[t]$ and $dg/dt \in k^*$ ($= k \setminus (0)$).*

Proof. ($\Rightarrow$) Certainly $k[t] \supset k[f, g] \supset k[f^p, g] = k[t]$. Moreover we have $t = F(f^p, g)$ for some suitable $F(X, Y) \in k[X, Y]$. Differentiating the above equation with respect to t, we get $1 = F_g \cdot dg/dt$ and $dg/dt \in k^*$.

($\Leftarrow$) We have $k[f^p, g] \supset k[f^p, g^p] = k[t^p]$. Moreover $g = ct + h(t^p)$, so, $t \in k[f^p, g]$.   q.e.d.

Applying the above proposition to Nagata's example we easily see that

$$(t^{p^2}, t^{mp} + t) \text{ defines a line}$$
$$\Longleftrightarrow \quad (t^p, t^{mp} + t) \text{ defines a line}$$
$$\Longleftrightarrow \quad (t, t^{mp} + t) \text{ defines a line.}$$

The last statement is self-evident.

We believe that Nagata's example is the principal one which illustrates the characteristic p phenomena. Hence we may formulate the following conjecture.

Conjecture 1. *Let k be an algebraically closed field of characteristic p. Let $x = f(t)$, $y = g(t) \in k[t]$ with $k[f(t), g(t)] = k[t]$. Then there exists an automorphism σ of $k[X, Y]$ such that*

(1) $\sigma(x) \in k[t^p]$,

(2) $(\frac{1}{p} \deg \sigma(x), \deg \sigma(y)) < (\deg x, \deg y)$.

For the ordering in (2), we may rearrange any pair of integers (m, n) by size, i.e., $m \leq n$ and then order pairs lexicographically.

The above Conjecture 1 means that Nagata's example is the principal char-acteristic p phenomena or the Frobenius map (i.e., taking the p-th power) is the only extra operation for characteristic p. Note that it trivially follows from our Proposition 1 that automorphisms and Frobenius maps can be used to build complicated embedded lines. Conjecture 1 implies the preceding natural pro-cesses are the only possible one. In other words we may use automorphisms and taking the p-th roots to decipher any embedded line to an axis. Hence Conjecture 1 may be used to classify any embedded line by the process to decipher it. We thus explain the title of this article. We wish to express our thanks to H. Kamat for clarifying some statements in this article.

§2. Expansions and Their Calculus.

To further our discussion let us restrict our attention to the fundamental case with the following irreducible parametrization.

$$x = t^{mp} + \text{lower terms} = \tau^{-mp} + \text{higher terms}$$
$$y = t^{np} + \text{lower terms} = \tau^{-np} + \text{higher terms}$$

where $\tau = t^{-1}$ and $(m, n) = 1$, say $p \nmid m$. We may apply the expansion technique (cf. Moh [6]) to rewrite

$$x = \Delta^{-m}$$
$$y = \Delta^{-n} + \cdots + c_i \Delta^i + \cdots + z$$

where $\Delta = \tau^p + \text{higher terms}$, and $p \nmid \text{ord } z$.

Proposition 2. (i) *Assume $m > n$. The contribution to the arithmetical genus of the point at ∞ is $[(mp - 1)(mp - np - 1) + q_2(p - 1)]/2$ where $q_2 = \text{ord}_\tau z - \text{ord}_\tau \Delta^{-n} = \text{ord}_\tau z + np$.*

(ii) *We have $k[x, y] = k[t]$ if and only if $\text{ord}_\tau z = (q_2^* - n)p - 1$ where $q_2^* = (q_2 + 1)/p = mn + (m - 1)(n - 1)/(p - 1)$ $(= mn + \bar{q})$.*

Proof. (i) This is well-known. However, a suitable reference cannot be located. We will provide a proof. The parameters at ∞ are

$$x_1 = x^{-1} = \Delta^m$$
$$y_1 = y \cdot x^{-1} = \Delta^{m-n} + \cdots + \Delta^m \cdot z.$$

Note that the q_2 at ∞ is invariant as

$$\text{ord}_\tau \Delta^m z - \text{ord}_\tau \Delta^{m-n} = q_2.$$

Using the notations of Moh [6], we may write the characteristic sequence of the plane curve at ∞ as $[mp, (m - n)p; q_2]$. After one blow-up, the characteristic sequence will be $[np, (m - n)p; q_2]$. After finite blow-ups it becomes $[m^*p, n^*p + q_2]$ with $m^* = 1$. In fact the number of characteristic pairs is reduced to one. We shall write it as $[m^*p, n^*p + q_2]$. In the one characteristic pair case the contribution to the arithmetical genus is

$$\frac{1}{2}(m^*p - 1)(n^*p + q_2 - 1) \quad (= \frac{1}{2}[(m^*p - 1)(n^*p - 1) + q_2(p - 1)]).$$

Note that the above formula can be easily proved by induction and one blow-up. Similarly assume the contribution to the genus for the singularity with $[np, (m - n)p; q_2]$ is

$$\frac{1}{2}[(np - 1)(mp - np - 1) + q_2(p - 1)].$$

Then the original contribution must be

$$\frac{1}{2}[(np - 1)(mp - np - 1) + q_2(p - 1)] + \frac{1}{2}(mp - np)(mp - np - 1)$$

$$= \frac{1}{2}[(mp - 1)(mp - np - 1) + q_2(p - 1)].$$

(ii) Assume $m > n$. We know $k[x, y] = k[t] \Leftrightarrow$ the curve is a non-singular rational curve with one place at $\infty \Leftrightarrow (mp - 1)(mp - 2)/2 = [(mp - 1)(mp - np - 1) + q_2(p - 1)]/2$ and the curve has only one place at ∞.

Solve the above numerical equation. We have

$$q_2 = \frac{(mp - 1)(np - 1)}{p - 1}$$

and

$$q_2^* = \frac{q_2 + 1}{p} = mn + \frac{(m - 1)(n - 1)}{p - 1}.$$

Remark. The general formula for the contribution to the arithmetical genus of a singularity with characteristic sequence $[m, n : q_2, \ldots, q_h]$ (cf. Moh [6], Abyhankar-Moh [1, 2]) is

$$\frac{1}{2}[(m - 1)(n - 1) + \sum_{i=2}^{h} q_2 \cdot (d_i - 1)]$$

where as usual $d_i = \gcd(m, n, q_2, \ldots, q_{i-1})$.

Using the above proposition we may rewrite

$$(1) \qquad \begin{cases} x = \Delta^{-m} \\ y = \Delta^{-n} + \cdots + c_i \Delta^i + \cdots + y_1 \Delta^{q_2^* - n} \end{cases}$$

where $y_1 = c\tau^{-1} +$ higher terms, $c \neq 0$.

Now given any element $h \in k((\tau))$ we may write uniquely

$$h = \sum_j \sum_{0 \leq i < p} c_{ij} y_1^i \Delta^j.$$

Definition 1. We shall call the above expansion of h the (Δ, y_1)-expansion of h.

Definition 2. Let $h = \sum_j \sum_{0 \leq i < p} c_{ij} y_1^i \Delta^j$ be the (Δ, y_1)-expansion of h. We shall define y_1^i-ord of h as $\min\{ (jp - i) ; c_{ij} \neq 0 \}$, where $0 \leq i < p$, if finite, otherwise ∞.

Proposition 3. Given x, y as in (1). Let r, s be integers. Then we have
(a)   if $p \nmid s$ then

$$y_1^i\text{-ord}(x^r y^s) - \text{ord}(x^r y^s)$$
$$\begin{cases} = i(q_2^* p - 1) & \forall\, i \leq \text{principal residue of } s \\ \geq p(q_2^* p - 1) & \forall\, i > \text{principal residue of } s; \end{cases}$$

(b)   if $p \mid s$ then $y_1^i\text{-ord}(x^r y^s) - \text{ord}(x^r y^s) \geq p(q_2^* p - 1)$.

Proof. Binomial theorem.

Proposition 4. Given $G = \sum_j \sum_{0 \leq i < p} c_{ij} y^i \Delta^j \in k((\tau))$ satisfying
(a) $y_1\text{-ord}\, G - \text{ord}\, G < q_2^* p - 1$
(b) $y_1^i\text{-ord}\, G - y_1\text{-ord}\, G \geq (i - 1)(q_2^* p - 1)$ for $0 < i < p$,
then we always have the following for $0 < l < p$ with $x^r y^s G^l = \sum d_{ij} y^i \Delta^j$:
(c) $y_1^i\text{-ord}(x^r y^s G^l) - \text{ord}(x^r y^s G^l) = i(y_1\text{-ord}\, G - \text{ord}\, G)$ $\forall\, 0 < i \leq l$.
(d)   Let $0 < i \leq l$ and $y_1^i\text{-ord}(x^r y^s G^l) = \gamma_{il} p - i$. Then

$$d_{i, \gamma_{il}} = \binom{l}{i} c_{1, \beta}^i c_{0, s}^{p - i}$$

with $y_1\text{-ord}\, G = \beta p - 1$ and $\text{ord}\, G_1 = sp$.
(e) $y_1^i\text{-ord}\, G^p - \text{ord}\, G^p \geq p(y_1\text{-ord}\, G - \text{ord}\, G)$.

Proof. Binomial theorem. Routine calculation.   q.e.d.

Remark. It is easy to see from the above propositions that the "gap" $= y_1\text{-ord}\, G - \text{ord}\, G$ is the essential ingredient of our computation.

§3. The Defining Equations.

We may study Conjecture 1 from the point of view of the defining equation of the embedded line. Let the parametrization of the embedded line be of the form (1). Then the defining equation of it is of the following form:

$$F(X, Y) = Y^{mp} - X^{np} + \cdots.$$

Let $G(X, Y)^p$ be the collection of all p-th power terms of $F(X, Y)$. Then

$$G(X, Y) = Y^m - X^n + \cdots,$$
(2)
$$F(X, Y) = G^p + \sum \alpha_i(X, Y) G^{p-i}$$

where $\deg_Y \alpha_i(X,Y) < m = \deg_Y G(X,Y)$ (cf. Abhyankar-Moh [1, 2], Moh [5]).

Definition 3. As in the above, $G(X,Y)$ is called the p-th pseudo-approximate root of $F(X,Y)$.

Example. Let us study the example of Nagata. The defining equation is

$$F(X,Y) = Y^{mp} + Y - X^{p^2}.$$

We have $G(X,Y) = Y^m - X^p$,

$$F(X,Y) = G^p + 0G^{p-1} + \cdots + 0G + Y.$$

Using the expansion technique of §2, we find

$$x = t^{mp} + t = \Delta^{-m},$$

$$\Delta = \tau^p - \frac{1}{m}\tau^{(m+1)p-1} + \cdots,$$

$$y = \Delta^{-p} - \frac{1}{m}\Delta^{mp-p-1} + y_1\Delta^{mp-p+(m-1)},$$

$$G(x,y) = y^m - x^p = \Delta^{-1} + y_1\Delta^{m-1} + \cdots.$$

We may observe that in our preceding computation we have $\alpha_i = 0$, $\forall\, i = 1$, $\ldots, p-1$. Note that in the case of characteristic zero, $\alpha_1 = 0$ means $G(X,Y)$ is the p-th approximate root of $F(X,Y)$. It hints Definition 3. For a further discussion, see our Theorem 2. Furthermore in the case of characteristic zero the approximate root will be of order positive in τ or negative in t if $m \nmid n$ and $n \nmid m$ and thus produces a contradiction. However in the case of characteristic p the p-th pseudo-approximate root $G(X,Y)$ may well be of order negative in τ as in Nagata's example and produces no direct contradiction.

Conjecture 2. *Let us use the above assumptions and notation. If $F(X,Y)$ defines an embedded line, then $\alpha_1 = \alpha_2 = \cdots = \alpha_{p-1} = 0$ in the equation (2).*

Let us clarify the (Δ, y_1)-expansion of $G(X,Y)$.

Proposition 5. *Let $G(X,Y)$ be as above and $\bar{q} = (m-1)(n-1)/(p-1)$. Then $G = a\Delta^{-\gamma} + \cdots$ where $a \neq 0$ and*

(1) $0 < \gamma < mn$,

(2) $y_1\text{-ord}\, G(x,y) = \dfrac{(m-1)(n-1)}{p-1}p - 1 = \bar{q}p - 1,$

$\qquad y_1\text{-ord}\, G(x,y) - \text{ord}\, G(x,y) < q_2^* p - 1,$

(3) $y_1^i\text{-ord}\, G(x,y) - y_1\text{-ord}\, G \geq (i-1)(q_2^* p - 1)$ *for* $0 < i < p$.

Proof. Let $G(X,Y) = Y^m - X^n - \sum c_{rs} X^r Y^s$ where $rm + ns < mn$ for $c_{rs} \neq 0$. Note that then we have $r < n$, $s < m$ and $\mathrm{ord}(x^r y^s) > -mnp$. It follows from Proposition 3 that $y_1^i\text{-ord}(x^r y^s) > -mnp + i(q_2^* p - 1)$ for $c_{rs} \neq 0$.

Let us prove our Proposition 5. Let us assume that $\gamma \leq 0$. Then $\mathrm{ord}_r(G(x,y) + c) > 0$ for a suitable c. In other words $\mathrm{ord}_t(G(x,y) + c) < 0$. Due to the fact that x, y, $G(x,y) + c$ are polynomials in t we are forced to conclude $G(x,y) + c = 0$ thus violating the basic assumption of the irreducibility of the parametrization of x, y or the defining equation being $F(X,Y)$. We conclude (1).

Since $p \nmid m$, it follows from Proposition 3 that

$y_1^i\text{-ord}\, y^m$

$$\begin{cases} = -mnp + i(q_2^* p - 1) & \forall\, i \leq \text{principal residue of } m \\ \geq -mnp + p(q_2^* p - 1) & \forall\, i > \text{principal residue of } m. \end{cases}$$

Combining this with the remark at the beginning of our proof, we see at once that

$$y_1\text{-ord}\, G(x,y) = \frac{(m-1)(n-1)}{p-1} p - 1 = \bar{q}p - 1$$

$y_1^i\text{-ord}\, G(x,y)$

$$\begin{cases} = -mnp + i(q_2^* p - 1) & \forall\, i \leq \text{principal residue of } m \\ \geq -mnp + i(q_2^* p - 1) & \forall\, i > \text{principal residue of } m. \end{cases}$$

Obviously we deduce (2) and (3).   *q.e.d.*

Remark. We may apply Proposition 4 to $G(x,y)$.

Theorem 1. *Let us use the above assumption and notation. Then Conjecture 2* $\Rightarrow \alpha_1 = \alpha_2 = \cdots = \alpha_{p-1} = 0$, $\mathrm{ord}\,\alpha_p(x,y) > \mathrm{ord}(x \cdot y) \Rightarrow$ *Conjecture 1.*

Proof. Let us give a weight m to x and n to y. Due to the restriction of $\deg_Y \alpha_i(X,Y) < m = \deg_Y G(X,Y)$, and $(m,n) = 1$, all terms in $\alpha_i(x,y)$ are of different weights. In other words after the substitution of $x = t^{mp} +$ (lower terms) $= \Delta^{-m}$, $y = t^{np} +$ (lower terms) $= \Delta^{-n} +$ (higher terms), there is no cancellation among the leading forms of different terms in a fixed $\alpha_i(x,y)$. Let us write

$$\alpha_p(X,Y) = bX^i Y^j + (\text{lower weighted terms}).$$

Clearly we have by Conjecture 2

$$G(x,y)^p = -\alpha_p(x,y).$$

Let us use the notation of Proposition 5. Comparing the orders of both sides of the above equation we conclude

$$\gamma p^2 = imp + jnp.$$

If $p|j$, then $p|i$. In this case X^iY^j is a p-th power and hence should be included in $G(X,Y)^p$. Thus we know $p\nmid j$.

We shall compare y_1-ord of both sides of the same equation. It follows from Proposition 5 that for the left hand side we have

$$y_1\text{-ord}\, G(x,y)^p > p(\frac{(m-1)(n-1)}{p-1}p - 1).$$

It follows from Proposition 3 that for the right hand side we have

$$y_1\text{-ord}\, \alpha_p(X,Y) = y_1\text{-ord}(X^iY^j) = -imp - jnp + (q_2^*p - 1).$$

We thus have the following inequality

$$-imp - jnp + (q_2^*p - 1) > p(\frac{(m-1)(n-1)}{p-1}p - 1)$$

or

$$-\gamma p^2 + (mn + \frac{(m-1)(n-1)}{p-1})p - 1 > p(\frac{(m-1)(n-1)}{p-1}p - 1)$$

or

$$mn + \frac{(m-1)(n-1)}{p-1}(p-1) > \gamma p$$

or

$$m + n > m + n - 1 > \gamma p = im + jn.$$

The last inequality means

$$\text{ord}\, \alpha_p(x,y) > \text{ord}(x \cdot y).$$

Clearly the above inequality means that in the expression of $\alpha_p(X,Y)$ there is no term that involves both X and Y. Moreover there cannot be both terms of the form X^i, Y^j with $i, j > 1$. It follows easily that

$$\alpha_p = \begin{cases} cx + h(y) & \text{or} \\ cy + h'(x). \end{cases}$$

We have either $\deg_t x \geq \deg_t y$ or $\deg_t y \geq \deg_t x$. In the first case $h'(x)$ must be linear and we may assume $\alpha_p = cx + h(y)$. In the second case $h(y)$ must be linear and we may assume $\alpha_p = cy + h'(x)$. Let us discuss the case $\deg_t x \geq \deg_t y$ and $\alpha_p = cx + h(y)$. If $c \neq 0$, then an automorphism $\sigma: y \mapsto y$, $x \mapsto cx + h(y)$ will satisfy $\sigma(x) = -G^p \in k[t^p]$. Note that the inequality $(m + n)p > \deg_t \alpha_p$ will imply

$$\deg \sigma(x)^{1/p} = \frac{1}{p}\deg \sigma(x) = \frac{1}{p}\deg \alpha_p < m = \deg x.$$

In other words

$$(\deg \sigma(x)^{1/p}, \deg \sigma(y)) < (\deg x, \deg y).$$

Suppose $c = 0$. Differentiating the following equation

$$G(x,y)^p = -\alpha_p = -h(y) = -y^j + \cdots$$

with respect to t we conclude at once that $y \in k[t^p]$. The case $\deg_t y \geq \deg_t x$ can be discussed verbatim. Hence Conjecture 1 has been proved.   q.e.d.

§4. $\alpha_1 = 0$.

In the last section we have shown that Conjecture 2 implies Conjecture 1. In this section we shall discuss Conjecture 2. In fact we shall indicate that Conjecture 2 is probably right by proving Theorem 2. For this purpose we need the following sequence of lemmas.

Definition 4. For non-negative integers u and v define

$$b_{u,v} = \left\{ \begin{array}{ll} \dbinom{u}{v} & \text{if } v \leq u \\ 0 & \text{if } v > u. \end{array} \right.$$

Also define

$$b_u = (b_{u,0}, \ldots, b_{u,s-1}) \quad \text{and} \quad c_u = (b_{u,1}, \ldots, b_{u,s-1}).$$

Here b_u is a vector with s components and c_u is a vector with $s - 1$ components.

Lemma 1. *Let $j_1, \ldots, j_s$ be non-negative integers with distinct residues modulo p (so that $s \leq p$). Then the s vectors $b_{j_1}, \ldots, b_{j_s}$ are linearly independent modulo p. Further, if $k_1, \ldots, k_{s-1}$ are non-negative integers with distinct non-zero residues modulo p, then the $s - 1$ vectors $c_{k_1}, \ldots, c_{k_{s-1}}$ are linearly independent modulo p.*

Proof. See Moh [7] Theorem 2.1 and its proof.

Lemma 2. *Let $\langle m, n \rangle$ be the semigroup generated by positive integers m and n where $(m, n) = 1$ as usual. Then $\gamma \in \langle m, n \rangle \Leftrightarrow mn - m - n - \gamma \notin \langle m, n \rangle$.*

Proof. This is a restatement of the well-known completely reflexive property of semigroup $\langle m, n \rangle$.

Lemma 3. *Let $j_1 > j_2 > \cdots > j_p \geq 0$ with $i_l m + j_l n = p\gamma$ for all $l = 1$, $\ldots, p$, $(m, n) = 1$, $p \nmid m$ and i_l, γ non-negative integers. Then $\gamma \in \langle m, n \rangle$.*

Proof. Due to the properties that $(m, n) = 1$ and $p \nmid m$, it follows that

$$j_1 = j_p + sm$$

where $s \geq p - 1$. Hence we have

$$i_1 m + j_1 n = i_1 m + j_p n + smn$$
$$= i_1 m + j_p n + (s - (p-1))mn + (p-1)mn = p\gamma$$

or

$$i_1 m + j_p n + (s - (p-1))mn + (p-1)(m+n)$$
$$+ (p-1)(mn - m - n - \gamma) = \gamma.$$

If $\gamma \notin \langle m, n \rangle$, Lemma 3 shows $mn - m - n - \gamma \in \langle m, n \rangle$. Then clearly the left hand side of the above equation is in $\langle m, n \rangle$. Thus $\gamma \in \langle m, n \rangle$. This is a contradiction. Hence $\gamma \in \langle m, n \rangle$. q.e.d.

We shall write the defining equation for the embedded line $F(x, y) = G^p + \sum \alpha_i G^{p-i} = G^p + H = 0$ as $G^p(x, y) = -H(x, y)$.

Lemma 4. $\alpha_1 = 0 \Leftrightarrow \deg_y H(x, y) < (p-1)m$.

Proof. Trivial.

We shall give the following *weight* to the variables x, y:

$$\mathrm{wt}(x) = m, \quad \mathrm{wt}(y) = n.$$

Let δ be the weight of $H(x, y)$. Let

$$H(x, y) = H_\delta(x, y) + \cdots$$

where $H_\delta(x, y)$ is the δ-weighted form of $H(x, y)$. Since both $F(x, y)$ and $G(x, y)^p$ share the same highest weighted form $(y^m - x^n)^p$, clearly we have $\delta < mnp$. It then follows that $H_\delta(x, y)$ contains at most p terms.

Lemma 5. *If* $\delta \geq (s-1)mn$, *then we must have*

$$y_1^i\text{-ord}\, H(x, y) > -\delta p + i(q_2^* p - 1)$$

for $i = 1, \ldots, s - 1$.

Proof. It follows from Proposition 3 that

$$y_1^i\text{-ord}(x^r y^s) \geq -\delta p + i(q_2^* p - 1)$$

where $x^r y^s$ is in $H(x, y)$. If the inequality in the statement is not valid we must have for some i,

$$y_1^i\text{-ord}\, H(x, y) = -\delta p + i(q_2^* p - 1).$$

On the other hand we have

$$G(x, y) = \alpha \Delta^{-\gamma} + \cdots + y_1 \Delta^{\bar{q}} + \cdots.$$

Hence we conclude by Proposition 4

$$y_1^i\text{-ord}\, G(x,y)^p \geq (\bar{q}p - 1)p$$

and for some i

$$-\delta p + i(q_2^* p - 1) \geq (\bar{q}p - 1)p,$$

so

$$(\bar{q}p - 1)p \leq -\delta p + i(q_2^* p - 1) \leq -imnp + i(q_2^* p - 1) \leq i(\bar{q}p - 1).$$

This is a contradiction. $\hfill$ q.e.d.

Lemma 6. $\delta = p\gamma$.

Proof. Otherwise we must have $\delta > p\gamma$ and the leading forms of terms in $H_\delta(x,y)$ cancel out.

Let $X^{i_1}Y^{j_1}, \ldots, X^{i_s}Y^{j_s}$ be the terms in $H_\delta(X,Y)$. Due to the observation that $j_l < mp$ and $p \nmid m$, it follows trivially that $s \leq p$ and $j_1, \ldots, j_s$ have all distinct residues modulo p. The expansion of $x^i y^j$ is as follows:

$$x^i y^j = \Delta^{-mi-nj}\Big(1 + \cdots + \binom{j}{1}y_1\Delta^{q_2^*} + \cdots$$

$$+ \binom{j}{2}y_1^2\Delta^{2q_2^*} + \cdots + \binom{j}{j}y_1^j\Delta^{jq_2^*} + \cdots\Big)$$

$$= \Delta^{-mi-nj}\Big(1 + \cdots + b_{j,1}y_1\Delta^{q_2^*} + \cdots + b_{j,s-1}y_1^{s-1}\Delta^{(s-1)q_2^*} + \cdots\Big)$$

with $mi + nj = \delta$ and the $b_{j,k}$ as in Definition 4. Considering the linear independence of the vectors $b_{j_1}, \ldots, b_{j_s}$ (by Lemma 1), we conclude that if the leading forms are cancelled out in $H_\delta(x,y)$, then for some i, $1 \leq i \leq r-1$, we have

$$y_1^i\text{-ord}\, H(x,y) = -\delta p + i(q_2^* p - 1).$$

This violates Lemma 5 because other conditions of it can be easily verified.
$\hfill$ q.e.d.

Lemma 7. *If $\delta \geq (s-1)mn$, then there are at least s terms in $H_\delta(X,Y)$.*

Proof. Same as the previous lemma.

Lemma 8. $\gamma \notin \langle m,n \rangle$, *the semigroup generated by m, n.*

Proof. Let $smn > \delta > (s-1)mn$ for some suitable $s \leq p$. It is easy to see that there are at most s terms in $H_\delta(X,Y)$. Hence there must be exactly s terms in $H_\delta(X,Y)$ by Lemma 7. If $\gamma = rm + sn \in \langle m,n \rangle$, then the term $(x^r y^s)^p$ is absorbed in $G(x,y)^p$. Thus $H_\delta(X,Y)$ will have less than s terms which is impossible. $\hfill$ q.e.d.

Remark. It follows from the above lemma that the number γ can be reached simply by considering $y^m - x^n = c\Delta^l + \cdots$ and then subtracting from $y^m - x^n$ a suitable $x^r y^s$ if $-l \in \langle m, n \rangle$. Continuing this process we must reach

$$y^m - x^n - cx^r y^s - \cdots = \Delta^{-\gamma} + \cdots + y\Delta^{\bar{q}} + \cdots$$

with $\gamma \notin \langle m, n \rangle$.

Now we are in a position to prove the following Theorem 2.

Theorem 2. *We always have* $\alpha_1 = 0$.

Proof. By Lemma 4 it suffices to prove $\deg_Y H(X, Y) < (p-1)m$. Assume the contrary, i.e.,

$$\deg_Y H(X, Y) \geq (p-1)m.$$

Then we have

$$\delta = \text{wt } H(X, Y) \geq (\deg_Y H(X, Y))n \geq (p-1)mn.$$

It follows from Lemma 6 that

$$\delta = p\gamma.$$

We have by Lemma 8

$$\gamma \notin \langle m, n \rangle$$

and by Lemma 3

$$\gamma \in \langle m, n \rangle.$$

Clearly we get a contradiction.   q.e.d.

Corollary. *If* $p = 2$, *then Conjecture 2 and Conjecture 1 are correct in the special case* $\deg_t x = pm$, $\deg_t y = pn$, $(m, n) = 1$.

Remark. In the characteristic zero case we may use $\alpha_1 = 0$ to define G as the p-th approximate root of F. In Theorem 2 we have shown that the pseudo-approximate root in characteristic p has the same formal property.

§5. Other Conjectures.

We have the following related conjectures.

Conjecture 3. *Let* k *be an algebraically closed field of characteristic* p. *Let* $F(X, Y)$ *be the defining equation of an embedded line, i.e.,*

$$k[X, Y]/(F(X, Y)) \overset{\rho}{\simeq} k[t].$$

Then there exists an automorphism σ *of* $k[X, Y]$ *such that*

(1) $k[X^p, Y^p, F(X, Y)] = k[\sigma(X)^p, \sigma(Y)],$

(2) $(\frac{1}{p}\deg \rho(\sigma(X)), \deg \rho(\sigma(Y))) < (\deg \rho(X), \deg \rho(Y))$.

Conjecture 4. *If $F(X,Y)$ defines an embedded line, then $F(X,Y) + \lambda$ defines an embedded line for all $\lambda \in k$.*

We believe that both Conjecture 3 and Conjecture 4 are equivalent to Conjecture 1 and Conjecture 2. For instance it is easy to see that Conjecture 2 implies Conjecture 4 in the following way: We may alternate taking the p-th root and applying automorphism to $F(X,Y)$ to reduce it to $X = 0$. So applying the same procedure to $F(X,Y) + \lambda$ we shall reduce it to

$$X + \lambda^{1/p^n} = 0.$$

Then it is clear that $F(X,Y) + \lambda$ defines an embedded line. On the other hand a proof of Conjecture 4 implying Conjecture 2 is rather long and will not be discussed here.

Note that Conjecture 4 is known to be true for characteristic zero. Hence Conjecture 4 if true can be used as the unified statement for all characteristics.

References

[1, 2] S. S. Abhyankar and T. T. Moh, Newton-Puiseux expansion and generalized Tschirnhausen transformation, I, II, J. reine angew. Math. **260**(1973), 47–83; **261**(1973), 29–54.

[3]   S. S. Abhyankar and T. T. Moh, Embedding of the line in the plane, J. reine angew. Math. **276**(1975), 148–166.

[4]   T. T. Moh, On the concept of approximate roots for algebra, J. Algebra **65**(1980), 347–360.

[5]   T. T. Moh, On two fundamental theorems for the concept of approximate roots, J. Math. Soc. Japan **34**(1980), 637–652.

[6]   T. T. Moh, On characteristic pairs of algebroid plane curves for characteristic p, Bull. Inst. Math. Academia Sinica **1**(1973), 75–91.

[7]   T. T. Moh, On the unboundedness of generators of prime ideals in power series rings of three variables, J. Math. Soc. Japan **26**(1974), 722–734.

[8]   M. Nagata, On automorphism group of $k[x,y]$, Lectures in Math., Kyoto Univ. **5**, Kinokuniya, Tokyo, 1972.

[9]   M. Nagata, A theorem of Gutwirth, J. Math. Kyoto Univ. **11**(1971), 149–154.

Tzuong-Tsieng Moh

Department of Mathematics
Purdue University
West Lafayette, IN 47907
U.S.A.

Algebraic Geometry and Commutative Algebra
in Honor of Masayoshi NAGATA
pp. 281–287 (1987)

A Cancellation Theorem for Projective Modules over Finitely Generated Rings

N. MOHAN KUMAR, M. Pavaman MURTHY and A. ROY

§1. Introduction.

Let A be a finitely generated ring of dimension $d \geq 2$ over $\mathbf{Z}$. In [9, Cor. 18.1, Th. 18.2] Vaserstein has shown that for $n \geq d + 1$, the group $E_n(A)$ generated by $n \times n$ elementary matrices over A acts transitively on unimodular elements in A. In particular stably free projective A-modules of rank $\geq d$ are free. In this paper, we show that if P, P' and Q are projective A-modules such that $P \oplus Q \cong P' \oplus Q$ and rank $P \geq d$, then $P \cong P'$ (see Cor. 2.5). We in fact show that for rank $P \geq d$, the group of elementary automorphism of $P \oplus A$ act transitively on unimodular elements in $P \oplus A$ (see Th. 2.4).

We also give some application of our main result to projective stable range of affine rings. For example we show that if A is an affine ring of dimension $d \geq 3$ over $\bar{\mathbf{F}}_p$, then A has projective stable range $\leq d$ (Th. 3.7).

Let A be a commutative ring and P a finitely generated projective A-module. We say that rank $P \geq r$ if for all $p \in \operatorname{Spec} A$, rank $P_p \geq r$. We say that the projective stable range of A is $\leq r$ (notation: $\operatorname{psr}(A) \leq r$) if for all finitely generated projective A-modules P of rank $\geq r$ and $(x, a) \in P \oplus A$ unimodular, we can find a $y \in P$ such that $x + ay$ is unimodular. We recall that A has stable range $\leq r$ (notation: $\operatorname{sr}(A) \leq r$) is defined same way as $\operatorname{psr}(A)$ but with P required to be free. For any A-module M and $x \in M$, we denote by $O_M(x)$ (sometimes just $O(x)$) the ideal $= \{ f(x) \mid f \in M^* = \operatorname{Hom}_A(M, A) \}$. For standard results in projective modules and algeblaic K-theory, we refer to [1]. By a finitely generated ring, we mean a commutative finitely generated $\mathbf{Z}$-algebra.

§2. Cancellation.

Let P be a projective module over a commutative ring A. If $\phi \in P^*$ (resp. $x \in P$), let us denote the elementary automorphsm of $P \oplus A$ sending (y, a) to $(y, a + \phi(y))$ (resp. (y, a) to $(y + ax, a)$) by E_ϕ^* (resp. E_x). Recall that $E(P, A)$ denotes

Received February 12, 1987.

the subgroup of $\mathrm{Aut}(P \oplus A)$ genarated by E_ϕ^* and E_x. Further, if I is an ideal of A, then $E(P, A; I)$ denotes the subgroup of $E(P, A)$ genarated by those E_ϕ^* and E_x which are identity modulo I.

It is easily verified that if J is another ideal of A, then the natural map

$$E(P, A; I) \longrightarrow E(P/JP, A/J; I + J/J)$$

is surjective.

If $\phi \in P^*$ and $x \in P$ are such that $\phi(x) = 0$, then $y \to y + \phi(y)x$ is an (unipotent) automorphism of P. Let us denote this automorphism by $\tau_{\phi,x}$. We have

$$\tau_{\phi,x} \oplus 1_A = E_{-x} E_{-\phi}^* E_x E_\phi^*$$

showing that $\tau_{\phi,x} \oplus 1_A \in E(P, A)$. Further, if $\phi \in IP^*$ or $x \in IP$, then $\tau_{\phi,x} \oplus 1_A \in E(P, A; I)$.

Recall that $\tau_{\phi,x}$ is called a transvection if ϕ is unimodular in P^* or x is unimodular in P.

We quote a result from [2, Prop. 4.1]:

Lemma 2.1. *Any transvection of P/IP can be lifted to an automorphism of P of the type $\tau_{\phi,x}$.*

For a commutative ring A, consider the following property:

$\mathcal{P}_r(A)$: *For all ideals I of A and projective A-modules P of rank $\geq r$, the group $E(P, A; I)$ acts transitively on the set of unimodular elements $(x, a) \in P \oplus A$ with the property $(x, a) \equiv (0, 1) \pmod{I}$.*

Remark 2.1. We recall that $\mathcal{P}_{d+1}$ is true if $\dim A = d$. [1, (3.4), p. 189].

Remark 2.2. Let I and J be ideals of A such that $I \cap J = 0$. Then

$$\mathcal{P}_r(A/I) + \mathcal{P}_r(A/J) \Longrightarrow \mathcal{P}_r(A).$$

We omit the proof.

Remark 2.3. $\mathcal{P}_r(A/\mathrm{nil}\, A) \Longrightarrow \mathcal{P}_r(A)$.

Proof. Let I be an ideal of A, P a projective A-module of rank $\geq r$, and let $(x, a) \in P \oplus A$ be a unimodular element with $(x, a) \equiv (0, 1) \pmod{I}$. We may assume that P is of constant rank > 0. Let "bar" denote "mod nil A". Since $\mathcal{P}_r(\bar{A})$ is true and $E(P \oplus A; I) \to E(\bar{P} \oplus \bar{A}; \bar{I})$ is surjective, we reduce to the case $(x, a) \equiv (0, 1) \pmod{I \cap \mathrm{nil}\, A}$. In particular, a is a unit in A. Let $a = 1 + n$, $n \in \mathrm{nil}\, A$. Applying $E_{-a^{-1}x}$ on (x, a) we further reduced to the case $x = 0$.

Next, we claim that given any $b \in 1 + I \cap \mathrm{nil}\, A$, $y \in P$, and $\phi \in P^*$, there exist an $\varepsilon \in E(P, A; An)$ such that $\varepsilon(0, b) = (0, b - n\phi(y))$. In fact we can take

$$\varepsilon = E_q E_{-p} E_\theta^* E_p$$

where $p = b^{-1}y$, $\theta = -n\phi$ and

$$q = (1 + \theta(p))^{-1}\theta(p)p.$$

Note that the first and the third factors belong to $E(P, A; An)$. Since P is of positive rank, there exist $y_1, \ldots, y_t \in P$ and $\phi_1, \ldots, \phi_t \in P^*$ such that $\sum \phi_i(y_i) = 1$. Using the claim proved above successively we see that $(0, a)$ can be mapped into $(0, 1 + n - n\phi_1(y_1) - \cdots - n\phi_t(y_t)) = (0, 1)$ by an element of $E(P, A; An)$.

We now state the main result.

Theorem 2.4. *Let A be a finitely generated ring of dimension d $(d \geq 2)$ and I an ideal of A. Let P be a projective A-module of rank $\geq d$ and $(x, a) \in P \oplus A$ a unimodular element such that $(x, a) \equiv (0, 1)$ (mod I).*

Then there exists $\varepsilon \in E(P \oplus A, I)$ such that $\varepsilon(x, a) = (0, 1)$.

Proof. In view of Remarks 2.1, 2.2, and 2.3, we may assume rank $P = d$ and that A is a domain. Further, we may assume $I \neq 0$.

Let $x_1, \ldots, x_d \in P$ such that $x_1 \wedge \cdots \wedge x_d \neq 0$. Let J denote the ideal $O(x_1 \wedge \cdots \wedge x_d)$. We have $\dim A/J < \dim A$.

Fix an integer $n \geq 2$ such that $n \neq 0$ in A. Let "tilde" denote "$\mathrm{mod}\, nIJ$". Since $\dim \tilde{A} \leq d - 1$ and rank $\tilde{P} = d$, by Remark 2.1 there is an $\tilde{\varepsilon} \in E(\tilde{P} \oplus \tilde{A}; \tilde{I})$ such that $\tilde{\varepsilon}(\tilde{x}, \tilde{a}) = (0, 1)$. Lift $\tilde{\varepsilon}$ to an $\varepsilon \in E(P + A; I)$. Replacing (x, a) by $\varepsilon(x, a)$, we may assume that $(x, a) \equiv (0, 1)$ (mod nIJ).

Thus we have ("bar" denoting "$\mathrm{mod}\, a$"):

(*) $\bar{P} = P/aP$ is free with $\bar{x}_1, \ldots, \bar{x}_d$ as a basis

(**) n is a unit in $\bar{A}$.

We claim that we can find $z_1, \ldots, z_d \in P$ such that $x_1 \wedge \cdots \wedge x_d = z_1 \wedge \cdots \wedge z_d$ and $\bar{z}_1 = \bar{x}$. If $d \geq 3$, then sr $\bar{A} \leq d - 1$ by ([9, Th. 17.2] or [7, Th. 9.1]). So there is a product of transvections of the free $\bar{A}$-modular $\bar{P}$, say $\bar{\sigma}$, such that $\bar{\sigma}(\bar{x}_1) = \bar{x}$. Lift $\bar{\sigma}$ to a $\sigma \in \mathrm{Aut}(P)$ by Lemma 2.0 and set $z_i = \sigma(x_i)$.

If $d = 2$, then sr $A \leq 3$, and since n is a unit in $\bar{A}$ and $\dim \bar{A} = 1$, we have $KSp_1(\bar{A}) = 0$ by [9, Th. 16.4]. So, by ([7, Cor. 9.10]), the map $SL_2(A) \to SL_2(\bar{A})$ is surjective. Now $\bar{P}$ is $\bar{A}$-free of rank 2 with $\bar{x}$ unimodular. So we can find $\bar{y} \in \bar{P}$ so that $\{\bar{x}, \bar{y}\}$ is a basis of $\bar{P}$. Further $\bar{y}$ can be chosen with the extra requirement that

$$\bar{\sigma}\begin{pmatrix} \bar{x}_1 \\ \bar{x}_2 \end{pmatrix} = \begin{pmatrix} \bar{x} \\ \bar{y} \end{pmatrix}$$

for a suitable $\bar{\sigma} \in SL_2(\bar{A})$. Lift $\bar{\sigma}$ to an $\sigma \in SL_2(A)$ and define z_1, z_2 by

$$\sigma\begin{pmatrix} x_1 \\ x_2 \end{pmatrix} = \begin{pmatrix} z_1 \\ z_2 \end{pmatrix}.$$

This completes the proof of the claim.

Let $z_1 = x + au$, $u \in P$. We have $1 - a \in IJ = IO(z_1 \wedge \cdots \wedge z_d) \subset IO(z_1)$. Let $1 - a = \phi(z_1)$ with $\phi \in IP^*$. Setting $\varepsilon = E_{(-x+(1-a)u)}E_{-u}E_\phi^* E_u$, we have

$\varepsilon(x, a) = (0, 1)$. We observe that E_ϕ^* and $E_{(-x+(1-a)u)}$ are identity mod I so that $\varepsilon \in E(P \oplus A; I)$.

Corollary 2.5. *Let A be a finitely generated ring of dimension d and let P_1, P_2, and Q be projective A-modules such that $P_1 \oplus Q \approx P_2 \oplus Q$. If rank $P_1 \geq d$, then $P_1 \approx P_2$.*

Proof. The case $d \leq 1$ is well-known and the case $d \geq 2$ follows from Theorem 2.4 by a standard argument.

§3. Projective stable ranges.

Theorem 3.1. *Let A be a noetherian ring of dimension 2, P a projective A-module of rank 2 containing a unimodular element, and let $(x, a) \in P \oplus A$ be unimodular. Assume furthur that for all ideals I of A, $E(P \oplus A; I)$ acts transitively on the set $\{ (z, b) \in P \oplus A \text{ unimodular} \mid (z, a) \equiv (0, 1) \pmod{I} \}$ and that if $\dim A/I \leq 1$, then $SK_1(A/I) = 0$.*
Then there is a $y \in P$ such that $x + ay$ is unimodular.

First we prove a lemma.

Lemma 3.2. *Let A be a noetherian ring of dimension d and P a projective A-module of rank $\geq d$. Assume that for all ideals J of A, $E(P \oplus A; J)$ acts transitively on the set $\{ (z, a) \in P \oplus A \text{ unimodular} \mid (z, b) \equiv (0, 1) \pmod{J} \}$. Let I be an ideal of A and let $\bar{\alpha} \in \mathrm{Aut}(P/IP)$. If $\bar{\alpha} = 0$ in $K_1(A/I)$ then $\bar{\alpha}$ can be lifted to an $\alpha \in \mathrm{Aut}(P/IP)$.*

Proof. Choose Q so that $P \oplus Q = A^N$ with N large enough to ensure $\bar{\alpha} \oplus 1_{\bar{Q}} \in E_N(\bar{A})$. Lift $\bar{\alpha} \oplus 1_{\bar{Q}}$ to a $\sigma \in \mathrm{Aut}(P \oplus Q)$. Then $\bar{\alpha} \oplus 1_{\bar{A}^N}$ lifts to an automorphism $\beta = \sigma \oplus 1_P$ of $P \oplus A^N$ $(= P \oplus Q \oplus P)$. We shall show the existence of α by downward induction on N.

We may assume $N = 1$. Let $e_1 = (0, 1) \in P \oplus A$. Then $\bar{\beta}(\bar{e_1}) = \bar{\alpha}(\bar{e_1}) = \bar{e_1}$. So $\beta(e_1) \equiv e_1 \pmod{I}$. By hypothesis, there is an $\varepsilon \in E(P \oplus A; I)$ such that $\varepsilon\beta(e_1) = e_1$. Changing β to $\varepsilon\beta$ we assume that $\beta(e_1) = e_1$. Now β induces an automorphism of $P = (P \oplus Ae_1)/Ae_1$ which lifts $\bar{\alpha}$. This proves the lemma.

Proof of Proposition 3.1. Let $\mathcal{P}_1, \ldots, \mathcal{P}_r$ be the minimal primes of ideals of A such that $a \notin \mathcal{P}_i$. Let $I = Aa + \cap_{i=1}^r \mathcal{P}_i$. Then $\dim A/I \leq 1$ and by hypothesis $SK_1(A/I) = 0$.

Let $z \in P$ be unimodular. With "bar" denoting "mod$I$", we have an automorphism $\bar{\alpha} \in SL(\bar{P})$ with $\bar{\alpha}(\bar{z}) = \bar{x}$. By Lemma 3.2, we can lift $\bar{\alpha}$ to an $\alpha \in \mathrm{Aut}(P)$. Then $\alpha(z)$ is unimodular in P and we can write $\alpha(z) \equiv x + ay \pmod{\cap \mathcal{P}_i}$ for some $y \in P$. Evidently $x + ay$ is unimodular modulo each $\mathcal{P}_i$, and also modulo the other minimal primes (since a belongs to them and (x, a) is unimodular in $P \oplus A$). Hence $x + ay$ is unimodular in P.

Theorem 3.3. *Let A be a finitely generated ring of dimension $d \geq 2$, P a projective A-module of rank $\geq d$ containing a unimodular element, and let $(x, a) \in P \oplus A$ be unimodular. Then there exists a $y \in P$ such that $x + ay$ is unimodular in P.*

Proof. We may assume P is constant rank. In view of [Bass stability theorem [1, (3.6) p. 185]], we may assume rank $P = d$. The case $d = 2$ follows from Theorem 3.1. If $d \geq 3$, we imitate the proof of Theorem 3.1 noting that $\bar{\alpha}$ can be chosen to be a product of transvecsions of $\bar{P}$ (Theorem 2.4) and applying Lemma 2.0 to lift $\bar{\alpha}$ to an automorphism α of P.

We need the following result from [3, Th. 1, p. 245].

Lemma 3.4. *Let A be a reduced affine ring of dimension d over an algebraically closed field. Let P be a projective A-module of rank d. Let $f \in P^*$ be such that Im f is a complete intersection ideal of height d. Then P has a unimodular element.*

(For proof, see the implication $\mathrm{Cl} \Rightarrow \mathrm{EE2}$ in [3, Th. 1, p. 245]).

Lemma 3.5. *Let A be a reduced affine ring of dimension d over an algebraically closed field. Let $I \subset A$ be a product of smooth maximal ideals of height d. There exists an $f \in I$ such that A/fA is regular and pure of dimension $d - 1$.*

Proof. Let $V(J) =$ singular locus of A. Then $I + J = A$ and hence $JI + I^2 = I$. Since I/I^2 is A/I-free of rank d, we can find $g_1, \ldots, g_d \in IJ$ such that the images of the g_i in I/I^2 form a basis for I/I^2. Let $h \in I^2$ such that $h - 1 \in J$. Put $f_1 = g_1 + h$. By 'Bertini arguments' (see [8, Th. 1.3 and Th. 1.4]) for general $a \in I^2 J$, $A/A(f_1 + a)$ is regular and pure of dimension $d - 1$.

Let $\bar{F}_p$ denote the algeblaic closure of the field of p elements.

Theorem 3.6. *Let A be a reduced affine ring of dimension d over $\bar{F}_p$. Suppose a) $d \geq 3$ or b) $d = 2$ and A is regular. Let $I \subset A$ be a product of smooth maximal ideals. Then I is a complete intersection.*

Proof. If $d = 2$, then by hypothesis A is regular and the proposition follows immediately from Rojtman's theorem ([6], [4]) and [5, Th. 3, p. 126]. If $d > 2$, by Lemma 3.5, there is an $f \in I$ such that A/fA is regular and pure of dimension $d - 1$. Hence by inductuon I/fA is generated by $d - 1$ elements. Hence I is a comlpete intersection.

Theorem 3.7. *Let A be an affine ring of dimension d over $\bar{F}_p$. Suppose one of the following conditions hold:*
a) $d = 2$ and A is regular
or

b) $d \geq 3$.

Then $\mathrm{psr}(A) \leq d$.

Proof. We may assume that A is reduced. In view of Th. 3.3, it suffuces
to show that if P is a projective A-module of rank d, then P has a unimodular
element. Let $f \in P^*$ be a generic section of P^* so that $\mathrm{Im}\, f = I$ is a product
of maximal ideals (neccesarily smooth) of height d. Then by Prop. 3.6, I is a
complete intersection. Hence by Lemma 3.4, P has a unimodular element.

Remark 3.8.   There exist two dimensional regular affine rings A over finite
fields which admit indecomposable projective modules of rank 2 (see [3]). For
such rings $\mathrm{sr}(A) = 2$ and $\mathrm{psr}(A) = 3$. When $d = 2$, we do not know if Prop. 3.6
or Th. 3.7 remain valid if the hypothesis that A is regular is dropped.

References

[1] H. Bass, Algebraic K-theory, Benjamin, N.Y. 1968

[2] S. M. Bhatwadeker and A. Roy, Some theorems about projective modules
over polynomial rings, J. Algebra 86 (1984), 150–158.

[3] N. Mohan Kumar, Some theorems on generation of ideals in affine algebras,
Comment. Math. Helvetici 59 (1984), 243–252.

[4] J. S. Milne, Zero cycles on algebraic varities in non-zero characteristic: Rojtman's theorem, Comp. math. 47 (1982), 271–287.

[5] M. P. Murthy and R. G. Swan, Vector bundles over affine surfaces, Invent.
Math. 36 (1976), 125–165.

[6] A. A. Rojtman, The torsion of the group of 0-cycles modulo rational equivalence, Ann. of Math. 46 (1978), 225–236.

[7] R. G. Swan, Serre's Problem, Conference on Commutative Algebra,1975,
Queen's papers on pure and applied Math. No. 42, Kingston, Ontatio,
Canada.

[8] R. G. Swan, A cancellation theorem for projective modules in the metastable
range, Invent. Math. 27 (1974), 23–43.

[9] L. N. Vaserstein and A. A. Suslin, Serre's problem on projective modules over
polynomial rings and algebraic K-theory, Math. USSR Izvestija 10 (1976),
937–1001.

N. Mohan Kumar

Tata Institute of Fundamental Research
Homi Bhabha Road
Bombay, 400005
India

M. Pavaman Murthy

Department of Mathematics
The University of Chicago
Chicago, IL 60637
U.S.A.

A. Roy

Tata Institute of Fundamental Research
Homi Bhabha Road
Bombay, 400005
India

Algebraic Geometry and Commutative Algebra
in Honor of Masayoshi NAGATA
pp. 289–311 (1987)

Semi-ampleness of the Numerically Effective Part
of Zariski Decomposition II

Atsushi MORIWAKI

§0. Introduction.

This paper is the supplement to [Mw1]. In [Mw1], we proved that the canonical ring of a projective variety with $\kappa \leq 2$ is finitely generated. We shall generalize this result to a Kähler manifold in this paper (cf. (3.3)). Of course, this problem is contained in the problem of minimal models. But very little is known about minimal models in Kähler case. Hence the auhtor thinks this result gives a weak evidence of the minimal model conjecture for Kähler manifolds.

§1 is a preliminary of this paper. In this section, we summarize several notations and known results. §2 is devoted to the elementary theory of Zariski decomposition. For lack of suitable numerical theory on complex manifolds, which was constructed by Nakayama [N1] in Kähler case, we shall make an ideal theorical approach to define Zariski decomposition (cf. (2.19)). The basic idea of this approach is due to Hironaka's idealistic exponent. Roughly speaking, first we define the limit of base loci of $H^0(X, \mathcal{O}_X(nD))$, which is denoted by $Z_X(D)$, and Zariski decompositon is defined by a good condition of $Z_X(D)$. Our definition of Zariski decomposition is different from Fujita's sense [F2]. But via Z_X, we can give some sufficient condition for Fujita's Zariski decomposition (cf. (2.33)). In §3, we treat the canonical ring of a compact complex manifold. Main result of this section is a generalization of [Mw1, Theorem (3.1)] to a Kähler manifold (cf. (3.3)). Wilson's example [W, Example 4.3] shows that Kählerianity is crucial for finite generation of the canonical ring of a compact complex manifold with dimension ≥ 4. But we can show the canonical ring of 3-fold is finitely generated by using recent Mori's result [Mo], Fujita's canonical bundle formula [F2] and (3.3) in this paper (cf. (3.5)).

§1. Preliminary.

(1.1) Let X be a normal complex variety of dimension d. We denote the group of Weil divisors (resp. Cartier divisors) by $Z_{d-1}(X)$ (resp. $\mathrm{Div}(X)$). For a field F (e.g., $\mathbf{Q}, \mathbf{R}, \ldots$), we set $Z_{d-1}(X)_F = Z_{d-1}(X) \otimes_{\mathbf{Z}} F$ and $\mathrm{Div}(X)_F =$

Received October 20, 1986.

$\mathrm{Div}(X) \otimes_{\mathbf{Z}} F$ and an element of $\mathbf{Z}_{d-1}(X)_F$ (resp. $\mathrm{Div}(X)_F$) is called an F-divisor (resp. F-Cartier divisor). For an element D of $\mathbf{Z}_{d-1}(X)_{\mathbf{R}}$, the rounding-down, rounding-up and fractional part of D is denoted by $[D]$, $\lceil D \rceil$ and $\{D\}$ respectively. Let D_1 and D_2 be $\mathbf{R}$-divisors. We say D_1 is $\mathbf{Q}$-linear equivalent to D_2, which is denoted by $D_1 \sim_{\mathbf{Q}} D_2$, if there are a positve integer d and a non-zero meromorphic function ϕ on X such that $d(D_1 - D_2) = (\phi)$. If we can take $d = 1$, simply we write $D_1 \sim D_2$. Let $\mathrm{Inv}(X)$ be the set of invetible sheaves on X. $\mathrm{Inv}(X)$ has the abelian group structure if we identify $L \otimes L^{-1}$ with $\mathcal{O}_X$ for any $L \in \mathrm{Inv}(X)$. This abelian group structure is canonically extended to $\mathrm{Inv}(X) \otimes \mathbf{R}$. For $L_1, L_2 \in \mathrm{Inv}(X) \otimes \mathbf{R}$, we denote the sum of L_1 and L_2 by $L_1 \otimes L_2$. If the natural homomorphism $\mathrm{Div}(X)_{\mathbf{Q}} \rightarrow \mathbf{Z}_{d-1}(X)_{\mathbf{Q}}$ is an isomorphism, we say X is $\mathbf{Q}$-factorial. Let ω_X be $(-d)$-th cohomology of the dualizing complex of X. It is well known that ω_X is reflexive. We say that X has only canonical singularities if there are a positive number d and a proper desingularization $f\colon Y \rightarrow X$ such that $(\omega_X^{\otimes d})^{**}$ is invertible and that $\omega_Y^{\otimes d} = f^*(\omega_X^{\otimes d})^{**} \otimes \mathcal{O}_Y(\sum_i a_i E_i)$ for some non-negative integers a_i, where $**$ is double dual and $\sum_i E_i$ is the exceptional locus of f. Let Δ be a $\mathbf{Q}$-divisor on X such that $[\Delta] = 0$. Furthermore, we say the pair (X, Δ) has only log-terminal singularities if there are a positive number d and a proper desingularization $f\colon Y \rightarrow X$ such that $d\Delta$ is an usual divisor, $(\omega_X^{\otimes d} \otimes \mathcal{O}_X(d\Delta))^{**}$ is invertible and that $\omega_X^{\otimes d} = f^*((\omega_X^{\otimes d} \otimes \mathcal{O}_X(d\Delta))^{**}) \otimes \mathcal{O}_Y(\sum_i a_i E_i)$ for some integers a_i with $a_i > -d$, where $\sum_i E_i$ is the support of the union of the exceptional locus of f and $f^{-1}(\mathrm{Supp}(\Delta))$.

(1.2) Here we assume X is compact. For a coherent reflexive sheaf L of rank one, we set $R(X, L) = \bigoplus_{m=0}^{\infty} H^0(X, (L^{\otimes d})^{**})$. A L-dimension $\kappa(X, L)$ of L is defined by

$$\kappa(X, L) = \begin{cases} -\infty & \text{if } R(X, L) = \mathbf{C}, \\ \text{trans. } \deg_{\mathbf{C}}(\text{quotient field of } R(X, L)) - 1 & \text{otherewise.} \end{cases}$$

Moreover, for an $\mathbf{R}$-divisor D, we set $R(X, D) = \bigoplus_{m=0}^{\infty} H^0(X, \mathcal{O}_X([mD]))$ and $\kappa(X, D)$, which is called D-dimension of D, is defined similarily as above. Note that if $\kappa(X, D) \geq 0$, there is an effective $\mathbf{R}$-divisor D' such that $D \sim_{\mathbf{Q}} D'$. D is called big if $\kappa(X, D) = \dim X$. For an invertible sheaf L on X, we set $I(X, L) = \mathrm{Image}(H^0(X, L) \otimes L^{-1} \rightarrow \mathcal{O}_X)$. For a $\mathbf{Q}$-Cartier divisor D on X, if there is a positive integer m such that mD is a Cartier divisor on X and $I(X, \mathcal{O}_X(mD)) = \mathcal{O}_X$, we say D is semi-ample.

(1.3) Let X be a compact Kähler manifold, $\mathrm{KC}(X)$ the Kähler cone of X and $\overline{\mathrm{KC}}(X)$ a closure of $\mathrm{KC}(X)$ in $H^{1,1}(X, \mathbf{R})$. Let D be an $\mathbf{R}$-divisor on X. We say D is quasi-nef if there is a bimeromorphic morphism $\mu\colon Y \rightarrow X$ of compact

Kähler manifolds such that the first chern class $c_1(\mu^*(D))$ of $\mu^*(D)$ is contained in $\overline{KC}(Y)$. For quasi-nef **R**-divisor D, we define

$$c(X, D) = \max\{d \geq 0 \mid (c_1(D))^d \in H^{d,d}(X, \mathbf{R}) \setminus \{0\}\}.$$

Then it is easy to see that $\kappa(X, D) \leq c(X, D)$ (cf. [N1, Proposition 2.10]). If $\kappa(X, D) = c(X, D)$, we say D is *good*. If a compact complex manifold is bimeromorphic to a compact Kähler manifold, we say this manifold is in class C according to Fujiki. The above notions can be defined on a manifold in class C by the same way. Since Lemma 2.8 in [N1] is valid for **R**-divisor, we have the following by similar arguement of [N1].

(1.3.1) Let D be a quasi-nef and good **R**-divisor on a compact complex manifold X in class C. Then there are a bimeromorphic morphism $\mu \colon Y \to X$ from a compact Kähler manifold Y to X, a fiber space $h \colon Y \to Z$ from Y to a non-singular projective variety Z and a nef and big **R**-divisor M on Z such that $\mu^*(D) \sim_{\mathbf{Q}} h^*(M)$.

Key tools of §2 are the following vanishing theorem (1.3.2) and non-vanishing theorem (1.3.3). We refer the reader to [K1] or [N1] for the notion concerning generalized normal crossing varieties.

(1.3.2) (cf. [N1, Theorem 3.1]) Let L be a quasi-nef and good **Q**-divisor on a compact Kähler manifold X and D an effective divisor on X. Assume that $\{L\}_{\text{red}}$ has only normal crossings and there is an injection $\mathcal{O}_X(D) \to \mathcal{O}_X(mL)$ for some positive integer m. Then the natural homomorphisms induced by $\mathcal{O}_X \to \mathcal{O}_X(D)$,

$$H^i(X, \mathcal{O}_X(\lceil L \rceil) \otimes \omega_X) \longrightarrow H^i(X, \mathcal{O}_X(\lceil L \rceil + D) \otimes \omega_X)$$

are injective for all $i \geq 0$.

(1.3.3) (cf. [N1, Theorem 5.4]) Let X be a compact Kähler generalized normal crossing variety, $f \colon X \to Z$ a surjective morphism form X to a projective variety Z and D an element of $\text{Div}_0(X)$. We assume the following.

(1) For all $n \geq 0$, every connected component of X_n is mapped surjectively to Z by f.

(2) There exists an element A of $\text{Div}_0(X) \otimes \mathbf{Q}$ such that the support of A is a generalized normal crossing divisor on X and $\lceil A \rceil \geq 0$.

(3) There is a nef Cartier divisor D_0 on Z such that $\mathcal{O}_X(dD) \simeq f^*(\mathcal{O}_Z(D_0))$ for some positive integer d.

(4) There is an ample Cartier divisor H_0 on Z such that $\mathcal{O}_X(q(D + A)) \otimes \omega_X^{-q} \simeq f^*(\mathcal{O}_Z(H_0))$ for some positive integer q.

Then there exist positive integers p and t_0 such that

$$H^0(X, \mathcal{O}_X(ptD + \lceil A \rceil)) \neq 0 \quad \text{for all integer } t \geq t_0.$$

(1.4) Here we review Fujita's canonical bundle formula (cf. [F2]). Let $f: X \to S$ be a proper fiber space of complex manifolds, which are not necessarily compact, such that a general fiber of f is an elliptic curve. This fibration is called an *elliptic fibration*. Set $\Sigma = \{ s \in S \mid f$ is not smooth over $s \}$. Then we have a J-invariant $J: S \setminus \Sigma \to \mathbf{P}^1$, which is meromorphic on S. Here we assume J is holomorphic on S by changing S. Set $S_0 = S \setminus \Sigma$, $X_0 = f^{-1}(S_0)$ and $f_0 = f|_{X_0}$. By using a cusp form of weight 6, we have an isomorphism $\lambda^0 : f_{0*}(\omega_{X_0}^{\otimes 12}) \simeq \omega_{S_0}^{\otimes 12}$. The problem is how λ^0 can be extended to S. Let $\{\Gamma_i\}_{i \in I}$ be a set of irreducible components of Σ such that $\operatorname{Codim}(\Gamma_i) = 1$. Let $\iota: D \to S$ be a general embedding of a unit disk D such that $\iota(D)$ meets Γ_i transversally, $D \times_S X \to D$ is smooth over D^* and that $D \times_S X$ is non-singular. And let $V \to D$ be a relative minimal model of $D \times_S X \to D$. Then it is easy to see that Kodaira type of singular fiber of $V \to D$ is independent of the choice of $\iota: D \to S$. Hence we can define α_i for each $i \in I$ by the following table.

Type	$_mI_b$	I^*	II	II^*	III	III^*	IV	IV^*
α	$1 - m^{-1}$	$1/2$	$1/6$	$5/6$	$1/4$	$3/4$	$1/3$	$2/3$

Let m_i be a multiplicity of fibers around Γ_i and m a positive integer such that $k = 12m$ is divisible by all m_i. Then $k\alpha_i$ is integer for every $i \in I$.

(1.4.1) (**Fujita's canonical bundle formula**) $(\lambda^0)^{\otimes m}$ can be extended to a homomorphism;

$$\lambda_m : f_*(\omega_X^{\otimes k}) \longrightarrow \omega_Y^{\otimes k} \otimes \mathcal{O}_S(\sum_{i \in I} k\alpha_i \Gamma_i) \otimes J^*(\mathcal{O}_{\mathbf{P}}(mp_\infty))$$

such that λ_m is an isomorphism in codimension one, where p_∞ is the infinite point of $\mathbf{P}^1$.

Fix m and set $Q = \omega_S^{\otimes k} \otimes \mathcal{O}_S(\sum_{i \in I} k\alpha_i \Gamma_i) \otimes J^*(\mathcal{O}_{\mathbf{P}}(mp_\infty))$. Let F be an effective divisor on X such that

$$\omega_X^{\otimes k} \otimes \mathcal{O}_X(-F) = \{\operatorname{Image}(f^* f_* \omega_X^{\otimes k} \longrightarrow \omega_X^{\otimes k})\}^{**}.$$

Then $f^* f_* \omega_X^{\otimes k} \to f^* Q$ induces an injective homomorpism $\omega_X^{\otimes k} \otimes \mathcal{O}_X(-F) \to f^* Q$. Therefore we have an effective divisor G on X such that $\omega_X^{\otimes k} \otimes \mathcal{O}_S(G - F) \simeq f^* Q$. Then we have;

(1.4.2) G is f-exceptional and λ_{tm} induces an isomorphism

$$\tilde{\lambda}_{tm} : f_*(\omega_X^{\otimes tk} \otimes \mathcal{O}_X(tG)) \simeq Q^{\otimes t}$$

for every $t \geq 0$.

Proof of (1.4.2).   Let U be the maximal open set of S such that λ_m is isomorphic over U. Then $\omega_X^{\otimes k} \otimes \mathcal{O}_X(-F) \to f^* Q$ is isomorphic over $f^{-1}(U)$.

Therefore $f(G)$ is contained in $X \setminus U$. Hence G is f-exceptional. An iso-morphism $\omega_X^{\otimes k} \otimes \mathcal{O}_X(G) \simeq f^* Q \otimes \mathcal{O}_X(F)$ induces $\eta_t \colon f_*(\omega_X^{\otimes tk} \otimes \mathcal{O}_X(tG)) \simeq Q^{\otimes t} \otimes f_*(\mathcal{O}_X(tF))$. Since G is f-exceptional, by (1.4.1), there is an injective homomorphism $\tilde{\lambda}_{tm} \colon f_*(\omega_X^{\otimes tk} \otimes \mathcal{O}_X(tG)) \to Q^{\otimes t}$. We have $j \circ \tilde{\lambda}_{tm} = \eta_t$ because $j \circ \tilde{\lambda}_{tm}$ is equal to η_t over S_0 and $Q^{\otimes t} \otimes f_*(\mathcal{O}_X(tF))$ is torsion free, where $j \colon Q^{\otimes t} \to Q^{\otimes t} \otimes f_*(\mathcal{O}_X(tF))$ is the natural inclusion. Therefore $\tilde{\lambda}_{tm}$ is an iso-morphism for every $t \geq 0$. $\qquad$ q.e.d.

(1.4.3) By the contruction of $\tilde{\lambda}_{tm}$, we have an isomorphism

$$\bigoplus_{t=0}^{\infty} H^0(\tilde{\lambda}_{tm}) \colon \bigoplus_{t=0}^{\infty} H^0(X, \omega_X^{\otimes tk} \otimes \mathcal{O}_X(tG)) \longrightarrow \bigoplus_{t=0}^{\infty} H^0(S, Q^{\otimes t})$$

as $\mathbf{C}$-algebra.

§2. Zariski decomposition.

In this section, we shall define Zariski decomposition by an ideal theorical approach and discuss elementary properties of Zariski decomposition. Though results of this section are almost valid for the relative case by slight modifica-tions, we shall treat only the absolute case for the simplicity.

(2.1) Let X be a normal complex variety. We set;

$$I(X) = \{\text{non-zero } \mathcal{O}_X\text{-coherent sub-sheaves of } \mathcal{M}_X\},$$

where $\mathcal{M}_X$ is the sheaf of meromorphic functions on X. A map $v \colon I(X) \to \mathbf{Z}$ is called a *valuation* on $I(X)$ if the following conditions are satisfied.

(1) $v(I \cdot J) = v(I) + v(J)$ for all $I, J \in I(X)$,

(2) $v(I + J) \geq \min\{v(I), v(J)\}$ for all $I, J \in I(X)$.

Let $\mu \colon Y \to X$ be a proper bimeromorphic morphism from a normal complex variety Y to X and Γ a prime divisor on Y. Set $U = (Y \setminus \mathrm{Sing}(Y)) \cap (\Gamma \setminus \mathrm{Sing}(\Gamma))$. Let y be a point of U, t a local equation of Γ at y and R a localization of $\mathcal{O}_{Y,y}$ with respect to $t\mathcal{O}_{Y,y}$. Then for an element I of $I(X)$, there is an integer v such that $IR = t^v R$. It is easy to see that this v is independent of choice of y and t. Hence we have a map $v \colon I(X) \to \mathbf{Z}$, which is clearly valuation on $I(X)$. We call this type of valuation a *geometric valuation* on $I(X)$ and set $\mathrm{RS}(X) = \{\text{geometric valuations on } I(X)\}$. Let C be a subvariety of X, $\mu \colon Y \to X$ be a normalized blowing-up along C and Γ a exceptional prime divisor of μ. We denote the valuation on $I(X)$ with respect to Γ by $v_{C,\Gamma}$. If $(C \setminus \mathrm{Sing}(C)) \cap (X \setminus \mathrm{Sing}(X)) \neq \phi$, there is the unique prime divisor Γ such that

$\mu(\Gamma) = C$. In this case, we write $v_{C,\Gamma}$ by v_C. Let E be an arbitrary set. We write the set of formal finite sum $\sum_{e \in E} n_e \cdot e$, $n_e \in \mathbf{Z}$, by $\mathbf{Z}[E]$. We can define the natural pairing;

$$(2.1.1) \qquad\qquad <\,,\,>: \mathbf{Z}[\mathrm{RS}(X)] \otimes_{\mathbf{Z}} \mathbf{Z}[\mathrm{I}(X)] \longrightarrow \mathbf{Z}$$

by $v(I)$. For elements a_1 and a_2 of $\mathbf{Z}[\mathrm{I}(X)]$, if $<v, a_1> \,=\, <v, a_2>$ for all $v \in \mathrm{RS}(X)$, we denote $a_1 \equiv a_2$. We set;

$$\mathrm{IE}(X) = \mathbf{Z}[\mathrm{I}(X)]/\equiv \,.$$

We write the class of I of $\mathrm{IE}(X)$ by $Cl(I)$.

(2.2) Next we give another description of $\mathrm{IE}(X)$. Let I be an element of $\mathrm{I}(X)$. The integral closure $\overline{I}$ of I is locally defined by

$$\overline{I}_x = \left\{ y \in \mathcal{M}_{X,x} \;\middle|\; \begin{array}{l} \text{There are a positive integer } n \text{ and} \\ a_i \in (I_x)^i \ (i = 1, \ldots, n) \text{ such that} \\ y^n + a_1 y^{n-1} + \cdots + a_{n-1} y + a_n = 0. \end{array} \right\}$$

for all $x \in X$. For I_1 and I_2 of $\mathrm{I}(X)$, we denote $I_1 \equiv' I_2$ if $\overline{I}_1 = \overline{I}_2$. Since $I_1 \cdot J_1 \equiv' I_2 \cdot J_2$ if $I_1 \equiv' I_2$ and $J_1 \equiv' J_2$, $\mathrm{I}(X)/\equiv'$ has the semi-module structure. Let $K(\mathrm{I}(X)/\equiv')$ be the Grothendieck module of $\mathrm{I}(X)/\equiv'$. Then it is easy to see that $K(\mathrm{I}(X)/\equiv') \simeq \mathrm{IE}(X)$ by the following proposition.

(2.3) Proposition (cf. [H, Theorem 7.5]) *Let I and J be elements of $\mathrm{I}(X)$. Then the following conditions are equivalent.*
  (1) *$\overline{I} = \overline{J}$.*
  (2) *There is a proper bimeromorphic morphism $\pi \colon W \to X$ from a normal complex varieties such that $I\mathcal{O}_W$ is invertible and $I\mathcal{O}_W = J\mathcal{O}_W$.*
  (3) *For every $v \in \mathrm{RS}(X)$, $v(I) = v(J)$.*

(2.4) Let $\mathrm{IE}^d(X)$ (resp. $\mathrm{IE}^{nd}(X)$) be a submodule of $\mathrm{IE}(X)$ generated by $Cl(I)$ such that $I^{**} = I$, $I \in \mathrm{I}(X)$ (resp. $I^{**} = \mathcal{O}_X$), where ** is double dual. Then it is easy to see that $\mathrm{IE}(X) = \mathrm{IE}^d(X) \oplus \mathrm{IE}^{nd}(X)$ if X is smooth. In this case, we denote the natural projection $\mathrm{IE}(X) \to \mathrm{IE}^d(X)$ by p^d. There is an injective homomorphism $\iota_X \colon \mathrm{Div}(X) \to \mathrm{IE}^d(X)$ defined by $\iota_X(D) = Cl(\mathcal{O}_X(D))$. If X is smooth, then ι_X is an isomorphism.

(2.5) For a field F (e.g., $\mathbf{Q}$, $\mathbf{R}$, ...), we set $\mathrm{IE}_F(X) = \mathrm{IE}(X) \otimes_{\mathbf{Z}} F$, $\mathrm{IE}_F^d(X) = \mathrm{IE}^d(X) \otimes_{\mathbf{Z}} F$ and $\mathrm{IE}_F^{nd}(X) = \mathrm{IE}^{nd}(X) \otimes_{\mathbf{Z}} F$. Then by (2.4), $\mathrm{IE}_F(X) = \mathrm{IE}_F^d(X) \oplus \mathrm{IE}_F^{nd}(X)$ if X is smooth. Note that $\mathrm{IE}_{\mathbf{Q}}(X)$ is nothing more than the Grothendieck module of {idealistic exponents in the sense of [H]} \ {zero ideal}. For an $\mathbf{R}$-divisor D on X, the image of D by $\iota_X \otimes 1 \colon \mathrm{Div}(X)_{\mathbf{R}} \to \mathrm{IE}_{\mathbf{R}}^d(X)$ is

denoted by $\iota_X(D)$ by an abuse of notation. The pairing of 2.1.1 is naturally extended to
$$< , >: \mathbf{R}[\mathrm{RS}(X)] \otimes_{\mathbf{R}} \mathrm{IE}_{\mathbf{R}}(X) \longrightarrow \mathbf{R}.$$
For $v \in \mathrm{RS}(X)$, $|<v, >|$ defines a semi-norm on $\mathrm{IE}_{\mathbf{R}}(X)$. Hence a family of semi-norms $\{|<v, >|\}_{v \in \mathrm{RS}(X)}$ gives a topological vector space structure on $\mathrm{IE}_{\mathbf{R}}(X)$. From now on, we fix this topology on $\mathrm{IE}_{\mathbf{R}}(X)$. Clearly this topology is Hausdorff. Let $\overline{\mathrm{IE}}_{\mathbf{R}}(X)$ be the completion of $\mathrm{IE}_{\mathbf{R}}(X)$ (cf. [T, I.5]). And let $\overline{\mathrm{IE}}_{\mathbf{R}}^d(X)$ (resp. $\overline{\mathrm{IE}}_{\mathbf{R}}^{nd}(X)$) be the closure of $\mathrm{IE}_{\mathbf{R}}^d(X)$ (resp. $\mathrm{IE}_{\mathbf{R}}^{nd}(X)$). Then $\overline{\mathrm{IE}}_{\mathbf{R}}(X) = \overline{\mathrm{IE}}_{\mathbf{R}}^d(X) \oplus \overline{\mathrm{IE}}_{\mathbf{R}}^{nd}(X)$ if X is smooth. In this case, by an abuse of notation, the projection $\overline{\mathrm{IE}}_{\mathbf{R}}(X) \to \overline{\mathrm{IE}}_{\mathbf{R}}^d(X)$ is denoted by p^d. We set $\overline{\mathrm{IE}}_{\mathbf{R}}^+(X) = \{ x \in \overline{\mathrm{IE}}_{\mathbf{R}} \mid <v,x> \geq 0 \text{ for all } v \in \mathrm{RS}(X) \}$. Then $\overline{\mathrm{IE}}_{\mathbf{R}}^+(X)$ is a closed and normal cone of $\overline{\mathrm{IE}}_{\mathbf{R}}^+(X)$, where normality means that if $x_1, x_2 \in \overline{\mathrm{IE}}_{\mathbf{R}}^+(X)$ and $x_1 + x_2 = 0$, $x_1 = x_2 = 0$. For elements $x_1, x_2 \in \overline{\mathrm{IE}}_{\mathbf{R}}(X)$, we write $x_1 \geq x_2$ if $x_1 - x_2 \in \overline{\mathrm{IE}}_{\mathbf{R}}^+(X)$.

(2.6) Let $f: Y \to X$ be a proper surjective morphism of normal complex varieties. Then we have a homomorphism $f^*: \mathrm{IE}_{\mathbf{R}}(X) \to \mathrm{IE}_{\mathbf{R}}(Y)$ defined by $f^*(Cl(I)) = Cl(I\mathcal{O}_Y)$.

(2.7) Proposition *Notation being as above, first we have*
(1) $f^*: \mathrm{IE}_{\mathbf{R}}(X) \to \mathrm{IE}_{\mathbf{R}}(Y)$ *is continuous.*
By (1), f^* can be extended to $\overline{\mathrm{IE}}_{\mathbf{R}}(X) \to \overline{\mathrm{IE}}_{\mathbf{R}}(Y)$ *which is denoted by f^* by an abuse of notation. Then, we have*
(2) $f^*: \overline{\mathrm{IE}}_{\mathbf{R}}(X) \to \overline{\mathrm{IE}}_{\mathbf{R}}(Y)$ *is injective.*

Proof. Let $\{a_n\}$ be a sequence in $\mathrm{IE}_{\mathbf{R}}(X)$ such that $\lim_{n \to \infty} a_n = 0$ and v an element of $\mathrm{RS}(Y)$. By flatting of morphism and normalization, we have the following commutative diagram satisfing conditions (a), (b) and (c).

$$
\begin{array}{ccc}
Y & \xleftarrow{\mu} & Y' \\
\downarrow{f} & & \downarrow{f'} \\
X & \xleftarrow{\tau} & X'
\end{array}
$$

(a) X' and Y' are normal complex varieties.
(b) μ and τ are proper bimeromorphic morphism and f' is proper and equi-dimensional.
(c) There is a prime divisor Γ on Y' such that v is determined by Γ.

Let Γ' be an image of Γ by f'. Since $<v, f^*(a_n)> = 0$ for all n in the case $\Gamma' = X$ and f' is equi-dimensional, we may assume Γ' is a prime divisor on X'. Let v' be a valuation on $\mathrm{I}(X)$ determined by Γ'. Let y' and x' be general points of Γ and Γ' such that $f'(y') = x'$, t and s local equations of Γ and Γ', and let R and R' be localization of $\mathcal{O}_{Y',y'}$ and $\mathcal{O}_{X',x'}$ with respect to $t\mathcal{O}_{Y',y'}$ and $s\mathcal{O}_{X',x'}$ respectively. We set $sR = t^c R$ for some positive integer c. Then clearly $<v', a_n> = c <v, f^*(a_n)>$. Therefore $\lim_{n\to\infty} <v, f^*(a_n)> = 0$. Hence f^* is continuous. Injectivity of f^* can be proved by the similar way as above.   q.e.d.

(2.8) Let X be a **Q**-factorial normal compact complex variety and D an effective **R**-divisor on X. Let $D = \sum_{i\in I} a_i D_i$ be an irreducible decomposition of D, and N a positive integer such that $N \cdot D_i \in \text{Div}(X)$ for all $i \in I$. We set $I_n = I(X, \mathcal{O}_X(N[nD]))$. (For the definition of $I(X, \mathcal{O}_X(N[nD]))$, see (1.2).) Then, considering a sequence $\{(1/n)Cl(I_n)\}_{n\in\mathbf{N}}$ in $\text{IE}_{\mathbf{R}}(X)$, we have the following lemma.

(2.9) **Lemma** $\{(1/n)Cl(I_n)\}_{n\in\mathbf{N}}$ *is a Cauchy sequence in* $\text{IE}_{\mathbf{R}}(X)$.

Proof. If we set $I_{n,m} = I(X, \mathcal{O}_X(N([nD]+[mD])))$, we have that

$$I_n \cdot I_m \subset I_{n,m},$$
$$I_{n,m} \cdot \mathcal{O}_X(N([nD]+[nD]-[(n+m)D])) \subset I_{n+m}.$$

Hence $a_n + a_m \geq a_{n+m}$ for all $v \in \text{RS}(X)$, where $a_n = <v, I_n> + <v, \iota_X(N([nD]))>$. Therefore by [PS, 98], the limit of a sequence $\{(1/n)a_n\}$ exists and its limit equals to $\inf_{n\to\infty} (1/n)a_n$. Since it is easy to see that

$$<v, (1/n)Cl(I_n)> = (1/n)a_n - <v, (1/n)\iota_X(N([nD]))>,$$
$$\lim_{n\to\infty} <v, (1/n)\iota_X(N([nD]))> = <v, \iota_X(ND)>,$$

the limit of the sequence $\{<v, (1/n)Cl(I_n)>\}$ exists.   q.e.d.

(2.9.1) The sequence $\{(1/(N\cdot n))Cl(I_n)\}$ has the limit in $\overline{\text{IE}}_{\mathbf{R}}(X)$, by this lemma. Since

$$I(X, \mathcal{O}_X(N[nD]))^{N'} \subset I(X, \mathcal{O}_X(NN'[nD])) \quad \text{and}$$
$$I(X, \mathcal{O}_X(NN'[nD])) \cdot \mathcal{O}_X(NN'[nD] - N[nN'D]) \subset I(X, \mathcal{O}_X(N[nN'D])),$$

$\lim_{n\to\infty} (1/(N\cdot n))Cl(I_n)$ is independent of the choice of N. Hence we denote its limit by $Z_X(D)$. Let D' be an **R**-divisor on X with non-negative D'-dimension. Then there is an effective **R**-divisor E on X such that $D' \sim_{\mathbf{Q}} E$. It is easy to see that $Z_X(E)$ is independent of the choice of E. Thus we can well define $Z_X(D')$ for an **R**-divisor D' with non-negative D'-dimension.

(2.10) Lemma Let D_i be an $\mathbf{R}$-divisor on X such that $\kappa(X, D_i) \geq 0$ ($i = 1, 2$). Then we have
(1) $Z_X(D_1 + D_2) \leq Z_X(D_1) + Z_X(D_2)$ and
(2) if D_1 is effective, $Z_X(D_1) \leq \iota_X(-D_1)$.

Proof. Clearly we may assume that D_i is effective ($i = 1, 2$). We set $I_{i,n} = I(X, \mathcal{O}_X(N[nD_i]))$ ($i = 1, 2$) and $J_n = I(X, \mathcal{O}_X(N[n(D_1 + D_2)]))$. Then

$$I_{1,n} \cdot I_{2,n} \subset I(X, \mathcal{O}_X(N([nD_1] + [nD_2]))),$$

$$I(X, \mathcal{O}_X(N([nD_1] + [nD_2]))) \cdot \mathcal{O}_X(N([nD_1] + [nD_2] - [n(D_1 + D_2)]))) \subset J_n.$$

Hence we have (1). (2) is obvious. *q.e.d.*

(2.11) Propositon Let $f: X \to Y$ be a fiber space of $\mathbf{Q}$-factorial compact normal complex spaces. Then $Z_X(f^*(D)) = f^*(Z_Y(D))$ for any $\mathbf{R}$-divisor D on Y with non-negative D-dimension.

Proof. Clearly we may assume D is effective. For a sufficiently large integer N, we set $I_n = I(X, \mathcal{O}_X(N[nf^*(D)]))$ and $I'_n = I(X, \mathcal{O}_X(Nf^*[nD]))$. Since $f_*\mathcal{O}_X = \mathcal{O}_Y$, we have $I'_n = I(Y, \mathcal{O}_Y(N[nD]))\mathcal{O}_X$. Hence $\lim_{n \to \infty} (1/n)Cl(I')_n = N \cdot f^*(Z_Y(D))$. On the other hand, there exists an effective divisor E on X and exists a positive integer n_0 such that $0 \leq [nf^*D] - f^*[nD] \leq E \leq f^*[n_0 D]$ for all $n \geq 0$. Thus we have the following.

$$I'_n \cdot \mathcal{O}_X(N(f^*[nD] - [nf^*D])) \subset I_n,$$
$$I_n \cdot \mathcal{O}_X(N([nf^*D] - f^*[nD] - E)) \subset I(X, \mathcal{O}_X(N(E + f^*[nD]))),$$
$$I(X, \mathcal{O}_X(N(E + f^*[nD]))) \cdot$$
$$\mathcal{O}_X(N(f^*[nD] + E - f^*[(n + n_0)D])) \subset I'_{n+n_0}.$$

Therefore for all $v \in \mathrm{RS}(X)$, $\lim_{n \to \infty} <v, (1/n)Cl(I_n)> = \lim_{n \to \infty} <v, (1/n)Cl(I'_n)>$. Hence $\lim_{n \to \infty} (1/n)Cl(I_n) = \lim_{n \to \infty} (1/n)Cl(I'_n)$. *q.e.d.*

(2.12) Proposition Let X be a $\mathbf{Q}$-factorial normal compact complex variety and D an effective $\mathbf{R}$-divisor on X. Then there are a bimeromorphic morphism $\mu: Y \to X$ from a compact complex manifold Y to X, a fiber space $g: Y \to W$ from Y to a smooth projective variety W with $\kappa(X, D) = \dim W$, and a big $\mathbf{R}$-divisor M on W such that $E = \mu^*(D) - g^*(M)$ is effective and $Z_Y(\mu^*(D)) = Z_Y(g^*(M)) + \iota_Y(-E)$. Furthermore, if D is a $\mathbf{Q}$-divisor, then M is also $\mathbf{Q}$-divisor.

Proof. For sufficiently large integer m, by resolution of singularities, flattening of morphism and normalization, we can construct the following commutative diagram satisfing conditions (1), (2) and (3).

$$
\begin{array}{ccccc}
X & \xleftarrow{\mu_1} & X_1 & \xleftarrow{\mu_2} & Y \\
\downarrow{\phi_m} & & \downarrow{g_1} & & \downarrow{g} \\
\text{Image}(\phi_m) = W_0 & \xleftarrow{\tau} & W & \xleftarrow{id_W} & W
\end{array}
$$

(1) Y and W are compact complex manifolds and X_1 is a normal compact complex variety.

(2) μ_1, μ_2 and τ is bimeromorphic morphisms and g_1 is a equi-dimensional fiber space.

(3) $H^0(X, \mathcal{O}_X([nD])) \subset \mathbf{C}(W_0)$ for all n.

Let M be the maximal $\mathbf{R}$-divisor on W such that $\mu_1^*(D) - g_1^*(M)$ is effective. Note that if D is a $\mathbf{Q}$-divisor, M is also $\mathbf{Q}$-divisor. To prove our proposition, it is sufficient to show that $H^0(W, \mathcal{O}_W([nM])) \to H^0(Y, \mathcal{O}_Y([\mu^*(nD)]))$ is bijective for every n, where $\mu = \mu_1 \circ \mu_2$. Let ψ be an element of $H^0(Y, \mathcal{O}_Y([\mu^*(nD)]))$. Then since $[\mu^*(nD)] - \mu^*([nD])$ is μ-exceptional, $(\psi)_Y + \mu^*([nD]) \geq 0$ on Y. Therefore by condition (3), $\psi \in \mathbf{C}(W)$. We set $(\psi)_W + nM = \sum_i n_i \Gamma_i$ on W, where $\Gamma_i's$ are prime divisors on W. Then $\mu_1^*(D) \geq g_1^*(M - \sum_i (n_i/n)\Gamma_i)$ since $(\psi)_{X_1} + \mu_1^*([nD]) \geq 0$. By using equi-dimensionality of g_1 and maximality of M, n_i is non-negative for all i. Hence $(\psi)_W + [nM] \geq 0$.   q.e.d.

(2.13) Lemma *Let X be a compact complex manifold, D an $\mathbf{R}$-divisor on X with non-negative D-dimension and C a curve on X. If $< v_C, Z_X(D)> = 0$, then $(D \cdot C) \geq 0$.*

Proof. Clearly we may assume that D is effective and C is smooth by embedded resolution of singularities. Let $\mu: Y \to X$ be a blowing-up along C and E an exceptional set of μ. Let $v_n = <v_C, I(X, \mathcal{O}_X([nD]))>$ and let D_n be a general element of $|[nD]|$. Then $\mu^*(D_n) = \overline{D}_n + v_n E$, where $\overline{D}_n$ is the strict transformation of D_n. Since E is projective, there is a very ample divisor A on E. Taking general element $A_1, \ldots, A_{e-1} \in |A|$, we set $F = A_1 \cap \cdots \cap A_{e-1}$ and $d = \deg(\mu|_F)$, where $e = \dim E$. Then

$$
0 \leq (\overline{D}_n \cdot F) = nd(D \cdot C) - d(\{nD\} \cdot C) - v_n(E \cdot F).
$$

Hence $(D \cdot C) \geq (1/n)(\{nD\} \cdot C) + (v_n/nd)(E \cdot F)$. Taking the limit of right hand, we have $(D \cdot C) \geq 0$.   q.e.d.

(2.14) Theorem *Let X be a $\mathbf{Q}$-factorial normal compact complex variety in class $\mathcal{C}$, and D an $\mathbf{R}$-Cartier divisor on X with non-negative D-dimension. Then the following are equivalent.*

(1) $Z_X(L) = 0$.

(2) L *is quasi-nef and good.*

Proof. Clearly we may assume that L is effective. First we suppose that $Z_X(L) = 0$. By Proposition (2.12), there are a bimeromorphic morphism $\mu\colon Y \to X$, a fiber space $g\colon Y \to W$ from Y to a non-singular projective variety W, and a big **R**-divisor M on W such that $E = \mu^*(L) - g^*(M)$ is effective and $Z_Y(\mu^*(L)) = Z_Y(g^*(M)) + \iota_Y(-E)$. Since $Z_Y(\mu^*(L)) = \mu^*(Z_Y(L)) = 0$, we have $\mu^*(L) = g^*(M)$ and $Z_Y(g^*(M)) = g^*(Z_W(M)) = 0$ by the normality of $\overline{\text{IE}}_{\mathbf{R}}^{+}(X)$. Due to Proposition (2.7), $Z_W(M) = 0$. Therefore we have M is nef and big by Lemma (2.13). Hence L is quasi-nef and good.

Next we suppose that L is quasi-nef and good. Then by (1.3.1), there are a bimeromorphic morphism $\mu\colon Y \to X$ from a compact complex manifold Y to X, a fiber space $g\colon Y \to W$ from Y to a non-singular projective variety W, and a nef and big **R**-divisor M on W such that $\mu^*(L) = g^*(M)$. Hence $\mu^*(Z_X(L)) = g^*(Z_W(M))$ by Proposition (2.11). Thus it is sufficient to prove the following lemma.

(2.15) Lemma *Let X be a non-singular projective variety and D a nef and big **R**-divisor on X. Then $Z_X(D) = 0$.*

Proof. Let H be an ample divisor on X. Since we can take a positive integer d such that $H^0(X, \mathcal{O}_X([dD] - H)) \neq 0$, we have a decomposition $D = A_0 + \Gamma_0$ suct that A_0 is an ample **Q**-divisor and Γ_0 is an effective **R**-divisor. Set $A_0 = \sum_{i \in I} a_i E_i$ and $\Gamma_0 = \sum_{i \in I} s_i E_i$. Since $A_0 + (1 - (1/n))\Gamma_0$ is ample for all $n \geq 1$, there are rational numbers $a_i^{(n)}$ such that $0 \leq a_i + (1 - (1/n))s_i - a_i^{(n)} \leq (1/n)$ ($i \in I$) and that $A_n = \sum_{i \in I} a_i^{(n)} E_i$ is ample. We set $\Gamma_n = \sum_{i \in I}(a_i + s_i - a_i^{(n)})E_i$. Let c_n's be positive integers such that $c_n A_n$ is very ample and $\lim_{n \to \infty} c_n = \infty$. Since $D = A_n + \Gamma_n$ and $I(X, \mathcal{O}_X(c_n A_n)) = \mathcal{O}_X$, we have

$$I(X, \mathcal{O}_X([c_n D])) \supset I(X, \mathcal{O}_X([c_n \Gamma_n])) \supset \mathcal{O}_X(-[c_n \Gamma_n]).$$

Hence for all $v \in \text{RS}(X)$,

$$0 \leq \, <v, (1/c_n)Cl(I(X, \mathcal{O}_X([c_n D])))> \, \leq \, <v, \iota_X(-\Gamma_n)>.$$

Thus $<v, Z_X(D)> = 0$ because $\lim_{n \to \infty} \Gamma_n = 0$.   *q.e.d.*

(2.16) Let D be an **R**-divisor on a compact complex manifold X such that $\kappa(X, D) \geq 0$. We consider a decomposition $D = P + N$ such that P is an **R**-divisor on X with non-negative P-dimension and N is an effective **R**-divisor. This decomposition is called a *sectional decomposition* if $Z_X(D) = Z_X(P) + \iota_X(-N)$. P (resp. N) is called the *positive part* (resp. *negative part*) of this decomposition. For any **R**-divisor D with non-negative D-dimension, it is easy to see that $p^d(Z_X(D)) \in \text{IE}_\mathbf{R}^d(X)$. Hence there is an effective $\mathbf{R}$-divisor N_c such that $p^d(Z_X(D)) = \iota_X(-N_c)$. We set $P_c = D - N_c$.

(2.17) Proposition *Notation being as in* (2.16), $D = P_c + N_c$ *is sectional decomposition.*

Proof. Clearly we may assume that D is effective. There is an effective divisor F_n such that $I(X, \mathcal{O}_X([nD]))^{**} = \mathcal{O}_X(-F_n)$. Set $F_n' = F_n + \{nD\}$. Then it is easy to see that $F_n' + F_m' \geq F_{n+m}'$ and $\inf_{n \to \infty} (1/n)F_n' = \lim_{n \to \infty} (1/n)F_n' = N_c$. Hence $nN_c \leq F_n'$. Therefore we have that $[nD] - F_n \leq [nP_c]$ for all $n \geq 0$. This implies that the natural homomorphism $H^0(X, \mathcal{O}_X([nP_c])) \to H^0(X, \mathcal{O}_X([nD]))$ is bijective for every $n \geq 0$. Thus $I(X, \mathcal{O}_X([nP_c])) \cdot \mathcal{O}_X([nD] - [nP_c]) = I(X, \mathcal{O}_X([nD]))$. So we have $Z_X(D) = Z_X(P_c) + \iota_X(-N_c)$.   q.e.d.

The sectional decomposition $D = P_c + N_c$ is called *canonical sectional decomposition* of D.

(2.18) Proposition *Notation being as above, the following are equivalent.*
(1) $D = P + N$ *is a sectional decomposition of D whose positive part is P.*
(2) $D = P + N$ *is a decomposition such that for some positive integer d, the natural homomorphism* $H^0(X, \mathcal{O}_X([ndP])) \to H^0(X, \mathcal{O}_X([ndD]))$ *is bijective for all $n \geq 0$.*

Proof. It is obvious that (2) implies (1). Hence we suppose (1). Let D' be a effective $\mathbf{R}$-divisor such that $D' \sim_\mathbf{Q} D$. Set $P' = D' - N$. Then there is a positive integer d such that $[ndD] \sim [ndD']$ and $[ndP] \sim [ndP']$ for all $n \geq 0$. Hence we may assume that D is effective. Let $D = P_c + N_c$ be the canonical sectional decomposition of D. Then $p^d(Z_X(P)) + \iota_X(-N) = \iota_X(-N_c)$. Hence $N_c \geq N$. Therefore by the proof of Proposition (2.17), $H^0(X, \mathcal{O}_X([nP])) \simeq H^0(X, \mathcal{O}_X([nD]))$ for all $n \geq 0$.   q.e.d.

(2.19) Let D be an $\mathbf{R}$-divisor on a $\mathbf{Q}$-factorial normal compact complex variety X such that $\kappa(X, D) \geq 0$. We say D has a quasi-Zariski decomposition if $Z_X(D)$ is an element of $\text{IE}_\mathbf{R}(X)$. Furthermore, we say D has a Zariski decomposition if $Z_X(D) \in \overline{\text{IE}}_\mathbf{R}^d(X)$ and X is smooth (cf. [C], [K2], [Mw1], [Mw2], [Mw3] and [Z]). Note that $Z_X(D) \in \overline{\text{IE}}_\mathbf{R}^d(X)$ if and only if $Z_X(D) \in \text{IE}_\mathbf{R}^d(X)$. Clearly if D has a *quasi-Zariski decomposition*, then there is a bimeromorphic morphism $\mu: Y \to X$ form a compact complex manifold Y to X such that $\mu^*(D)$ has a *Zariski decomposition*. But the author does not know whether the converse holds (cf. Proposition (2.25)). Cutkosky example in [C] shows that $Z_X(D)$ is not necessarily an element of $\text{IE}_\mathbf{Q}^d(X)$ even if $D \in \text{Div}(X)$ and $Z_X(D) \in \text{IE}_\mathbf{R}^d(X)$. The numerical characterization of Zariski decomposition is the following.

(**2.20**) **Proposition** *Let X be a compact Kähler manifold and D an* **R**-*divisor with non-negative D-dimension. Then the following are equivalent.*

(1) *D has a Zariski decomposition.*

(2) *D has a sectional decomposition $D = P + N$ such that its positive part P is quasi-nef and good.*

Proof. First we suppose (1). Let $D = P_c + N_c$ be the canonical decomposition of D. Then $Z_X(P_c) = 0$ since $Z_X(D) \in \mathrm{IE}^d_{\mathbf{R}}(X)$. Hence by Theorem (2.14), P_c is quasi-nef and good. Next we suppose (2). Because $Z_X(P) = 0$ by Theorem (2.14), $Z_X(D) = \iota_X(-N) \in \mathrm{IE}^d_{\mathbf{R}}(X)$.   q.e.d.

The fundamental conjecture of Zariski decomposition is the following.

(**2.21**) **Conjecture** *For any* **R**-*divisor D on a* **Q**-*factorial normal compact complex variety X such that $\kappa(X, D) \geq 0$, there is a bimeromorphic morphism $\mu\colon Y \to X$ from a compact complex manifold X to Y such that $\mu^*(D)$ has a Zariski decomposition.*

A partial answer for this conjecture is

(**2.22**) **Proposition** *Let X be a* **Q**-*factorial normal compact complex variety and D an* **R**-*divisor on X such that $0 \leq \kappa(X, D) \leq 2$. Then there is a bimeromorphic morphism $\mu\colon Y \to X$ such that $\mu^*(D)$ has a Zariski decomposition. Moreover, if D is a* **Q**-*divisor, then $Z_Y(\mu^*(D)) \in \mathrm{IE}^d_{\mathbf{Q}}(Y)$.*

Proof. By Proposition (2.12), there are a bimeromorphic morphism $\mu\colon Y \to X$ from a compact complex manifold Y to X, a fiber space $g\colon Y \to W$ form Y to a non-singular projective variety W and a big **R**-divisor M on W such that $E = \mu^*(D) - g^*(M)$ is effective and $Z_Y(\mu^*(D)) = g^*(Z_W(M)) + \iota_Y(-E)$. Furthermore, if D is a **Q**-divisor, then M is a **Q**-divisor. Hence it is sufficient to see the following lemma.

(**2.23**) **Lemma** *Let W be a non-singular projective variety and M a big* **R**-*divisor on W. Assume that $\dim W \leq 2$. Then M has a Zariski decomposition and if M is a* **Q**-*divisor, $Z_W(M) \in \mathrm{IE}^d_{\mathbf{Q}}(W)$.*

Proof. Clearly we may assume that $\dim W = 2$. Let $M = P_c + N_c$ be the canonical sectional decomposition of D. Then for all integral curve C, $< v_C, Z_W(P_c) > = 0$. Hence by Lemma (2.13), P_c is nef. Therefore $Z_W(P_c) = 0$ by Theorem (2.14). Thus $Z_W(M) = \iota_W(-N_c) \in \mathrm{IE}^d_{\mathbf{R}}(W)$. In the case when M is a **Q**-divisor, rationality of Zariski decomposition is due to Zariski [Z]. q.e.d.

(2.24) Remark By the same proof of Proposition (2.22), the conjecture (2.21) can be reduced to the case where X is a non-singular projective variety and D is a big $\mathbf{R}$-divisor.

(2.25) Propositon Let D be an $\mathbf{R}$-divisor (resp. $\mathbf{Q}$-divisor) on a compact complex manifold X such that $\kappa(X, D) \geq 0$. Then the following are equivalent.

(1) $R(X, D)$ is a finitely generated $\mathbf{C}$-algebra.

(2) $Z_X(D) \in \mathrm{IE}_{\mathbf{R}}(X)$ (resp. $\mathrm{IE}_{\mathbf{Q}}(X)$) and there is a bimeromorphic morphism $\mu: Y \to X$ of compact complex manifolds such that the positive part of the canonical sectional decomposition of $\mu^*(D)$ is a semi-ample $\mathbf{Q}$-divisor.

(3) There is a bimeromorphic morphism $\mu: Y \to X$ of compact complex manifolds such that $Z_Y(\mu^*(D)) \in \overline{\mathrm{IE}}^d_{\mathbf{R}}(Y)$ (resp. $\overline{\mathrm{IE}}^d_{\mathbf{Q}}(Y)$) and the poitive part of the canonical sectional decomposition of $\mu^*(D)$ is a semi-ample $\mathbf{Q}$-divisor.

Proof. It is well known that (3) implies (1) (cf.[F1]) and it is trivial that (2) implies (3). Now we suppose (1). Then there is a positive integer d such that $\mathrm{Sym}^n(H^0(X, \mathcal{O}_X([dD]))) \to H^0(X, \mathcal{O}_X([ndD]))$ is surjective for all $n \geq 0$. For any non-negative integer k, we set $I_k = I(X, \mathcal{O}_X([kD]))$. Then it is easily checked that $(I_d)^n \cdot \mathcal{O}_X(n[dD]) \subset I_{nd} \cdot \mathcal{O}_X([ndD])$. Hence we have the following commutative diagram.

$$
\begin{array}{ccccc}
\mathrm{Sym}^n(H^0(X, \mathcal{O}([dD]))) \otimes \mathcal{O}_X & \longrightarrow & (I_d)^n \cdot \mathcal{O}(n[dD]) & \longrightarrow & 0 \\
\downarrow & & \downarrow & & \\
H^0(X, \mathcal{O}_X([ndD])) \otimes \mathcal{O}_X & \longrightarrow & I_{nd} \cdot \mathcal{O}_X([ndD]) & \longrightarrow & 0 \\
\downarrow & & & & \\
0 & & & &
\end{array}
$$

Therefore $(I_d)^n \cdot \mathcal{O}_X(n[dD]) = I_{nd} \cdot \mathcal{O}_X([ndD])$. Thus we get $Z_X(D) = (1/d)Cl(I_d) + \iota_X(-(1/d)\{dD\})$. We take a bimeromorphic morphism $\mu: Y \to X$ such that $I_d \cdot \mathcal{O}_Y$ is invertible and we write $I_d \cdot \mathcal{O}_Y = \mathcal{O}_Y(-N_d)$ for some effective divisor N_d on Y. Then $P_d = \mu^*([dD]) - N_d$ is base point free. Since

$$d \cdot Z_Y(\mu^*(D)) = Cl(I_d \cdot \mathcal{O}_Y) + \iota_Y(-\mu^*(\{dD\})) = \iota_Y(-N_d - \mu^*(\{dD\})),$$

$dP_c = \mu^*(dD) - N_d - \mu^*(\{dD\}) = P_d$, where P_c is the positive part of the canonical sectional decomposition of $\mu^*(D)$. Therefore P_c is a semi-ample $\mathbf{Q}$-divisor.   q.e.d.

(2.26) Corollary Let D be an on a compact complex manifold X with non-negative D-dimension. Then we have;

(1) if $\kappa(X, D) = 0$, $R(X, D)$ is finitely generated, and

(2) if $\kappa(X, D) = 1$, $R(X, D)$ is finitely generated if and only if for some bimeromorphic morphism $\mu: Y \to X$ of compact complex manifolds, there is

a sectional decomposition $\mu^*(D) = P + N$, *which is not necessarily canonical, such that the positive part P is a* **Q**-*divisor.*

Proof. (1) is obvious by Proposition (2.12) and Proposotion (2.25). To prove (2), we may assume D is effective. "only if" is clear by Proposition (2.25). Now we suppose that for some bimeromorphic morphism $\mu: Y \to X$ of compact complex manifolds, there is a sectional decomposition $\mu^*(D) = P + N$ such that the positive part P is a **Q**-divisor. Due to Proposition (2.12), there are a bimeromorphic morphism $\pi: V \to Y$ of compact complex manifolds, a fiber space $g: V \to C$ from V to a smooth curve C, and a big **Q**-divisor M on C such that $E = \pi^*(P) - g^*(M)$ is effective and $Z_V(\pi^*(P)) = Z_V(g^*(M)) + \iota_V(-E)$. Since M is ample, the positive part of the canonical sectional decomposition of $(\mu \circ \pi)^*(D)$ is $g^*(M)$. Therefore $R(X, D)$ is finitely generated by Proposition (2.25).   q.e.d.

(2.27)   Finally we give some sufficient condition for Fujita's Zariski decomposition [F2] via Z_X (cf. Proposition (2.33)). Let X be a smooth projective variety and D an **R**-divisor on X. We say D is *pseudo-effective* if for any ample **R**-divisor A on X, $\kappa(X, D + A) \geq 0$. Note that D is nef, then D is quasi-effective. In pervious version of this paper, we mainly treated effective divisors. But by Nakayama's letter, we find our treatment can be extended to the case of pseudo-effective divisors.

(2.28) Lemma   *Let D be a pseudo-effective* **R**-*divisor on a non-singular projective variety X and A an ample* **R**-*divisor. Then we have the following.*
(1) $\lim_{\epsilon \downarrow 0} Z_X(D + \epsilon A)$ *exists.*
(2) *If A' is another ample* **R**-*divisor,* $\lim_{\epsilon \downarrow 0} Z_X(D + \epsilon A) = \lim_{\epsilon \downarrow 0} Z_X(D + \epsilon A')$.

Proof.   For positive numbers $\epsilon \geq \epsilon'$, by Lemma (2.10) and Theorem (2.14),

$$\begin{aligned} Z_X(D + \epsilon A) &= Z_X(D + \epsilon' A + (\epsilon - \epsilon')A) \\ &\leq Z_X(D + \epsilon' A) + Z_X((\epsilon - \epsilon')A) \\ &\leq Z_X(D + \epsilon' A). \end{aligned}$$

Hence to show (1), it is sufficient to see that $\{< v, Z_X(D + \epsilon A)>\}_{\epsilon > 0}$ is bounded for all $v \in \mathrm{RS}(X)$. Changing birational model of X, we may assume that there is a prime divisor Γ such that $v_\Gamma = v$. We set

$$a_\epsilon = < v, Z_X(D + \epsilon A)>,$$

$$b_\epsilon = \sup \left\{ < v, \mathcal{O}(-F)> \, \middle| \, \begin{array}{l} F \text{ is an effective } \mathbf{R}\text{-divisor such that} \\ F \sim_\mathbf{Q} D + \epsilon A \end{array} \right\}.$$

For any $\mathbf{R}$-divisor $F \sim_{\mathbf{Q}} D + \epsilon A$,

$$<v, \mathcal{O}(-F)> \cdot (\Gamma \cdot A^{\dim X - 1}) \leq (D + \epsilon A \cdot A^{\dim X - 1}).$$

Therefore, $b_\epsilon < \infty$. On the other hand, it is easy to see that $a_\epsilon \leq b_\epsilon$ and $b_{\epsilon'} \leq b_\epsilon$ for any positive numbers $\epsilon' \leq \epsilon$. Thus we have (1).

Let A' be another ample $\mathbf{R}$-divisor. To see (2), it is sufficient to check that $\lim_{\epsilon \downarrow 0} Z_X(D + \epsilon A) \leq \lim_{\epsilon \downarrow 0} Z_X(D + \epsilon A')$. Since A is an ample $\mathbf{R}$-divisor, there are a positive integer d and an effective $\mathbf{R}$-divisor E such that $dA \sim_{\mathbf{Q}} A' + E$. Hence by Lemma (2.10),

$$Z_X(D + \epsilon A) = Z_X(D + (\epsilon/d)(A' + E)) \leq Z_X(D + (\epsilon/d)A') + \iota_X((-\epsilon/d)E).$$

Taking the limits of the above inequalities, we have our desired inequality.

$$q.e.d.$$

(2.29) By this lemma, we set $\tilde{Z}_X(D) = \lim_{\epsilon \downarrow 0} Z_X(D + \epsilon A)$. By the similar argument of Lemma (2.13), we can easily check that D is nef if and only if $\tilde{Z}_X(D) = 0$.

(2.30) Lemma *Notation being as above, let D be a big $\mathbf{R}$-diviosr. Then, we have the following.*

(1) $Z_X(D) = \tilde{Z}_X(D)$.

(2) *If D' is another big $\mathbf{R}$-divisor which is numerically equivlent to D, then $Z_X(D) = Z_X(D')$.*

(3) *Let $N_c = \sum_{i \in I} n_i N_i$ be the negative part of the canonical sectional decomposition of D. Then $\{N_i\}_{i \in I}$ is linearly independent in Neron-Severi group.*

Proof. Let A be an ample divisor on X. By the definition of $\tilde{Z}_X(D)$, it is obvious that $\tilde{Z}_X(D) \leq Z_X(D)$. There are a positive integer d and an effective $\mathbf{R}$-divisor E such that $dD \sim_{\mathbf{Q}} A + E$, where M is a nef and big divisor. Then $Z_X(D + \epsilon(A + E)) = (1 + d\epsilon)Z_X(D)$. Hence we have $(1 + \epsilon d)Z_X(D) \leq Z_X(D + \epsilon A) + \iota_X(-\epsilon E)$. Therefore $Z_X(D) \leq \tilde{Z}_X(D)$. Hence we get (1). Since $(D - D') + \epsilon A$ is ample, we have;

$$
\begin{aligned}
Z_X(D) &= \lim_{\epsilon \downarrow 0} Z_X(D + \epsilon(2A)) \\
&= \lim_{\epsilon \downarrow 0} Z_X((D' + \epsilon A) + ((D - D') + \epsilon A)) \\
&\leq \lim_{\epsilon \downarrow 0} Z_X(D' + \epsilon A) \\
&= Z_X(D').
\end{aligned}
$$

Hence we obtain (2). To see (3), it is sufficient to see that if $\sum a_i N_i$ is numerically equivalent to $\sum b_i N_i$ for non-negative numbers a_i, b_i, then $a_i = b_i$.

Considering ta_i and tb_i for sufficiently small $t > 0$, we may assume that $a_i \leq n_i$ and $b_i \leq n_i$. Let P_c be the positive part of the canonical sectional decomposition, $D_a = P_c + \sum a_i N_i$ and $D_b = P_c + \sum b_i N_i$. Then by (2), $Z_X(D_a) = Z_X(D_b)$. On the other hand, by Proposition (2.17), $Z_X(D_a) = Z_X(P_c) + \iota_X(-\sum a_i N_i)$ and $Z_X(D_b) = Z_X(P_c) + \iota_X(-\sum b_i N_i)$. Therefore, we have (3). q.e.d.

By Lemma (2.30), there is an effective **R**-divisor N_F such that $p^d(\tilde{Z}_X(D)) = \iota_X(-N_F)$. Set $P_F = D - N_F$. Then we get;

(2.31) Proposition *Notation being as above, we have*

$$\tilde{Z}_X(D) = \tilde{Z}_X(P_F) + \iota_X(-N_F).$$

Proof. Let A be an ample **R**-divisor on X. We set $\iota_X(-N_\epsilon) = p^d(Z_X(D + \epsilon A))$ for some effective **R**-divisor N_ϵ. Clearly $E_\epsilon = N_F - N_\epsilon$ is effective and $\lim_{\epsilon \downarrow 0} E_\epsilon = 0$. Set $P_\epsilon = D + \epsilon A - N_\epsilon = P_F + \epsilon A + E_\epsilon$, $E_\epsilon = \sum_{j \in J} a_{\epsilon,j} E_j$, $E = \sum_{j \in J} E_j$, and $a_\epsilon = \max_{j \in J}\{a_{\epsilon,j}\}$. Then $\lim_{\epsilon \downarrow 0} a_\epsilon = 0$ and $E_\epsilon \leq a_\epsilon E$. Since A is ample, there are a positive integer d and an effective **R**-divisor E' such that $dA \sim_{\mathbf{Q}} E + E'$. If we set $\delta_\epsilon = \epsilon + da_\epsilon$, we have

$$
\begin{aligned}
Z_X(P_F + \delta_\epsilon A) &= Z_X(P_\epsilon + a_\epsilon(E + E') - E_\epsilon) \\
&\leq Z_X(P_\epsilon) + \iota_X(E_\epsilon - a_\epsilon(E + E')).
\end{aligned}
$$

Taking limits of the above inequlities, we get;

$$\tilde{Z}_X(P_F) \leq \lim_{\epsilon \downarrow 0} Z_X(P_\epsilon) = \tilde{Z}_X(D) - \iota_X(-N_F),$$

by Lemma (2.28). On the other hand, since $Z_X(P_\epsilon) \leq Z_X(P_F + \epsilon A) + \iota_X(-E_\epsilon)$, we obtain $\lim_{\epsilon \downarrow 0} Z_X(P_\epsilon) \leq \tilde{Z}_X(P_F)$. q.e.d.

(2.32) Notation being as above, we assume $\tilde{Z}_X(D) \in \overline{\mathrm{IE}}_{\mathbf{R}}^d(X)$. There is an effective **R**-divisor N_F such that $\tilde{Z}_X(D) = \iota_X(-N_F)$. Set $P_F = D - N_F$. Then we have the following proposition (The property of this proposition is nothing more than the definition of Fujita's Zariski decomposition.).

(2.33) Proposition *In the situation (2.32), P_F is nef and for any birational morphism $\mu: Y \to X$ of non-singular projective varieties and any effective **R**-divisor N on Y, if $\mu^*(D) - N$ is nef, then $N - \mu^*(N_F)$ is effective.*

Proof. Using Proposition (2.11), it is easy to see that $\mu^*(\tilde{Z}_X(D)) = \tilde{Z}_Y(\mu^*(D))$. Hence we may assume $Y = X$. Since $\tilde{Z}_X(P_F) = 0$ by Proposition (2.31), P_F is nef. Set $P = D - N$. Then by Lemma (2.10),

$$\tilde{Z}_X(D) \leq \tilde{Z}_X(P) + \iota_X(-N) = \iota_X(-N).$$

Therefore $N_F \leq N$.   q.e.d.

§3. Canonical rings.

In this section, we shall treat the canonical ring of a compact complex manifold.

(3.1) Theorem *Let X be a compact Kähler manifold and Δ be a $\mathbf{Q}$-divisor on X such that $[\Delta] = 0$ and Δ_{red} has only simple normal crossings. Assume that $\kappa(X, \omega_X \otimes \mathcal{O}_X(\Delta)) \geq 0$ and $Z_X(\omega_X \otimes \mathcal{O}_X(\Delta)) \in \mathrm{IE}_{\mathbf{Q}}(X)$. Then the log-canonical ring $\bigoplus_{m=0}^{\infty} H^0(X, \omega_X^{\otimes m} \otimes \mathcal{O}_X([m\Delta]))$ is finitely generated.*

Proof. Taking a sequence of permissible blowing-ups with respect to Δ_{red}, we may assume $Z_X(\omega_X \otimes \mathcal{O}_X(\Delta)) \in \mathrm{IE}_{\mathbf{Q}}^d(X)$. Let D be an effective $\mathbf{Q}$-divisor such that $\omega_X \otimes \mathcal{O}_X(\Delta) \sim_{\mathbf{Q}} \mathcal{O}_X(D)$ and $D = P_c + N_c$ the canonical sectional decomposition. Then since P_c is quasi-nef and good, by changing a model of X, if necessarily, we may assume that there are a fiber space $h \colon X \to Z$ from X to a smooth projective variety Z, and a nef and big $\mathbf{Q}$-divisor M on Z such that $P_c \sim_{\mathbf{Q}} h^*(M)$. Let d be a positive integer such that $d\Delta$, dP_c and dM are divisors. By Proposition (2.25), it is sufficient to see that M is semi-ample. Hence if we can prove the following, the proof of our theorem is complete.

(3.1.1) For all $n \geq 0$, if the base locus of ndM, say $\mathrm{Bs}(Z, ndM)$, is not empty, then there are an irreducible subvariety C_n of Z and a positive integer p such that C_n is contained in $\mathrm{Bs}(Z, ndM)$ but is not in $\mathrm{Bs}(Z, mpndM)$ for all $m >> 0$.

Take a birational morphism $\pi \colon Z' \to Z$ of smooth projective varieties such that $L = \mathrm{Image}(H^0(Z', \pi^*(\mathcal{O}_Z(ndM))) \otimes \mathcal{O}_{Z'} \longrightarrow \pi^*(\mathcal{O}_Z(ndM)))$ is invertible. Then $\pi^*(\mathcal{O}_Z(ndM)) = L \otimes \mathcal{O}_{Z'}(E)$ for some effective divisor E on Z'. Since M is nef and big, there is an effective divisor Γ on Z' such that $\pi^*(M) - \delta\Gamma$ is ample for $0 < \delta << 1$. Take a bimeromorphic morphism $\mu \colon Y \to X$ of compact Kähler manifolds such that the following conditions are satisfied.

(1) There is a morphism $g \colon Y \to Z'$ such that $h \circ \mu = \pi \circ g$.

$$\begin{array}{ccc} Y & \xrightarrow{g} & Z' \\ \downarrow{\mu} & & \downarrow{\pi} \\ X & \xrightarrow{h} & Z \end{array}$$

(2) There is a simple normal crossing divisor $\sum_{i\in I} F_i$ on Y.

(3) $\omega_Y = \mu^*(\omega_X \otimes \mathcal{O}(\Delta)) \otimes \mathcal{O}_Y(\sum_{i\in I} e_i F_i)$ for some rational numbers e_i with $e_i > -1$.

(4) $\mu^*(N_c) = \sum_{i\in I} f_i F_i$ for some non-negative rational numbers f_i.

(5) $g^*(E) = \sum_{i\in I} r_i F_i$ for some non-negative integers r_i.

(6) $g^*(\Gamma) = \sum_{i\in I} b_i F_i$ for some non-negative integers b_i.

For $0 < \delta \ll 1$, we set;

$$a_i = e_i + f_i \quad (i \in I),$$
$$c = \min_{i\in I}\{(a_i + 1 - \delta b_i)/r_i\},$$
$$I_0 = \{i \in I \mid cr_i = a_i + 1 - \delta b_i\}.$$

Then $c > 0$. Changing a model of Y, if necessarily, we can choose $M_1 \in |q(\pi^*(M) - \delta\Gamma)|$ for a sufficiently large integer q satisfying the following conditions (7) and (8). (See [Mw1, (2.12) and (2.13)].)

(7) $g^*(M_1) = \sum_{i\in I} s_i F_i$ for some non-negative integers s_i.

For sufficiently small δ', we set;

$$c' = \min_{i\in I}\{(a_i + 1 - \delta b_i)/r_i\},$$
$$I_1 = \{i \in I \mid c'(r_i + \delta' s_i) = a_i + 1 - \delta b_i\} \subset I_0,$$
$$A = \sum_{i\in I\setminus I_1} (-c'(r_i + \delta' s_i) + a_i - \delta b_i)F_i,$$
$$B = \sum_{i\in I_1} F_i.$$

(8) Every non-empty intersection of F_i's with $i \in I_1$ maps surjectively to $g(B)$ by g.

Since

$$g^*(\mathcal{O}_{Z'}((1 - c'\delta' q)(\pi^*(M) - \delta\Gamma)) \otimes L^{\otimes c'} \otimes \mathcal{O}_{Z'}((mnd - c'nd - 2)\pi^*(M)))$$
$$\sim_{\mathbf{Q}} g^*\pi^*(\mathcal{O}_Z(mndM)) \otimes \mathcal{O}_Y(A - B) \otimes \omega_Y^{-1}$$

and $g^*(E) \supset B$, by (1.3.2),

$$H^1(Y, g^*\pi^*(\mathcal{O}_Z(mndM)) \otimes \mathcal{O}_Y(\lceil A\rceil - B))$$
$$\longrightarrow H^1(Y, g^*\pi^*(\mathcal{O}_Z(mndM)) \otimes \mathcal{O}_Y(\lceil A\rceil))$$

is injective if $0 < \delta' \ll 1$ and $m \gg 0$. On the other hand, by (1.3.3), there is a positive integer p such that

$$H^0(B, (g^*\pi^*(\mathcal{O}_Z(mpndM)) \otimes \mathcal{O}_Y(\lceil A\rceil))|_B) \neq 0.$$

if $m \gg 0$. Hence B is not contained in $\mathrm{Bs}(Y, g^*\pi^*(mpndM) + \lceil A \rceil)$. Since

$$\mu_*(\lceil A \rceil) \leq \mu_*(\lceil \sum_{i \in I} e_i F_i \rceil + \lceil \mu^*(N_c) \rceil) = \lceil -\Delta \rceil + \lceil N_c \rceil = \lceil N_c \rceil,$$

by properties of sectional decomposition,

$$H^0(Y, g^*\pi^*(\mathcal{O}_Z(mpndM))) \longrightarrow H^0(Y, g^*\pi^*(\mathcal{O}_Z(mpndM)) \otimes \mathcal{O}_Y(\lceil A \rceil))$$

is bijective if $m \gg 0$. Therefore B is not contained in $\mathrm{Bs}(Y, g^*\pi^*(mpndM))$. This implies that $(\pi \circ g)(B)$ is not contained in $\mathrm{Bs}(Z, mpndM)$.   q.e.d.

(3.2) **Remark** To see $Z_X(\omega_X \otimes \mathcal{O}_X(\Delta)) \in \mathrm{IE}_{\mathbf{Q}}(X)$, it is sufficient to see that $Z_X(\omega_X \otimes \mathcal{O}_X(\Delta)) \in \mathrm{IE}_{\mathbf{R}}(X)$. Indeed we can prove that $Z_X(\omega_X \otimes \mathcal{O}_X(\Delta)) \in IE_{\mathbf{R}}(X)$ implies $Z_X(\omega_X \otimes \mathcal{O}_X(\Delta)) \in \mathrm{IE}_{\mathbf{Q}}(X)$. For an algebraic case, see [K2], [Mw2] or [Mw3].

By Proposition (2.22) and Theorem (3.1), we have;

(3.3) **Theorem** *Let (X, Δ) be a normal compact complex variety with only log-terminal singularities. Assume that X is in class C and $\kappa(X, \omega_X \otimes \mathcal{O}_X(\Delta)) \leq 2$. Then the log-canonical ring $\bigoplus_{m=0}^{\infty} H^0(X, \{\omega_X^{\otimes m} \otimes \mathcal{O}_X([m\Delta])\}^{**})$ is finitely generated.*

(3.4) **Remark** Wilson [W, Example 4.3] gave an example of 4-dimensional compact non-Kähler manifold X such that $\kappa(X) = 2$ and the canonical ring of X is not finitely generated.

By the above remark, Kählerianity is essential to see that canonical ring is finitely generated for a compact complex manifofd whose dimension ≥ 4. But for 3-fold, combining recent Mori's result [Mo], Fujita's canonical bundle formula (cf.(1.4)) and Theorem (3.3), we have the following theorem.

(3.5) **Theorem** *Let X be a three dimensional normal compact complex variety with only canonical singularities. Then the canonical ring $R(X, \omega_X)$ of X is finitely generated.*

Proof. By the existence of minimal model of a projective 3-fold, the case $\kappa(X) = 3$ is clear (cf.[Mo]). On the other hand, if $\kappa(X) \leq 1$, the problem is obvious by Proposition (2.26). Hence we may assume that $\kappa(X) = 2$. In this case, we follow Fujita's arguement [F2]. Due to Iitaka's theory, changing a bimeromorphic model of X, flattening of a morphism and resolution of singularities, if necessarily, we have an elliptic fibration $f: X \to Y$ such that the following conditions (1) - (7) are satisfied.

(1) There is a commutative diagram;

$$
\begin{array}{ccccc}
M & \xleftarrow{\pi} & V & \xleftarrow{\rho} & X \\
\downarrow g & & \downarrow f' & & \downarrow f \\
T & \xleftarrow{\mu} & S & \xleftarrow{id_S} & S
\end{array}
$$

(2) T and S is a smooth projective surfaces.

(3) M and X are 3-dimensional compact complex manifolds and V is a compact complex variety.

(4) g and f are elliptic fibrations and f' is flat.

(5) π, ρ and μ are bimeromorphic morphisms.

(6) There is a normal crossing divisor $E = \sum_{i \in I} E_i$ on S such that f is smooth over $S \setminus E$.

(7) J-invariant $J \colon S \to \mathbf{P}^1$ is holomorphic.

Then by Fujita's canonical bundle formula (cf. (1.4)), there exist an effective $\mathbf{Q}$-divisor G on X and an effective $\mathbf{Q}$-divisor Σ on S such that

(8) G is f-exceptional, $[\Sigma] = 0$ and Σ_{red} has only normal crossings and that

(9) there is an positive integer d such that $\bigoplus_{t=0}^{\infty} H^0(X, \omega_X^{\otimes td} \otimes \mathcal{O}_X(tG))$ is isomorphic to $\bigoplus_{t=0}^{\infty} H^0(X, \omega_S^{\otimes td} \otimes \mathcal{O}_S(td\Sigma))$ as $\mathbf{C}$-algebra.

Since f' is flat and G is f-exceptional, G is $(\pi \circ \rho)$-exceptional. Hence the canonical ring $\bigoplus_{t=0}^{\infty} H^0(X, \omega_X^{\otimes td})$ is isomorphic to $\bigoplus_{t=0}^{\infty} H^0(X, \omega_X^{\otimes td} \otimes \mathcal{O}_X(tG))$. Therefore by Theorem (3.3), the canonical ring of X is finitely generated. q.e.d.

Next we consider singularities of $Proj(R(X, \omega_X))$.

(3.6) Theorem *Let X be a 3-dimensional normal compact complex variety with only canonical singularities. Then for the canonical model of X, say $S = Proj(R(X, \omega_X))$, there is an effective Weil divisor Δ on S such that (S, Δ) has only log-terminal singularities.*

Proof. If $\kappa(X) \neq 2$, this theorem is obvious. Hence we may assume that $\kappa(X) = 2$. In this case, by the arguement of Theorem (3.5), it is sufficient to show the following proposition.

(3.7) Proposition *Let X be a non-singular projective variety and Σ a $\mathbf{Q}$-divisor on X such that $[\Sigma] = 0$ and Σ_{red} has only normal crossings. Assume that $\kappa(X, K_X + \Sigma) = \dim X$ and $R(X, K_X + \Sigma)$ is finitely generated. Then, there is an effective Weil divisor Δ such that (S, Δ) has only log-terminal singularities, where $S = Proj(R(X, K_X + \Sigma))$.*

Proof. Since $R(X, K_X + \Sigma)$ is finitely generated, changing a model of X, if necessarily, we may assume the following.

(1) There is a birational morphism $\mu\colon X \to S$.

(2) $K_X + \Sigma$ has a Zariski decomposition with rational coefficients $K_X + \Sigma = P + N$.

(3) There is an ample $\mathbf{Q}$-Cartier divisor H on S such that $P = \mu^*(H)$.

(4) There is a normal crossing divisor $E = \sum_{i \in I} E_i$ such that $\Sigma = \sum_{i \in I} a_i E_i$ and $N = \sum_{i \in I} n_i E_i$.

Let d be a positive integer such that dP is a usual divisor. Set $A = N - \Sigma$. Then $\lceil A \rceil l$ is effective and there is a positve integer m such that $\lceil A \rceil + E_i \leq mN$ for all $i \in I$ with $n_i \neq 0$. Since $ndP + A - K_X = (nd-1)P$, $H^1(X, \mathcal{O}_X(ndP + \lceil A \rceil)) = 0$ for $n > 1$ by Kawamata-Viehweg vanishing theorem. Let E_i be a component of N. By the definition of Zariski decomposition, $H^0(X, \mathcal{O}_X(ndP + \lceil A \rceil)) \simeq H^0(X, \mathcal{O}_X(ndP + \lceil A \rceil + E_i))$ if $nd \geq m$. Hence we have $H^0(E_i, \mathcal{O}_{E_i}(ndP + \lceil A \rceil + E_i)) = 0$ for $nd > m$. If E_i is not μ-exceptional, $H^0(E_i, \mathcal{O}_{E_i}(ndP + \lceil A \rceil + E_i)) \neq 0$ for sufficiently large n because $\mathcal{O}_{E_i}(P)$ is nef and big. Therefore, N is μ-exceptional. Set $\Delta = \mu_*(\Sigma)$. Then taking direct images of $K_X + \Sigma = P + N$, we have $K_S + \Delta = H$. Hence $K_S + \Delta$ is a $\mathbf{Q}$-Cartier divisor on S and $K_X - \mu^*(K_S + \Delta) = N - \Sigma = \sum_{i \in I}(n_i - a_i)E_i$. q.e.d.

(3.8) Remark Nakayama [N2] proved the same assertion for a compact Kähler manifold, when the canonical ring is finitely generated.

References

[C] Cutkosky, D., Zariski decomposition of divisors on algebraic varieties, Duke Math. J., 53, (1986), 149–156.

[F1] Fujita, T., Semipositive line bundles, J. Fac. Sci. Univ. Tokyo, Sect. IA Math. 30, (1983), 353–378.

[F2] Fujita,T., Zariski decomposition and canonical ring of elliptic threefolds, J. Math. Soc. Japan, 38, No.1, (1986), 19–37.

[H] Hironaka, H., Introduction to the theory of infinitely near singular points, Memorias de Mat. del Inst. "Jorge Juan", 28, Consejo Superior de Investigaciones Cientificas, Madrid, (1974).

[K1] Kawamata, Y., Pluricanonical systems on minimal algebraic varieties, Inv. Math., 79, (1985), 567-588.

[K2] Kawamata, Y., The Zariski decomposition of log-canonical divisor, preprint, Tokyo, (1985).

[N1] Nakayama, N., On the lower semi-continuity of the plurigenera, to appear in Advanced Studies in Pure Math., Proc. of Symp. of Alg. Geom. at Sendai, (1985).

[N2] Nakayama, N., The singularity of the canonical model of Kähler manifold, preprint, Tokyo, (1986).

[Mo] Mori, S., Flip conjecture and the existence of minimal models for 3-folds, to appear.

[Mw1] Moriwaki, A., Semi-ampleness of the numerically effective part of Zariski decomposition, to appear in J. Math. Kyoto Univ.

[Mw2] Moriwaki, A., Zariski decomposition on higher dimensional algebraic varieties, Master Thesis, Kyoto Univ., (1986) (in Japanese).

[Mw3] Moriwaki, A., Relative Zariski decomposition on higher dimensional algebraic varieties, Proc. Japan Acad., 62, Ser A, (1986), 108-111.

[PS] Pólya, G. and Szegö, G., Aufgaben und Lehrsätze aus der Analysis 1, Die Grundlehren der Mathematischen Wissenschaften in Einzeldarstellungen, Band XIX, Springer Verlag, (1925).

[T] Treves, F., Topological vector spaces, distributions and kernels, Academic Press, (1967).

[W] Wilson, P.M.H., On the canonical ring of algebraic varieties, Compositio Math. 43, (1981), 365-385.

[Z] Zariski, O., The theorem of Riemann-Roch for high multiples of an effective divisor on an algebraic surface, Ann. of Math., 76, (1962), 560-615.

Atsushi MORIWAKI

Department of Mathmatics
Faculty of Science
Kyoto University
Kyoto, 606
Japan

Algebraic Geometry and Commutative Algebra
in Honor of Masayoshi NAGATA
pp. 313–355 (1987)

On the Moduli of Todorov Surfaces

David R. Morrison[*]

Let Z be a canonical surface, that is, a compact complex surface with only rational double points as singularities whose canonical divisor K_Z is ample. A theorem of Bombieri [5] says that if $c_1^2(Z) \geq 10$ and $p_g(Z) \geq 6$ then the bicanonical map $\phi_{|2K_Z|}$ is either birational, or maps Z two to one onto a (birationally) ruled surface. On the other hand, a recent theorem of Francia [13] says that if $\chi(\mathcal{O}_Z) \geq 2$ then the linear system $|2K_Z|$ is free; in particular, the birational image has dimension 2. It is then natural to study the canonical surfaces Z with $\chi(\mathcal{O}_Z) \geq 2$ whose bicanonical map is not birational, but whose bicanonical image is a non-ruled surface.

The first observation about such surfaces rests on the classical consequence of Clifford's theorem that a non-ruled surface X of degree d in $\mathbf{P}^{n-1}$ must satisfy $d \geq 2n - 4$, with equality if and only if X is a K3 surface with rational double points (cf. [2], for example). If X is the bicanonical image of a canonical surface Z and the bicanonical map has degree m, then the degree of X is $4c_1^2(Z)/m$ while $n = h^0(2K_Z) = \chi(\mathcal{O}_Z) + c_1^2(Z)$. If we assume that $m \geq 2$, X is non-ruled, and $\chi(\mathcal{O}_Z) \geq 2$ we get a chain of inequalities

$$2c_1^2(Z) \geq 4c_1^2(Z)/m \geq 2\chi(\mathcal{O}_Z) + 2c_1^2(Z) - 4 \geq 2c_1^2(Z)$$

which imply that $m = \chi(\mathcal{O}_Z) = 2$ and X is a K3 surface with rational double points. More generally, we may consider canonical surfaces Z with $\chi(\mathcal{O}_Z) = 2$ for which the bicanonical map factors through a finite map of degree 2 $Z \longrightarrow X$ with X a K3 surface with rational double points; it is these which we call *Todorov surfaces*.

The first example of canonical surfaces whose bicanonical map has degree 2 and whose bicanonical image is a K3 surface were given by Todorov [41], as counterexamples for the Torelli problem. Earlier examples of Todorov surfaces in the more general sense were given by Kunev [20] for the same purpose: Kunev's surfaces have $p_g = c_1^2 = 1$ and $q = 0$ so that the bicanonical map has degree 4 onto $\mathbf{P}^2$; when that map is Galois, there is an intermediate K3 surface.

Our goal here is to give an explicit description of the moduli spaces of Todorov surfaces. We show that Todorov surfaces from 11 irreducible families, which are characterized by the order of the 2-torsion subgroup of $\operatorname{Pic}(Z)$

---

[*]Research supported by a National Science Foundation Postdoctoral Fellowship, and by a fellowship from the Japan Society for the Promotion of Science.
Received June 12, 1986.

and by $c_1^2(Z)$. (Examples of surfaces belonging to each of the 11 families were essentially constructed by Todorov [41].) For each family, we give a description of the (coarse) moduli space in the form $\mathcal{V}/\Gamma$, where $\mathcal{V}$ is an open subset of a projective bundle over a Hermitian symmetric space D, and Γ is an arithmetic group acting on D and on $\mathcal{V}$. The main tools (which were also employed by Todorov) are the global Torelli theorem [36] and the surjectivity of the period map [19] for algebraic $K3$ surfaces. Our description of the moduli spaces is more precise than the one given by Todorov: the inaccuracies in appendix 1 of [41] arise in part from Todorov's failure to allow the surfaces to acquire rational double points. One of the themes of this paper is therefore the systematic extension of certain known results about smooth surfaces to include the case of surfaces with rational double points.

It is possible for a Todorov surface to have deformations which are not Todorov surfaces. We have not considered those deformations here, but the larger families which include the Todorov surfaces have been studied in the case $c_1^2(Z) = 1$ or 2 by several people, including Catanese [9], [10], [11], Catanese-Debarre [12], Oliverio [34], Todorov [40] and Usui [42], [43], [44].

We begin with some combinatorial preliminaries in section 1; we have borrowed an idea from Beauville [3] and phrased this section in the language of binary linear codes. The coding theory is applied to the study of ordinary double points on $K3$ surfaces in section 2; we obtain a characterization of these surfaces (theorem (2.5)) which generalizes a result of Nikulin [29] about Kummer surfaces. In sections 3 and 4 we study double covers of surfaces in the presence of rational double points; at the end of section 3 we give a modification of the classical "canonical resolution" process for desingularizing the double cover of a smooth surface branched along a singular curve, which has the advantage of producing the *minimal* desingularization of the double cover. Todorov surfaces are introduced in section 5, and their moduli spaces are studied in section 7. (Further information about the moduli spaces can be found in our related joint paper with M.-H. Saito [27].) Section 6 and the appendix constitute a technical digression concerning embeddings of certain even integral symmetric bilinear forms (called *lattices* in the text); we assume throughout the paper a general familiarity on the part of the reader with the theory of such embeddings (as presented in [31] or [24], for example). The paper closes with a few miscellaneous remarks about Todorov surfaces.

A few words about notation. When S is a smooth surface, we regard the first Chern class as defining maps $c_1 : \mathrm{Pic}(S) \longrightarrow H^2(S, \mathbf{Z})$ and $c_1 : \mathrm{Div}(S) \otimes \mathbf{Q} \longrightarrow H^2(S, \mathbf{Z})$; we also regard the Néron-Severi group $\mathrm{NS}(S)$ as the image in $H^2(S, \mathbf{Z})$ of $\mathrm{Pic}(S)$ under c_1. If $\Gamma \in \mathrm{Div}(S) \otimes \mathbf{Q}$ has the property that $c_1(\Gamma) \in H^2(S, \mathbf{Z})$ and the denominators of the coefficients of Γ are relatively prime to the orders of all torsion elements in $\mathrm{Pic}(S)$, we write $\mathcal{O}_S(\Gamma)$ for the unique line bundle such that $c_1(\Gamma) = c_1(\mathcal{O}_S(\Gamma))$. In particular, when S is simply connected, $\mathcal{O}_S(\Gamma)$ is defined for any $\Gamma \in \left(\mathrm{Div}(S) \otimes \mathbf{Q}\right) \cap c_1^{-1}(H^2(S, \mathbf{Z}))$.

We also use a certain amount of "modern" terminology: a linear system is *free* if it has neither fixed components nor base points; a line bundle $\mathcal{L}$ on a normal surface X is *nef* if $\mathcal{L} \cdot C \geq 0$ for all curves C on X, and is *big* if $\mathcal{L} \cdot \mathcal{L} > 0$; and a birational morphism $\pi : X \longrightarrow Y$ is *non-discrepant* [1] if $\pi^*(K_Y) = K_X$. A non-discrepant birational morphism $\pi : X \longrightarrow Y$ between surfaces with rational double points is called a *partial desingularization*.

I would like to thank the members of Kyoto-Osaka algebraic geometry seminar for many stimulating discussions during the course of this work.

§1. Equidistant binary linear codes.

A *binary linear code* $\mathcal{C} = (V \subset \mathbf{F}_2^I)$ is a finite set I together with a vector subspace V of the $\mathbf{F}_2$-vector space $\mathbf{F}_2^I$ of maps from I to $\mathbf{F}_2$. (We allow $I = \phi$ and use the convention that $\mathbf{F}_2^\phi$ is a zero-dimensional vector space.) An isomorphism between two binary linear codes $\mathcal{C} = (V \subset \mathbf{F}_2^I)$ and $\mathcal{D} = (W \subset \mathbf{F}_2^J)$ is an isomorphism of sets $\sigma : I \longrightarrow J$ such that the induced map $\sigma^* : \mathbf{F}_2^J \longrightarrow \mathbf{F}_2^I$ restricts to an isomorphism of $\mathbf{F}_2$-vector spaces $\sigma^*|_W : W \overset{\sim}{\longrightarrow} V$. The group of automorphisms of $\mathcal{C}$ is denoted by $\mathrm{Aut}(\mathcal{C})$.

If $\phi \in \mathbf{F}_2^I$, the *weight* of ϕ is the natural number $\#\{i \in I | \phi(i) = 1\}$. A binary linear code $\mathcal{C} = (V \subset \mathbf{F}_2^I)$ is *equidistant* if all nonzero elements of V have the same weight; this common weight is called the *distance* of the code. [2]

The easiest examples of binary linear codes are the *trivial codes* $\mathcal{T}_k = (\{0\} \subset \mathbf{F}_2^{\{1,\dots,k\}})$ for $k \geq 0$. More interesting examples are given by the following construction (cf. [3]): let W be an $\mathbf{F}_2$-vector space of dimension $\alpha < \infty$, and let $W^* = \mathrm{Hom}_{\mathbf{F}_2}(W, \mathbf{F}_2)$ be the dual space. Regarding elements of W^* as maps of sets produces a binary linear code $\mathcal{U}_\alpha = (W^* \subset \mathbf{F}_2^W)$. Moreover, since $\phi(0) = 0$ for all $\phi \in W^*$, we may regard W^* as a vector subspace of $\mathbf{F}_2^{W \setminus \{0\}}$, producing a second code $\mathcal{C}_\alpha = (W^* \subset \mathbf{F}_2^{W \setminus \{0\}})$. It is easily seen that $\mathcal{C}_\alpha$ and $\mathcal{U}_\alpha$ are equidistant with distance $2^{\alpha - 1}$.

Closely related to these is another binary linear code $\mathcal{D}_{\alpha+1} = (V \subset \mathbf{F}_2^W)$ defined as follows (cf. [3]): V is spanned by the image of W^* in $\mathbf{F}_2^W$ and by the element $\phi_0 \in \mathbf{F}_2^W$, where $\phi_0(w) = 1$ for all $w \in W$. $\mathcal{D}_{\alpha+1}$ is not equidistant, for ϕ_0 has weight 2^α while all other nonzero elements of V have weight $2^{\alpha-1}$.

There are several constructions which can be used to produce new codes from given ones.

(i) If $(V \subset \mathbf{F}_2^I)$ is a binary linear code and J is a subset of I, let $V_J = \{\phi \in V \mid \phi|_{I \setminus J} \equiv 0\}$ and call the resulting binary linear code $(V_J \subset \mathbf{F}_2^J)$ the *linear subcode* associated to J. For example, the linear subcode of $\mathcal{D}_{\alpha+1}$ associated to $W \setminus \{0\}$ is the binary linear code $\mathcal{C}_\alpha$.

---

[1] Those who have read Reid's footnote [38; p. 133] will recognize the present author as a linguistic conservative.

[2] This terminology derives from the *Hamming distance* (cf. [4; p.16]) between two elements $\phi, \psi \in \mathbf{F}_2^I$, which is defined to be $\#\{i \in I | \phi(i) \neq \psi(i)\}$; $\mathcal{C}$ is equidistant with distance ω if and only if the Hamming distance between every pair of distinct elements is ω.

(ii) If $C = (V \subset \mathbf{F}_2^I)$ and $\mathcal{D} = (W \subset \mathbf{F}_2^J)$ are binary linear codes, let $C \oplus \mathcal{D}$ denote the *direct sum code* $(V \oplus W \subset \mathbf{F}_2^{I \cup J})$. For example, $\mathcal{U}_\alpha$ is isomorphic to the direct sum $C_\alpha \oplus T_1$, with the isomorphism given by $W \cong (W \setminus \{0\}) \cup \{0\}$.

(iii) If $C = (V \subset \mathbf{F}_2^I)$ is a binary linear code and $\sigma : J \longrightarrow I$ is an arbitrary map, we call $\sigma^*(C) = (\sigma^*(V) \subset \mathbf{F}_2^J)$ the *pullback code*.

(iv) As a special case of (iii), if n is a natural number and $\sigma : J \longrightarrow I$ satisfies $\#\{j \in J | \sigma(j) = i\} = n$ for each $i \in I$, we call $\sigma^*(C)$ the *n-fold repetition of C* and denote it by C^n.

The pullback construction (iii) enables us to regard $\mathcal{U}_\alpha$ as a "universal binary linear code of dimension α", in the following way. Given a binary linear code $C = (V \subset \mathbf{F}_2^I)$, we let $W = \operatorname{Hom}(V, \mathbf{F}_2)$ and define a tautological map $\tau : I \longrightarrow W$ by $\tau(i)(\phi) = \phi(i)$ for all $\phi \in V$. Then τ^* establishes an isomorphism between W^* and V, so that the original binary linear code C is the pullback by τ of the code $(W^* \subset \mathbf{F}_2^W)$; this latter code is isomorphic to U_α.

Theorem (1.1). *Let $C = (V \subset \mathbf{F}_2^I)$ be an equidistant binary linear code with distance ω, let $W = \operatorname{Hom}(V, \mathbf{F}_2)$ with $\tau : I \longrightarrow W$ the tautological map, and let $\alpha = \dim V$ and $k = \#(I)$. Then $2^\alpha | 2\omega$, $k \geq (\omega / 2^{\alpha-1})(2^\alpha - 1)$, and for all nonzero $w \in W$, $\#\{i \in I | \tau(i) = w\} = \omega / 2^{\alpha-1}$. In particular, if we let $l = k - (\omega / 2^{\alpha-1})(2^\alpha - 1)$, then*

$$C \cong (C_\alpha)^{\omega/2^{\alpha-1}} \oplus T_l.$$

Proof. For each $w \in W$, let

$$a_w = \#\{i \in I | \tau(i) = w\}.$$

It suffices to show that $a_0 = l$ and that for $w \neq 0$, $a_w = \omega / 2^{\alpha-1}$, for in that case, writing $\mathcal{U}_\alpha = C_\alpha \oplus T_\alpha$ we will have $\tau^*(C_\alpha) = (C_\alpha)^{\omega/2^{\alpha-1}}$ and $\tau^*(T_1) = T_l$.

For each $\phi \in V$ we have

$$\{i \in I | \phi(i) = 1\} = \bigcup_{\substack{w \in W \\ w(\phi)=1}} \{i \in I | \tau(i) = w\}.$$

Thus if $\phi \neq 0$ we get $\sum_{w(\phi)=1} a_w = \omega$ and $\sum_{w(\phi)=0} a_w = k - \omega$ by the assumption of equidistance, while if $\phi = 0$ these sums are 0 and k respectively. We combine these as

$$(*) \qquad \sum_w (-1)^{w(\phi)} a_w = \begin{cases} k - 2\omega & \text{if } \phi \neq 0 \\ k & \text{if } \phi = 0 \end{cases}.$$

Let $A_{w,\phi} = (-1)^{w(\phi)}$. The matrix $A = (A_{w,\phi})_{w \in W, \phi \in V}$ is a *Hadamard matrix* (cf. [4; p. 144]): this means that $AA^T = diag(2^\alpha, \ldots, 2^\alpha)$, as is easily verified.

If we compose the matrix of equations $(*)$ with A^T, we find for a fixed $w_0 \in W$,

$$2^\alpha a_{w_0} = \sum_\phi (-1)^{w_0}(\phi) \sum_w (-1)^{w(\phi)} a_w = \begin{cases} k - (k - 2\omega) & \text{if } w_0 \neq 0 \\ k + (2^\alpha - 1)(k - 2\omega) & \text{if } w_0 = 0 \end{cases}$$

which immediately implies $a_0 = k - (\omega/2^{\alpha-1})(2^\alpha - 1)$ and $a_{w_0} = \omega/2^{\alpha-1}$ for $w_0 \neq 0$, as required.   Q.E.D.

Corollary (1.2). *Let β be a natural number, let $\mathcal{C} = (V \subset \mathbf{F}_2^I)$ be a binary code, let $\alpha = \dim V$, and let $k = \#(I)$. Then $\mathcal{C}$ is isomorphic to a linear subcode of $\mathcal{D}_{\beta+1}$ if and only if*
 (i) *either $\mathcal{C}$ is equidistant with distance $2^{\beta-1}$ or $\mathcal{C} \cong \mathcal{D}_{\beta+1}$, and*
 (ii) $2^{\beta-\alpha}(2^\alpha - 1) \leq k \leq 2^\beta - \beta + \alpha - 1$.
Moreover, given α and k satisfying (ii) with $\alpha \leq \beta + 1$, there is a linear subcode $\mathcal{C} = (V \subset \mathbf{F}_2^I)$ of $\mathcal{D}_{\beta+1}$ with $\alpha = \dim V$ and $k = \#(I)$.

Proof. First suppose that $\mathcal{C}$ is isomorphic to the linear subcode of $\mathcal{D}_{\beta+1}$ associated to $J \subset W$. If $J = W$ then (i) and (ii) are immediate. If $J \subset W\backslash\{w_0\}$, consider the affine translation T of W defined by $T(w) = w + w_0$; it is easy to check that $T \in \operatorname{Aut} \mathcal{D}_{\beta+1}$. Since $T(w_0) = 0$, by replacing J by $T(J)$ we may assume that $J \subset W \setminus \{0\}$. But in that case, $\mathcal{C} \cong (V_J \subset \mathbf{F}_2^J)$ is a linear subcode of $\mathcal{C}_\beta$ and hence equidistant with distance $2^{\beta-1}$, proving (i).

By theorem (1.1), $k \geq 2^{\beta-\alpha}(2^\alpha - 1)$; on the other hand, since $V_J = \{\phi \in W^* | \phi|_{W\backslash J} \equiv 0\}$ we see that the linear span of the subset $W \setminus J$ in W has dimension $\beta - \alpha = \operatorname{codim} V_J$. In particular, $W \setminus J$ must contain at least $\beta - \alpha + 1$ points, so that $2^\beta - k \geq \beta - \alpha + 1$, finishing the proof of (ii).

Conversely, suppose that $\mathcal{C}$ satisfies (i) and (ii). If $\mathcal{C} \cong \mathcal{D}_{\beta+1}$ the statement is immediate, so we may assume without loss of generality that $\mathcal{C}$ is equidistant with distance $2^{\beta-1}$. By theorem (1.1), $\mathcal{C} \cong (\mathcal{C}_\alpha)^{2^{\beta-\alpha}} \oplus \mathcal{T}_l$, where $l = k - 2^{\beta-\alpha}(2^\alpha - 1)$; in particular, $\alpha \leq \beta$. We will exhibit a linear subcode of $\mathcal{C}_\beta$ which is isomorphic to this latter code.

Let $0, w_1, \ldots, w_{\beta-\alpha}$ be $\beta - \alpha + 1$ points of W in general position, and let $K = W \setminus \{0, w_1, \ldots, w_{\beta-\alpha}\}$. By theorem (1.1), there is an isomorphism $\sigma : (V_K \subset \mathbf{F}_2^K) \xrightarrow{\sim} (\mathcal{C}_\alpha)^{2^{\beta-\alpha}} \oplus \mathcal{T}_m$, where $m = 2^{\beta-\alpha} - \beta + \alpha - 1$. Let L be the subset of K mapping to $(\mathcal{C}_\alpha)^{2^{\beta-\alpha}}$ under σ. Since $\#(L) = 2^{\beta-\alpha}(2^\alpha - 1) \leq k \leq 2^\beta - \beta + \alpha - 1 = \#(K)$, there is a set J with $\#(J) = k$ such that $L \subset J \subset K$. But now σ induces an isomorphism $(V_J \subset \mathbf{F}_2^J) \cong (\mathcal{C}_\alpha)^{2^{\beta-\alpha}} \oplus \mathcal{T}_l$, which proves the converse.

Note that we have proved the last statement as well, in our explicit construction of $(V_J \subset \mathbf{F}_2^J)$, once we observe that when $\alpha = \beta + 1$ the only solution of (ii) is $k = 2^\beta$.   Q.E.D.

We can use theorem (1.1) to compute the automorphism group of an equidistant code.

Lemma (1.3). *Let* $C = (V \subset \mathbf{F}_2^I) \cong C_\alpha^{\omega/2^{\alpha-1}} \oplus T_l$ *be an equidistant binary linear code with distance* ω, *where* $\alpha = \dim V$, $k = \#(I)$ *and* $l = k - (\omega/2^{\alpha-1})(2^\alpha - 1)$. *Then there is an exact sequence:*

$$1 \longrightarrow (\mathfrak{S}_{\omega/2^{\alpha-1}})^{2^\alpha-1} \times \mathfrak{S}_l \longrightarrow \operatorname{Aut} C \longrightarrow \operatorname{GL}(\alpha, \mathbf{F}_2) \longrightarrow 1$$

where $\mathfrak{S}_n$ *denotes the symmetric group on* n *letters.*

Proof. Let $\tau : I \longrightarrow W = \operatorname{Hom}(V, \mathbf{F}_2)$ be the tautological map. If $\sigma \in \operatorname{Aut} C$ then σ induces a linear automorphism $\sigma^* \in \operatorname{GL}(V)$, and this in turn induces $\sigma^{**} \in \operatorname{GL}(W)$; we thus get a homomorphism $\operatorname{Aut} C \longrightarrow \operatorname{GL}(W)$. The kernel of this homomorphism is easily identified: $\sigma^{**} = 1_W$ if and only if $\sigma^* = 1_V$, and in that case, for every $i \in I$ and $\phi \in V$ we have

$$\tau(\sigma(i))(\phi) = \phi(\sigma(i)) = \sigma^*(\phi)(i) = \phi(i) = \tau(i)(\phi)$$

so that $\tau(\sigma(i)) = \tau(i)$. (In other words, σ acts as a permutation of each of the fibers of τ.) The same computation shows that any permutation σ of I which acts as a permutation of each fibers of τ acts as the identity on V and so lies in the kernel of $\operatorname{Aut} C \longrightarrow \operatorname{GL}(W)$. Since τ has $2^\alpha - 1$ fibers of cardinality $\omega/2^{\alpha-1}$ and 1 fiber of cardinality l, we see that the kernel of $\operatorname{Aut} C \longrightarrow \operatorname{GL}(W)$ is exactly $(\mathfrak{S}_{\omega/2^{\alpha-1}})^{2^\alpha-1} \times \mathfrak{S}_l$.

It remains to show that the homomorphism $\operatorname{Aut} C \longrightarrow \operatorname{GL}(W)$ is surjective. Fix a nonzero $w_0 \in W$ and for each nonzero $w \in W$ choose an isomorphism

$$f_w : \{i \in I | \tau(i) = w\} \xrightarrow{\sim} \{i \in I | \tau(i) = w_0\}.$$

(This is possible because each of these sets has cardinality $\omega/2^{\alpha-1}$.) Given $\gamma \in \operatorname{GL}(W)$, define σ by

$$\sigma(i) = \begin{cases} i & \text{if } \tau(i) = 0 \\ f_{\gamma(w)}^{-1} \circ f_w(i) & \text{if } \tau(i) = w \neq 0. \end{cases}$$

Then $\sigma \in \operatorname{Aut} C$ and $\sigma^{**} = \gamma$.   *Q.E.D.*

We will also need the automorphism groups of the codes $\mathcal{D}_{\alpha+1}$.

Lemma (1.4). *If* $\mathcal{D}_{\alpha+1} = (V \subset \mathbf{F}_2^W)$ *then* $\operatorname{Aut}(\mathcal{D}_{\alpha+1})$ *is the affine general linear group* $AGL(W)$.

Proof. As we noted in the proof of corollary (1.2), if $w_0 \in W$ then the affine translation $T : w \longmapsto w + w_0$ lies in $\operatorname{Aut}(\mathcal{D}_{\alpha+1})$. Moreover it is easy to see that $\operatorname{Aut}(\mathcal{U}_\alpha) \subset \operatorname{Aut}(\mathcal{D}_{\alpha+1})$. Since $\operatorname{Aut}(\mathcal{U}_\alpha) = \operatorname{GL}(W)$ by the previous lemma, we have $AGL(W) \subset \operatorname{Aut}(\mathcal{D}_{\alpha+1})$.

On the other hand, for any $\gamma \in \operatorname{Aut}(\mathcal{D}_{\alpha+1})$ after composing with an affine translation we may assume $\gamma(0) = 0$. But then $\gamma \in \operatorname{Aut}(\mathcal{U}_\alpha) = \operatorname{GL}(W)$. Thus, $\operatorname{Aut}(\mathcal{D}_{\alpha+1}) \subset AGL(W)$.   *Q.E.D.*

§2. K3 surfaces with ordinary double points.

A K3 *surface with rational double points* is a compact complex surface X, all of whose singular points are rational double points, such that $q(X) = 0$ and the canonical divisor K_X is trivial. (This is a slight generalization of the usual definition.) We say that X has *ordinary double points* if all singular points of X have type A_1.

Let Σ be a K3 surface with ordinary double points, let $\pi : S \longrightarrow \Sigma$ be the minimal desingularization, and let E_Σ be the subgroup of $\mathrm{Div}(S)$ generated by $\{\pi^{-1}(P) \mid P \in \mathrm{Sing}\,\Sigma\}$. The *double point lattice of* Σ is the saturation $L_\Sigma = (E_\Sigma \otimes \mathbf{Q}) \cap c_1^{-1}(H^2(S, \mathbf{Z}))$ of E_Σ in the free abelian group $(\mathrm{Div}(S) \otimes \mathbf{Q}) \cap c_1^{-1}(H^2(S, \mathbf{Z}))$. The intersection form gives L_Σ the structure of an even lattice, and determines an adjoint map $\mathrm{ad} : L_\Sigma \longrightarrow L_\Sigma^* = \mathrm{Hom}(L_\Sigma, \mathbf{Z})$. (We shall usually suppress the adjoint map, and regard L_Σ as a subgroup of L_Σ^*.) There is then a chain of embeddings (cf. [31]) $E_\Sigma \subset L_\Sigma \subset L_\Sigma^* \subset E_\Sigma^* \subset (E_\Sigma \otimes \mathbf{Q})$; since E_Σ^*/E_Σ is a 2-elementary group, we see that $L_\Sigma \subset E_\Sigma \otimes (1/2)\mathbf{Z}$ and hence that $[L_\Sigma : E_\Sigma]$ is a power of 2. We call the integer $\alpha = \log_2[L_\Sigma : E_\Sigma]$ the 2-*index of* Σ.

The basis $\{\pi^{-1}(P) \mid P \in \mathrm{Sing}\,\Sigma\}$ of E_Σ furnishes a natural identification of $\mathrm{Hom}(E_\Sigma, \mathbf{Z}/2\mathbf{Z})$ with $\mathbf{F}_2^{\mathrm{Sing}\,\Sigma}$. We let $f : L_\Sigma \longrightarrow \mathbf{F}_2^{\mathrm{Sing}\,\Sigma}$ denote the composite homomorphism

$$L_\Sigma \hookrightarrow \mathrm{Hom}(E_\Sigma, \mathbf{Z}) \longrightarrow \mathrm{Hom}(E_\Sigma, \mathbf{Z}/2\mathbf{Z}) \xrightarrow{\sim} \mathbf{F}_2^{\mathrm{Sing}\,\Sigma}$$

The kernel of f is exactly E_Σ, and f induces a binary linear code $\mathcal{C}_\Sigma = (\mathrm{Im}\,f \subset \mathbf{F}_2^{\mathrm{Sing}\,\Sigma}$, called the *double point code of* Σ. Since $\mathrm{Im}\,f \cong L_\Sigma/E_\Sigma$, the 2-index of Σ is $\dim(\mathrm{Im}\,f)$. Note that the double point lattice L_Σ can be recovered from the double point code $\mathcal{C}_\Sigma$ via the natural identification of $\mathbf{F}_2^{\mathrm{Sing}\,\Sigma}$ with $\mathrm{Hom}(E_\Sigma, \mathbf{Z}/2\mathbf{Z})$: if for $\phi \in \mathbf{F}_2^{\mathrm{Sing}\,\Sigma}$ we let

$$\Gamma_\phi = \frac{1}{2} \sum_{\substack{P \in \mathrm{Sing}\,\Sigma \\ \phi(P)=1}} \pi^{-1}(P) \in E_\Sigma \otimes \mathbf{Q},$$

then L_Σ is generated by E_Σ and $\{\Gamma_\phi \mid \phi \in \mathrm{Im}\,f\}$.

Important examples of K3 surfaces with ordinary double points are furnished by the Kummer surfaces: by definition, a *Kummer surface* is a surface of the form $\Sigma = T/(\pm 1)$, where T is a complex torus of complex dimension 2, so that a Kummer surface always has 16 ordinary double points[3]: the images of the 2-torsion subgroup of T. D. B. Fuks (cf. [36; appendix to section 5]), V. V. Nikulin [29], and A. Beauville [3] have shown that the double point code of a Kummer surface Σ is isomorphic to the binary linear code $\mathcal{D}_5$ described in section 1; identifying $\mathrm{Sing}\,\Sigma$ with the 2-torsion subgroup of T gives $\mathrm{Sing}\,\Sigma$ the

---

[3]Some authors use the term "Kummer surface " to denote the minimal desingularization of $T/(\pm 1)$.

structure of an $\mathbf{F}_2$-vertor space of dimension 4, and applying the construction of $\mathcal{D}_5$ to that vector space produces the double point code of Σ.

We can generalize these examples in the following way: let Σ_0 be a Kummer surface and let $\Sigma \longrightarrow \Sigma_0$ be a partial desingularization. There is a natural inclusion $\mathrm{Sing}\,\Sigma \subset \mathrm{Sing}\,\Sigma_0$ which induces an inclusion $E_\Sigma \subset E_{\Sigma_0}$. Hence $L_\Sigma = L_{\Sigma_0} \cap (E_\Sigma \otimes \mathbf{Q})$, and $\mathcal{C}_\Sigma$ is the linear subcode of $\mathcal{C}_{\Sigma_0}$ associated to the subset $\mathrm{Sing}\,\Sigma \subset \mathrm{Sing}\,\Sigma_0$. If $k = \#(\mathrm{Sing}\,\Sigma)$ and α is the 2-index of Σ, corollary (1.2) then implies that $0 \leq \alpha \leq 5$ and $2^{4-\alpha}(2^\alpha - 1) \leq k \leq \alpha + 11$.

For each α and k such that $0 \leq \alpha \leq 5$ and $2^{4-\alpha}(2^\alpha - 1) \leq k \leq \alpha + 11$, let us fix once and for all a partial desingularization $\Sigma_{\alpha,k}$ of a Kummer surface Σ_0 such that $k = \#(\mathrm{Sing}\,\Sigma_{\alpha,k})$ and the 2-index of $\Sigma_{\alpha,k}$ is α, and let us define $E_{\alpha,k} = E_{\Sigma_{\alpha,k}}$, $L_{\alpha,k} = L_{\Sigma_{\alpha,k}}$, and $\mathcal{C}_{\alpha,k} = \mathcal{C}_{\Sigma_{\alpha,k}}$. Such a $\Sigma_{\alpha,k}$ is easy to construct, since corollary (1.2) guarantees the existence of a subset $J_{\alpha,k} \subset \mathrm{Sing}\,\Sigma_0$ whose linear subcode has the correct properties; we simply blow up the points in $\mathrm{Sing}\,\Sigma_0 \backslash J_{\alpha,k}$. For later convenience, we choose Σ_0 to be the Kummer surface of a principally polarized abelian surface whose endomorphism ring is $\mathbf{Z}$.

Theorem (2.1). *Let Σ be a K3 surface with ordinary double points, let $k = \#(\mathrm{Sing}\,\Sigma)$ and let α be the 2-index of Σ. Then*

(i) $0 \leq \alpha \leq 5$ *and* $2^{4-\alpha}(2^\alpha - 1) \leq k \leq \alpha + 11$, *and*

(ii)   *there exists an isomorphism of binary linear codes $\mathcal{C}_\Sigma \xrightarrow{\sim} \mathcal{C}_{\alpha,k}$ which induces an isometry $L_\Sigma \xrightarrow{\sim} L_{\alpha,k}$.*

Proof. By a theorem of V. V. Nikulin [29; theorem 1 and corollary 1] if $k \geq 16$ then Σ is a Kummer surface. In particular, by the result of Fuks-Nikulin-Beauville noted before, $\mathcal{C}_\Sigma \cong \mathcal{D}_5$, while $\mathcal{C}_{5,16} \cong \mathcal{D}_5$ as well. A map $\mathrm{Sing}\,\Sigma \longrightarrow \mathrm{Sing}\,\Sigma_{5,16}$ inducing the composite isomorphism of binary linear codes must clearly give an isometry $L_\Sigma \xrightarrow{\sim} L_{5,16}$.

We may thus assume that $k \leq 15$ so that Σ is not a Kummer surface. Write $\mathcal{C}_\Sigma = (V \subset \mathbf{F}_2{}^{\mathrm{Sing}\,\Sigma})$; if $\phi \in V$ then

$$c_1(\Gamma_\phi) = \frac{1}{2} \sum_{\substack{P \in \mathrm{Sing}\,\Sigma \\ \phi(P)=1}} c_1(\pi^{-1}(P)) \in H^2(S, \mathbf{Z}),$$

where $\pi : S \longrightarrow \Sigma$ is the minimal desingularization. We now recall the following fundamental fact, proved for example in [29; lemma 3] or [24; lemma 3.3].

Lemma (2.2). *If $C_1, \ldots C_n$ are smooth disjoint rational curves in a smooth K3 surface S such that $c_1((1/2) \sum C_i) \in H^2(S, \mathbf{Z})$, then $n = 0, 8$ or 16.*

Applying this lemma to our situation, we have $n \leq k < 16$ so that $n = 0$ or 8; this implies that the weight of each nonzero $\phi \in V$ is 8 so that $\mathcal{C}_\Sigma$ is equidistant with distance 8. Now by theorem (1.1), $2^\alpha \mid 16$ (so that $\alpha \leq 4$) and $k \geq 2^{4-\alpha}(2^\alpha - 1)$.

Let N_Σ be the orthogonal complement of $c_1(L_\Sigma)$ in $H^2(S, \mathbf{Z})$. Then by a standard argument [31], $N_\Sigma^*/N_\Sigma \cong L_\Sigma^*/L_\Sigma$, while L_Σ^*/L_Σ is a subquotient of $E_\Sigma^*/E_\Sigma \cong (\mathbf{Z}/2\mathbf{Z})^k$ with $\#(E_\Sigma^*/E_\Sigma) = 2^{2\alpha} \cdot \#(L_\Sigma^*/L_\Sigma)$. We conclude that $N_\Sigma^*/N_\Sigma \cong (\mathbf{Z}/2\mathbf{Z})^{k-2\alpha}$, but since $\mathrm{rank}(N_\Sigma) = 22-k$ we see that $k-2\alpha \leq 22-k$, or in other words, $k \leq \alpha + 11$.

We may now apply corollary (1.2) to conclude that $\mathcal{C}_\Sigma$ is isomorphic to a linear subcode of $\mathcal{D}_5$, and thus (by theorem (1.1)), $\mathcal{C}_\Sigma \cong \mathcal{C}_{\alpha,k}$. But as in the previous case, such an isomorphism induces an isometry $L_\Sigma \xrightarrow{\sim} L_{\alpha,k}$. *Q.E.D.*

We will several additional results about K3 surfaces with rational double points, and their minimal desingularizations.

Proposition (2.3). *Let S be a smooth K3 surface, let $C_1, \cdots C_k$ be smooth rational curves on S with $C_i \cdot C_j = -2\delta_{ij}$, let σ be a permutation of $\{1, \ldots, k\}$, and let ϕ be an automorphism of S. Suppose that $\phi(C_i) = C_{\sigma(i)}$, and that for all $x \in H^2(S, \mathbf{Z})$ with $x \cdot C_i = 0$ for all i we have $\phi^*(x) = x$. If $k \leq 15$, then ϕ and σ are trivial.*

Proof. ϕ^* acts trivially on $H^{2,0}(S)$, so that $\phi^*(\omega) = \omega$ for any holomorphic 2-form ω on S, that is, ϕ is a *symplectic* automorphism of S; these have been classified by Nikulin [30].

It suffices to prove the proposition for ϕ of prime order $p \geq 2$. In that case, Nikulin's analysis shows that ϕ has $24(p+1)^{-1}$ isolated fixed points on S which become A_{p-1} singularities on the quotient S/ϕ; moreover, the fixed part of $H^2(S, \mathbf{Z})$ under the action of ϕ has rank $22 - 24(p+1)^{-1}(p-1)$.

Let the cycle decomposition of σ consist of l fixed elements and m p-cycles, so that $k = l+mp$ and $m \leq k/p$. The fixed part of $H^2(S, \mathbf{Z})$ under the action of ϕ then has rank $(22-k)+l+m = 22+m(1-p)$; hence $22 - 24(p+1)^{-1}(p-1) = 22 + m(1-p)$ so that $m = 24(p+1)^{-1}$. But now since $k \leq 15$,

$$\frac{24}{p+1} \leq \frac{k}{p} \leq \frac{15}{p}$$

which is not possible for $p \geq 2$. *Q.E.D.*

Notice that in contrast to proposition (2.3), the translation by a point of order 2 on a complex torus T induces an automorphism of order 2 on the minimal desingularization S of the Kummer surface $T/(\pm 1)$ which permutes the 16 exceptional curves non-trivially, and acts as the identity on their orthogonal complement in $H^2(S, \mathbf{Z})$.

We turn now to the problem of characterizing the set of smooth K3 surfaces which are minimal desingularizatons of K3 surfaces with ordinary double points. The result (theorem (2.5) below), generalizes a theorem of Nikulin [29; theorem 3] in the case of Kummer surfaces.

Let S be a smooth K3 surface, and let $C_1, \ldots, C_n$ be a collection of smooth rational curves on S whose intersection form is negative-definite. We define the *root system spanned by* $C_1, \ldots, C_n$ to be the set

$$R = R(C_1, \ldots, C_n) = \{\textstyle\sum a_i C_i \mid a_i \in \mathbf{Z}, \ (\textstyle\sum a_i C_i)^2 = -2\} \subset \mathrm{Div}(S).$$

R satisfies the usual axioms of a root system, with a change in sign on the quadratic form (cf. [6]). In particular, for $E_1, E_2 \in R$ we have $|E_1 \cdot E_2| \leq 2$, with equality if and only if $E_1 = \pm E_2$.

The *Weyl group of the root system* R is the subgroup $W(R) \subset \mathrm{Aut}(\mathrm{Div}(S))$ generated by the reflections s_E for all $E \in R$, where

$$s_E(D) = D + (E \cdot D)E$$

for $D \in \mathrm{Div}(S)$.

Proposition (2.4). *Let R be a root system on S, let $E_1, \ldots, E_k$ be a collection of effective divisors on S such that $E_i \in R$ for each i, and $E_i \cdot E_j = -2\delta_{ij}$. Suppose that for each $I \subset \{1, \ldots, k\}$ with $|I| = 4$, $(1/2)c_1(\sum_{i \in I} E_i) \notin H^2(S, \mathbf{Z})$. Then there exists some $w \in W(R)$ such that $w(E_i)$ is an effective irreducible divisor (and hence a smooth rational curve) for each i.*

Proof. Let $C_1, \ldots, C_n$ span the root system R. For any effective $E \in R$ we write $E = \sum a_i C_i$ and define the *length of* E to be the positive integer $l(E) = \sum a_i$. Thus $l(E) = 1$ if and only if E is irreducible.

Suppose that $E_1, \ldots, E_{r-1}$ are irreducible, and let $l = l(E_r)$. By a double induction on r and l, it suffices to show; if $l > 1$, there is some $w \in W(R)$ such that $w(E_1), \ldots, w(E_{r-1})$ are effective and irreducible, $w(E_r)$ is effective, and $l(w(E_r)) < l$.

We thus assume $l \geq 2$, so that E_r is reducible. There must be some component C of E_r with $C \cdot E_r < 0$; since $E_r \neq \pm C$, we have $E_r \cdot C = -1$. Since $E_i \cdot E_r = 0$ for $i < r$, we see that C does not coincide with any such E_i. Thus, since C and E_i are both effective, $E_i \cdot C = 0$ or 1 for each $i < r$.

Re-order the set $\{E_1, \ldots, E_{r-1}\}$ so that $E_i \cdot C = 1$ for $1 \leq i \leq m$, $E_i \cdot C = 0$ for $m + 1 \leq i \leq r - 1$. Let $w_0 = s_C$; one easily computes

$$w_0(E_i) = \begin{cases} E_i + C & \text{if } 1 \leq i \leq m \\ E_i & \text{if } m + 1 \leq i \leq r - 1 \\ E_r - C & \text{if } i = r. \end{cases}$$

$E_r - C$ is effective since C is a component of E_r; hence, if $m = 0$ then $w = w_0$ is the desired element of $W(R)$.

If $m \geq 1$, since $E_1 \cdot (E_r - C) = -1$, E_1 must be a component of $E_r - C$ so that $E_r - C - E_1$ is effective. Let $w_1 = s_{E_1} w_0$ and compute again:

$$w_1(E_i) = \begin{cases} C & \text{if } i = 1 \\ E_i + C + E_1 & \text{if } 2 \leq i \leq m \\ E_i & \text{if } m + 1 \leq i \leq r - 1 \\ E_r - C - E_1 & \text{if } i = r. \end{cases}$$

If $m = 1$, $w = w_1$ is the desired element.

If $m \geq 2$ then $E_2 \cdot (E_r - C - E_1) = -1$ so that $E_r - C - E_1 - E_2$ is effective. Moreover, $C \cdot (E_r - C - E_1 - E_2) = -1$ so that $E_r - 2C - E_1 - E_2$ is also effective. Let $w_2 = s_C s_{E_1} w_1$ and make one final computation:

$$w_2(E_i) = \begin{cases} E_2 & \text{if } i = 1 \\ E_1 & \text{if } i = 2 \\ E_i + 2C + E_1 + E_2 & \text{if } 3 \leq i \leq m \\ E_i & \text{if } m + 1 \leq i \leq r - 1 \\ E_r - 2C - E_1 - E_2 & \text{if } i = r. \end{cases}$$

If $m = 2$, $w = w_2$ is the desired element. The case $m \geq 3$, however, will lead to a contradiction: in the case, $E_3 \cdot (E_r - 2C - E_1 - E_2) = -2$ and both curves are effective, which implies $E_r - 2C - E_1 - E_2 = E_3$. But then $(1/2)c_1(E_1 + E_2 + E_3 + E_r) = c_1(C + E_1 + E_2 + E_3) \in H^2(S, \mathbf{Z})$, contrary to our assumptions. Q.E.D.

Theorem (2.5). *Let S be a smooth K3 surface. The following are equivalent:*

(i) *There exists a K3 surface Σ with k ordinary double points whose 2-index is α and a map $\pi : S \longrightarrow \Sigma$ which is the minimal desingularization of Σ.*

(ii) *There exists a primitive embedding of the lattice $L_{\alpha,k}$ into $NS(S)$.*

Proof. Since $NS(S)$ is the image of the first Chern class map, (i) $\Rightarrow$ (ii) follows immediately from the definition of the double point lattice L_Σ and theorem (2.1). To show that (ii) $\Rightarrow$ (i), let $\phi : L_{\alpha,k} \hookrightarrow NS(S)$ be a primitive embedding, and let $\{\pi^{-1}(P) \mid P \in \operatorname{Sing} \Sigma_{\alpha,k}\}$ be the natural basis of $E_{\alpha,k}$.

Let $NA(S) = \{ x \in H^{1,1}(S) \cap H^2(S, \mathbf{R}) \mid x \cdot x > 0$ and $x \cdot C > 0$ for every effective curve C on $S \}$, let V be the component of $\{x \in H^{1,1}(S) \cap H^2(S, \mathbf{R}) \mid x \cdot x > 0\}$ which contains $NA(S)$ and let W be the subgroup of $\operatorname{Aut} H^2(S, \mathbf{Z})$ generated by the reflections s_δ associated to all $\delta \in NS(S)$ with $\delta \cdot \delta = -2$. Choose some $\kappa \in ((\operatorname{Im} \phi)^\perp \otimes \mathbf{R}) \cap V$; the standard facts about W (cf. [45] or [33; proposition 1.10]) imply that there is some $w_1 \in W$ such that $w_1(\kappa) \in \overline{NA(S)}$.

For each $P \in \operatorname{Sing} \Sigma_{\alpha,k}$, $w_1\phi(\pi^{-1}(P))$ is an element of $NS(S)$ with self-intersection -2, so by Riemann-Roch, $\pm w_1\phi(\pi^{-1}(P))$ is the class of an effective divisor E_P. We let w_2 be the product of all reflections $s_{w_1\phi(\pi^{-1}(P))}$ such that $-w_1\phi(\pi^{-1}(P))$ is effective; then $w_2 w_1 \phi(\pi^{-1}(P)) = c_1(E_P)$ is effective for each P. Since $w_2 w_1(\kappa) = w_1(\kappa) \in \overline{NA(S)}$, we see that $w_2 w_1(\kappa) \cdot C \geq 0$ for every effective curve C on S.

Write $E_P = \sum a_{i,P} C_{i,P}$ with $C_{i,P}$ irreducible and $a_{i,P} > 0$. Since $w_2 w_1(\kappa) \cdot E_P = \kappa \cdot \phi(\pi^{-1}(P)) = 0$ while $w_2 w_1(\kappa) \cdot C_{i,P} \geq 0$ we see that $w_2 w_1(\kappa) \cdot C_{i,P} = 0$ for each i and P. Hence, for each irreducible component $C_{i,P}$ of E_P, $c_1(C_{i,P})$ lies in $(w_2 w_1(\kappa)^{\perp}) \cap NS(S)$.

Since $(w_2 w_1(\kappa)^{\perp}) \cap NS(S)$ has a negative-definite intersection form, the elements of self-intersection -2 in that set form a root system R. We apply proposition (2.4) to this root system and the set of effective curves $\{E_P \mid P \in \operatorname{Sing} \Sigma_{\alpha,k}\}$ and obtain some $w_3 \in W(R)$ such that $w_3(E_P)$ is an effective irreducible divisor for each P. Contracting these smooth rational curves $w_3(E_P)$ gives the desired surface Σ.   Q.E.D.

We will need one additional lemma about root systems on K3 surfaces.

Lemma (2.6). *Let R be a root system on the smooth K3 surface S, and let $C_1, \ldots C_k$ be distinct irreducible curves in R such that for all $D \in R$, $(\sum C_i) \cdot D \in 2\mathbf{Z}$. If w is an element of the Weyl group $W(R)$ such that $w(C_i)$ is effective and irreducible for each i, then there is a permutation σ of $\{1, \ldots, k\}$ such that $w(C_i) = C_{\sigma(i)}$.*

Proof. We first claim that for any $w_1 \in W(R)$, $(1/2)(w_1(\sum C_i) - \sum C_i)$ is contained in the $\mathbf{Z}$-linear span L of R. We write w_1 as a product of reflections in elements of R, and prove our claim by induction on the number of reflections.

If w_1 is the identity, there is nothing to prove. Otherwise, write $w_1 = s_E w_2$ for some $E \in R$; by induction hypothesis, $(1/2)(w_2(\sum C_i) - \sum C_i) \in L$. Now

$$w_2\left(\sum C_i\right) \cdot E = \left(\sum C_i\right) \cdot w_2^{-1}(E) \in 2\mathbf{Z}$$

so that

$$\frac{1}{2}\left(s_E w_2\left(\sum C_i\right) - w_2\left(\sum C_i\right)\right) = \frac{1}{2}\left(w_2\left(\sum C_i\right) \cdot E\right) E \in L.$$

Hence

$$\begin{aligned}
\frac{1}{2}\left(w_1\left(\sum C_i\right) - \sum C_i\right) &= \frac{1}{2}\left(s_E w_2\left(\sum C_i\right) - w_2\left(\sum C_i\right)\right) \\
&\quad + \frac{1}{2}\left(w_2\left(\sum C_i\right) - \sum C_i\right) \in L,
\end{aligned}$$

proving the claim.

We apply this when $w_1 = w$, and find that

$$\frac{1}{2}\left(\sum w(C_i) - \sum C_i\right) \in L.$$

But $\{C_1, \ldots, C_k\}$ is part of a basis of L (a complete basis is formed by all irreducible effective curves in R), and each $w(C_i)$ either coincides with some C_j, or is also part of the basis. Since there is a linear relation among $\{C_i\} \cup \{w(C_i)\}$, each $w(C_i)$ must coincide with some C_j and the lemma follows.   Q.E.D.

§3. Double covers of surfaces with ordinary double points.

Let W be a normal compact complex surface, S be a smooth compact complex surface, and $\rho : W \to S$ be a proper finite holomorphic map of degree 2. There is a natural involution [4] $j : W \to W$ comuting with ρ whose eigenspace decomposition induces a splitting $\rho_* \mathcal{O}_W \cong \mathcal{O}_S \oplus \mathcal{M}^{-1}$ for some line bundle $\mathcal{M}$ on S. If Δ is the branch locus of ρ (a reduced Cartier divisor on S) then $\mathcal{O}_S(\Delta) = \mathcal{M}^{\otimes 2}$; note that Δ is the image of the fixed point set of j.

Conversely, given a reduced Cartier divisor Δ on S and a line bundle $\mathcal{M}$ such that $\mathcal{O}_S(\Delta) = \mathcal{M}^{\otimes 2}$, there is a standard construction for a double cover W : in each sufficiently small open set $U \subset S$, one chooses a local section $y \in \Gamma(U, \mathcal{M}^{\otimes 2})$ which vanishes on $\Delta \cap U$ and defines $\rho^{-1}(U) = \{x \in \Gamma(U, \mathcal{M}) \mid x^2 = y\}$. The double cover $\rho : W \to S$ is then uniquely specified by the data $(S, \mathcal{M}, \Delta)$, and is called the *double cover of S branched on Δ with associated line bundle $\mathcal{M}$.*

If Q is a smooth point of a surface S, ξ is a projectivized tangent vector to S at Q, and Δ is a curve on S containing Q, Δ is said to have an *infinitely near m-ple point at Q in direction ξ* if Δ has multiplicity m at Q, and on the blowup $\tilde{S}$ of S at Q, the proper transform of Δ has multiplicity m at the point of the exceptional divisor corresponding to ξ. If $\rho : W \to S$ is a double cover branched on the reduced Cartier divisor Δ, then W has only rationaly double points if and only if Δ has neither infinitely near triple points nor points of multiplicity greater than three.

The construction outlined above is quite well known (more details can be found in [16; section 2] or [35; section II]); we wish to generalize it to the case in which the quotient may have ordinary double points.

Let Y be a normal compact complex surface, let $j : Y \to Y$ be an involution, and let $\sigma : Y \to \Sigma$ be the quotient by j. We call j an *ODP-involution* if all singularities of Σ are ordinary double points, and for each $P \in \operatorname{Sing} \Sigma, \sigma^{-1}(P)_{\text{red}}$ is a smooth point of Y which is an isolated fixed point of j. (Note that when Y is smooth, any involution of Y is an ODP-involution.) Given such an involution, we let $\pi : S \to \Sigma$ be the minimal desingularization, let $W = S \times_\Sigma Y$ be the fiber product, and let $\rho : W \to S$ and $\gamma : W \to Y$ be the natural projections. The map γ is simply the blowup of the set $\cup_{P \in \operatorname{Sing} \Sigma} \sigma^{-1}(P)_{\text{red}}$, and each curve $\pi^{-1}(P)$ is a component of the branch locus of ρ. Moreover, since each point $\sigma^{-1}(P)_{\text{red}}$ is an *isolated* fixed point of j, each such component is disjoint from the rest of the branch locus. We may thus write the branch locus of ρ in the form $\Delta = \pi^*(B) + \sum_{P \in \operatorname{Sing} \Sigma} \pi^{-1}(P)$ for some reduced Cartier divisor B on Σ with $B \cap \operatorname{Sing} \Sigma = \phi$.

Conversely, suppose that Σ is a surface with ordinary double points, $\pi : S \to \Sigma$ is the minimal desingularization and $(\mathcal{M}, B)$ is a pair satisfying:

Condition (3.1). *$\mathcal{M}$ is a line bundle on S and B is a reduced Cartier*

---

[4] *An involution is an automorphism of degree 2.*

divisor on Σ *such that* $B \cap \operatorname{Sing} \Sigma = \phi$ *and* $\mathcal{O}_S(\pi^*(B) + \sum_{P \in \operatorname{Sing} \Sigma} \pi^{-1}(P)) = \mathcal{M}^{\otimes 2}$.

Then we define the *double cover of* Σ *branched on* $B \cup \operatorname{Sing} \Sigma$ *with associated line bundle* $\mathcal{M}$ as follows: first form the double cover $\rho : W \to S$ branched on $\Delta = \pi^*(B) + \sum P \in \operatorname{Sing} \Sigma \pi^{-1}(P)$ with associated line bundle $\mathcal{M}$. For each $P \in \operatorname{Sing} \Sigma$, $(1/2)\rho^{-1}(\pi^{-1}(P))$ is an exceptional curve of the first kind on W, and these curves are mutually disjoint; let Y be the surface obtained by contracting these curves. The induced meromorphic map $Y - - \to \Sigma$ extends to a holomorphic map $\sigma : Y \to \Sigma$, which is the desired double cover. It is easy to check that $\sigma : Y \to \Sigma$ is uniquely specified by the data $(\Sigma, \mathcal{M}, B)$ and that this gives a one to one correspondence between ODP-involutions and such triples. As in the previous case, Y has only rational double points if and only if B has neither infinitely near triple points nor points of mulitiplicity greater than three.

Lemma (3.2). *Let* $\sigma : Y \to \Sigma$ *be the quotient by an ODP-involution, and let* B *be the divisorial part of the branch locus of* σ. *Then* $2K_Y = \sigma^*(2K_\Sigma + B)$.

Proof. Let $\pi : S \to \Sigma$ be the minimal desingularization, and let $W = S \times_\Sigma Y$ be the fiber product with projections $\rho : W \to S$ and $\gamma : W \to Y$. ρ is branched on $\Delta = \pi^*(B) + \sum_{P \in \operatorname{Sing} \Sigma} \pi^{-1}(P)$ so that by the usual canonical divisor formula for a double cover, we have

$$K_W = \rho^* K_S + \frac{1}{2} \rho^* \Delta$$

and hence

$$2K_W = \rho^*(2K_S) + \rho^* \pi^*(B) + \sum P \in \operatorname{Sing} \Sigma \rho^* \pi^{-1}(P).$$

On the other hand, by the canonical divisor formula for a blowup applied to γ, we have

$$K_W = \rho^* K_Y + \sum_{P \in \operatorname{Sing} \Sigma} \gamma^{-01} \sigma^{-1}(P).$$

Since $2\gamma^{-1}\sigma^{-1}(P) = \rho^* \pi^{-1}(P)$ and $\rho^* \pi^*(B) = \gamma^* \sigma^*(B)$, comparing these shows that

$$\gamma^*(2K_Y) = \rho^*(2K_S) + \gamma^* \sigma^*(B).$$

But now since π is non-discrepant,

$$\rho^*(2K_S) = \rho^* \pi^*(2K_\Sigma) = \gamma^* \sigma^*(2k_\Sigma)$$

which implies $2K_Y = \sigma^*(2K_\Sigma) + \sigma^*(B)$ as claimed.   Q.E.D.

We now recall the following definition from [25]: let Σ' be a complex surface, let $Q \in \Sigma'$ be a smooth point and let ξ be a projectivized tangent vector at Q. Choose local coordinates y and z at Q such that $\xi(z) = 0$. The *directed blowup of Q in direction ξ of weight* 1 *is the blowup* $\Sigma \to \Sigma'$ *of the ideal generated by* y *and* z^2. The inverse image of Q under such a blowup is a smooth rational curve on Σ; Σ has an ordinary double point at one point of this curve, but is otherwise smooth in a neighborhood of the exceptional curve.

Lemma (3.3). *Let* Σ' *be a compact complex surface with ordinary double points, let* $Q \in \Sigma'$ *be a smooth point, and let* ξ *be a projectivized tangent vector at* Q. *Let* $\delta : \Sigma \to \Sigma'$ *be the ordinary blowup of* Q *or the directed blowup of* Q *in direction* ξ *of weight* 1, *let* C *be the exceptional divisor of* δ *and let* $Y \to \Sigma$ *be the double cover branched along* $B \cup \text{Sing} \Sigma$ *with associated line bundle* $\mathcal{M}$, *where* $(\mathcal{M}, B)$ *satisfies condition* (3.1). *Then* $B' = \delta(B)$ *is a reduced Cartier divisor on* Σ' *and there exists a unique line bundle* $\mathcal{M}'$ *on the minimal desingularization* S' *of* Σ' *such that*

(i) $(\mathcal{M}', B')$ *satisfies condition* (3.1),

(ii) *if* $Y' \to \Sigma'$ *is the double cover of* Σ' *branched on* $B' \cup \text{Sing} \Sigma'$ *with associated line bundle* $\mathcal{M}'$, *and* $\overline{Y} = \Sigma \times_{\Sigma'} Y'$ *is the fiber product, then the normalization of* $\overline{Y}$ *is* Y, *and*

(iii) *if* δ *is an ordinary blowup then for some* $m \geq 0$, B' *has a point of mulitiplicity* m *at* Q *and*

$$\delta^*(B') = B + 2[m/2]C$$

while if δ *is a directed blowup then for some* $n \geq 0$, B' *has an infinitely near* $(2n+1)$*-ple point at* Q *in direction* ξ *and*

$$\delta^*(B') = B + (4n+2)C.$$

Proof. We prove this in case δ is a directed blowup, leaving the easier (and well-known) case of an ordinary blowup to the reader. Let C be the exceptional divisor of δ, let R be the singular point of Σ contained in C, and let $E = \pi^{-1}(R)$ where $\pi : S \to \Sigma$ is the minimal desingularization. $2C$ is then a Cartier divisor on Σ, and $\pi^*(2C) = 2\tilde{C} + E$ for some $\tilde{C}$ which is an exceptional curve of the first kind on S; the induced map $\tilde{\delta} : S \to S'$ blows down $\tilde{C}$ and E. Since Σ' is smooth at Q, the Weil divisor B' (which is Cartier on $\Sigma' \setminus Q \cong \Sigma \setminus C$) must be Cartier on all of Σ', and $B' \cap \text{Sing} \Sigma = \phi$.

Let $\mathcal{M}' = \tilde{\delta}_*(\mathcal{M})$; there are then nonnegative integers m, n and r such that $\tilde{\delta}^*(\mathcal{M}') = \mathcal{M} \otimes \mathcal{O}_S(m\tilde{C} + nE)$ and $\tilde{\delta}^*(B') = B + rC$. Intersecting the first equation with E and with $\tilde{C}$ shows that $\mathcal{M} \cdot E = -m + 2n$ and $\mathcal{M} \cdot \tilde{C} = m - n$. From the second equation we deduce $\pi^* \delta^*(B') = \pi^*(B) + r\tilde{C} + (r/2)E$; again intersecting with E and with $\tilde{C}$ we see that $\pi^*(B) \cdot E = 0$ and $\pi^*(B) \cdot \tilde{C} = r/2$. By (3.1), $\mathcal{M}^{\otimes 2} = \mathcal{O}_S(\pi^*(B) + E + \sum_{P \in \text{Sing} \Sigma} \pi^{-1}(P))$, so that $2(\mathcal{M} \cdot E) = \pi^*(B) \cdot e - 2$ and $2(\mathcal{M} \cdot \tilde{C}) = \pi^*(B) \cdot \tilde{C} + 1$. Combining these computations, we see that $r = 4n + 2$ and $m = 2n + 1$.

Now

$$(\tilde{\delta}^*(\mathcal{M}'))^{\otimes 2} = \mathcal{O}_S(\pi^*(B) + E + \sum_{\substack{P \in \text{Sing } \Sigma \\ P \neq R}} \pi^{-1}(P) + 2(2n+1)\tilde{C} + 2nE)$$

$$= \mathcal{O}_S(\pi^*\delta^*(B') + \sum_{\substack{P \in \text{Sing } \Sigma \\ P \neq R}} \pi^{-1}(P)),$$

which implies that $(\mathcal{M}', B')$ satisfies (3.1). Moreover, the pullback $\overline{Y}$ of the double cover Y' is a double cover of Σ branched on

$$\delta^*(B') + \sum_{\substack{P \in \text{Sing } \Sigma \\ P \neq R}} \pi^{-1}(P) = B + (4n+2)C + \sum_{\substack{P \in \text{Sing } \Sigma \\ P \neq R}} \pi^{-1}(P);$$

pulling this branch locus back to S gives

$$\pi^*\delta^*(B') = \pi^*(B) + E + \sum_{\substack{P \in \text{Sing } \Sigma \\ P \neq R}} \pi^{-1}(P) + 2((2n+1)\tilde{C} + nE)$$

so that the normalization is the double cover branched on

$$\pi^*(B) + E + \sum_{\substack{P \in \text{Sing } \Sigma \\ P \neq R}} \pi^{-1}(P),$$

and coincides with the original double cover.

Finally, note that if $\alpha : S \to \overline{S}$ is the contraction of $\tilde{C}$ and $\beta : \overline{S} \to S'$ the contraction of $\alpha(E)$, then $\beta^*(B') = \alpha(\pi^*(B)) + (2n+1)\alpha(E)$ while $\alpha^*(\alpha(\pi^*(B))) = \pi^*(B) + (2n+1)\tilde{C}$, so that B' has an infinitely near $(2n+1)$-ple point in direction ξ (corresponding to $\tilde{C} \cap E$), as required.   Q.E.D.

As an application of lemma (3.3), we exhibit a method for finding the minimal desingularization of a singularity arising from a double cover of a smooth surface branched along a singular curve. One standard method of desingularizing in this case is the "canonical resolution" process (cf. [16;section 2] or [1; section III.7]): given a double cover $Y_1 \to S_1$ with branch locus B_1 such that S_1 is smooth, one blows up a singular point of B_1 via $S_2 \to S_1$ and defines $Y_2 \to S_2$ to be the normalization of the fiber product $S_2 \times_{S_1} Y_1$; repeating this process eventually desingularizes Y_1.

Our variant of this, which produces the minimal desingularization, is as follows. If we are inductively given a surface Σ_k with ordinary double point and a (normal) double cover $Y_k \to \Sigma_k$ branched on $B_k \cap \text{Sing } \Sigma_k$, we choose some $Q \in \Sigma_k$ which is a sinular point of B_k. (If no such Q exists, then Y_k is smooth and we stop.) If Q is an infinitely near $(2n+1)$-ple point of B_k in direction ξ, let $\Sigma_{k+1} \to \Sigma_k$ be the directed blowup (of weight 1) of Σ_k at Q in direction ξ;

otherwise, let $\Sigma_{k+1} \to \Sigma_k$ be the ordinary blowup of Σ_k at Q. In either case, we get a new (noemal) double cover $Y_{k+1} \to \Sigma_{k+1}$ branched along $B_{k+1} \cup \operatorname{Sing} \Sigma_{k+1}$ if we choose B_{k+1} according to (3.3)(iii) (setting $B = B_{k+1}$ and $B' = B_k$), so that we may continue the process.

To see that this process terminates and produces the minimal desingularization of the original surface Y_1, let $Y \to Y_1$ be the minimal desingularization, let j be the induced ODP-involution on Y and let $\Sigma = Y/j$. The natural map $\Sigma \to \Sigma_1$ is a birational morphism between surfaces with ordinary double points and $\Sigma_1 = S_1$ is smooth; theorem 1.4 of [25] (cf. also [39]) then implies that $\Sigma \to \Sigma_1$ can be factored as a composite $\Sigma = \Sigma_n \to \Sigma_{n-1} \to \cdots \to \Sigma_1$ of ordinary blowups and directed blowups of weight 1. If we use descending induction and lemma (3.3), we see that there are double covers $Y_k \to \Sigma_k$ branched on $B_k \cup \operatorname{Sing} \Sigma_k$ with Y_{k+1} the normalization of the fiber product $\Sigma_{k+1} \times_{\Sigma_k} Y_k$. If some $\Sigma_{k+1} \to \Sigma_k$ were an ordinary blowup at either an infinitely near $(2n+1)$-ple point of B_k or a point not in $\operatorname{Sing} B_k$, then Y would have an exceptional curve of the first kind mapping to $\operatorname{Sing} Y_1$; hence, such blowups are not allowed, and the process described above is the one which produces the minimal desingularization.

§4. Involutions on canonical surfaces.

A *canonical surface* is a complex surface Z with rational double points whose canonical divisor K_Z is ample. We call the involution $j : Z \to Z$ an *RDP-involution* if Z is a canonical surface, j is an involution, and the quotient Z/j has rational double points. To each RDP-involution $j : Z \to Z$ we associate a triple $(X, \mathcal{M}, B)$ as follows: $X = A/j$, B (which is a Weil divisor on X) is the divisorial part of the branch locus of the quotient map $Z \to X$, and if $\mu : S \to X$ is the minimal desingularization, $W = S \times_X Z$ is the fiber product and $\eta : W \to S$ is the projection, then $\mathcal{M}$ is the line bundle on S such that $\eta_* \mathcal{O}_W = \mathcal{O}_S \oplus \mathcal{M}^{-1}$.

Theorem (4.1). *The association described above gives a one to one correspondence between isomorphism classes of RDP-involutions and isomorphism classes of triples $(X, \mathcal{M}, B)$ with the following properties:*

(i) *X is a surface with rational double points, $\mathcal{M}$ is a line bundle on the minimal desingularization $\mu : S \to X$, and B is a Cartier divisor on X such that $2K_X + B$ is ample*

(ii) *there exist smooth disjoint rational curves $C_1, \ldots, C_k$ on S such that*

$$\mathcal{M}^{\otimes 2} = \mathcal{O}_S(\mu^*(B) + \sum C_i)$$

(iii) *$\mu^*(B)$ is disjoint from the curves C_i, and is a reduced divisor with neither infinitely near triple points nor points of multiplicity greater than three.*

Note that property (ii) determines a natural partial desingularization ν : $\Sigma \to X$, where $\pi : S \to \Sigma$ is the contraction of the curves $C_1, \ldots, C_k$; we call this the *distinguished partial desingularization* of X.

Proof. Let $j : Z \to Z$ be an RDP-involution, and let $\varepsilon : \tilde{Z} \to Z$ be the minimal desingularization. The involution j lifts to an involution on $\tilde{Z}$ (which we shall again denote by j); we let $\tilde{X} = \tilde{Z}/j$ and let $\phi : \tilde{X} \to X$ be the induced map.

ϕ is a birational morphism between surfaces with rational double points, so we may apply theorem (1.4) of [25] (cf. also [39]). Specialized to the present situation in which $\tilde{X}$ has only ordinary double points, that theorem guarantees that ϕ can be factored as

$$\tilde{X} = \Sigma_1 \xrightarrow{\psi_1} \Sigma_2 \longrightarrow \cdots \xrightarrow{\psi_{n-1}} \Sigma_n = \Sigma \xrightarrow{\nu} X$$

where each Σ_i has ordinary double points, each ψ_i is either an ordinary blowup or a directed blowup of weight 1, and ν is a partial desingularization. (The referee points out that each ψ_i is actually an ordinary blowup, as follows from the analysis at the end of section 3.)

Let $Y_1 = \tilde{Z}$; since Y_1 is smooth, the map $Y_1 \to \Sigma_1$ is the quotient by an ODP-involution, and is therefore the double cover of Σ_1 brached on $B_1 \cup \operatorname{Sing} \Sigma_1$ with associated line bundle $\mathcal{M}_1$ for some $(\mathcal{M}_1, B_1)$ satisfying (3.1). Let $B_{i+1} = \psi_i(B_i)$ for each 1. By induction using lemma (3.3), we see that there is a sequence of double covers Y_1 of Σ_1 brached on $B_i \cup \operatorname{Sing} \Sigma_i$ with associated line bundles $\mathcal{M}_i$ fitting into a diagram

$$
\begin{array}{ccccccccccc}
Y_1 & \longrightarrow & Y_2 & \longrightarrow & \cdots & \longrightarrow & Y_n & = & Y & \xrightarrow{\beta} & Z \\
\downarrow & & \downarrow & & & & \downarrow & & \sigma\downarrow & & \tau\downarrow \\
\Sigma_1 & \longrightarrow & \Sigma_2 & \longrightarrow & \cdots & \longrightarrow & \Sigma_n & = & \Sigma & \xrightarrow{\nu} & X
\end{array}
$$

in which each square is the normalization of the Cartesian product; the induction also shows that B_i is a Cartier divisor disjoint from $\operatorname{Sing} \Sigma_i$.

Let C be a component of the exceptional locus of ν. Then $\sigma^*(C)$ is contracted by β; since Z is canonical, this implies that $\sigma^*(C) \cdot K_Y = 0$ and hence, by lemma (3.2), $C \cdot (2K_\Sigma + B_n) = 0$. On the other hand, since ν is non-discrepant $C \cdot K_\Sigma = 0$, so that $C \cdot B_n = 0$. Thus, $\nu(B_n)$ is Cartier divisor on X with $\nu^*(\nu(B_n)) = B_n$. Since $mB_n - \nu^*(mB)$ is supported on the exceptional locus of ν for any integer m for which mB is Cartier, we conclude that $B = \nu(B_n)$ is Cartier and $\nu^*(B) = B_n$.

Properties (ii) and (iii) now follow from the description of ODP-involutions given in section 3 (applied to $\sigma : Y \to \Sigma$), once we set $\mathcal{M} = \mathcal{M}_n$. It remains to show that $2K_X + B$ is ample. But $\nu^*(2K_X + B) = 2K_\Sigma + B_n$ while $\sigma^*(2K_\Sigma + B_n) = 2K_Y$ by lemma (3.2). Hence, since β is non-discrepant, $2K_\Sigma + B_n$ is nef; moreover, $C \cdot (2K_\Sigma + B_n) = 0$ if and only if $\sigma^*(C) \cdot K_Y = 0$. Since Z is canonical, this holds if and only if $\sigma^*(C)$ is contracted by β, that is, if and only if C is contracted by ν. It follows that $2K_X + B = \nu(2K_\Sigma + B_n)$ is ample.

Conversely, if we are given $(X, \mathcal{M}, B)$, we let $\pi : S \to \Sigma$ be the contraction of the curves $C_1, \ldots, C_k$ and let $\nu : \Sigma \to X$ be the induced non-discrepant map. Properties (ii) and (iii) guarantee the existence of a double cover $\sigma : Y \to \Sigma$ branched on $\nu^*(B) \cup \operatorname{Sing} \Sigma$ with associated line bundle $\mathcal{M}$; we let Z be the canonical model of Y, which exists since $K_Y = \sigma^*(2K_\Sigma + \nu^*(B)) = \sigma^* \nu^*(2K_X + B)$ is nef and big. The argument in the preceding paragraph now shows that (since $2K_X + B$ is ample), the rational map $Z \dashrightarrow X$ is a finite morphism (since curves are contracted by $\beta : Y \to Z$ if and only if their images are contracted by $\nu : \Sigma \to X$.)
$\hfill$ Q.E.D.

§5. Todorov surfaces.

Let Z be a canonical surface, let $j : Z \to Z$ be an RDP-involution and suppose that $X = Z/j$ is a K3 surface with rational double points. As in section 4, we associate a triple $(X, \mathcal{M}, B)$ to j, which determines a distinguished partial desingularization $\nu : \Sigma \to X$ in which Σ has only ordinary double points. Note that since K_X is trivial, B is an ample Cartier divisor on X. Moreover, since the minimal desingularization S of Σ is simply connected,

$$\mathcal{M} = \mathcal{O}_S(\frac{1}{2}\mu^*(B) + \frac{1}{2} \sum_{P \in \operatorname{Sing} \Sigma} \pi^{-1}(P)).$$

Lemma (5.1). *Let X be a K3 surface with rational double points, B be an ample Cartier divisor on X, $\mu : S \to X$ be the minimal desingularization and $\nu : \Sigma \to X$ be a partial disingularization with induced map $\pi : S \to \Sigma$ such that Σ has only ordinary double points and*

$$\frac{1}{2}c_1(\mu^*(B) + \sum_{P \in \operatorname{Sing} \Sigma} \pi^{-1}(P)) \in H^2(S, \mathbf{Z}).$$

Then

(i) If $B^2 = 2$, $\#(\operatorname{Sing} \Sigma) = 1$ or 9, and there is a point $Q \in \operatorname{Sing} \Sigma$ such that

$$\frac{1}{2}c_1(\sum_{\substack{P \in \operatorname{Sing} \Sigma \\ P \neq Q}} \pi^{-1}(P)) \in H^2(S, \mathbf{Z}),$$

then the linear system $|\nu^*(B)|$ has a single base point, at Q.

(ii) In all other cases, the linear system $|B|$ is free.

Proof. $\mu^*(B)$ is a nef and big divisor on the smooth K3 surface S, so that by a theorem of Mayer [22], either the linear system $|\mu^*(B)|$ is free (and hence $|B|$ is free), or $\mu^*(B) \equiv nE + C$ for some elliptic pencil $|E|$ on S, and some section C of $|E|$, where $n = (1/2)(1 + \mu^*(B)^2) \geq 2$. In the latter case; C is the unique fixed component of $|\mu^*(B)|$ and we compute:

$$E \cdot (\mu^*(B) + \sum_{P \in \text{Sing } \Sigma} \pi^{-1}(P)) = 1 + \sum_{P \in \text{Sing } \Sigma} E \cdot \pi^{-1}(P).$$

Now $(1/2)c_1(\mu^*(B) + \sum_{P \in \text{Sing } \Sigma} \pi^{-1}(P)) \in H^2(S, \mathbf{Z})$ so that this intersection number must be even; hence, there is some $Q \in \text{Sing } \Sigma$ with $E \cdot \pi^{-1}(Q) \geq 1$. For each Q, we have

$$0 = \pi^{-1}(Q) \cdot \mu^*(B) = \pi^{-1}(Q) \cdot (nE + C) \geq n + \pi^{-1}(Q) \cdot C$$

so that $\pi^{-1}(Q) = C$ and $n = 2$. Thus, $|\nu^*(B)|$ has a unique base point at $Q = \pi(C)$, and $B^2 = \mu^*(B)^2 = 2$.

To finish the proof of (ii), note that when $|\mu^*(B)|$ is not free,

$$\frac{1}{2}c_1(\sum_{\substack{P \in \text{Sing } \Sigma \\ P \neq Q}} \pi^{-1}(P)) = \frac{1}{2}c_1(\mu^*(B) + \sum_{P \in \text{Sing } \Sigma} \pi^{-1}(P)) - c_1(E + C)$$

lies in $H^2(S, \mathbf{Z})$. Moreover, by lemma (2.2),

$$\#(\text{Sing } \Sigma \backslash \{Q\}) = 0, \quad 8, \text{or} \quad 16;$$

since $\#(\text{Sing } \Sigma) \leq 16$, we see that $\#(\text{Sing } \Sigma) = 1$ or 9.

It remains to show that when the hypotheses of (i) are satisfied, $|\mu^*(B)|$ cannot be free. Since $\mu^*(B)$ is nef and big, Mayer's vanishing theorem [22] implies $H^1(\mu^*(B)) = 0$, so that $h^0(\mu^*(B)) = 3$ by Riemann-Roch. Let $Q \in \text{Sing } \Sigma$ be the point such that

$$\frac{1}{2}c_1(\sum_{\substack{P \in \text{Sing } \Sigma \\ P \neq Q}} \pi^{-1}(P)) \in H^2(S, \mathbf{Z}),$$

and define

$$\mathcal{E} = \mathcal{O}_S(\frac{1}{2}(\mu^*(B) + \sum_{P \in \text{Sing } \Sigma} \pi^{-1}(P)) - \frac{1}{2} \sum_{\substack{P \in \text{Sing } \Sigma \\ P \neq Q}} \pi^{-1}(P) - \pi^{-1}(Q)).$$

Then $\mathcal{E}^{\otimes 2} = \mathcal{O}_S(\mu^*(B) - \pi^{-1}(Q))$ and $\mathcal{E} \cdot \mathcal{E} = 0$. If $\mathcal{E}^{-1} \otimes \Omega_S^2 = \mathcal{E}^{-1}$ were effective, we would have

$$1 = h^0(\mathcal{O}_S(\pi^{-1}(Q))) = h^0((\mathcal{E}^{-1})^{\otimes 2} \otimes \mathcal{O}_S(\mu^*(B))) \geq h^0(\mathcal{O}_S(\mu^*(B))) = 3,$$

a contradiction. Hence, $H^2(\mathcal{E}) = H^0(\mathcal{E}^{-1} \otimes \Omega_S^2)^* = 0$, so that by Riemann-Roch, $h^0(\mathcal{E}) \geq 2$. In particular, $\mathcal{E} = \mathcal{O}_S(E)$ for some effective divisor E with $2E \equiv \mu^*(B) - \pi^{-1}(Q)$. Now $h^0(E) \geq 2$ implies that $h^0(2E) \geq 3$; since

$h^0(2E + \pi^{-1}(Q)) = 3$, we see that $\pi^{-1}(Q)$ is a fixed component of $|2E + \pi^{-1}(Q)| = |\mu^*(B)|$, which is therefore not free. $Q.E.D.$

The following computation of the invariants of a double cover of a K3 surface is essentially due to Todorov [41].

Theorem (5.2). *Let Z be a canonical surface and let $j : Z \to Z$ be an RDP-involution such that Z/j is a K3 surface with rational double points. Let $(X, \mathcal{M}, B)$ be the associated triple, let $\nu : \Sigma \to X$ be the distinguished partial desingularization, let $k = \#(\mathrm{Sing}\,\Sigma)$, let α be the 2-index of Σ, and let $d = c_1^2(Z) = (1/2)B^2$. Then*
(i) *the linear system $|B|$ is free, and $h^0(B) = d + 2$,*
(ii) *$p_g(Z) = (d - k)/4 + 3$ and $q(Z) = 0$,*
(iii) *$h^0(2K_Z) = d + (d - k)/4 + 4$, and*
(iv) *if $\tilde{Z}$ is the minimal desingularization of Z, then the 2-torsion subgroup of $\mathrm{Pic}(\tilde{Z})$ has order 2^α.*

Proof. Since $\nu^*(B) \cap \mathrm{Sing}\,\Sigma = \emptyset$, $|\nu^*(B)|$ cannot have a base point at a point of $\mathrm{Sing}\,\Sigma$, so $|B|$ is free by lemma (5.1). Let $\mu : S \to X$ and $\epsilon : \tilde{Z} \to Z$ be the minimal desingularizations. Since B and K_Z are ample on X and Z respectively, $\mu^*(B)$ and $\epsilon^*(K_Z) = K_{\tilde{Z}}$ are nef and big on S and $\tilde{Z}$ respectively. By the Kodaira-Ramanujam vanishing theorem[5] [37], keeping in mind that $K_S = 0$, we have
$$h^1(\mu^*(B)) = h^2(\mu^*(B)) = h^1(2K_{\tilde{Z}}) = h^2(2K_{\tilde{Z}}) = 0$$
so that by Riemann-Roch,
$$h^0(B) = h^0(\mu^*(B)) = \chi(\mathcal{O}_S) + \frac{1}{2}(\mu^*(B))^2 = 2 + d$$
and
$$h^0(2K_Z) = h^0(2K_{\tilde{Z}}) = \chi(\mathcal{O}_{\tilde{Z}}) + c_1^2(\tilde{Z}) = \chi(\mathcal{O}_{\tilde{Z}}) + d.$$
This proves (i), and shows that (iii) is a consequence of (ii).

To prove (ii) and (iv), note that by Bertini's theorem there is some $B_1 \in |B|$ such that $\nu^*(B_1)$ is smooth and disjoint from $\mathrm{Sing}\,\Sigma$. If we deform B to B_1 the double cover Z deforms to some surface Z_1 ; since Z and Z_1 both have rational double points, $p_g(Z) = p_g(Z_1)$ and $q(Z) = q(Z_1)$. Moreover, by the theory of "simultaneous resolution" of rational double points [7], $\tilde{Z}$ is diffeomorphic to $\tilde{Z}_1$. Since the 2-torsion subgroup of $\mathrm{Pic}(\tilde{Z})$ is a topological invariant, these subgroups of $\mathrm{Pic}(\tilde{Z})$ and $\mathrm{Pic}(\tilde{Z}_1)$ coinside. It thus suffices to prove (ii) and (iv) in the case that $\nu^*(B)$ is smooth, which we now assume.

Let $\sigma : Y \to \Sigma$ be the double cover branched on $\nu^*(B) \cup \mathrm{Sing}\,\Sigma$, and let $W = S \times_\Sigma Y$ be the fiber product with projection maps $\rho : W \to S$ and $\gamma : W \to Y$.

---
[5]In the cases in which we use it, this theorem was proved earlier by Mayer [22] and Kodaira [18].

Since $\nu^*(B)$ is smooth, Y coincides with the minimal desingularization $\tilde{Z}$ of Z. The map γ is the blowup of k points on Y, so that $c_1^2(W) = c_1^2(Y) - k$; since $h^0(2K_W) = h^0(2K_Y)$, Riemann-Roch immediately yields $h^1(2K_W) = k$.

On the other hand, since $\mathcal{M}^{\otimes 2} = \mathcal{O}_S(\mu^*(B) + \sum_{P \in \mathrm{Sing}\,\Sigma} \pi^{-1}(P))$, we have $\mathcal{M}^{\otimes 2} \cdot \pi^{-1}(P) < 0$ for each $P \in \mathrm{Sing}\,\Sigma$; hence, $\sum_{P \in \mathrm{Sing}\,\Sigma} \pi^{-1}(P)$ is contained in the fixed locus of $|\mathcal{M}^{\otimes 2}|$ so that $h^0(\mathcal{M}^{\otimes 2}) = h^0(\mu^*(B)) = d + 2$. Since $\mathcal{M}^{\otimes 2} \cdot \mathcal{M}^{\otimes 2} = 2d - 2k$ and $h^2(\mathcal{M}^{\otimes 2}) = 0$, Riemann-Roch now shows that $h^1(\mathcal{M}^{\otimes 2}) = k$.

Now $\sigma_* \mathcal{O}_W = \mathcal{O}_S \oplus \mathcal{M}^{-1}$ and $\sigma_* \mathcal{O}_W(2K_W) = \sigma_* \sigma^* \mathcal{M}^{\otimes 2} = \mathcal{M}^{\otimes 2} \oplus \mathcal{M}$ so that

$$h^1(\mathcal{O}_W) = h^1(\mathcal{O}_S) + h^1(\mathcal{M}^{-1})$$

and

$$h^1(2K_W) = h^1(\mathcal{M}^{\otimes 2}) + h^1(\mathcal{M}).$$

Since $h^1(\mathcal{O}_S) = 0$ and $K_S = 0$, Serre duality implies that $h^1(\mathcal{O}_W) = h^1(\mathcal{M} \otimes K_S) = h^1(\mathcal{M})$. But now

$$q(Z) = h^1(\mathcal{O}_W) = h^1(\mathcal{M}^{\otimes 2}) - h^1(2K_W) = k - k = 0.$$

To compute the geometric genus, we compute the topological Euler characteristic of W . The branch curve of $\rho : W \to S$ has topological Euler characteristec $2 - 2g(\mu^*(B)) + 2k = 2k - 2d$, while S has topological Euler characteristic 24, so that

$$c_2(W) = 2 \cdot 24 - (2k - 2d) = 48 + 2d - 2k.$$

Thus,

$$12\chi(\mathcal{O}_W) = c_1^2(W) + c_2(W) = 48 + 3d - 3k$$

which immediately implies that $p_g(Z) = p_g(W) = (d - k)/4 + 3$, proving (ii).

To prove (iv), note that $\rho : W \to S$ is a double cover of smooth surfaces branched along $\alpha + 1$ smooth disjoint irreducible curves (the inverse images of B and $\mathrm{Sing}\,\Sigma$). Since $\mathrm{Pic}(S)$ has no 2-torsion, by [3; lemme 2] the 2-torsion subgroup of $\mathrm{Pic}(W)$ (which coincides with the 2-torsion subgroup of $\mathrm{Pic}(\tilde{Z})$) has order 2^α. Q.E.D

A *Todorov surface* is a canonical surface Z with $\chi(\mathcal{O}_Z) = 2$ which has an involution $j : Z \to Z$ shch that $X = Z/j$ is a K3 surface with rational double points. If Z is a Todorov surface and $\nu : \Sigma \to X$ is the distinguished partial desingularization, theorem (5.2) shows that $p_g(Z) = 1$ and $q(Z) = 0$. We call $\alpha = \log_2 |2\text{-torsion in } \mathrm{Pic}(\tilde{Z})|$ and $k = c_1^2(Z) + 8$ the *fundamental invariants* of Z : by theorem (5.2) , α is the 2-index of Σ and $k = \#(\mathrm{Sing}\,\Sigma)$. Since $c_1^2(Z) > 0$, we have $k \geq 9$, while lemma (5.1) and theorem (5.2) (i) show that $(\alpha, k) \neq (1, 9)$. In addition, theorem (2.1) implies that $0 \leq \alpha \leq 5$ and

$2^{4-\alpha}(2^\alpha - 1) \le k \le \alpha + 1$. Combining all of these conditions, we find that there are 11 possible values for (α, k):

$$(\alpha, k) \in \{(0,9), (0,10), (0,11), (1,10), (1,11), (1,12), (2,12),$$
$$(2,13), (3,14), (4,15), (5,16)\}.$$

Examples of Todorov surfaces with each of these possible values for (α, k) can in principle be given by a method due to Tokorov [41], as follows: embed the Kummer surface Σ_0 (which was chosen in section 2) as a quartic surface in $\mathbf{P}^3$, and let $J_{\alpha,k} \subset \text{Sing}\,\Sigma_0$ be the subset such that $\Sigma_{\alpha,k} \to \Sigma_0$ is the blowup of $\text{Sing}\,\Sigma_0 \backslash J_{\alpha,k}$. If there is a quadric surface $Q \subset \mathbf{P}^3$ containing $\text{Sing}\,\Sigma_0 \backslash J_{\alpha,k}$ such that the proper transform B of $Q \cap \Sigma_0$ on $\Sigma_{\alpha,k}$ is smooth and disjoint from $\text{Sing}\,\Sigma_{\alpha,k}$, then double cover of $\Sigma_{\alpha,k}$ branched of $B \cup \text{Sing}\,\Sigma_{\alpha,k}$ will be a Todorov surface with fundamental invariants (α, k) . To guarantee that such a Q exists, one must take some care in selecting the sets $J_{\alpha,k}$, to ensure that the points in $\text{Sing}\,\Sigma_0 \backslash J_{\alpha,k}$ satisfy appropriate "general position" conditions with respect to linear systems of quadric surfaces in $\mathbf{P}^3$.

We will not pursue this method here[6], but instead will appeal to the surjectivity of the period map for K3 surfaces to show in section 7 that there is a nonempty irreducible family of Todorov surfaces with fundamental invariants (α, k) for each of the 11 possible values of (α, k) .

We need a few additional properties of Todorov surfaces.

Lemma (5.3). *Let* Z *be a Todorov surface, and let* $j : Z \to Z$ *be an involution such that* $X = Z/j$ *is a K3 surface with rational double points. Then*

(i)   *the bicanonical map* $\phi_{|2K_Z|}$ *is a morphism,*

(ii) $\phi_{|2K_Z|}$ *factors through the quotient* $\tau : Z \to X$, *and*

(iii) *the involution* j *is uniquely determined by* Z .

Proof. (i) By theorem (5.2)(i), $|B|$ is free; hence, $|2K_Z| = |\tau^*(B)|$ is free so that $\phi_{|2K_Z|}$ is a morphism. (Alternatively, we could use a theorem of Francia [13] which guarantees that any canonical surface with $p_g = 1$ and $q = 0$ has a free bicanonical system).

(ii) By parts (i) and (iii) of theorem (5.2), $h^0(B) = d + 2 = h^0(2K_Z)$ so that $\phi_{|2K_Z|}$ factors through $\tau : Z \to X$. (Compare Persson [35; prop. 3.1].)

(iii) By Mayer's analysis of free linear systems on K3 surfaces [22], $\phi_{|B|}$ either embeds X , or maps X two to one onto a rational surface U . In the first case, $\deg \phi_{|2K_Z|} = 2$, so that there can be only one involution of Z through which $\phi_{|2K_Z|}$ factors. In the second case, $\deg \phi_{|2K_Z|} = 4$; by Galois theory for a biquadratic extension, if there were a second involution i through which

---

[6]The reader who wishes to try the exercise of selecting the sets $J_{\alpha,k}$ appropriately will need to know that the subsets of $\text{Sing}\,\Sigma_0$ of cardinality 6 which lie in a hyperplane in $\mathbf{P}3$ occurring in section 1 of [41] and chapter 6 of [15] coincide with the subsets J occurring in proposition (6.1)(ii) below.

$\phi_{|2K_Z|}$ factored, the map $\phi_{|2K_Z|} : Z \to U$ would be a Galois cover with Galois group $\mathbf{Z}/2\mathbf{Z} \times \mathbf{Z}/2\mathbf{Z}$ and all involutions through which $\phi_{|2K_Z|}$ factored would be contained in G. The natural representation $G \to \mathrm{Aut}(H^0(K_Z)) \cong \mathbf{C}^*$ would then be non-trivial (since $p_g(U) = 0$), but any involution whose quotient is a K3 surface would lie in the kernel of that representation. Since G is generated by such involutions, we get a contradiction; hence, the involution j is unique. Q.E.D

The free linear system $|B|$ on X is called *hyperelliptic* when $\phi_{|B|}$ maps X two to one onto a rational surface U; this happens exactly when the bicanonical map of the corresponding Todorov surface has degree 4. The following lemma gives a sufficient condition for this to happen; a partial converse will be proved in section 7.

Lemma (5.4). *Let Z be a Todorov surface with fundamental invariants (α, k). If $(\alpha, k) = (0, 9)$ or $(1, 10)$ then the bicanonical map $\phi_{|2K_Z|}$ has degree 4 and the linear system $|B|$ on $X = Z/j$ is hyperelliptic.*

Proof. If $k = 9$, then $B^2 = 2$ and $h^0(2K_Z) = h^0(B) = 3$; it follows that $\phi_{|2K_Z|}$ and $\phi_{|B|}$ are maps of degree 4 and 2 respectively onto $\mathbf{P}^2$.

If $(\alpha, k) = (1, 10)$, let $\nu : \Sigma \to X$ be the distinguished partial desingularization, and $\pi : S \to \Sigma$ be the minimal desingularization. There are points $P_1, P_2 \in \mathrm{Sing}\,\Sigma$ such that

$$\frac{1}{2}c_1\Big(\sum_{\substack{P \in \mathrm{Sing}\,\Sigma \\ P \neq P_1, P_2}} \pi^{-1}(P)\Big) \in H^2(S, \mathbf{Z}).$$

This implies that $(1/2)c_1(B - \pi^{-1}(P_1) - \pi^{-1}(P_2)) \in H^2(S, \mathbf{Z})$ as well. Let $\mathcal{E} = \mathcal{O}_S((1/2)(B - \pi^{-1}(P_1) - \pi^{-1}(P_2)))$. Then $\mathcal{E} \cdot \mathcal{E} = 0$ so by Riemann-Roch $\mathcal{E}$ or $\mathcal{E}^{-1}$ is effective. Since $B \cdot \mathcal{E} = 2$ and B is nef, $\mathcal{E}^{-1}$ cannot be effective; thus, $\mathcal{E} = \mathcal{O}_S(E)$ for some effective divisor E, and the linear system $|E|$ has (projective) dimension at least 1.

Let $|E_0|$ be the moving part of $|E|$. Since B is nef, $B \cdot (E - E_0) \geq 0$ so that $0 \leq B \cdot E_0 \leq 2$. If $B \cdot E_0 = 0$ then the Hodge index theorem implies $E_0^2 < 0$, but a curve of negative self-intersection on a K3 surface cannot move. Thus, $B \cdot E_0 = 1$ or 2. But then every smooth curve $B' \in |B|$ has a linear system $|E_0 \cap B'|_{B'}$ of dimension at least 1 and degree 1 or 2. Since B' is not rational, the degree must be 2 and B' must be hyperelliptic. Hence, $\phi_{|B|}$ (which induces the canonical map on B') must have degree 2. Q.E.D

§6. Embeddings of Todorov lattices.

Let $(\Sigma, \mathcal{L})$ be a pair consisting of K3 surface Σ with ordinary double points and a line bundle $\mathcal{L}$ on Σ with $\mathcal{L} \cdot \mathcal{L} = 2(\# \mathrm{Sing}\,\Sigma) - 16 > 0$ such that if $\pi : S \to \Sigma$

is the minimal desigularization, then

$$\frac{1}{2}c_1(\pi^*(\mathcal{L}) \otimes \mathcal{O}_S(\sum_{P \in \text{Sing}\, \Sigma} \pi^{-1}(P))) \in H^2(S, \mathbf{Z}).$$

(Such a pair can be obtained from the quotient $X = Z/j$ of a Todorov surface by taking $\nu : \Sigma \to X$ to be the distinguished partial desingularization, and $\mathcal{L} = \mathcal{O}_\Sigma(\nu^*(B))$.) For such a pair $(\Sigma, \mathcal{L})$, we let $M_{\Sigma,\mathcal{L}}$ be the saturation in $H^2(S, \mathbf{Z})$ of the lattice generated by $c_1(L_\Sigma)$ and $c_1(\mathcal{L})$.

Let $\Sigma_{\alpha,k}$ be the fixed partial desingularization of a Kummer surface Σ_0 with 2-index α and k ordinary double points which was chosen in section 2. Recall that we chose Σ_0 to be the Kummer surface of a principally polarized abelian surface A_0, so that there is a map $f : A_0 \to \Sigma_0$ and a line bundle $\mathcal{N}_0$ on Σ_0 such that $f^*\mathcal{N}_0 = \Theta_0^{\otimes 2}$, where Θ_0 is the principal polarization on A_0.

Let $\pi_0 : S_0 \to \Sigma_0$ be the minimal desingularization, and let $\pi_{\alpha,k} : S_0 \to \Sigma_{\alpha,k}$ and $\nu_{\alpha,k} : \Sigma_{\alpha,k} \to \Sigma_0$ be the natural maps. If we let

$$\mathcal{L}_{\alpha,k} = \nu_{\alpha,k}^*(\mathcal{N}_0^{\otimes 2}) \otimes \mathcal{O}_{\Sigma_{\alpha,k}}(\sum_{\substack{P \in \text{Sing}\, \Sigma_0 \\ P \notin \text{Sing}\, \Sigma_{\alpha,k}}} \nu_{\alpha,k}^{-1}(P)),$$

then $\mathcal{L}_{\alpha,k} \cdot \mathcal{L}_{\alpha,k} = 2k - 16$. Moreover, since Σ_0 is a Kummer surface,

$$\frac{1}{2}c_1(\pi_{\alpha,k}^*(\mathcal{L}_{\alpha,k}) \otimes \mathcal{O}_{s_0}(\sum_{P \in \text{Sing}\, \Sigma_{\alpha,k}} \pi_{\alpha,k}^{-1}(P)) =$$

$$c_1(\pi_0^*\mathcal{N}_0) + \frac{1}{2}c_1(\sum_{P \in \text{Sing}\, \Sigma_0} \pi_0^{-1}(P)) \in H^2(S_0, \mathbf{Z}).$$

Thus, when $k \geq 9$ the pair $(\Sigma_{\alpha,k}, \mathcal{L}_{\alpha,k})$ satisfies the hypotheses above. We define $M_{\alpha,k} = M_{\Sigma_{\alpha,k}, \mathcal{L}_{\alpha,k}}$, and call this a *Todorov lattice* when $k \geq 9$ and $(\alpha, k) \neq (1, 9)$.

If $(\Sigma, \mathcal{L})$ is a pair satisfing the hypotheses above, we let $\lambda = c_1(\pi^*\mathcal{L})$, let $e_P = c_1(\pi^{-1}(P))$ for each $P \in \text{Sing}\, \Sigma$, and let $\mu = (1/2)(\lambda + \sum_{P \in \text{Sing}\, \Sigma} e_P) \in H^2(S, \mathbf{Z})$.

Proposition (6.1). *Let $(\Sigma, \mathcal{L})$ be a pair satisfying the hypotheses above, let α be the 2-index of Σ, and let $k = \#(\text{Sing}\, \Sigma)$.*

(i) *If $(\alpha, k) \neq (5, 16)$ then $M_{\Sigma,\mathcal{L}}$ is generated by $c_1(L_\Sigma)$ and μ.*

(ii) *If $(\alpha, k) = (5, 6)$, then $M_{\Sigma,\mathcal{L}}$ is generated by $c_1(L_\Sigma)$, μ, and an element of the form $(1/4) + (1/2)\sum_{P \in J} e_P$, where $J \subset \text{Sing}\, \Sigma$ is a collection of 6 points such that for any choice of $\mathbf{F}_2$-vector space structure on $\text{Sing}\, \Sigma$ compatible with the code $\mathbf{C}_\Sigma$, every hyperplane in $\text{Sing}\, \Sigma$ contain either 2 or 4 points of J.*

(iii) *There is an isomorphism of binary linear codes $\mathbf{C}_\Sigma \xrightarrow{\sim} \mathbf{C}_{\alpha,k}$ such that the induced isometry $c_1(L_\Sigma) \xrightarrow{\sim} c_1(L_{\alpha,k})$ extends to an isometry $M_{\Sigma,\mathcal{L}} \xrightarrow{\sim} M_{\alpha,k}$ which sends $c_1(\pi^*\mathcal{L})$ to $c_1(\pi^*\mathcal{L}_{\alpha,k})$.*

Proof. Let $\tilde{M}_{\Sigma,\mathcal{L}}$ be the lattice generated by $c_1(L_\Sigma)$ and μ. We first claim that if $(\alpha, k) = (5, 16)$, then $M_{\Sigma,\mathcal{L}}$ is strictly larger than $\tilde{M}_{\Sigma,\mathcal{L}}$. For in the case, $c_1(L_\Sigma)$ contains $1/2 \sum_{P \in \mathrm{Sing}\,\Sigma} e_P$ so that $\tilde{M}_{\Sigma,\mathcal{L}}$ is generated by $c_1(L_\Sigma)$ and $\nu = \mu - 1/2 \sum_{P \in \mathrm{Sing}\,\Sigma} e_P$. Now $\nu \cdot \nu = 4$ and $\nu \cdot e_P = 0$ for each P, so $\tilde{M}_{\Sigma,\mathcal{L}}$ splits as an orthogonal direct sum $< \nu > \oplus c_1(L_\Sigma)$, where $< \nu >$ denotes the span of ν. This implies that $\tilde{M}^*_{\Sigma,\mathcal{L}}/\tilde{M}_{\Sigma,\mathcal{L}}$ is isomorphic to $\mathbf{Z}/4\mathbf{Z} \oplus L^*_\Sigma/L_\Sigma$. Now $E^*_\Sigma/E_\Sigma \cong (\mathbf{Z}/2\mathbf{Z})^{16}$ while $\log_2[L_\Sigma : E_\Sigma] = \alpha = 5$ so that $L^*_\Sigma/L_\Sigma \cong (\mathbf{Z}/2\mathbf{Z})^6$; in particular, the finite abelian group $\tilde{M}^*_{\Sigma,\mathcal{L}}/\tilde{M}_{\Sigma,\mathcal{L}}$ has 7 generators. But since $\tilde{M}_{\Sigma,\mathcal{L}}$ has rank 17, this impiles that $\tilde{M}_{\Sigma,\mathcal{L}}$ admits no primitive embedding into K3 lattice Λ (which has rank 22). But $M_{\Sigma,\mathcal{L}}$ is primitively embedded in $H^2(S, \mathbf{Z}) \cong \Lambda$; hence, $M_{\Sigma,\mathcal{L}} \neq \tilde{M}_{\Sigma,\mathcal{L}}$.

Returning to the general case, let x be an element of $M_{\Sigma,\mathcal{L}}$ and write

$$x = \frac{a}{2k - 16}\lambda + \frac{1}{2}\sum b_P e_P$$

with $a, b_P \in \mathbf{Q}$. Now

$$
\begin{aligned}
x \cdot e_P &= -b_P \\
x \cdot \mu &= \frac{1}{2}a - \frac{1}{2}\sum b_P \\
\frac{1}{2}x \cdot x &= \frac{a^2}{4(k - 8)} - \frac{1}{4}\sum (b_P)^2
\end{aligned}
$$

and all of these quantities must be integers. In paticular, a and each b_P are integers as is $a^2/(k - 8)$. Since $1 \leq k - 8 \leq 8$, there are three cases:

(1) $a \equiv 0 \bmod k - 8$
(2) $k = 12$ and $a \equiv 2 \bmod 4$
(3) $k = 16$ and $a \equiv 4 \bmod 8$.

In the first case, write $a = (k - 8)c$ with $c \in \mathbf{Z}$. Then

$$x - c\mu = \frac{1}{2}\sum (b_P - c)e_P$$

is in the saturation of $c_1(E_\Sigma)$, and hence lies in $c_1(L_\Sigma)$, so that $x \in \tilde{M}_{\Sigma,\mathcal{L}}$. In the second case, write $a = 4c + 2$ with $c \in \mathbf{Z}$. Then $a \in 2\mathbf{Z}$ so that $\sum b_P \in 2\mathbf{Z}$; on the other hand,

$$\frac{a^2}{4(k - 8)} - \frac{1}{4} \in \frac{1}{2}\mathbf{Z}$$

so that $(1/4)\sum(b_P)^2 \notin (1/2)\mathbf{Z}$ a contradiction. Thus, whenever $k \leq 15$ we have $M_{\Sigma,\mathcal{L}} = \tilde{M}_{\Sigma,\mathcal{L}}$, which proves (i). (Recall that $k = 16$ implies $\alpha = 5$).

We now assume $(\alpha, k) = (5, 16)$; if x falls in the case (1) above then $x \in \tilde{M}_{\Sigma,\mathcal{L}}$. If x falls in case (3), write $a = 8c + 4$ with $c \in \mathbf{Z}$, and let $J = \{P \mid$

b_P is odd}. Then

$$x - c\mu - \sum \left[\frac{b_P - c}{2}\right] e_P = \frac{1}{4}\lambda + \frac{1}{2} \sum_{P \in J} e_P$$

so that element of $M_{\Sigma,\mathcal{L}} \setminus \tilde{M}_{\Sigma,\mathcal{L}}$ is congruent mod $\tilde{M}_{\Sigma,\mathcal{L}}$ to an element of the form $y_J = (1/4)\lambda + (1/2)\sum_{P \in J} e_P$. Moreover, if $y_{J'} = (1/4)\lambda + (1/2)\sum_{P \in J'} e_P$ is another such element, then their difference is

$$\frac{1}{2} \sum_{P \in J} e_P - \frac{1}{2} \sum_{P \in J'} e_P \in c_1(L_\Sigma).$$

Thus, $[M_{\Sigma,\mathcal{L}} : \tilde{M}_{\Sigma,\mathcal{L}}] = 2$.

Fix a subset $J \subset \text{Sing}\,\Sigma$ such that y_J generates $M_{\Sigma,\mathcal{L}} \setminus \tilde{M}_{\Sigma,\mathcal{L}}$; by adding $(1/2)\sum_{P \in \text{Sing}\,\Sigma} e_P$ if necessary, we may assume $\#(J) \geq 8$. Now $(1/2)y_J \cdot y_J = (1/2) - (1/4)\#(J)$ so that $\#(J) \equiv \text{mod}4$. If $J = \{Q, R\}$ contains of 2 points, let $H \subset \text{Sing}\,\Sigma$ be a hyperplane (with respect to some fixed $\mathbf{F}_2$-vector space structre) which contains Q but does not contain R. Then $(1/2)\sum e_P \in c_1(L_\Sigma)$ so that $y_{J'} = y_J + (1/2)\sum_{P \notin H} e_P - e_R \in M_{\Sigma,\mathcal{L}}$, where $J' = \{Q\} \cup (H \setminus \{R\})$. But then $\#(J') = 8 \not\equiv \text{mod}4$, which is a contradiction. Thus, $\#(J) = 6$; moreover, for any J' with $y_{J'} \in M_{\Sigma,\mathcal{L}}$ we have $\#(J') = 6$ or 10.

To finish the proof of (ii), let H be any hyperplane in $\text{Sing}\,\Sigma$ (so that $(1/2)\sum_{P \notin H} e_P \in c_1(L_\Sigma)$) and let $J' = (H \cap J) \cup (\text{Sing}\,\Sigma \setminus (H \cup J))$. Then

$$y_{J'} = y_J = \frac{1}{2}\sum_{P \notin H} e_P - \sum_{P \notin J \setminus H} e_P \in M_{\Sigma,\mathcal{L}}$$

while $\#(J') = 2 + 2\#(H \cap J)$. Thus, $\#(H \cap J) = 2$ or 4, proving (ii).

Finally, by theorem (2.1) there is an isomorphism of binary linear codes $\mathcal{C}_\Sigma \xrightarrow{\sim} \mathcal{C}_{\alpha,k}$ which induces an isometry $L_\Sigma \xrightarrow{\sim} L_{\alpha,k}$; this clearly extends to an isometry $\tilde{M}_{\Sigma,\mathcal{L}} \xrightarrow{\sim} \tilde{M}_{\Sigma_{\alpha,k},\mathcal{L}_{\alpha,k}}$. When $k \leq 15$, (iii) now follows from (i). In the remaining case $(\alpha, k) = (5, 16)$ we must show: for any two subsets J, J' of $\text{Sing}\,\Sigma_{5,16}$ satisfying the condition in (ii), there is an automorphism σ of the binary linear code $\mathcal{C}_{5,16} \cong \mathcal{D}_5$ such that $\sigma(J) = J'$. Recall that by lemma (1.4), $\text{Aut}(\mathcal{D}_5) = \text{AGL}(\text{Sing}\,\Sigma_{5,16})$.

Fix a vector space structure on $\text{Sing}\,\Sigma_{5,16}$ with zero-vector P_0, fix a basis $P_1, \ldots, P_4$ of the vector space, and let $P_5 = P_1 + P_2 + P_3 + P_4$ (vector space addition). We will show that any J satisfying (ii) can be mapped to $\{P_0, \ldots, P_5\}$ by an affine linear transformation. (It is easy to see that $\{P_0, \ldots, P_5\}$ satisfies (ii).) First, choose an affine transformation T so that the zero-vector P_0 lies in $T(J)$. Let $R_1, \ldots, R_4$ be 4 elements of $T(J)$ distinct from P_0; since no hyperplane contains 5 or 6 elements of $T(J)$, the set $\{P_0, R_1, \ldots, R_4\}$ is not contained in a hyperplane; hence, $R_1, \ldots, R_4$ is a basis of $\text{Sing}\,\Sigma_{5,16}$. Let γ be the linear transformation such that $\gamma(R_i) = P_i$ for $i = 1, \ldots, 4$ for some point Q, $Q \neq P_i$.

For each $i = 1, \ldots, 4$ consider the hyperplane H_i spanned by the set of 4 point $\{P_j \mid j \neq i\}$. Since $\#(H_i \cap J) = 2$ or 4 we see thst $Q \notin H_i$. But the only point of $\mathrm{Sing}\,\Sigma_{5,16}$ not lying in $H_1 \cup \cdots \cup H_4$ is P_5; hence $Q = P_5$ and $\gamma \circ T(J) = \{P_0, \ldots, P_5\}$.   Q.E.D.

Recall that a *finite quadratic form* is a finite abelian group G together with a map $q : G \to \mathbf{Q}/2\mathbf{Z}$ such that (1) $q(nx) = n^2 q(x)$ for $x \in G$, $n \in \mathbf{Z}$ and (2) $b(x, y) = (1/2)(q(x + y) - q(x) - q(y))$ defines a bilinear map of $\mathbf{Z}$-modules $b : G \times G \to \mathbf{Q}/\mathbf{Z}$. We call a finite quadratic form *special* if $2x = 0$ impiles $q(x) \in \mathbf{Z} \bmod 2\mathbf{Z}$ for all $x \in G$. Note that if G is a 2-elementary group and $x_1, \ldots, x_n$ is a generating set for G, then q is special if and only if $q(x_i) \in \mathbf{Z} \bmod 2\mathbf{Z}$ for all $i = 1, \ldots, n$.

If a and l are natural numbers with $2|al$ and $(a, l) = 1$, we let z_l^a denote the finite quadratic form $(\mathbf{Z}/l\mathbf{Z}, q)$ where $q(x) = a/l \bmod 2\mathbf{Z}$ for some generator x of $\mathbf{Z}/l\mathbf{Z}$.

The intersection form on the Todorov lattice $M_{\alpha,k}$ induces an adjoint map $\mathrm{ad} : M_{\alpha,k} \to M_{\alpha,k}^* = \mathrm{Hom}(M_{\alpha,k}, \mathbf{Z})$; we define $G_{\alpha,k}$ to be the cokernel of this map. $G_{\alpha,k}$ is a finite abelian group which inherits the structure of a finite quadratic form $(G_{\alpha,k}, q_{\alpha,k})$ (called the *discriminat-form*) from $M_{\alpha,k}$ by the following procedure: for each $x \in M_{\alpha,k}^*$ there is some $n \in \mathbf{Z}$ and some $y \in M_{\alpha,k}$ such that $\mathrm{ad}(y) = nx$; one defines $q_{\alpha,k}(x) = (1/n^2) y \cdot y \bmod 2\mathbf{Z}$.

Proposition (6.2). *Let $M_{\alpha,k}$ be a Todorov lattice.*

(i) *There is a special 2-elementary quadratic form $(G'_{\alpha,k}, q'_{\alpha,k})$ of rank $2s$ and an orthogonal direct sum decomposition $(G_{\alpha,k}, q_{\alpha,k}) \cong z_l^a \oplus (G'_{\alpha,k}, q'_{\alpha,k})$, where*

$$(2s, l, a) = \begin{cases} (k - 2\alpha - 1, k - 8, 2) & \text{if } k \text{ is odd} \\ (k - 2\alpha, 2k - 16, k - 7) & \text{if } k \text{ is even, } k < 16 \\ (4, 4, 1) & \text{if } (\alpha, k) = (5, 16) \end{cases}$$

Note that since $k \geq 9$ and (α, k) satisfies $(2.1)(i)$, $2s \geq 4$ in all cases.

(ii) *If $(\alpha, k) \neq (5, 16)$ and $\sigma \in \mathrm{Aut}(\mathcal{C}_{\alpha,k})$, consider the automorphism of $M_{\alpha,k}$ defined by $\lambda \mapsto \lambda$, $e_p \mapsto e_{\sigma(P)}$. If the induced automorphism of $G_{\alpha,k}$ coincides with the action of -1, then $(\alpha, k) = (0, 9)$ or $(1, 10)$.*

Proof. (i) We first treat the case $(\alpha, k) = (5, 16)$. Let A_0 be the principally polarized abelian surface used to construct $\Sigma_0 = \Sigma_{5,16}$, let $f : A_0 \to \Sigma_0$ be the quotient map and let $\pi_0 : S_0 \to \Sigma_0$ be the minimal desigularization. Since $\mathrm{End}(A_0) = \mathbf{Z}$, the principal polarization Θ_0 generates $\mathrm{NS}(A_0)$, and $\Theta_0 \cdot \Theta_0 = 2$. Let $T(A_0)$ be the transcendental lattice, that is, the orthogonal complement of $\mathrm{NS}(A_0)$ in $H^2(A_0, \mathbf{Z})$. The Kneser-Nikulin uniqueness theorem [31;corollary 1.13.3] immediately implies that $T(A_0) \cong U^{\oplus 2} \oplus T'$ (cf. [24;corollary 2.6]) where U is the hyperbolic plane and T' is infinite cyclic with generator y such that $y \cdot y = -2$.

f induces a map $f_* : T(A_0)(2) \to H^2(S_0, \mathbf{Z})$ preserving quadratic forms (where $L(2)$ denote multiplication of the quadratic form L by 2) whose image has finite index in $T(S_0)$. If $n = [T(S_0) : T(A_0)(2)]$ then the discriminant of $T(S_0)$ is $n^2 \cdot disc(U(2)^{\oplus 2} \oplus T'(2)) = -2^6 n^2$. On the other hand, the discriminant of $NS(S_0) = M_{5,16}$ is $(1/4)disc(\tilde{M}_{5,16}) = (1/4) \cdot 4disc(L_{5,16}) = 2^6$. Since these discriminants have the same absolute value, we must have $n = 1$.

This implies that $(G_{5,16}, q_{5,16})$ is isomorphic to the discriminant-form of $T(A_0)(2)$ by an isomorphism reversing the sign on the quadratic form. Let $(G'_{5,16}, q'_{5,16})$ be the inverse image of the discriminant-form $U(2)^{\oplus 2}$ under this isomorphism, and let x be the inverse image of $\text{ad}((1/4)y')$ where y' is the image of y in $T(A_0)(2)$. Then $(G'_{5,16}, q'_{5,16})$ is a special 2-elementary form of rank 4, and $q_{5,16}(x) = (-1/16)y' \cdot y' = 1/4$ so that the subgroup $< x >$ of $G_{5,16}$ generated by x is isomorphic to z_4^1. The orthogonal direct sum decomposition $T(A_0)(2) \cong U(2)^{\oplus 2} \oplus T'(2)$ induces an orthogonal direct sum decomposition $(G_{5,16}, q_{5,16}) \cong (G'_{5,16}, q'_{5,16}) \oplus < x >$, proving (i) in this case.

We assume for the rest of the proof that $k \leq 15$ so that $M_{\alpha,k} = \tilde{M}_{\alpha,k}$ by proposition (6.1). We abbreviate $q_{\alpha,k}$ by q, and let b be the associated bilinear form. Let $\hat{M}_{\alpha,k}$ be the span of μ and $\{e_P \mid P \in \text{Sing}\,\Sigma_{\alpha,k}\}$, and let μ^* and $\{e_P^*\}$ be the dual basis of $\hat{M}_{\alpha,k}^* = \text{Hom}(\hat{M}_{\alpha,k}, \mathbf{Z})$. We will first compute $\hat{G}_{\alpha,k} = \hat{M}_{\alpha,k}^* / \text{ad}(\hat{M}_{\alpha,k})$.

Fix a point $Q \in \text{Sing}\,\Sigma_{\alpha,k}$ and define $\xi_P = e_Q^* - e_P^* \in \hat{M}_{\alpha,k}^*$. μ^* and $\{e_P^*\}$ generate $\hat{G}_{\alpha,k}$, with relations:

$$\text{ad}(\mu) = -4\mu^* - \sum_P e_P^* = -4\mu^* - ke_Q^* + \sum_P \xi_P$$
$$\text{ad}(e_P) = -\mu^* - 2e_P^* = -\mu^* - 2e_Q^* + 2\xi_P.$$

We fix a second point $R \neq Q$ in $\text{Sing}\,\Sigma_{\alpha,k}$, and consider the basis of $\hat{M}_{\alpha,k}$ consisting of e_Q, $\mu - 4e_Q$, $-2\mu + \sum_P e_P + (8 - k)e_Q$, $\{e_P - e_Q \mid P \neq Q, R\}$. Then one easily computes (using the fact that $\xi_Q = 0$):

$$\text{ad}(e_Q) = -\mu^* - 2e_Q^*$$
$$\text{ad}(\mu - 4e_Q) = (8 - k)e_Q^* + \sum_{P \neq Q} \xi_P$$
$$\text{ad}\left(-2\mu + \sum_P e_P + (8 - k)e_Q\right) = (2k - 16)e_Q^*$$
$$\text{ad}(e_P - e_Q) = 2\xi_P$$

This implies that $\hat{G}_{\alpha,k}$ is generated by e_Q^* and $\{\xi_P \mid P \neq Q\}$, subject to the relations $(2k - 16)e_Q^* \equiv 2\xi_P \equiv 0$ and $\sum_{P \neq Q} \xi_P \equiv (k - 8)e_Q^*$. This presentation also makes it easy to compute the quadratic form on $\hat{G}_{\alpha,k}$. For example, $b(\xi_{P_1}, \xi_{P_2}) = (1/4)(e_{P_1} - e_Q) \cdot (e_{P_2} - e_Q) \equiv (-1/2) \mod \mathbf{Z}$ if $P_1 \neq P_2, P_i \neq Q$. The complete results of the computation are the following:

$$b(\xi_{P_1}, \xi_{P_2}) \equiv \frac{1}{2} \mod \mathbf{Z} \quad \text{if } P_1 \neq P_2,\, P_i \neq Q$$

$$b(\xi_P, e_Q^*) \equiv \frac{1}{2} \bmod \mathbf{Z} \quad \text{if } P \neq Q$$

$$q(\xi_P) \equiv 1 \bmod 2\mathbf{Z} \quad \text{if } P \neq Q$$

$$q(e_Q^*) \equiv \frac{9 - k}{2k - 16} \bmod 2\mathbf{Z}.$$

Let V denote the subgroup $\mathrm{ad}(M_{\alpha,k})/\mathrm{ad}(\hat{M}_{\alpha,k})$ of $\hat{G}_{\alpha,k}$, so that $G_{\alpha,k} = V^\perp/V$. Since $M_{\alpha,k}/\hat{M}_{\alpha,k} = L_{\alpha,k}/E_{\alpha,k}$, every element of V has the form $\mathrm{ad}(c_1(\Gamma_\phi))$, where $\phi \in \mathbf{F}_2{}^{\mathrm{Sing}\,\Sigma_{\alpha,k}}$ belongs to the double point code of $\Sigma_{\alpha,k}$ and $c_1(\Gamma_\phi) = (1/2)\sum_{P \in \mathrm{Sing}\,\Sigma_{\alpha,k}} e_P$. If we let $I = I_\phi = \{P \in \mathrm{Sing}\,\Sigma_{\alpha,k} \mid \phi(P) = 1\}$ then $\#(I) = 0$ or 8 and

$$\mathrm{ad}(c_1(\Gamma_\phi)) = \mathrm{ad}(4e_Q) - \mathrm{ad}\left(\frac{1}{2}\sum_{P \in I}(e_Q - e_P)\right) \equiv \sum_{P \in I} \xi_P \bmod \mathrm{ad}(\hat{M}_{\alpha,k}).$$

For such sets I, let $\eta_I = \sum_{P \in I} \xi_P$.

Suppose that k is odd. Then the 2-Sylow subgroup $\hat{G}'_{\alpha,k}$ of $\hat{G}_{\alpha,k}$ is generated by $(k - 8)e_Q^*$ and $\{\xi_P\}$, while the sum of the remaining Sylow subgroups is generated by $x = 2e_Q^*$. The Sylow decomposition then induces an orthogonal direct sum decomposition $(\hat{G}_{\alpha,k}, q_{\alpha,k}) \cong (\hat{G}'_{\alpha,k}, q'_{\alpha,k}) \oplus <x>$, where $<x>$ denotes the subgroup generated by x. Moreover, since

$$q(x) = 4q(e_Q^*) \equiv 2/(k - 8) \bmod 2\mathbf{Z},$$

we have $<x> \cong z_{k-8}^2$.

Since each η_I belongs to $\hat{G}'_{\alpha,k}$ we have $V \subset \hat{G}'_{\alpha,k}$; if we define $\tilde{V}^\perp$ to be the orthogonal complement of V in $\hat{G}'_{\alpha,k}$ and $G'_{\alpha,k} = \tilde{V}^\perp/V$, then there is an orthogonal direct sum decomposition $(G_{\alpha,k}, q_{\alpha,k}) \cong (G'_{\alpha,k}, q'_{\alpha,k}) \oplus <x>$. From the values of q and b on the generators of $\hat{G}'_{\alpha,k}$ it follows that $(\hat{G}'_{\alpha,k}, q'_{\alpha,k})$ is a special 2-elementary form; since $G'_{\alpha,k}$ is a subquotient of $\hat{G}'_{\alpha,k}$, $(G'_{\alpha,k}, q'_{\alpha,k})$ has the same property. Finally, by computing the discriminant, we see that $G'_{\alpha,k}$ has rank $k - 2\alpha - 1$, proving (i) when k is odd.

Suppose now that k is even, $k < 16$. We choose a subset J of $\mathrm{Sing}\,\Sigma_{\alpha,k}$, not containing Q, in the following manner: if $\alpha = 0$, let J be any 3 points of $\Sigma_{\alpha,k}$ distinct from Q. If $\alpha = 1$ and $\eta_I \in V$ is the nontrivial element, let J consist of 2 points in I and 1 point not in I with $Q \notin J$; this is possible since $k \geq 10$ so that $\#(\mathrm{Sing}\,\Sigma_{\alpha,k} \setminus I) \geq 2$. If $\alpha = 2$ or 3, let W be an $\mathbf{F}_2$-vector space of dimension α let $\tau : \mathrm{Sing}\,\Sigma_{\alpha,k} \to W$ be a surjective map such that $\tau^*(\mathcal{U}_\alpha)$ is the double point code of $\Sigma_{\alpha,k}$; such maps exist by theorem (1.1). Choose J so that $Q \notin J$, $J \cap \tau^{-1}(0) = \emptyset$, and for each nonzero $w \in W$, $J \cap \tau^{-1}(w)$ consists of a single point.

The subset J has the following properties: $\#(J) \equiv 3 \bmod 4$, and for each $I \subset \mathrm{Sing}\,\Sigma_{\alpha,k}$ such that $\eta_I \in V$; the set $I \cap J$ contains an even number of points.

We let $f_J = e_Q^* + \sum_{P \in J} \xi_P$. The properties of J enable us to compute:

$$b(f_J, \xi_P) \equiv \begin{cases} \dfrac{1}{2} \bmod \mathbf{Z} & \text{if } P \in J \\ 0 \bmod \mathbf{Z} & \text{if } P \notin J, \, P \neq Q \end{cases}$$

$$b(f_J, (k-8)e_Q^*) \equiv \frac{1}{2} \bmod \mathbf{Z}$$

$$q(f_J) \equiv \frac{9-k}{2k-16} - 3 \equiv \frac{k-7}{2k-16} \bmod 2\mathbf{Z}.$$

The subgroup $< f_J >$ of $(G_{\alpha,k}, q_{\alpha,k})$ generated by f_J is thus isomorphic to Z_{2k-16}^{k-7}.

Let $\zeta_P = (k-8)e_Q^* + \xi_P$, and let $\hat{G}'_{\alpha,k}$ be the subgroup of $\hat{G}_{\alpha,k}$ generated by $\{\zeta_P \mid P \in J\}$ and $\{\xi_P \mid P \notin J\}$. We claim there is an orthogonal direct sum decomposition $(\hat{G}_{\alpha,k}, q_{\alpha,k}) = (\hat{G}'_{\alpha,k}, q'_{\alpha,k}) \oplus < f_J >$. It is clear that $\hat{G}'_{\alpha,k} \perp f_J$; we must show that $\hat{G}'_{\alpha,k} + < f_J > = \hat{G}_{\alpha,k}$. Since k is even, $k - 7$ is a unit mod $(2k - 16)$ so that $e_Q^* \in \hat{G}'_{\alpha,k} + < f_J >$. But then $\xi_P \in \hat{G}'_{\alpha,k} + < f_J >$ for each P, so that $\hat{G}'_{\alpha,k} + < f_J > = \hat{G}_{\alpha,k}$.

It remains to show that $\hat{G}'_{\alpha,k}$ is a special 2-elementary form and that $V \subset \hat{G}'_{\alpha,k}$; the rest of the proof is then the same as in the case of odd k. For the first statement, we simply compute

$$q(\zeta_P) \equiv \frac{(k-8)(9-k)}{2} + 2(k-8)(\frac{1}{2}) + 1 \bmod 2\mathbf{Z}$$

so that $q(\zeta_P) \in \mathbf{Z} \bmod 2\mathbf{Z}$. For the second statement, if $\eta_I \in V$ let $\#(I \cap J) = 2m$; then

$$\eta_I \equiv \sum_{P \in I \backslash I} \xi_P + \sum_{P \in J \backslash I} +2m(k-8)e_Q^* = \sum_{P \in I \backslash J} \xi_P + \sum_{P \in J \backslash I} \zeta_P \in \hat{G}'_{\alpha,k};$$

hence $V \subset \hat{G}'_{\alpha,k}$, proving (i).

(ii) Let $\sigma \in \text{Aut}(\mathbf{C}_{\alpha,k})$ be an element whose induced action on $G_{\alpha,k}$ coincides with the action of -1. The natural action of σ sends e_Q^* to $e_Q^* - \xi_{\sigma(Q)}$ and sends ξ_P to $\xi_{\sigma(P)} - \xi_{\sigma(Q)}$.

Fisrt, suppose that k is odd. Since σ acts as -1 on $G_{\alpha,k}$, $e_Q^* + e_Q^* - \xi_{\sigma(Q)} \equiv 0$ in $G_{\alpha,k}$; this implies that $2e_Q^* - \xi_{\sigma(Q)} \in V$. In particular, since every element of V has order 2, we have $4e_Q^* \equiv 0$ in $\hat{G}_{\alpha,k}$. Now the order of e_Q^* in $\hat{G}_{\alpha,k}$ is $2k - 16$; hence, $2k - 16$ divides 4. Since $k \geq 9$ is odd, this implies that $k = 9$; α must then be 0 since $M_{\alpha,k}$ is a Todorov lattice.

Suppose instead that k is even; by hypothesis, $k < 16$. Let $J \subset \text{Sing } \Sigma_{\alpha,k}$ be the subset chosen in the proof of (i). Since $\#(J)$ is odd, any $\sigma \in \text{Aut}(\mathbf{C}_{\alpha,k})$ sends f_J to $f_{\sigma(J)}$. If σ acts as -1 on $G_{\alpha,k}$, we see as above that $f_J + f_{\sigma(J)} = 2f_J - \sum_{P \in J} \xi_P + \sum_{P \in \sigma(J)} \xi_P \in V$. Since each ξ_P and every element of V has order 2, this implies that $4f_J \equiv 0$ in $\hat{G}_{\alpha,k}$. Now f_J has order $2k - 16$, so that $2k - 16$ divides 4 as before; since k is even, we find that $k = 10$.

The construction of J now implies that $\#(J) = 3$. Thus, $\#(J \cup \sigma(J) \cup \{Q\}) \leq 7$ so that $\sum_{P \in J} \xi_P - \sum_{P \in \sigma(J)} \xi_P \neq \sum_P \xi_P = 2e_Q^*$. Hence, $f_J + f_{\sigma(J)} = 2e_Q^* - \sum_{P \in J} \xi_P + \sum_{P \in \sigma(J)} \xi_P$ is a nontrivial element of V; this implies that $\alpha \geq 1$. Since $M_{\alpha,k}$ is a Todorov lattice, $(\alpha, k) = (1, 10)$.   Q.E.D.

Let L be a nondegenerate lattice of signature (r_+, r_-), and let $O(L)$ be its orthogonal group. There are natural homomorphisms det : $O(L) \to \{\pm 1\}$ and spin : $O(L) \to \{\pm 1\}$, where det γ is the determinant of γ and spin γ is its real spinor norm. We define

$$O_-(L) = \{\gamma \in O(L) \mid \det \gamma = \text{spin } \gamma\}.$$

This group can be given the following geometric interpretation (cf. [21]). A (positive) sign structure on L is a choice of one of the connected components of the set of oriented r_+-planes in $L \otimes \mathbf{R}$ on which the form is positive definite; the sign structure containing the oriented plane ν is denoted by $[\nu]$. If $[\nu]$ is a sign structure on L, then

$$O_-(L) = \{\gamma \in O(L) \mid \gamma([\nu]) = [\nu]\}.$$

Theorem (6.3). Let $M_{\alpha,k}$ be a Todorov lattice and let Λ be the K3 lattice.
(i) There is a primitive embedding $\phi : M_{\alpha,k} \hookrightarrow \Lambda$.
(ii) If ϕ_1, $\phi_2 : M_{\alpha,k} \hookrightarrow \Lambda$ are primitive emdeddings, then there is some $\gamma \in O_-(\Lambda)$ such that $\gamma \circ \phi_1 = \phi_2$.

Proof. Statement (i) follows from the definition of $M_{\alpha,k}$, for $M_{\alpha,k} \subset H^2(\Sigma_{\alpha,k}, \mathbf{Z}) \cong \Lambda$ is a primitive sublattice.

To check statement (ii), we use a variant of a theorem of Nikulin [33;theorem 1.6] which is proved as theorem (A.1) in an appendix to this paper. Let $M = \phi_1(M_{\alpha,k})$ and $K = M^{\perp}$. K has signature $(2, 19 - k)$ and $k \leq 16$ so that K is indefinite and rank$(K) \geq 3$. To check hypothesis (iii) of theorem (A.1), note that $G_M \cong G_K$ so that G_M can be generated by a set containing at most rank(K) elements. Moreover, by proposition (6.2), for each $p \neq 2$ the p-Sylow subgroup of G_M can be generated by a single element, and the 2-Sylow subgroup of G_M has an orthogonal direct sum decomposition of the form $G_1 \oplus G_2$ where G_2 is a special 2-elementary form of rank at least 4.

If we now let $M_1 = \phi_2(M_{\alpha,k})$, by theorem (A.1) there is some $\gamma \in O_-(\Gamma)$ such that $\gamma|_M = \phi_2 \circ \phi_1^{-1}$, in other words, $\gamma \circ \phi_1 = \phi_2$.

§7. The moduli of Todorov surfaces.

Let Z be a Todorov surface with fundamental invariants (α, k). By lemma (5.3)(iii), the RDP involution $j : Z \to Z$ whose quotient is a K3 surface is uniquely determined by Z; this implies that there is a one to one correspondence between isomorphism classes of Todorov surfaces with these invariants, and isomorphism classes of triples $(X, \mathcal{M}, B)$ satisfying the conditions in theorem (4.1) in which X is a K3 surface with rational double points, $B^2 = 2k - 16$, and the distinguished partial desingularization Σ of X has k singular points and 2-index α.

Suppose that (α, k) is one of the 11 possible values of the fundamental invariants of a Todorov surface. A *K3 surface of Todorov type* (α, k) is a triple $(X, \mathcal{L}, \Sigma)$ consisting of a K3 surface X with rational double points, an ample line bundle $\mathcal{L}$ on X with $\mathcal{L} \cdot \mathcal{L} = 2k - 16$, and a partial desingularization $\nu : \Sigma \to X$ with k ordinary double points which has 2-index α, such that if $\pi : S \to \Sigma$ and $\mu = \nu \circ \pi : S \to X$ are the minimal desingularizations then $(1/2)c_1(\mu^*(\mathcal{L}) \otimes \mathcal{O}_S(\sum_{P \in \mathrm{Sing}\,\Sigma} \pi^{-1}(P))) \in H^2(S, \mathbf{Z})$. There is a natural transformation from triples $(X, \mathcal{M}, B)$ as in the preceding paragraph to K3 surfaces of Todorov type (α, k) given by setting $\mathcal{L} = \mathcal{O}_X(B)$ and letting Σ be the distinguished partial desingularization; we will first study the moduli of these latter surfaces. Note that when $(X, \mathcal{L}, \Sigma)$ is a K3 surface of Todorov type (α, k), $(\Sigma, \nu^*\mathcal{L})$ is a pair of the type considered in section 6.

Let us fix once and for all a primitive embedding of the Todorov lattice $M_{\alpha,k}$ into the K3 lattice Λ (these exist by theorem (6.3)(i)), and identify $M_{\alpha,k}$ with its image in Λ. We let $N_{\alpha,k}$ be the orthogonal complement of $M_{\alpha,k}$ in Λ, and fix a positive sign structure $[\nu_{\alpha,k}]$ on $N_{\alpha,k}$. The *period space* $D_{\alpha,k}$ is then defined by

$$D_{\alpha,k} = \{\omega \in \mathbf{P}(N_{\alpha,k} \otimes \mathbf{C}) \mid \omega \cdot \omega = 0,\ \omega \cdot \bar{\omega} > 0,\ \mathrm{Re}(\omega) \wedge \mathrm{Im}(\omega) \in [\nu_{\alpha,k}]\}$$

(The last condition ensures that $D_{\alpha,k}$ is connected.) The *integral automorphism group* of $D_{\alpha,k}$ is the group $\mathrm{Aut}_{\mathbf{Z}}(D_{\alpha,k}) = O_-(N_{\alpha,k})/(\pm 1)$.

Let $(X, \mathcal{L}, \Sigma)$ be a K3 surface of Todorov type (α, k), and let $\mu : S \to X$ and $\pi : S \to \Sigma$ be the minimal desingularizations. A *special marking* of $(X, \mathcal{L}, \Sigma)$ is an isometry $\phi : H^2(S, \mathbf{Z}) \to \Lambda$ together with an isomorphism of codes $\psi : C_\Sigma \to C_{\alpha,k}$ such that $c_1(\mu^*(\mathcal{L})) = \phi^{-1}(\lambda)$, $c_1(\pi^{-1}(P)) = \phi^{-1}(e_{\psi(P)})$ for each $P \in \mathrm{Sing}\,\Sigma$, and $\mathrm{Re}(\phi(\omega)) \wedge \mathrm{Im}(\phi(\omega)) \in [\nu_{\alpha,k}]$ for any nonzero holomorphic 2-form ω on S, where we have regarded $\phi(\omega)$ as an element of $N_{\alpha,k} \otimes \mathbf{C}$.

Lemma (7.1). *Every K3 surface of Todorov type (α, k) has a special marking.*

Proof. Let $M_{\Sigma,\mathcal{L}}$ be the saturation of the lattice generated by $c_1(L_\Sigma)$ and $c_1(\mu^*(\mathcal{L}))$ as in section 6; by proposition (6.1), there is an isomorphism of codes $\psi : C_\Sigma \to C_{\alpha,k}$ and an isometry $\eta : M_{\Sigma,\mathcal{L}} \to M_{\alpha,k}$ such that $\eta^{-1}(\lambda) = c_1(\mu^*(\mathcal{L}))$ and $\eta^{-1}(e_{\psi(P)}) = c_1(\pi^{-1}(P))$. If we choose an isometry $\phi' : H^2(S, \mathbf{Z}) \to \Lambda$ such that $\phi'[c_1(\mu^*(\mathcal{L})) \wedge \mathrm{Re}(\omega) \wedge \mathrm{Im}(\omega)] = [\lambda \wedge \nu_{\alpha,k}]$ for a nonzero holomorphic 2-form ω on S, then $\phi' \circ \eta^{-1}$ is a primitive embedding of $M_{\alpha,k}$ into Λ. By theorem (6.3)(ii), there is some $\gamma \in O_-(\Lambda)$ with $\gamma \circ \phi' \circ \eta^{-1} = 1_{M_{\alpha,k}}$. Let $\phi = \gamma \circ \phi'$; then $\phi^{-1}(\lambda) = c_1(\mu^*(\mathcal{L}))$ and $\phi^{-1}(e_{\psi(P)}) = c_1(\pi^{-1}(P))$ for each $P \in \mathrm{Sing}\,\Sigma$. Since γ

preserves the sign structure $[\lambda \wedge \nu_{\alpha,k}]$ on Λ,

$$[\lambda \wedge \mathrm{Re}(\phi(\omega)) \wedge \mathrm{Im}(\phi(\omega))] = \phi[c_1(\mu^*(\mathcal{L}) \wedge \mathrm{Re}(\omega)\mathrm{Im}(\omega)] = [\lambda \wedge \nu_{\alpha,k}]$$

so that $\mathrm{Re}(\phi(\omega)) \wedge \mathrm{Im}(\phi(\omega)) \in [\nu_{\alpha,k}]$ as sign structures on $N_{\alpha,k}$.   Q.E.D.

Note that if $\gamma \in O(\Lambda)$ satisfies $\gamma(\lambda) = \lambda$ and $\gamma(M_{\alpha,k}) = M_{\alpha,k}$ then $\gamma|_{N_{\alpha,k}}$ preserves the sign structure $[\nu_{\alpha,k}]$ if and only if $\gamma \in O_-(\lambda)$. Thus, if (ϕ,ψ) is a special marking of $(X,\mathcal{L},\Sigma)$, it follows directly from the definitions that (ϕ',ψ') is a special marking of the same $(X,\mathcal{L},\Sigma)$ if and only if $(\phi',\psi') = (\gamma \circ \phi, \sigma \circ \psi)$ for some $(\gamma,\sigma) \in \tilde{\Gamma}_{\alpha,k}$, where

$$\tilde{\Gamma}_{\alpha,k} = \{(\gamma,\sigma) \in O_-(\Lambda) \times \mathrm{Aut}(C_{\alpha,k}) | \gamma(\lambda) = \lambda, \ \gamma(e_P) = e_{\sigma(P)}\}.$$

If $(\gamma,\sigma) \in \tilde{\Gamma}_{\alpha,k}$ then $\gamma|_{N_{\alpha,k}}$ preserves the sign structure $[\nu_{\alpha,k}]$ so that $\gamma|_{N_{\alpha,k}}$ acts on $D_{\alpha,k}$; we define $\Gamma_{\alpha,k} = \mathrm{Image}(\tilde{\Gamma}_{\alpha,k} \to \mathrm{Aut}_{\mathbf{Z}}(D_{\alpha,k}))$ The set of special markings of $(X,\mathcal{L},\Sigma)$ thus determines a point in $D_{\alpha,k}/\Gamma_{\alpha,k}$.

We recall some definitions from [23]. Let X be a K3 surface with rational double points and let $\mu : S \to X$ be the minimal desingularization. The *root system of X* is the root system $R(X)$ spanned by the curves on S which are contracted by μ; the Weyl group of this root system is denoted by $W(X)$, and called the *Weyl group of X*. $W(X)$ acts on $H^2(S,\mathbf{Z})$, and the group of invariants $H^2(S,\mathbf{Z})^{W(X)}$ is denoted by $I^2(X)$. $I^2(X)$ coinincides with the orthogonal complement of $c_1(R(X))$, and inherits the structure of a lattice from the intersection form on $H^2(S,\mathbf{Z})$. A *marking* of X is an embedding of lattices $\phi_0 : I^2(X) \to \Lambda$ for which there exist extensions $\phi : H^2(S,\mathbf{Z}) \to \Lambda$ such that ϕ is an isometry and $\phi|_{I^2(X)} = \phi_0$.

If $(X,\mathcal{L},\Sigma)$ is a K3 surface of Todorov type (α,k), a special marking (ϕ,ψ) of $(X,\mathcal{L},\Sigma)$ induces a marking $\phi_0 = \phi|_{I^2(X)} : I^2(X) \to \Lambda$ of X with the property that the image of ϕ_0 is contained in the span of λ and $N_{\alpha,k}$.

Proposition (7.2). *Let X be a K3 surface with rational double points, let $\mathcal{L}$ be an ample line bundle on X, and let $\phi_0 : I^2(X) \to \Lambda$ be a marking such that $\phi_0(c_1(\mathcal{L})) = \lambda$, the image of ϕ_0 is contained in the span of λ and $N_{\alpha,k}$, and $\mathrm{Re}(\tilde{\phi}(\omega)) \wedge \mathrm{Im}(\tilde{\phi}(\omega)) \in [\nu_{\alpha,k}]$ for any nonzero holomorphic 2-form ω on the minimal desingularization S of X and any extension $\tilde{\phi} : H^2(S,\mathbf{Z}) \to \Lambda$ of ϕ_0. Then there is a partial desingularization Σ of X such that $(X,\mathcal{L},\Sigma)$ is a K3 surface of Todorov type (α,k), and a special marking (ϕ,ψ) of $(X,\mathcal{L},\Sigma)$ such that $\phi|_{I^2(X)} = \phi_0$. Moreover, the partial desingularization Σ is uniquely determined by $X,\mathcal{L}$ and ϕ_0.*

Proof. Let $\tilde{\phi} : H^2(S,\mathbf{Z}) \to \Lambda$ be any extension of ϕ_0. By composing $\tilde{\phi}$ with reflection in some of the classes e_P if necessary, we may assume that each $\tilde{\phi}^{-1}(e_P)$ is the first Chern class of an effective divisor E_P. Since e_P is orthogonal to both λ and $N_{\alpha,k}$, E_P belongs to the root system $R(X)$.

The structure of the Todorov lattice $M_{\alpha,k}$ guarantees that for each $I \subset$ Sing $\Sigma_{\alpha,k}$ with $|I| = 4$ we have $(1/2)c_1(\sum_{P \in I} E_P) \notin H^2(S, \mathbf{Z})$. But now by proposition (2.4), there is some $w \in W(X)$ such that $C_P = w(E_P)$ is effective and irreducible for each $P \in$ Sing $\Sigma_{\alpha,k}$. If we let $\pi : S \to \Sigma$ be the contraction of the curves C_P for $P \in$ Sing $\Sigma_{\alpha,k}$, let $\phi = \tilde{\phi} \circ w^{-1}$ and define $\psi : C_\Sigma \to C_{\alpha,k}$ by $\psi^{-1}(P) = \pi(C_P) \in$ Sing Σ, then $(X, \mathcal{L}, \Sigma)$ is a K3 surface of Todorov type (α, k) and (ϕ, ψ) is a special marking such that $\phi|_{I^2(X)} = \phi_0$.

To prove the last statement, suppose that $(X, \mathcal{L}, \Sigma_i)$ is a K3 surface of Todorov type (α, k) with a special marking (ϕ_i, ψ_i) such that $\phi_i|_{I^2(X)} = \phi_0$ for $i = 1, 2$. The proof of the weakly polarized global Torelli theorem [23; p. 319] then implies that there is some $w \in W(X)$ with $\phi_2 = \phi_1 \circ w$.

For each $Q \in$ Sing Σ_2, let $C_Q = \pi_2^{-1}(Q)$. Then $C_Q \in R(X)$, and for any $D \in R(X)$ we have

$$D \cdot \sum_{Q \in \text{Sing } \Sigma_2} C_Q = D \cdot (\mu_2^*(\mathcal{L}_2) \otimes \mathcal{O}_S(\sum_{Q \in \text{Sing } \Sigma_2} C_Q)) \in 2\mathbf{Z}.$$

Moreover, if $P = \psi_1^{-1}\psi_2(Q)$ then

$$c_1(\pi_1^{-1}(P)) = \phi_1^{-1}(e_{\psi_2(Q)}) = w\phi_2^{-1}(e_{\psi_2(Q)}) = wc_1(C_Q)$$

which implies that $w(C_Q) = \pi_1^{-1}(P)$ is effective and irreducible for each $Q \in$ Sing Σ_2. By lemma (2.6), there is some permuation σ of Sing Σ_2 such that $w(C_Q) = C_{\sigma(Q)}$. But this implies that $\psi_2 = \sigma \circ \psi_1$; hence, Σ_1 and Σ_2 are obtained by contracting the same set of curves on S, that is, $\Sigma_1 = \Sigma_2$. Q.E.D.

In spite of proposition (7.2), the choice of a partial desingularization Σ and the notion of a special marking of $(X, \mathcal{L}, \Sigma)$ are by no means superfluous when studying the moduli of K3 surfaces of Todorov type (α, k). The special markings are needed to allow us to describe the group $\Gamma_{\alpha,k}$ in an efficient way; the necessity of the choice of Σ will be discussed in section 8.

We are now in a position to prove:

Theorem (7.3). $D_{\alpha,k}/\Gamma_{\alpha,k}$ *is a coarse moduli space for K3 surfaces of Todorov type* (α, k).

Proof. Suppose that $(X_i, \mathcal{L}_i, \Sigma_i)$ for $i = 1, 2$ and K3 surfaces of Todorov type (α, k) which are assigned to the same point $x \in D_{\alpha,k}/\Gamma_{\alpha,k}$, and let $\mu_i : S_i \to X_i$ and $\pi_i : S_i \to \Sigma_i$ be the minimal desingularizations. If we choose some $\omega \in D_{\alpha,k}$ which maps to x in $D_{\alpha,k}/\Gamma_{\alpha,k}$, then there are special markings (ϕ_i, ψ_i) of $(X, \mathcal{L}_i, \Sigma_i)$ such that $\phi_i^{-1}(\omega) \in H^{2,0}(S)$ and $\phi_i^{-1}(\lambda) = c_1(\mu_i^*(\mathcal{L}_i))$ for $i = 1, 2$. In particular, $\phi = \phi_2^{-1} \circ \phi_1 : H^2(S_1, \mathbf{Z}) \to H^2(S_2, \mathbf{Z})$ is an isometry preserving the Hodge structure and $\phi(c_1(\mu_1^*(\mathcal{L}_1))) = c_1(\mu_2^*(\mathcal{L}_2))$. Let $\phi_0 = \phi|_{I^2(X_1)}$. Since $\mathcal{L}_i$ is ample on X_i, by the weakly polarized global Torelli theorem [23; p. 319] (cf. also [36]) there is an isomorphism $\phi : X_2 \xrightarrow{\sim} X_1$ such that $\phi^*|_{I^2(X_1)} = \phi_0$ and $\phi^*(\mathcal{L}_1) = \mathcal{L}_2$. By proposition (7.2), ϕ also induces an isomorphism $\bar{\phi} : \Sigma_2 \xrightarrow{\sim}$

Σ_1, and hence an isomorphism between the triples $(X_i, \mathcal{L}_i, \Sigma_i)$ for $i = 1, 2$. We conclude that the natural map from the moduli space of such triples to $D_{\alpha,k}/\Gamma_{\alpha,k}$ is injective.

To prove that the map is surjective, take an arbitrary point $\omega \in D_{\alpha,k}$. We use the surjectivity of the period map for algebraic K3 surfaces in the form given in [23; p. 325] (and due essentially to Kulikov [19]): there is a K3 surface with rational double points X, an ample line bundle $\mathcal{L}$ on X and a marking $\phi_0 : I^2(X) \to \Lambda$ with $\phi_0^{-1}(\lambda) = c_1(\mathcal{L})$ such that if $\mu : S \to X$ is the minimal desingularization then $\phi^{-1}(\omega) \in H^{2,0}(S)$ for any extension $\phi : H^2(S, \mathbf{Z}) \to \Lambda$ of ϕ_0; note that the image of ϕ_0 lies in the span of λ and $N_{\alpha,k}$. By proposition (7.2), there is a (unique) partial desingularization Σ of X and a special marking (ϕ, ψ) of $(X, \mathcal{L}, \Sigma)$ such that $\phi|_{I^2(X)} = \phi_0$. $(X, \mathcal{L}, \Sigma)$ is thus assigned to the image of ω in $D_{\alpha,k}/\Gamma_{\alpha,k}$.   Q.E.D.

Further details about the structure of the moduli space $D_{\alpha,k}/\Gamma_{\alpha,k}$, and in particular about the "period map" $D_{\alpha,k}/\Gamma_{\alpha,k} \to D_{\alpha,k}/\mathrm{Aut}_{\mathbf{Z}}(D_{\alpha,k})$, can be found in our joint paper with M.-H. Saito [27].

Let $(X, \mathcal{L}, \Sigma)$ be a K3 surface of Todorov type (α, k). The set of Todorov surfaces whose associated K3 surface is $(X, \mathcal{L}, \Sigma)$ is parametrized by the set of divisors B with $\mathcal{L} = \mathcal{O}_X(B)$ which satisfy:

Condition (7.4). $\mu^*(B)$ *is disjoint from* $\pi^{-1}(\mathrm{Sing}\,\Sigma)$ *and is a reduced divisor with neither infinitely near triple points nor points of multiplicity greater than three.*

We may thus describe the moduli space of Todorov surfaces in the following way. Using the procedure discussed in [8] and [23; pp. 320-321] of glueing together local deformations, we construct a universal marked family $f : \mathcal{X}_{\alpha,k} \to D_{\alpha,k}$ of K3 surfaces with rational double points and a relatively ample line bundle $\mathbf{L}_{\alpha,k}$ on $\mathcal{X}_{\alpha,k}$ whose first Chern class in each fiber is mapped to λ by the marking, where the markings of the fibers are required to have images in the span of λ and $N_{\alpha,k}$, and to send the sign structure induced from the Hodge structure on the fiber to $[\nu_{\alpha,k}]$. Let $\mathcal{V}_{\alpha,k}$ be the open subset of the projective bundle $\mathbf{P}(R^0 f_* \mathbf{L}_{\alpha,k}) \to D_{\alpha,k}$ consisting of all sections whose divisor satisfies condition (7.4). The action of the group $\tilde{\Gamma}_{\alpha,k}$ on $D_{\alpha,k}$ then extends to an action on $\mathcal{V}_{\alpha,k}$.

Theorem (7.5). $\mathcal{V}_{\alpha,k}/\tilde{\Gamma}_{\alpha,k}$ *is a coarse moduli space for Todorov surfaces with fundamental invariants* (α, k), *and each fiber of the natural map* $\mathcal{V}_{\alpha,k}/\tilde{\Gamma}_{\alpha,k} \to D_{\alpha,k}/\Gamma_{\alpha,k}$ *is connected, and nonempty. In particular, since* $D_{\alpha,k}/\Gamma_{\alpha,k}$ *is connected, the set of Todorov surfaces with fixed fundamental invariants forms a nonempty irreducible family.*

Proof. Most of this has been proved in the discussion above; we must still consider the structure of the fibers of the map $\mathcal{V}_{\alpha,k}/\tilde{\Gamma}_{\alpha,k} \to D_{\alpha,k}/\Gamma_{\alpha,k}$. It suffices to show that each fiber of $\mathcal{V}_{\alpha,k} \to D_{\alpha,k}$ is connected and nonempty.

Let $(X, \mathcal{L}, \Sigma)$ be a K3 surface of Todorov type (α, k). By Riemann-Roch, $\mathcal{L} = \mathcal{O}_X(B)$ for some effective divisor B which moves in a linear system $|B|$ of dimension $k - 8 \geq 1$. Since $(\alpha, k) \neq (1, 9)$, lemma (5.1) guarantees that $|B|$ is free; but then by Bertini's theorem, there is some $B' \in |B|$ which is smooth and disjoint from Sing X. B' satisfies condition (7.4), so that the set of divisors in $|B|$ satisfying (7,4) (which coincides with the fiber of $\mathcal{V}_{\alpha,k} \to D_{\alpha,k}$) is a nonempty Zariski-open subset. Q.E.D.

When $(\alpha, k) \neq (5, 16)$, we can give a bit more information about the map $\mathcal{V}_{\alpha,k}/\tilde{\Gamma}_{\alpha,k} \to D_{\alpha,k}/\Gamma_{\alpha,k}$ by analyzing the action of $\tilde{\Gamma}_{\alpha,k}$ more carefully.

Lemma (7.6). *Suppose that $(\alpha, k) \neq (5, 16)$.*
(i) The natural map $\tilde{\Gamma}_{\alpha,k} \to O_-(N_{\alpha,k})$ is injective.
(ii) -1 is in the image of this map if and only if $(\alpha, k) = (0, 9)$ or $(1, 10)$.

Proof. Let $(X, \mathcal{L}, \Sigma)$ be a K3 surface of Todorov type (α, k) whose minimal desingularization S has Picard number $k + 1$. (This implies that $\mathrm{NS}(S) = M_{\Sigma,\mathcal{L}}$ and $\Sigma = X$.) Let (ϕ, ψ) be a special marking of $(X, \mathcal{L}, \Sigma)$, and for $(\gamma, \sigma) \in \tilde{\Gamma}_{\alpha,k}$ let $\tilde{\gamma} = \phi^{-1} \circ \gamma \circ \phi$ and $\tilde{\sigma} = \psi^{-1} \circ \sigma \circ \psi$.
(i) Suppose that $(\gamma, \sigma) \in \mathrm{Ker}(\tilde{\Gamma}_{\alpha,k} \to O_-(N_{\alpha,k}))$. Then $\tilde{\gamma}$ is an automorphism of $H^2(S, \mathbf{Z})$ preserving the intersection form and the Hodge structure, $\tilde{\gamma}$ acts as the identity on $\Gamma^2(X)$, and $\tilde{\gamma}(\pi^{-1}(P)) = \pi^{-1}(\tilde{\sigma}(P))$. Since $\tilde{\gamma}$ preserves $c_1(\mu^*(\mathcal{L}))$ and maps each $\pi^{-1}(P)$ to an effective curve $\pi^{-1}(\tilde{\sigma}(P))$, by the weakly polarized global Torelli theorem there is an automorphism $\phi : S \to S$ such that $\phi^* = \tilde{\gamma}$. Now ϕ and $\tilde{\sigma}$ satisfy the hypotheses of proposition (2.3); since $k \leq 15$, ϕ and $\tilde{\sigma}$ must be trivial, so that γ and σ are trivial as well.
(ii) Suppose that $(\gamma, \sigma) \in \tilde{\Gamma}_{\alpha,k}$ maps to -1 in $O_-(N_{\alpha,k})$. Then $\gamma|_{N_{\alpha,k}}$ acts as -1, while $\gamma|_{M_{\alpha,k}}$ fixes λ and permutes the e_P's according to σ. If $g : M_{\alpha,k}^*/M_{\alpha,k} \to N_{\alpha,k}^*/N_{\alpha,k}$ is the natural isomorphism, then $g^{-1} \circ (\gamma|_{N_{\alpha,k}})^* \circ g = (\gamma|_{M_{\alpha,k}})^*$ mod $M_{\alpha,k}$; this implies that the permutation σ acts as -1 on $G_{\alpha,k} = M_{\alpha,k}^*/M_{\alpha,k}$. By proposition (6.2)(ii), $(\alpha, k) = (0, 9)$ or $(1, 10)$.
Conversely, if $(\alpha, k) = (0, 9)$ or $(1, 10)$, let $|B|$ be the linear system on X with $\mathcal{L} = \mathcal{O}_X(B)$. By lemma (5.4), $|B|$ is hyperelliptic so that there is an involution $\phi : X \to X$ with $\phi^*(\mathcal{L}) = \mathcal{L}$ and X/ϕ rational. The rationality of X/ϕ implies that $\phi^*(\omega) = -\omega$ for any holomorphic 2-form ω on S; since $\mathrm{NS}(S) = M_{\Sigma,\mathcal{L}}$, this implies that ϕ^* acts as -1 on all of $M_{\Sigma,\mathcal{L}}^\perp$. Now if we let $\gamma = \phi \circ \phi^* \circ \phi^{-1}$ and $\sigma = \psi \circ (\phi|_{\mathrm{Sing} X}) \circ \psi^{-1}$ then $(\gamma, \sigma) \in \tilde{\Gamma}_{\alpha,k}$ and γ acts as -1 on $\phi(M_{\Sigma,\mathcal{L}}^\perp) = N_{\alpha,k}$. Q.E.D.

Notice that the last paragraph of the proof above applies when $(\alpha, k) \neq (0, 9)$ or $(1, 10)$ to provide a partial converse to lemma (5.4).

Corollary (7.7). *Let $(X, \mathcal{L}, \Sigma)$ be a K3 surface of Todorov type (α, k) with $(\alpha, k) \neq (0, 9)$, $(1, 10)$, or $(5, 16)$, and suppose that the minimal desingularization of X has Picard number $k + 1$. Then the linear system $|B|$ on X associated to $\mathcal{L}$ is not hyperelliptic.*

As an application of lemma (7.6), we prove:

Corollary (7.8). *If* $(\alpha, k) \neq (5, 16)$, *then the natural action of* $\tilde{\Gamma}_{\alpha,k}$ *on* $\mathcal{V}_{\alpha,k}$ *factors through the projection* $\tilde{\Gamma}_{\alpha,k} \to \Gamma_{\alpha,k}$. *In particular,* $\mathcal{V}_{\alpha,k}/\Gamma_{\alpha,k}$ *is the coarse moduli space for Todorov surfaces with fundamental invariants* (α, k).

Proof. If $(\alpha, k) \neq (0, 9)$ or $(1, 10)$ then $\Gamma_{\alpha,k} = \tilde{\Gamma}_{\alpha,k}$ by lemma (7.6) and there is nothing to prove. If $(\alpha, k) = (0, 9)$ or $(1, 10)$ and $(X, \mathcal{L}, \Sigma)$ is generic, then lemma (7.6) implies that the unique element of $\mathrm{Ker}(\tilde{\Gamma}_{\alpha,k} \to \mathrm{Aut}_{\mathbf{Z}}(D_{\alpha,k}))$ induces the hyperelliptic involution on X. Since $\phi_{|B|}$ factors through the quotient by that hyperelliptic involution, the induced action on $\mathbf{P}H^\circ(\mathcal{L})$ is trivial. Hence, the action of the kernel on $\mathcal{V}_{\alpha,k}$ is trivial.   *Q.E.D.*

Corollary (7.8) implies that for $(\alpha, k) \neq (5, 16)$, the Stein factorization of the "period map" $\mathcal{V}_{\alpha,k}/\Gamma_{\alpha,k} \to D_{\alpha,k}/\mathrm{Aut}_{\mathbf{Z}}(D_{\alpha,k})$ is nothing other than $\mathcal{V}_{\alpha,k}/\Gamma_{\alpha,k} \to D_{\alpha,k}/\Gamma_{\alpha,k} \to D_{\alpha,k}/\mathrm{Aut}_{\mathbf{Z}}(D_{\alpha,k})$. We compute the degree of the finite part of the Stein factorization in [27].

§8. Concluding remarks.

1. The reader will have notice that much of the technical difficulty in studying the moduli of Todorov surfaces derives from the necessity of choosing a partial desingularization Σ when discussing K3 surfaces of Todorov type (α, k). Of course, if we had been willing to consider only general Todorov surfaces (for example, ones for which $\Sigma = X$), this difficulty would have been eliminated, But in studying *all* Todorov surfaces, we cannot eliminate the choice of Σ, as we now give an example to show.

Let C_1 and C_2 be conics and let L_1 and L_2 be lines in $\mathbf{P}^2$, meeting transversely (pairwise). The sextic curve $C = C_1 + C_2 + L_1 + L_2$ can then be written as a sum of two cubics in two different ways: $C = (C_1 + L_1) + (C_2 + L_2)$ and $C = (C_1 + L_2) + (C_2 + L_1)$. The base points of the two associated pencils of cubics give two distinguished subsets S_1 and S_2 of $\mathrm{Sing}\, C$ with $\#(S_i) = 9$. Let X be the double cover of $\mathbf{P}^2$ branched along C, and let $\mathcal{L}$ be the pullback of $\mathcal{O}_{\mathbf{P}^2}(1)$. X is a K3 surface with 13 ordinary double points; if we let Σ_i be the partial desingularization of X obtained by blowing up all points of $\mathrm{Sing}\, X$ not in the inverse image of S_i for $i = 1, 2$, then $(X, \mathcal{L}, \Sigma_i)$ is a K3 surface of Todorov type $(0, 9)$. In particular, the set of isomorphism classes of pairs $(X, \mathcal{L})$ contains less information than the set of isomophism classes of triples $(X, \mathcal{L}, \Sigma)$.

2. Let $(X, \mathcal{L}, \Sigma)$ be a K3 surface of Todorov type (α, k) and let $|B|$ be the linear system with $\mathcal{L} = \mathcal{O}_X(B)$. Suppose that $B_t \in |B|$ is a family of divisors parametrized by the unit disk Δ such that B_t satisfies condition (7.4) for $t \neq 0$, but B_0 does not satisfy (7.4). (For example, if $\Sigma = X$ we could require that B_0 pass through a singular point of X.) The corresponding family Z_t of Todorov surfaces over the punctured disk Δ^* exhibits a phenomenon first

observed by Friedman [14]: it is a family of regular[7] surfaces of general type with trivial monodromy which can be completed to a family over the disk Δ, but *cannot* be completed to such a family with a smooth central fiber. For if the central fiber were smooth, that fiber would also be a Todorov surface with the same fundamental invariants and its branch locus would be B_0, a contradiction. (In fact, the natural central fiber contains a component birational to either an elliptic surface, or a Todorov surface with different invariants.)

3. In [26], we showed that any algebraic surface Z with geometric genus 1 has an "associated K3 surface": one whose transcendental lattice is isomorphic to that of Z by an isomorphism preserving the intersection forms and (integral) Hodge structures. We then asked whether there is an algebraic cycle on the product of Z with its associated K3 surface which realizes this isomorphism. (The existence of such a cycle is predicted by the Hodge conjecture.) Somewhat surprisingly, if Z is a Todorov surface, the K3 surface $X = Z/j$ is *not* the associated K3 surface of Z! This happens because the natural map $Z \to X$ multiplies the intersection form by 2 (cf. the proof of proposition (6.2) in the case $(\alpha, k) = (5, 16)$). To construct the desired algebraic cycle, we must find a double cover of X (or some surface birational to X) which is also a K3 surface.

Such a double cover is easily found when $\alpha > 0$: the double cover of Σ branched along a set $I \subset \text{Sing}\,\Sigma$ such that $(1/2)c_i(\sum_{P \in I} \pi^{-1}(P)) \in \text{NS}(S)$ is a K3 surface Y. If $\alpha = 0$ (so that $k = 9, 10$, or 11) we must work somewhat harder to find Y, and we will only sketch the construction when Z is generic. In that case, let $\text{Sing}\,\Sigma = \{P_1, \ldots, P_k\}$, let Q_i be the image of P_i in $\mathbf{P}^{k-7}$ under the map $\phi_{|B|}$, and let $H \subset \mathbf{P}^{k-7}$ be the hyperplane spanned by the $k - 7$ points $Q_8, Q_9, \ldots, Q_K$. When Z is generic the inverse image of H on X is an irreducible rational curve whose proper transform C on S is smooth. If $C_i = \pi^{-1}(P_i)$, then $C, C_1, \ldots, C_7$ are smooth disjoint rational curves on S such that $1/2c_1(C + C_1 + \cdots + C_7) \in \text{NS}(S)$. The double cover $\tilde{Y}$ of S branched along $C + C_1 + \cdots + C_7$ is then birational to a K3 surface Y.

The graph of the pair of degree 2 rational maps $Z \to X$ and $Y \dashrightarrow X$ now gives a cycle on the product $Z \times Y$ which induces an isomorphism $T(Z) \otimes \mathbf{Q} \to T(Y) \otimes \mathbf{Q}$ preserving intersection forms and Hodge structures. Since this isomorphism might not be defined over $\mathbf{Z}$, Y might not itself be the associated K3 surface W of Z. However, by a theorem of Mukai [28] (combined with a result of Nikulin [32] in the case $(\alpha, k) = (0, 9)$ in which the Picard number of S may be 10), there is an algebraic cycle in the product $Y \times W$ inducing an isomorphism $T(Y) \times \mathbf{Q} \to T(W) \otimes \mathbf{Q}$ such that the composite isomorphism $T(Z) \times \mathbf{Q} \to T(W) \otimes \mathbf{Q}$ preserves intersection forms and Hodge structures, and is defined over $\mathbf{Z}$. We may then "compose" the algebraic cycles on $Z \times Y$ and $Y \times W$ as in [26] to get a cycle on $Z \times W$ with the desired property.

---

[7]We may even assume that the surfaces are simply connected, if we take $(\alpha, k) = (0, 9)$ or $(0, 10)$; cf. [9], [41].

Appendix

In this appendix we modify a theorem of Nikulin [30] concerning embeddings into an even unimodular lattice L to cover the case of equivalence of embedding under the subgroup $O_-(L)$ of the orthogonal group $O(L)$ of L (which was defined in section 6).

We denote the discriminant-form of a lattice L by (G_L, q_L). If G is a finite abelian group, G_p denotes the p-Sylow subgroup of G, and $l(G)$ denotes the minimum number of generators of G.

Theorem (A.1). *Let L be an even integral unimodular symmetric bilinear form, let M and M_1 be nondegenerate primitive sublattices of L, let $\phi : M \to M_1$ be an isometry and let K be the orthogonal complement of M. Suppose that*
  (i) *the bilinear form is indefinite when restricted to K,*
  (ii) $\operatorname{rank}(K) \geq 3$, *and*
  (iii) *either*
       (a) $l(G_M) \leq \operatorname{rank}(K) - 2$, *or*
       (b) $l(G_{M_P}) \leq \operatorname{rank}(K) - 2$ *for all $p \neq 2$, $l(G_{M_2}) = \operatorname{rank}(K)$, and there is an orthogonal direct sum decomposition*

$$(G_{M_2}, q_{M_2}) \cong (G_1, q_1) \oplus (G_2, q_2)$$

such that (G_2, q_2) is a special 2-elementary form of rank at least 2.
  Then there is some $\phi \in O_-(L)$ such that $\phi|_M = \phi$.

Proof. We modify slightly the proof of theorems 1.2 and 1.6 in [30], following the notation there. Note first that our hypothesis (iii) implies that conditions (1.5) and (1.6) of [30] hold (cf. [31]; theorem 1.14.2]; the point is that a special 2-elementary form of rank at least 2 must contain $u_+^{(2)}(2)$ or $v_+^{(2)}(2)$ as a direct summand, in the notation of [31]); hence, we may apply the argument of [30].

We begin as Nikulin does, by invoking the version of Witt's theorem over local rings due to Kneser [17] to guarantee the existence of p-adic isometries $\phi_p \in O(L \otimes \mathbf{Z}_p)$ such that $\phi_p|_{M \otimes \mathbf{Z}_p} = \phi \otimes \mathbf{Z}_p$. Suppose that $\det \phi_2 = -1$. If there is some $\alpha \in O_-(L)$ with $\det \alpha = -1$, we may proceed as at the bottom of p. 77, [8] replacing M_1, ϕ and Φ by $\alpha(M_1)$, $\alpha \circ \phi$ and $\alpha \circ \Phi$ to reduce to the case $\det \phi_2 = 1$. To see that such an α exists, note that L, being even, indefinite, and unimodular, contains elements x and y on which the bilinear form has matrix $\begin{pmatrix} 0 & 1 \\ 1 & 0 \end{pmatrix}$; α may be taken to be the reflection in $x - y$.

We may thus assume that $\det \phi_2 = 1$. We now proceed as on pp. 77-78, making the following modification of formula (1.7) on p. 78: define $O_{\mathbf{Q}}^{++} = \{\psi \in O_{\mathbf{Q}}^+ \mid \Theta(\psi) > 0\}$. Then

(*) $O_{\mathbf{A}}^+ = O_{\mathbf{A}}' \cdot O_{\mathbf{Q}}^{++} \cdot H_{\mathbf{A}}^+(K).$

---

[8]Page numbers refer to the English translation of [30].

This is proved the same way as formula (1.7): if $\mathbf{Q}^{*+}$ denotes the positive rational numbers, then

$$[O_\mathbf{A}^+ : O_\mathbf{A}' \cdot O_\mathbf{Q}^{++} \cdot H_\mathbf{A}^+(K)] = [I_\mathbf{Q} : \mathbf{Q}^{*+} \cdot \Theta(H_\mathbf{A}^+(K)]$$

so that (*) follows from the fact (proved on pp. 78-79) that under our hypotheses, $\Theta(H^+(K_p)) \subset \mathbf{Z}_p^* \cdot \mathbf{Q}_p^{*2}$.

Thus, when formula (1.7) is applied in the proof (near the top of p. 80) we may assume that $\Theta(\psi) > 0$ in the decomposition $\phi_p = \phi_p' \circ \psi \circ \alpha_p$. Hence, at the end of the proof we have

$$\Theta(\phi) = \Theta(\phi' \circ \psi) = \Theta(\psi) > 0$$

while $\det \phi = 1$ so that $\phi \in O_-(L)$. *Q.E.D.*

References

[1] W. Barth, C. Peters, and A. Van de Ven, Compact Complex Surfaces, Berlin-Heidelberg-New York-Tokyo: Springer-Verlag (1984).

[2] A. Beauville, "L'application canonique pour les surfaces de type général", Invent. Math. **55** (1978) 121-140.

[3] A. Beauville, "Sur le nombre maximum de points doubles d'une surface dans $\mathbf{P}^3(\mu(5) = 31)$", Journées de géométrie algébrique (juillet 1979), Alphen aan den Rijn: Sijthoff & Noordhoff (1980) 207-215.

[4] I. F. Blake and R. C. Mullin, An Introduction to Algebraic and Combinatorial Coding Theory, New York-San Francisco-London: Academic Press (1976).

[5] E. Bombieri, "Canonical models of surfaces of general type", Publ. Math. IHES **42** (1973) 171-219.

[6] N. Bourbaki, Groupes et algèbres de Lie Chap. IV, V, VI, Paris: Hermann (1968).

[7] E. Brieskorn, "Singular elements of semi-simple groups", Proc. Int. Cong. Math. Nice, t. **2** (1970) 279-284.

[8] D. Burns, Jr. and M. Rapoport, "On the Torelli problems for Kählerian K3 surfaces", Ann. scient. Éc. Norm. Sup. (4) **8** (1975) 235-274.

[9] F. Catanese, "Surfaces with $K^2 = p_g = 1$ and their period mapping", Algebraic Geometry (Proceedings, Copenhagen 1978), Lecture Notes in Math. **732**, Berlin-Heidelberg-New York: Springer-Verlag(1979) 1-29.

[10] F. Catanese, "The moduli and the global period mapping of surfaces with $K^2 = p_g = 1$: a counterexample to the global Torelli problem:", Compositio Math. **41** (1980) 401-414.

[11] F. Catanese, "On the period map of surfaces with $K^2 = \chi = 2$", Classification of Algebraic and Analytic manifolds, Progress in Math. **39**, Boston-Basel-Stuttgart: Birkhäuser (1983) 27-43.

[12] F. Catanese and O. Debarre, "Surfaces with $K^2 = 2$, $p_g = 1$, $q = 0$", preprint.

[13] P. Francia, "The bicanonical map for surfaces of general type", preprint.

[14] R. Friedman, "A degenerating family of quintic surfaces with trivial monodromy", Duke Math. J. **50** (1983) 203-214.

[15] P. Griffiths and J. Harris, Principles of Algebraic Geometry, New York-Chichester-Brisbane-Toronto: John Wiley & Sons (1978).

[16] E. Horikawa, "On deformations of quintic surfaces", Invent. Math. **31** (1975) 43-85.

[17] M. Keneser, "Witts Satz für quadratischer Formen über lokalen Ringen", Nachr. Akad. Wiss. Göttingen Math.-Phys. K1. II (1972) 195-203.

[18] K. Kodaira, "Pluricanonical systems on algebraic surfaces of general type", J. Math. Soc. Japan **20**(1968) 170-192.

[19] V. S. Kulikov, "Epimorphicity of the period mapping for surfaces of type K3" (Russian), Uspehi Mat. Nauk. **32**: 4 (1977) 257-258.

[20] V. T. Kunev, "An example of a simply-connected surface of general type for which the local Torelli theorem does not hold" (Russian), C. R. Ac. Bulg. Sc. **30** (1977) 323-325.

[21] E. Looijenga and J. Wahl, "Quadratic functions and smoothing surface singularities", Topology **25** (1986) 261-291.

[22] A. Mayer, "Families of K3 surfaces", Nagoya Math. J. **48** (1972) 1-17.

[23] D. R. Morrison, "Some remarks on the moduli of K3 surfaces", Classification of Algebraic and Analytic Manifolds, Progress in Math. **39**, Boston-Basel-Stuttgart: Birkhäuser (1983) 303-332.

[24] D. R. Morrison, "On K3 surfaces with large Picard number", Invent. Math. **75** (1984) 105-121.

[25] D. R. Morrison, "The birational geometry of surfaces with rational double points", Math. Ann. **271** (1985) 415-438.

[26] D. R. Morrison, "Isogenies between algebraic surfaces with geometric genus one", Tokyo J. Math. **10** (1987) 179-187.

[27] D. R. Morrison and M.-H. Saito, "Cremona transformations and degrees of period maps for K3 surfaces with ordinary double points", Algebraic Geometry, Sendai 1985, Adv. Stud. Pure Math. **10** (1987) 477-513.

[28] S. Mukai, "On the moduli space of bundles on K3 surfaces, I", Vector Bundles on Algebraic Varieties, Tata Institute of Fundamental Reserch (1987) 341-413.

[29] V. V. Nikulin, "On Kummer surfaces", Izv. Akad. Nauk SSSR **39** (1975) 278-293; Math. USSR Izvestija **9** (1975) 261-275.

[30] V. V. Nikulin, "Finite automorphism groups of Kähler K3 surfaces", Trudy Mosk. Mat. Ob. **38** (1979) 75-137; Trans. Moscow Math. Soc. **38** (1980) 71-135.

[31] V. V. Nikulin, "Integral symmetric bilinear forms and some of their applications", Izv. Akad. Nauk. SSSR **43** (1979) 111-177; Math. USSR Izvestija **14** (1980) 103-167.

[32] V. V. Nikulin, Letter to the author, 15 November 1984.

[33] A. Ogus, "A crystalline Torelli theorem for supersingular K3 surfaces", Arithmetic and Geometry: Papers Dedicated to I. R. Shafarevich (vol. II: Geometry), Progress in Math. **36**, Boston-Basel-Stuttgart: Birkhäuser (1983) 361-394.

[34] P. Oliverio, "On the period map for surfaces with $K_S^2 - 2$, $p_g = 1$, $q = o$ and torsion $\mathbf{Z}/2\mathbf{Z}$", Duke Math. J. **50** (1983) 561-571.

[35] U. Persson, "Double coverings and surfaces of general type", Algebraic Geometry (Proceedings, Tromsø, Norway 1977), Lecture Notes in Math. **687**, Berlin-Heidelberg-New York: Springer-Verlag (1978) 168-195.

[36] I. Piateckii-Shapiro and I. R. Shafarevich, "A Torelli theorem for algebraic surfaces of type K3", Izv. Akad. Nauk. SSSR **35** (1971) 530-572; Math. USSR Izvestija **5** (1971) 547-587.

[37] C. P. Ramanujam, "Remarks on the Kodaira vanishing theorem", J. Indian Math. Soc. **36** (1972) 41-51.

[38] M. Reid, "Minimal models of canonical 3-folds", Algebraic Varieties and Analytic Varieties, Adv. Studies in Pure Math. 1, Amsterdam-New York-Oxford: North-Holland and Tokyo: Kinokuniya (1983) 131-180.

[39] F. Sakai, "Normal Gorenstein surfaces and blowing ups", Saitama Math. J. **1** (1983) 29-35.

[40] A. N. Todorov, "Surfaces of general type with $p_g = 1$ and $(K, K) = 1$, I", Ann. scient. Éc. Norm. Sup. (4) **13** (1980) 1-21.

[41] A. N. Todorov "A construction of surfaces with $p_g = 1$, $q = 0$ and $2 \leq (K^2) \leq 8$: counterexamples of the global Torelli theorem", Invent. Math. **63** (1981) 287-304.

[42] S. Usui, "Period map of surfaces with $p_g = c_1^2 = 1$ and K ample", Mem. Fac. Sci. Kochi Univ. (Math.) **3** (1981) 37-73.

[43] S. Usui, "Torelli theorem for surfaces with $p_g = c_1^2 = 1$ and K ample and with certain type of automorphism", Compositio Math. **45** (1982) 293-314.

[44] S. Usui, "Period map of surfaces with $p_g = 1$, $c_1^2 = 2$ and $\pi_1 = \mathbf{Z}/2\mathbf{Z}$", Mem. Fac. Sci. Kochi Univ. (Math.) **5** (1984) 15-26; 103-104.

[45] E. B. Vinberg, "Discrete groups generated by reflections in Lobačevskii spaces", Mat. Sbornik **72** (1967) 471-488; Math. USSR Sbornik **1** (1967) 429-444.

David. R. MORRISON

Department of Mathematics
Duke University
Durham, NC 27706
U.S.A.

Algebraic Geometry and Commutative Algebra
in Honor of Masayoshi NAGATA
pp. 357–377 (1987)

Curves, K3 Surfaces and Fano 3-folds of Genus ≤ 10

Shigeru MUKAI*

A pair (S, L) of a K3 surface S and a pseudo-ample line bundle L on S with $(L^2) = 2g - 2$ is called a (polarized) K3 surface of genus g. Over the complex number field, the moduli space $\mathcal{F}_g$ of those (S, L)'s is irreducible by the Torelli type theorem for K3 surfaces [12]. If L is very ample, the image S_{2g-2} of $\Phi_{|L|}$ is a surface of degree $2g - 2$ in $\mathbf{P}^g$ and called the projective model of (S, L), [13]. If $g = 3, 4, 5$ and (S, L) is general, then the projective model is a complete intersection of $g - 2$ hypersurfaces in $\mathbf{P}^g$. This fact enables us to give an explicit description of the birational type of $\mathcal{F}_g$ for $g \leq 5$. But the projective model is no more complete intersection in $\mathbf{P}^g$ when $g \geq 6$. In this article, we shall show that a general K3 surface of genus $6 \leq g \leq 10$ is still a complete intersection in a certain homogeneous space and apply this to the discription of birational type of $\mathcal{F}_g$ for $g \leq 10$ and the study of curves and Fano 3-folds. The homogeneous space X is the quotient of a simply connected semi-simple complex Lie group G by a maximal parabolic subgroup P. For the positive generator $\mathcal{O}_X(1)$ of $\mathrm{Pic}X \cong \mathbf{Z}$, the natural map $X \to \mathbf{P}(H^0(X, \mathcal{O}_X(1)))$ is a G-equivariant embedding and the image coincides with the G-orbit $G \cdot \bar{v}$, where v is a highest weight vector of the irreducible representation $H^0(X, \mathcal{O}_X(1))^\vee$ of G. For each $6 \leq g \leq 10$, G and the representation $U = H^0(X, \mathcal{O}_X(1))$ are given as follows:

	g	6	7	8	9	10
	G	SL(5)	Spin(10)	SL(6)	Sp(3)	exceptional of type G_2
	$\dim G$	24	45	35	21	14
(0.1)	U	$\wedge^2 V^5$	half spinor representation	$\wedge^2 V^6$	$\wedge^3 V^6 / \sigma \wedge V^6$	adjoint representation
	$\dim U$	10	16	15	14	14
	$\dim X$	6	10	8	6	5

where V^i denotes an i-dimensional vector space and $\sigma \in \wedge^2 V^6$ is a non-degenerate 2-vector of V^6.

---
*Partially supported by SFB 40 Theoretische Mathematik at Bonn and Educational Projects for Japanese Mathematical Scientists.
Received April 7, 1987.

In the case $7 \leq g \leq 10$, dim U is equal to $g + n - 1$, $n = \dim X$. X is of degree $2g - 2$ in $\mathbf{P}(U) \cong \mathbf{P}^{g+n-2}$ and the anticanonical (or 1st Chern) class of X is $n - 2$ times hyperplane section (cf. (1.5)). Hence a smooth complete intersection of $X = X_{2g-2}$ and $n - 2$ hyperplanes is a K3 surface of genus g. (This has been known classically in the case $g = 8$ and is first observed by C. Borcea [1] in the case $g = 10$.)

Theorem 0.2. *If two K3 surfaces S and S' are intersections of X_{2g-2} ($7 \leq g \leq 10$) and g-dimensional linear subspaces P and P', respectively, and if $S \subset P$ and $S' \subset P'$ are projectively equivalent, then P and P' are equivalent under the action of $\bar{G}$ on $\mathbf{P}(U)$, where $\bar{G}$ is the quotient of G by its center.*

By the theorem there exists a nonempty open subset Ξ of the Grassmann variety $G(n - 2, U)$ of $n - 2$ dimensional subspaces of U such that the natural morphism $\Xi/\bar{G} \to \mathcal{F}_g$ is injective. For each $7 \leq g \leq 10$, it is easily checked that $\dim \Xi/\bar{G} = 19 = \dim \mathcal{F}_g$. Hence the morphism is birational.

Corollary 0.3. *The generic K3 surface of genus $7 \leq g \leq 10$ is a complete intersection of $X_{2g-2} \subset \mathbf{P}(U)$ and a g-dimensional linear subspace in a unique way up to the action $\bar{G}$ on $\mathbf{P}(U)$. In particular, the moduli space $\mathcal{F}_g$ is birationally equivalent to the orbit space $G(n - 2, U)/\bar{G}$.*

In the case $g = 6$, the generic K3 surface is a complete intersection of X, a linear subspace of dimension 6 and a quadratic hypersurface in $\mathbf{P}(U) \cong \mathbf{P}^9$. We have a similar result on the uniqueness of this expression of the K3 surface (see (4.1)). In the proof of these results, special vector bundles, instead of line bundles in the case $g \leq 5$, play an essential role. For instance, the generic K3 surface (S, L) of genus 10 has a unique (up to isomorphism) stable rank two vector bundle with $c_1(E) = c_1(L)$ and $c_2(E) = 6$ on it and the embedding of S into $X = G/P$ is uniquely determined by this vector bundle E.

The following is the table of the birational type of $\mathcal{F}_g$ for $g \leq 10$:

(0.4)

genus	2	3	4
birational type	$\mathbf{P}(S^6U^3)/\mathrm{PGL}(3)$	$\mathbf{P}(S^4U^4)/\mathrm{PGL}(4)$	$\mathbf{P}(U^{30})/\mathrm{SO}(5)$

5	6	7
$G(3, S^2U^6)/\mathrm{PGL}(6)$	$(U^{13} \oplus U^9)/\mathrm{PGL}(2)$	$G(8, U^{16})/\mathrm{PSO}(10)$

8	9	10
$G(6, \wedge^2 V^6)/\mathrm{PGL}(6)$	$G(4, U^{14})/\mathrm{PSp}(3)$	$G(3, \mathfrak{g})/\bar{G}_2$

where U^d is a d-dimensional irreducible representation of the universal covering group.

Corollary 0.5. $\mathcal{F}_g$ *is unirational for every $g \leq 10$.*

By [5], there exists a Fano 3-fold V with the property $\text{Pic} V \cong \mathbf{Z}(-K_V)$ and $(-K_V)^3 = 22$. The moduli space of these Fano 3-folds are unirational by their description in [5]. The generic K3 surface of genus 12 is an anticanonical divisor of V and hence $\mathcal{F}_{12}$ is also unirational.

Problem 0.6. Describe the birational types, *e.g.*, the Kodaira dimensions, of the 19-dimensional varieties $\mathcal{F}_g$ for $g \gg 0$. Are they of general type?

If (S, L) is a K3 surface of genus g, then every smooth member of $|L|$ is a curve of genus g. Conversely if C is a smooth curve of genus $g \geq 2$ on a K3 surface, then $\mathcal{O}_S(C)$ is pseudo-ample and $(S, \mathcal{O}_S(C))$ is a K3 surface of the same genus as C. In the case $g \leq 9$, the generic curve lies on a K3 surface, that is , the natural rational map

$$\phi_g : \mathcal{P}_g = \bigcup_{(S,L) \in \mathcal{F}_g} |L| \dashrightarrow \mathcal{M}_g = (\text{the moduli space of curves of genus } g)$$

is generically surjective (§6). The inequality $\dim \mathcal{M}_g \leq \dim \mathcal{P}_g = 19 + g$ holds if and only if $g \leq 11$ and ψ_{11} is generically surjective ([10]). But in spite of $\dim \mathcal{M}_{10} = 27 < \dim \mathcal{P}_{10} = 29$, we have

Theorem 0.7. *The generic curve of genus 10 cannot lie on a K3 surface.*

Proof. Let $\mathcal{F}'_{10}$ (resp. $\mathcal{M}'_{10}$) be the subset of $\mathcal{F}_{10}$ (resp. $\mathcal{M}_{10}$) consisting of K3 surfaces (resp. curves) of genus 10 obtained as a complete intersection in the homogeneous space $X_{18}^5 \subset \mathbf{P}(\mathfrak{g})$. $\mathcal{M}'_{10}$ has a dominant morphism from a Zariski open subset U of $G(4, \mathfrak{g})/\bar{G}$. Since the automorphism of a curve of genus ≥ 2 is finite, the stabilizer group is finite for every 4-dimensional subspace of $\mathfrak{g}$ which gives a smooth curve of genus 10. Hence we have $\dim \mathcal{M}'_{10} \leq \dim U = \dim G(4, \mathfrak{g}) - \dim G = 26 < \dim \mathcal{M}_{10}$. On the other hand $\mathcal{F}'_{10}$ contains a dense open subset of $\mathcal{F}_{10}$ by Theorem 0.2. Hence the image of ψ_{10} is contained in the closure of $\mathcal{M}'_{10} = \psi_{10}(\mathcal{F}'_{10})$ and ψ_{10} is not generically surjective. q.e.d.

Remark 0.8. Every curve of genus 10 has g_{12}^4, a 4-dimensional linear system of degree 12. If C is a general linear section of the homogeneous space $X_{18} \subset \mathbf{P}^{13}$, then every g_{12}^4 of C embeds C into a quadric hypersurface in $\mathbf{P}^4$. But if C is the generic curve of genus 10, then the image $C_{12} \subset \mathbf{P}^4$ embedded by any g_{12}^4 is not contained in any quadratic hypersurface. This fact gives an alternate proof of the theorem.

In the case $7 \leq g \leq 10$, a Fano 3-fold $V_{2g-2} \subset \mathbf{P}^{g+1}$ is obtained as a complete intersection of the homogeneous space X_{2g-2}^n and a linear subspace of codimension $n - 3$ in $\mathbf{P}(U) = \mathbf{P}^{n+g-2}$. By the Lefschetz theorem, the Fano 3-fold $V = V_{2g-2}$ has the property $\mathrm{Pic} V \cong \mathbf{Z}(-K_V)$. The existence of such V has been known classically but was shown by totally different construction ([6]). Theorem 0.2 holds for Fano 3-folds, too.

Theorem 0.9. *Let V_{2g-2} and V'_{2g-2} ($7 \leq g \leq 10$) be two Fano 3-folds which are complete intersections of the homogeneous space $X_{2g-2}^n \subset \mathbf{P}^{n+g-2}$ and linear subspaces of codimension $n-3$. If V_{2g-2} and V'_{2g-2} are isomorphic to each other, then they are equivalent under the action of $\bar{G}$.*

We note that, by [1], the families of Fano 3-folds in the theorem is locally complete in the sense of [7].

The original version of this article was written during the author's stay at the Max-Planck Institüt für Mathematik in Bonn 1982 and at the Mathematics Institute in University of Warwick 1982-3. He expresses his hearty thanks to both institutions for their hospitality. He also thanks Mrs. Kozaki for nice typing into LaTeX.

Conventions. Varieties and vector spaces are considered over the complex number field $\mathbf{C}$. For a vector space or a vector bundle E, its dual is denoted by $E^\vee$. For a vector space V, $G(r, V)$ (resp. $G(V, r)$) is the Grassmann variety of r-dimensional subspaces (resp. quotient spaces) of V. $G(1, V)$ and $G(V, 1)$ are denoted by $\mathbf{P}_*(V)$ and $\mathbf{P}(V)$, respectively.

§1. Preliminary

We study some properties of the Cayley algebra $\mathcal{C}$ over $\mathbf{C}$. $\mathcal{C}$ is an algebra over $\mathbf{C}$ with a unit 1 and generated by 7 elements e_i, $i \in \mathbf{Z}/7\mathbf{Z}$. The multiplication is given by

$$(1.1) \qquad \begin{array}{l} e_i{}^2 = -1 \text{ and } e_i e_{i+a} = -e_{i+a} e_i = e_{i+3a} \\ \text{for every } i \in \mathbf{Z}/7\mathbf{Z} \text{ and } a = 1, 2, 4. \end{array}$$

The algebra $\mathcal{C}$ is not associative but alternative, *i.e.*, $x(xy) = x^2 y$ and $(xy)y = xy^2$ hold for every $x, y \in \mathcal{C}$. Let $\mathcal{C}_0$ be the 7-dimensional subspace of $\mathcal{C}$ generated by $e_i, i \in \mathbf{Z}/7\mathbf{Z}$ and U the subspace of $\mathcal{C}_0$ spanned by $\alpha = e_3 + \sqrt{-1} e_5$ and $\beta = e_6 - \sqrt{-1} e_7$. It is easily checked that $\alpha^2 = \beta^2 = \alpha\beta = \beta\alpha = 0$, *i.e.*, U is totally isotropic with respect to the multiplication of $\mathcal{C}$. Moreover, U is maximally totally isotropic with respect to the multiplication of $\mathcal{C}$, *i.e.*, if $xU = 0$ or $Ux = 0$, then x belongs to U. Let q be the quadratic form $q(x) = x^2$ on $\mathcal{C}_0$ and b the associated symmetric bilinear form. $b(x, y)$ is equal to $xy + yx$ for every x and $y \in \mathcal{C}_0$. Let V be the subspace of $\mathcal{C}_0$ of vectors orthogonal to U with respect to q (or b). Since U is totally isotropic with respect to q, V contains U and the quotient V/U carries the quadratic form $\bar{q}$.

Lemma 1.2. $x'(xy) = b(x,y)x' - b(x',y)x + y(x'x)$ *for every* x, x' *and* $y \in C_0$.

Proof. By the alternativity of C, we have $u(vw) + v(uw) = (uv + vu)w$. Hence, if u and v belongs to C_0, then we have $u(vw) + v(uw) = b(u,v)w$. So we have

$$
\begin{aligned}
x'(xy) &= x'(b(x,y) - yx) = b(x,y)x' - x'(yx) \\
&= b(x,y)x' - (b(x',y)x - y(x'x)) \\
&= b(x,y)x' - b(x',y)x + y(x'x).
\end{aligned}
$$

q.e.d.

If $x \in U$ and $y \in V$, then $U(xy) = 0$ by the above lemma and hence xy belongs to U. Hence the right multiplication homomorphism $R(y)$, $x \mapsto xy$, by $y \in V$ maps U into itself. Since $R(x)$ is zero on U if and only if $x \in U$, R gives an injective homomorphism $\bar{R} : V/U \to \mathrm{End}(U)$.

Proposition 1.3. (1) $\bar{R}(\bar{x})^2 = \bar{q}(\bar{x}) \cdot \mathrm{id}$ *for every* $\bar{x} \in V/U$, *and*

(2) $\bar{R}$ *is an isomorphism onto* $sl(U)$, *the vector space consisting of trace zero endomorphisms of* U.

Proof. (1) follows immediately from the alternativity of C. It is easy to check the following fact: if r is an endomorphism of a 2-dimensional vector space and if r^2 is a constant multiplication, then either r itself is a constant multiplication or the trace of r is equal to zero. Hence by (1), $\bar{R}(\bar{x})$ is a constant multiplication or belongs to $sl(U)$, for every $\bar{x} \in V/U$. Therefore, $\bar{R}(V/U)$ is contained in the 1-dimensional vector space consisting of constant multiplications of U or contained in the 3-dimensional vector space $sl(U)$. Since the quadratic form q is nondegenerate on V/U, the former is impossible and $\bar{R}(V/U)$ coincides with $sl(U)$. q.e.d.

Let G be the automorphism group of the Cayley algebra C. It is known that G is a simple algebraic group of type G_2. The automorphisms which map U onto itself form a maximal parabolic subgroup P of G. The subspace spanned by e_1, e_2 and e_4 (resp. by $e_3 - \sqrt{-1}\,e_5$ and $e_6 + \sqrt{-1}\,e_7$) can be identified with $sl(U)$ (resp. $U^\vee$) by $\bar{R}$ (resp. b). C is isomorphic to $\mathbf{C} \oplus U \oplus sl(U) \oplus U^\vee$ and if $f \in \mathrm{GL}(U)$, then $1 \oplus f \oplus \mathrm{ad}(f) \oplus {}^t f$ is an automorphism of the Cayley algebra C. Hence the maximal parabolic subgroup P contains $\mathrm{GL}(U)$ and $X = G/P$ can be identified with the set of 2-dimensional subspaces of C_0 which are equivalent to U under the action of $G = \mathrm{Aut}\,C$.

Let $\mathcal{U}$ be the maximally totally isotropic universal subbundle of $C_0 \otimes \mathcal{O}_X$: the fibre $\mathcal{U}_x \subset C_0$ at x is the 2-dimensional subspace corresponding to $x \in X$. Let $\mathcal{V}$ be the subsheaf of $C_0 \otimes \mathcal{O}_X$ consisting of the germs of sections which are orthogonal to $\mathcal{U}$ with respect to the bilinear form $b \otimes 1$ on $C_0 \otimes \mathcal{O}_X$. $\mathcal{V}$ is a rank 5 subbundle of $C_0 \otimes \mathcal{O}_X$ and contains $\mathcal{U}$ as a subbundle. The quotient bundle

$(\mathcal{C}_0 \otimes \mathcal{O}_X)/\mathcal{V}$ is isomorphic to $\mathcal{U}^\vee$ by $b \otimes 1$ and $\mathcal{V}/\mathcal{U}$ has a quadratic form $\overline{q \otimes 1}$ induced by $q \otimes 1$ on $\mathcal{C}_0 \otimes \mathcal{O}_X$. By Proposition 1.3, we have

Proposition 1.4. *The right multiplication induces an isomorphism $\bar{R}$ from $\mathcal{V}/\mathcal{U}$ onto the vector bundle $sl(\mathcal{U})$ of trace zero endomorphisms of $\mathcal{U}$ and $\bar{R}(\bar{x})^2$ is equal to $\overline{(q \otimes 1)}(\bar{x}) \cdot \mathrm{id}$ for every $\bar{x} \in \mathcal{V}/\mathcal{U}$.*

Next we shall compute the anticanonical class of X and the degree of $\mathcal{O}_X(1)$, the ample generator of $\mathrm{Pic}X$, and show some vanishings of the cohomology groups of homogeneous vector bundles $\mathcal{U}(i)$ and $(S^2\mathcal{U})(i)$ etc.

Let G be a simply connected semi-simple algebraic group and P a maximal parabolic subgroup of G. Fixing a Borel subgroup B in P, the Lie algebra $\mathfrak{g}$ of G is the direct sum of $\mathfrak{b}$ and 1-dimensional eigenspaces $\mathfrak{g}^\beta$, where β runs over all negative roots. If we choose a suitable root basis Δ, then there exists a positive root $\alpha \in \Delta$ such that $\mathfrak{p}$ is equal to the direct sum of $\bigoplus \mathfrak{g}^\gamma$ and $\mathfrak{b}$, where γ runs over all positive roots which are linear combinations of the roots in $\Delta \setminus \{\alpha\}$ with nonnegative coefficients. A positive root β is said to be *complementary* if $\mathfrak{g}^\beta \cap \mathfrak{p} = 0$ or equivalently if β cannot be expressed as a linear combination of the roots in $\Delta \setminus \{\alpha\}$ with nonnegative coefficients.

Proposition 1.5. (Borel-Hirzebruch [2]) *Let G, P, Δ and α be as above and L the positive generator of $\mathrm{Pic}(G/P)$. Then we have*

(1) *the quotient $\mathfrak{g}/\mathfrak{p}$ is isomorphic to $\bigoplus_{\beta \in R_P} \mathfrak{g}^\beta$, where R_P is the set of positive complementary roots. In particular, $\dim(G/P)$ is equal to the cardinality n of R_P,*

(2) *$(L^n) = n! \prod_{\beta \in R_P} \frac{(\beta, w)}{(\beta, \rho)}$, where w is the fundamental weight corresponding to α (or L) and ρ is a half of the sum of all positive roots, and*

(3) *the sum of all $\beta \in R_P$ is r times ρ for some positive integer r and $c_1(G/P)$ (or the anticanonical class of G/P) is equal to r times $c_1(L)$.*

A homogeneous vector bundle on G/P is obtained from a representation of P and hence from that of reductive part G_0 of P. Note that the weight spaces of G and G_0 are naturally identified.

Theorem 1.6. (Bott [3]) *Let E be a homogeneous vector bundle over G/P induced by an irreducible representation of the reductive part of P. Let γ be the highest weight of the representation and ρ a half of the sum of all positive roots of G. Then we have*

(1) *if $(\gamma + \rho, \beta) = 0$ for a positive root β, then $H^i(G/P, E)$ vanishes for every i, and*

(2) *let i_0 be the number of positive roots β with $(\gamma + \rho, \beta)$ negative (i_0 is called the index of E). Then $H^i(G/P, E) = 0$ for all i except for i_0 and $H^{i_0}(G/P, E)$ is an irreducible G-module.*

Returning to our first situation, our variety X is the quotient of the exceptional Lie group G of type G_2 by a maximal parabolic subgroup P. The root system G_2 has two root basis α_1 and α_2 with different lengths and the root α corresponding to P in the above manner is the longer one, say α_2. The line bundle $L = \mathcal{O}_X(1)$ and the vector bundle $\mathcal{U}^\vee(1)$ on X come from the representation with the highest weights $w_1 = 3\alpha_1 + 2\alpha_2$ and $w_2 = 2\alpha_1 + \alpha_2$, respectively, which are the fundamental weights of G. Since $\mathcal{U}$ is of rank 2 and $\wedge^2\mathcal{U} \cong \mathcal{O}_X(1)$, $\mathcal{U}^\vee$ is isomorphic to $\mathcal{U}(1)$. ρ is equal to $w_1 + w_2$ and the inner products of ρ, w_1, w_2 and the 6 positive roots are as follows:

	α_1	$3\alpha_1 + \alpha_2$	$2\alpha_1 + \alpha_2$	$3\alpha_1 + 2\alpha_2$	$\alpha_1 + \alpha_2$	α_2
ρ	1	6	5	9	4	3
w_1	0	3	3	6	3	3
w_2	1	3	2	3	1	0

By (1.5), X has dimension 5, $c_1(X) = 3c_1(L)$ and has degree

$$(L^5) = 5!\frac{3 \cdot 3 \cdot 6 \cdot 3 \cdot 3}{6 \cdot 5 \cdot 9 \cdot 4 \cdot 3} = 18$$

in $\mathbf{P}^{13}$. The homogeneous vector bundles $(S^m\mathcal{U})(n)$ comes from the irreducible representation with the highest weight $mw_1 + (n - m)w_2$. Applying (1.6), we have

Proposition 1.7. *The cohomology groups of* $\mathcal{U}(n), (S^2\mathcal{U})(n)$ *and* $(S^3\mathcal{U})(n)$ *are zero except for the following cases:*

(1) $H^0(X, \mathcal{U}(n))$ *for* $n \geq 1, H^0(X, (S^2\mathcal{U})(n))$ *for* $n \geq 2$ *and* $H^0(X, (S^3\mathcal{U})(n))$ *for* $n \geq 3$,

(2) $H^1(X, (S^3\mathcal{U})(1))$ *and* $H^4(X, (S^3\mathcal{U})(-1))$, *and*

(3) $H^5(X, \mathcal{U}(n)), H^5(X, (S^2\mathcal{U})(n))$ *and* $H^5(X, (S^3\mathcal{U})(n))$ *for* $n \leq -3$.

Let S be a smooth K3 surface which is a complete intersection of 3 members of $|\mathcal{O}_X(1)|$. By using the Koszul complex

$$(1.8) \quad 0 \longrightarrow \mathcal{O}_X(-3) \longrightarrow \mathcal{O}_X(-2)^{\oplus 3} \longrightarrow \mathcal{O}_X(-1)^{\oplus 3} \longrightarrow \mathcal{O}_X \longrightarrow \mathcal{O}_S \longrightarrow 0,$$

we have

Lemma 1.9. *If* E *is a vector bundle on* X *and if* $H^{i+j}(X, E(-j)) = 0$ *for every* $0 \leq j \leq 3$, *then* $H^i(S, E|_S) = 0$.

Since $\mathcal{U}$ is of rank 2, $sl(\mathcal{U})$ is isomorphic to $S^2\mathcal{U} \otimes (\det \mathcal{U})^{-1} \cong (S^2\mathcal{U})(1)$. By Proposition 1.7 and Lemma 1.9, we have

Proposition 1.10. *Let S be as above. Then $H^i(S, sl(\mathcal{U})|_S)$ vanishes for every i, $H^1(S, (sl\mathcal{U})(n)|_S)$ vanishes for every n and $\mathcal{U}|_S$ or $(S^3\mathcal{U})(1)|_S$ has no nonzero global sections.*

§2. Proof of Theorem 0.2 in the case $g = 10$

Let A be a 3-dimensional subspace of $H^0(X, L)$ and S_A the intersection of $X = X_{18}$ and the linear subspace $\mathbf{P}(H^0(X, L)/A)$ of $\mathbf{P}(H^0(X, L))$. Let L_A and U_A be the restrictions of L and $\mathcal{U}$ to S_A, respectively. Let Ξ be the subset of the Grassmann variety $G(3, H^0(X, L))$ consisting of A's such that S_A are smooth K3 surfaces and that the vector bundles U_A are stable with respect to the ample line bundles L_A.

Proposition 2.1. Ξ *is a nonempty open subset of $G(3, H^0(X, L))$.*

Proof. U_A is a rank 2 bundle and $\det U_A \cong L_A^{-1}$. By Moishezon's theorem [9], $\operatorname{Pic} S_A$ is generated by L_A if A is general. Since $H^0(S_A, U_A) = 0$ by Proposition 1.10, U_A is stable if A is general. Since the stableness is an open condition [8], we have our proposition.   q.e.d.

In this section we shall prove the following:

(2.2) *If two 3-dimensional subspaces A and B belong to Ξ and if the polarized K3 surfaces (S_A, L_A) and (S_B, L_B) are isomorphic to each other, then S_A and S_B, and hence A and B, are equivalent under the action of G.*

Let $\varphi : S_A \xrightarrow{\sim} S_B$ be an isomorphism such that $\varphi^* L_B \cong L_A$.

Step I. There is an isomorphism $\beta : U_A \xrightarrow{\sim} \varphi^* U_B$.

Proof. Since $c_1(U_A) = -c_1(L_A)$ and $c_1(U_B) = -c_1(L_B)$, the first Chern classes of U_A and $\varphi^* U_B$ are same. Since (S_B, U_B) is a deformation of (S_A, U_A), U_B and U_A have the same second Chern number. Hence the two vector bundles $\mathcal{H}om_{\mathcal{O}_S}(U_A, \varphi^* U_B)$ and $\mathcal{E}nd_{\mathcal{O}_S}(U_A)$ have the same first Chern class and the same second Chern number. Therefore, by the Riemann-Roch theorem and Proposition 1.10, we have

$$
\begin{aligned}
\chi(\mathcal{H}om_{\mathcal{O}_S}(U_A, \varphi^* U_B)) &= \chi(\mathcal{E}nd_{\mathcal{O}_S}(U_A)) \\
&= \chi(\mathcal{O}_{S_A}) + \chi(sl(U_A)) = 2.
\end{aligned}
$$

By the Serre duality, we have

$$
\begin{aligned}
\dim \operatorname{Hom}_{\mathcal{O}_S}(U_A, \varphi^* U_B) &+ \dim \operatorname{Hom}_{\mathcal{O}_S}(\varphi^* U_B, U_A) \\
&\geq \chi(\mathcal{H}om_{\mathcal{O}_S}(U_A, \varphi^* U_B)) = 2.
\end{aligned}
$$

Hence there is a nonzero homomorphism from U_A to $\varphi^* U_B$ or vice versa. Since U_A and $\varphi^* U_B$ are stable vector bundles and have the same slope, the nonzero homomorphism is an isomorphism.   q.e.d.

Step II. There is an isomorphism $\gamma : C_0 \xrightarrow{\sim} C_0$ (as **C**-vector spaces) such that the following diagram is commutative:

$$
\begin{array}{ccc}
U_A & \xrightarrow{\beta} & \varphi^* U_B \\
\cap & & \cap \\
C_0 \otimes \mathcal{O}_{S_A} & \xrightarrow{\gamma \otimes 1} & C_0 \otimes \mathcal{O}_{S_A} = \varphi^*(C_0 \otimes \mathcal{O}_{S_B})
\end{array}
$$

Proof. Let γ_0 be the dual map of

$$\mathrm{Hom}(\beta, \mathcal{O}_{S_A}) : \mathrm{Hom}_{\mathcal{O}_S}(\varphi^* U_B, \mathcal{O}_{S_A}) \longrightarrow \mathrm{Hom}_{\mathcal{O}_S}(U_A, \mathcal{O}_{S_A}).$$

Claim: The inclusion $U_A \subset C_0 \otimes \mathcal{O}_{S_A}$ induces an isomorphism $\mathrm{Hom}(C_0, \mathbf{C}) \xrightarrow{\sim} \mathrm{Hom}_{\mathcal{O}_S}(U_A, \mathcal{O}_{S_A})$.

Let $\mathcal{K}$ be the dual of the quotient bundle $(C_0 \otimes \mathcal{O}_X)/\mathcal{U}$ on X. The natural map from $\mathrm{Hom}(C_0, \mathbf{C})$ to $\mathrm{Hom}_{\mathcal{O}_X}(\mathcal{U}, \mathcal{O}_X)$ is an isomorphism because both are irreducible G-modules. Hence both $H^0(X, \mathcal{K})$ and $H^1(X, \mathcal{K})$ are zero. By the exact sequence

$$0 \longrightarrow \mathcal{K} \longrightarrow C_0^\vee \otimes \mathcal{O}_X \xrightarrow{\alpha} \mathcal{U}^\vee \longrightarrow 0$$

and Proposition 1.7, we have $H^i(X, \mathcal{K}(-i)) = H^{i+1}(X, \mathcal{K}(-i)) = 0$ for $i = 1, 2$ and 3. Hence by Lemma 1.9, both $H^0(S, \mathcal{K}|_S)$ and $H^1(S, \mathcal{K}|_S)$ are zero and we have our claim.

By the claim and by applying the claim to $\varphi^* U_B \subset \varphi^*(C_0 \otimes \mathcal{O}_{S_B})$, we have a homomorphism $\gamma : C_0 \longrightarrow C_0$ such that the following diagram

$$
\begin{array}{ccc}
C_0 & \xrightarrow{\gamma} & C_0 \\
\downarrow \wr & & \downarrow \wr \\
\mathrm{Hom}_{\mathcal{O}_S}(U_A, \mathcal{O}_{S_A})^\vee & \xrightarrow{\gamma_0} & \mathrm{Hom}_{\mathcal{O}_S}(\varphi^* U_B, \mathcal{O}_{S_A})^\vee
\end{array}
$$

is commutative. Since β is an isomorphism, γ_0 and γ are isomorphisms and γ enjoys our requirement. q.e.d.

Step III. There is an isomorphism $\gamma : C_0 \xrightarrow{\sim} C_0$ (as **C**-vector spaces) such that $(\gamma \otimes 1)(U_A) = \varphi^* U_B \subset C_0 \otimes \mathcal{O}_X$ and $x^2 = \gamma(x)^2$ for every $x \in C_0$.

Proof. Take an isomorphism γ which satisfies the requirement of Step II. Put $q(x) = x^2$ and $q'(x) = \gamma(x)^2$. Then q and q' are quadratic forms on C_0 and both $q \otimes 1$ and $q' \otimes 1$ are identically zero on U_A. Hence replacing γ by some multiple by a nonzero constant if necessary, we have our assertion by the following:

Claim: The quadratic forms Q on C_0 such that $(Q \otimes 1)|_{U_A} = 0$ form at most one dimensional vector space.

Let $\mathcal{N}$ be the kernel of the homomorphism $S^2\alpha : S^2\mathcal{C}_0 \otimes \mathcal{O}_X \longrightarrow S^2\mathcal{U}^\vee$. Since $S^2\mathcal{C}_0$ is a sum of two irreducible G-modules of dimension 1 and 27 and since $H^0(S^2\alpha)$ is a homomorphism of G-modules, we have dim $H^0(X,\mathcal{N}) =$ dim $\mathrm{Ker} H^0(S^2\alpha) = 1$. By the exact sequence

$$H^{i-1}(X, S^2\mathcal{U}^\vee(-n)) \longrightarrow H^i(X,\mathcal{N}(-n)) \longrightarrow H^i(X, S^2\mathcal{C}_0 \otimes \mathcal{O}_X(-n))$$

and Proposition 1.7, $H^i(X,\mathcal{N}(-i))$ is zero for every $i = 1,2$ and 3. Hence by the Koszul complex (1.8), the restriction map $H^0(X,\mathcal{N}) \longrightarrow H^0(S,\mathcal{N}|_S)$ is surjective and we have dim $H^0(S,\mathcal{N}|_S) \leq$ dim $H^0(X,\mathcal{N}) = 1$, which shows our claim.   q.e.d.

Step IV. There is an isomorphism $\gamma : \mathcal{C}_0 \xrightarrow{\sim} \mathcal{C}_0$ such that $(\gamma \otimes 1)(U_A) = \varphi^*U_B, x^2 = \gamma(x)^2$ for every $x \in \mathcal{C}_0$ and $(\gamma \otimes 1)(xy) = ((\gamma \otimes 1)(x))((\gamma \otimes 1)(y))$ for every $x \in U_A$ and $y \in V_A$.

Proof. Take an isomorphism γ which satisfies the requirements of Step III. Then $\gamma \otimes 1$ maps V_A onto $\varphi^*V_B \subset \mathcal{C}_0 \otimes \mathcal{O}_X$ and induces an isomorphism $\Gamma : V_A/U_A \longrightarrow \varphi^*(V_B/U_B)$ which is compatible with the quadratic forms on V_A/U_A and V_B/U_B. Let $r_A : V_A/U_A \longrightarrow sl(U_A)$ be the restriction of $\bar{R} : \mathcal{V}/\mathcal{U} \longrightarrow sl(\mathcal{U})$ to S_A. Consider the following diagram:

$$\begin{array}{ccc} V_A/U_A & \xrightarrow{r_A} & sl(U_A) \\ \Gamma \downarrow & & \downarrow \mathrm{ad}(\gamma \otimes 1) \\ \varphi^*(V_B/U_B) & \xrightarrow{\varphi^* r_B} & \varphi^* sl(U_B) \end{array}$$

The vector bundles $sl(U_A)$ and $sl(U_B)$ have the quadratic forms $f \mapsto (\mathrm{tr} f^2)/2$ and all the homomorphisms in the above diagram are isomorphisms compatible with the quadratic forms by Proposition 1.4. If g is an automorphism of $sl(U_A)$ and preserves the quadratic form, then g or $-g$ comes from an automorphism of U_A because $H^1(S, \mathbf{Z}/2\mathbf{Z}) = 0$. Since every endomorphism of U_A is a constant multiplication, g is equal to $\pm\mathrm{id}$. Therefore, the above diagram is commutative up to sign. Hence, for γ or $-\gamma$, the above diagram is commutative. Since $xy = r_A(\bar{y})(x)$ for every $x \in U_A$ and $y \in V_A$, γ or $-\gamma$ satisfies our requirements, where $\bar{y} \in V_A/U_A$ is the image of $y \in V_A$.   q.e.d.

We shall show that, for the isomorphism γ in Step IV, $\tilde{\gamma} = 1 \oplus \gamma : \mathcal{C}_0 \longrightarrow \mathcal{C}^0$ satisfies $\tilde{\gamma}(xy) = \tilde{\gamma}(x)\tilde{\gamma}(y)$ for every $x,y \in \mathcal{C}_0$. If $x,y \in \mathcal{C}_0$, then $xy + yx$ is equal to $b(x,y)$, where $b(x,y)$ is the inner product associated to the quadratic form q. Hence the *real* part of xy is equal to $b(x,y)/2$, that is, $xy - b(x,y)/2$ belongs to $\mathcal{C}_0$. Since γ preserves the quadratic form q, $\tilde{\gamma}(x,y)$ and $\tilde{\gamma}(x)\tilde{\gamma}(y)$ have the same *real* part, that is, their difference belongs to $\mathcal{C}_0$. Put $\delta(x,y) = \tilde{\gamma}(x,y) - \tilde{\gamma}(x)\tilde{\gamma}(y)$ for every $x,y \in \mathcal{C}_0$. $\delta : \mathcal{C}_0 \otimes \mathcal{C}_0 \longrightarrow \mathcal{C}_0$ is skew-symmetric and $\delta \otimes 1$ is identically zero on $U_A \otimes V_A \subset \mathcal{C}_0 \otimes \mathcal{C}_0 \otimes \mathcal{O}_{S_A}$.

Step V. $\delta \otimes 1$ is identically zero on $V_A \otimes V_A \subset \mathcal{C}_0 \otimes \mathcal{C}_0 \otimes \mathcal{O}_{S_A}$.

Proof. Since $\delta \otimes 1$ is skew-symmetric and identically zero on $U_A \otimes V_A$, $\delta \otimes 1$ induces a skew-symmetric form $\bar{\delta}$ on V_A/U_A. Since V_A/U_A is isomorphic to $sl(U_A)$, $\wedge^2(V_A/U_A)^\vee$ is also isomorphic to $sl(U_A)$ and has no nonzero global sections. Hence $\bar{\delta}$ is zero and $\delta \otimes 1$ is identically zero on $V_A \otimes V_A$. *q.e.d.*

Step VI. Every homomorphism f from V_A to U_A is zero.

Proof. Since V_A/U_A is isomorphic to $sl(U_A)$, there are no nonzero homomorphisms from V_A/U_A to $\mathcal{O}_{S_A}$. Hence V_A/U_A cannot be a subsheaf of $C_0 \otimes \mathcal{O}_{S_A}$. Therefore, the exact sequence $0 \longrightarrow U_A \longrightarrow V_A \longrightarrow V_A/U_A \longrightarrow 0$ does not split. Hence the restriction $f|_{U_A} : U_A \longrightarrow U_A$ of f to U_A is not an isomorphism. Since every endomorphism of U_A is a constant multiplication, $f|_{U_A}$ is zero and f induces a homomorphism $\bar{f} : V_A/U_A \longrightarrow U_A$. Since $V_A/U_A \cong sl(U_A)$, we have

$$
\begin{aligned}
\mathrm{Hom}_{\mathcal{O}_S}(V_A/U_A, U_A) &\cong H^0(S_A, sl(U_A) \otimes U_A) \\
&\cong H^0(S_A, U_A \oplus (S^3 U_A) \otimes L_A).
\end{aligned}
$$

Hence by Proposition 1.10, $\bar{f}$ is zero and f is also zero. *q.e.d.*

Step VII. δ is zero.

Proof. Let T be the cokernel of the natural injection $\wedge^2 V_A \longrightarrow \wedge^2 C_0 \otimes \mathcal{O}_{S_A}$. Since $\delta \otimes 1$ belongs to $\mathrm{Hom}_{\mathcal{O}_S}(T, C_0 \otimes \mathcal{O}_{S_A})$, it suffices to show that $\mathrm{Hom}_{\mathcal{O}_S}(T, \mathcal{O}_{S_A})$ is zero. There is an exact sequence

$$
0 \longrightarrow V_A \otimes E_A \longrightarrow T \longrightarrow \overset{2}{\bigwedge} E_A \longrightarrow 0,
$$

where E_A is the quotient bundle $(C_0 \otimes \mathcal{O}_{S_A})/V_A$ and isomorphic to $U_A^\vee$ by the bilinear form b associated to q. By Step VI, we have $\mathrm{Hom}_{\mathcal{O}_S}(V_A \otimes E_A, \mathcal{O}_{S_A}) \cong \mathrm{Hom}_{\mathcal{O}_S}(V_A, U_A) = 0$. Since $\wedge^2 E_A$ is an ample line bundle, $\mathrm{Hom}_{\mathcal{O}_S}(\wedge^2 E_A, \mathcal{O}_{S_A})$ is zero. Therefore, by the above exact sequence, $\mathrm{Hom}_{\mathcal{O}_S}(T, \mathcal{O}_{S_A})$ is zero. *q.e.d.*

By Step VII, $1 \oplus \gamma$ is an automorphism of the Cayley algebra $\mathcal{C}$. The automorphism of $X_{18} = G/P$ induced by $1 \oplus \gamma$ maps S_A onto S_B. Hence we have (2.2) and, in particular, Theorem 0.2.

§3. Generic K3 surfaces of genus 7, 8, and 9

The proof of Theorem 0.2 in the case $g = 7, 8$, and 9 is very similar to and rather easier than the case $g = 10$ dealt in the previous sections. The $(24 - 2g)$-dimensional homogeneous spaces $X = X_{2g-2} \subset \mathbf{P}^{22-g}$ ($g = 7, 8$ and 9) are also generalized Grassmann variety as in the case $g = 10$. In the case $g = 8$, $X_{14} \subset \mathbf{P}^{14}$ is the Grassmann variety $G(V, 2)$ of 2-dimensional quotient spaces of a 6-dimensional vector space V embedded into $\mathbf{P}(\wedge^2 V)$ by the Plücker coordinates. In the case $g = 9$, $X \subset \mathbf{P}^{13}$ is the Grassmann variety of 3-dimensional totally isotropic quotient spaces of a 6-dimensional vector space V with a nondegenerate skew-symmetric tensor $\sigma \in \wedge^2 V$, where a quotient $f : V \to V'$ is totally isotropic with respect to σ if $(f \otimes f)(\sigma)$ is zero in $V' \otimes V'$. The embedding $X_{16} \subset \mathbf{P}^{13}$ is the linear hull of the composite of the natural embedding $X \subset G(V, 3)$ and the Plücker embedding $G(V, 3) \subset \mathbf{P}(\wedge^3 V)$. In the case $g = 7$, $X \subset \mathbf{P}^{15}$ is a 10-dimensional spinor variety. Let V be a 10-dimensional vector space with a non-degenerate second symmetric tensor. The subset of $G(V, 5)$ consisting of 5-dimensional totally isotropic quotient spaces of V has exactly two connected components, one of which is our spinor variety X. The pull-back of the tautological line bundle $\mathcal{O}_{\mathbf{P}}(1)$ by the composite $X \hookrightarrow G(V, 5) \hookrightarrow \mathbf{P}(\wedge^5 V)$ is twice the positive generator L of $\operatorname{Pic} X$ and the vector space $H^0(X, L)$ is a half spinor representation of $\operatorname{Spin}(V)$, the universal covering groups of $\operatorname{SO}(V)$. In each case, X is a compact hermitian symmetric space and the anticanonical class of X is equal to $\dim X - 2$ times the positive generator L of $\operatorname{Pic} X$ (Proposition 1.5 and [2] §16). Moreover, by Proposition 1.5 and an easy computation, we have that the embedded variety $X \hookrightarrow \mathbf{P}(H^0(X, L))$ has degree $2g - 2$. Hence every smooth complete intersection of X and a linear subspace of codimension $n - 2$ (resp. $n - 3$, $n - 1$) is the projective (resp. canonical, anticanonical) model of a K3 surface (resp. curve, Fano 3-fold) of genus g.

Each homogeneous space $X = X_{2g-2}$ has a natural homogeneous vector bundle $\mathcal{E}$ on it. In the case $g = 8$, we have the exact sequence

$$(3.1) \qquad 0 \longrightarrow \mathcal{F} \longrightarrow V \otimes \mathcal{O}_X \xrightarrow{\alpha} \mathcal{E} \longrightarrow 0,$$

where $\mathcal{E}$ (resp. $\mathcal{F}$) is the universal quotient (resp. sub-) bundle and is of rank 2 (resp. 4). In the case $g = 7$ (resp. 9), we have the exact sequence

$$(3.2) \qquad 0 \longrightarrow \mathcal{E}^{\vee} \longrightarrow V \otimes \mathcal{O}_X \xrightarrow{\alpha} \mathcal{E} \longrightarrow 0,$$

where $\mathcal{E}$ is the universal maximally totally isotropic quotient bundle with respect to $\sigma \otimes 1 \in V \otimes V \otimes \mathcal{O}_X$ and is of rank 5 (resp. 3).

Theorem 0.2 is a consequence of the openness of the stability condition and the following:

Theorem 3.3. *Let S and S' be two K3 surfaces which are complete intersections of $X_{2g-2} \subset \mathbf{P}^{22-g}$ ($g = 7, 8$ and 9) and linear subspaces R and R', respectively. Then we have*

(1) *if R is general, then the vector bundle $\mathcal{E}|_S$ is stable with repsect to $\mathcal{O}_S(1)$, the restriction of $L = \mathcal{O}_X(1)$ to S, and*

(2) *if $\mathcal{E}|_S$ and $\mathcal{E}|_{S'}$ are stable with respect to $\mathcal{O}_S(1)$ and $\mathcal{O}_{S'}(1)$ and if $S \subset R$ and $S' \subset R'$ are projectively equivalent, then R and R' are equivalent under the action of G on X.*

For the proof we need the following property of the vector bundle $E = \mathcal{E}|_S$.

Proposition 3.4. *Let S be a complete intersection of $X = X_{2g-2} \subset \mathbf{P}^{22-g}$ and a g-dimensional linear subspace and E the restriction of $\mathcal{E}$ to S. Then we have*

(1) $H^i(S, sl(E)) = 0$ *for every i,*

(2) *the homomorphism $H^0(\alpha) : V \to H^0(S, E)$ is an isomorphism,*

(3) *in the case $g = 7$ (resp. 9), the kernel of the homomorphism $H^0(S^2\alpha) : S^2V \to H^0(S, S^2E)$ (resp. $H^0(\wedge^2\alpha) : \wedge^2 V \to H^0(S, \wedge^2 E)$) is 1-dimensional and generated by $\sigma \otimes 1$, and*

(4) *in the case $g = 7$ (resp. 8, resp. 9), $E(-1), (\wedge^2 E)(-1), (\wedge^3 E)(-2)$ or $(\wedge^4 E)(-2)$ (resp. $E(-1)$, resp. $E(-1)$ or $(\wedge^2 E)(-1)$) has no nonzero global sections.*

We prove the proposition in the case $g = 7$. The other cases are similar. According to [4], we take $\alpha_i = e_i - e_{i+1}$, $1 \le i \le 4$, and $\alpha_5 = e_4 + e_5$ as a root basis of SO(10). The positive roots are $e_i \pm e_j, i < j$ and the conjugacy class of the maximal parabolic subgroup P corresponds to α_5 (or α_4). The homogeneous vector bundles $\mathcal{O}_X(1), \wedge^i \mathcal{E}, sl(\mathcal{E})$ and $S^2 \mathcal{E}$ are induced by the irreducible representations of the reductive part of P with the highest weights $\frac{1}{2}(e_1 + \cdots + e_5)$, $e_1 + \cdots + e_i$, $e_1 - e_5$ and $2e_1$, respectively. The half ρ of the sum of positive roots is equal to $4e_1 + 3e_2 + 2e_3 + e_4$. Applying Bott's theorem, we have

Lemma 3.5. $(g = 7)$ *The cohomology groups of $\mathcal{E}(n), (\wedge^2\mathcal{E})(n), (sl\,\mathcal{E})(n)$ and $(S^2\mathcal{E})(n)$ vanish except for the following cases:*

(1) $H^0(X, \mathcal{E}(n)), H^0(X, (\wedge^2\mathcal{E})(n)), H^0(X, (S^2\mathcal{E})(n))$ *for $n \ge 0$ and $H^0(X, (sl\,\mathcal{E})(n))$ for $n \ge 1$,*

(2) $H^9(X, (\wedge^2\mathcal{E})(-8))$, *and*

(3) $H^{10}(X, \mathcal{E}(n)), H^{10}(X, (sl\,\mathcal{E})(n))$ *for $n \le -9$ and $H^{10}(X, (\wedge^2\mathcal{E})(m)), H^{10}(X, (S^2\mathcal{E})(m))$ for $m \le -10$.*

Remark 3.6. In the above case $g = 7$, the 10 roots $e_i + e_j$, $1 \le i < j \le 5$, are complementary to P. Their sum is equal to $4(e_1 + \cdots + e_5)$ and this is 8 times the fundamental weight w. By Proposition 1.5, the self intersection number of $\mathcal{O}_X(1)$ is equal to

$$10! \prod_{\beta \in R_P} \frac{(\beta, w)}{(\beta, \rho)} = 10! \prod_{0 \le i < j \le 4} (i+j)^{-1} = 12.$$

Hence X is a 10-dimensional variety of degree 12 in $\mathbf{P}^{15}$ and the anticanonical class is 8 times the hyperplane section.

Proof of Proposition 3.4 (in the case $g = 7$): S is a complete intersection of 8 members of $|\mathcal{O}_X(1)|$. Hence, if $\mathcal{A}$ is a vector bundle on X and $H^{i+a}(X, \mathcal{A}(-a))$ vanishes for every $0 \le a \le 8$, then so does $H^i(S, \mathcal{A}|_S)$.

(1 and 4)   (1) and the vanishings of $H^0(S, E(-1))$ and $H^0(S, (\wedge^2 E)(-1))$ follow immediately from Lemma 3.5. Since $\wedge^5 \mathcal{E}$ is isomorphic to $\mathcal{O}_X(2)$, $\wedge^k \mathcal{E}$ is isomorphic to $(\wedge^{5-k} \mathcal{E})^\vee \otimes \mathcal{O}_X(2)$. Hence by the Serre duality and Lemma 3.5, we have

$$
\begin{aligned}
H^i(X, (\wedge^3 \mathcal{E})(-2 - i)) &\cong H^{10-i}(X, (\wedge^3 \mathcal{E})^\vee(2 + i - 8))^\vee \\
&\cong H^{10-i}(X, (\wedge^2 \mathcal{E})(i - 8))^\vee = 0
\end{aligned}
$$

and

$$
\begin{aligned}
H^i(X, (\wedge^4 \mathcal{E})(-2 - i)) &\cong H^{10-i}(X, (\wedge^4 \mathcal{E})^\vee(2 + i - 8))^\vee \\
&\cong H^{10-i}(X, \mathcal{E}(i - 8)))^\vee = 0,
\end{aligned}
$$

for every $0 \leq i \leq 8$. Therefore, $(\wedge^3 E)(-2)$ or $(\wedge^4 E)(-2)$ has no nonzero global sections.

(2) By the Serre duality, $H^i(X, E^\vee(-i))$ and $H^{i+1}(X, E^\vee(-i))$ are isomorphic to $H^{10-i}(X, \mathcal{E}(i - 8))^\vee$ and $H^{9-i}(X, \mathcal{E}(i - 8))^\vee$, respectively and both are zero for every $0 \leq i \leq 8$, by Lemma 3.5. Hence both $H^0(S, E^\vee)$ and $H^1(S, E^\vee)$ vanish. Therefore, by the exact sequence (3.2), we have (2).

(3) Let $\mathcal{K}$ be the kernel of the homomorphism $S^2\alpha : S^2V \otimes \mathcal{O}_X \to S^2\mathcal{E}$. We have the exact sequence

$$
0 \longrightarrow \mathcal{K} \longrightarrow S^2 V \otimes \mathcal{O}_X \longrightarrow S^2 \mathcal{E} \longrightarrow 0.
$$

The G-module S^2V is isomorphic to the direct sum of an irreducible G-module of dimension 54 and a trivial G-module generated by σ. Hence the G-module $H^0(X, \mathcal{K}) = \text{Ker } H^0(S^2\alpha)$ is 1-dimensional and generated by σ. By Lemma 3.5 and the Kodaira vanishing theorem, $H^{i-1}(X, (S^2\mathcal{E})(-i))$ and $H^i(X, \mathcal{O}_X(-i))$ are zero. Hence by the above exact sequence, $H^i(X, \mathcal{K}(-i))$ vanishes for every $1 \leq i \leq 8$. By using the Koszul complex, we have that the restriction map $H^0(X, \mathcal{K}) \to H^0(S, \mathcal{K}|_S)$ is surjective. Therefore, the kernel of $H^0(S^2\alpha|_S)$ is at most 1-dimensional. It is clear that the kernel contains $\sigma \otimes 1$. Hence we have (3).   q.e.d.

Proof of Theorem 3.3:   Let S (resp. S') be a K3 surface which is a complete intersection of X and a linear subspace P (resp. P') and E (resp. E') the restriction of $\mathcal{E}$ to S (resp. S'). If P is general, then Pic S is generated by $\mathcal{O}_S(1)$ and, by (4) of Proposition 3.4, E is stable. Hence we have i). Assume that S and S' are isomorphic to each other as polarized surfaces and that E and E' are stable. By (1) of Proposition 3.4 and the same argument as Step I in §2, E and E' are isomorphic to each other. By (2) of Proposition 3.4, we have an isomorphism $\beta : V \xrightarrow{\sim} V'$ and a commutative diagram

$$
\begin{array}{ccccc}
V \otimes \mathcal{O}_S & \xrightarrow{\alpha|_S} & E & \longrightarrow & 0 \\
{\scriptstyle \beta \otimes 1} \downarrow {\scriptstyle \wr} & & \downarrow {\scriptstyle \wr} & & \\
V' \otimes \mathcal{O}_{S'} & \xrightarrow{\alpha|_{S'}} & E' & \longrightarrow & 0.
\end{array}
$$

Hence, in the case $g = 8$, S and S' are equivalent under the action of $\mathrm{GL}(V)$. In the case $g = 7$ or 9, by (3) of Proposition 3.4, $S^2\beta$ maps σ to $a\sigma$ for a nonzero constant a. Hence, replacing β by $a^{1/2}\beta$, we may assume that $S^2\beta$ preserves σ. Hence S and S' are equivalent under the action of $\mathrm{SO}(V, \sigma)$ or $\mathrm{Sp}(V, \sigma)$. q.e.d.

§4. Generic K3 surface of genus 6

A K3 surface of genus 6 is obtained as a complete intersection in the Grassmann variety $G(2, V^5)$ of 2-dimensional subspaces in a fixed 5-dimensional vector space V^5. $G(2, V^5)$ is embedded into $\mathbf{P}^9$ by Plücker coordinates and has degree 5. A smooth complete intersection $X_5 \subset \mathbf{P}^6$ of $G(2, V^5)$ and 3 hyperplanes in $\mathbf{P}^9$ is a Fano 3-fold of index 2 and degree 5. A smooth complete intersection X_5 and a quadratic hypersurface in $\mathbf{P}^6$ is an anticanonical divisor of X_5 and is a K3 surface of genus 6. The isomorphism class of X_5 does not depend on the choice of 3 hyperplanes and X_5 has an action of $\mathrm{PGL}(2)$ (see below).

Theorem 4.1. *Let S and S' be two general smooth complete intersections of X_5 and a quadratic hypersurface in $\mathbf{P}^6$. If $S \subset \mathbf{P}^6$ and $S' \subset \mathbf{P}^6$ are projectively equivalent, then they are equivalent under the action of $\mathrm{PGL}(2)$ on X_5.*

All the Fano 3-folds of index 2 and degree 5 are unique up to isomorphism [5]. There are several ways to describe the Fano 3-folds. The following is most convenient for our purpose: Let V be a 2-dimensional vector space and $f \in S^6V$ an invariant polynomial of an octahedral subgroup of $\mathrm{PGL}(V)$. f is equal to $xy(x^4 - y^4)$ for a suitable choice of a basis $\{x, y\}$ of V. Then the closure X_5 of the orbit $\mathrm{PGL}(V) \cdot \bar{f}$ in $\mathbf{P}_*(S^6V) := (S^6V - \{0\})/\mathbf{C}^*$ is a Fano 3-fold of index 2 and degree 5, [11]. $H^0(X_5, \mathcal{O}_X(2))$ is generated by $H^0(X_5, \mathcal{O}_X(1)) = S^6V$, [5], and has dimension $\frac{1}{2}(-K_X)^3 + 3 = 23$. Hence the kernel A of the natural map $S^2H^0(X, \mathcal{O}_X(1)) \longrightarrow H^0(X, \mathcal{O}_X(2))$ is a 5-dimensional $\mathrm{SL}(V)$-invariant subspace. As an $\mathrm{SL}(V)$-module, $S^2H^0(X, \mathcal{O}_X(1))$ is isomorphic to $S^2(S^6V) \cong S^{12}V \oplus S^8V \oplus S^4V \oplus \mathbf{1}$. Hence we have

Proposition 4.2. (1) $H^0(X_5, \mathcal{O}_X(-K_X))$ *is isomorphic to* $S^{12}V \oplus S^8V \oplus \mathbf{1}$ *as $\mathrm{SL}(V)$-module, and*

(2) *the vector space A of quadratic forms which vanish on $X_5 \subset \mathbf{P}^6$ is isomorphic to S^4V as $\mathrm{SL}(V)$-module.*

There is a non-empty open subset Ξ of $|-K_X|$ and a natural morphism $\Xi/\mathrm{PGL}(V) \longrightarrow \mathcal{F}_6$. Both the target and the source are of dimension 19 and the morphism is birational by the theorem. Hence by the proposition we have

Corollary 4.3. *The generic K3 surface of genus 6 can be embedded into X_5 as an anticanonical divisor in a unique way up to the action of $\mathrm{PGL}(V)$. In particular, the moduli space $\mathcal{F}_6$ is birationally equivalent to the orbit space $(S^{12}V \oplus S^8V)/\mathrm{PGL}(V)$.*

First we need to show that $\mathrm{PGL}(V)$ is the full automorphism group of X_5:

Proposition 4.4. *The automorphism group $\mathrm{Aut}\,X_5$ of X_5 is connected and the natural homomorphism $\mathrm{PGL}(V) \longrightarrow \mathrm{Aut}\,X_5$ is an isomorphism.*

Proof. There is a 2-dimensional family of lines on $X_5 \subset \mathbf{P}^6$ and a 1-dimensional subfamily of lines ℓ of special type, *i.e.*, lines such that $N_{\ell/X} \cong \mathcal{O}(1) \oplus \mathcal{O}(-1)$. The union of all lines of special type is a surface and has singularities along a rational curve C. C is the image of the 6-th Veronese embedding of $\mathbf{P}(V) \cong \mathbf{P}^1$ into $\mathbf{P}(S^6V)$. C is invariant under the action of $\mathrm{Aut}\,X_5$. Every automorphism of X_5 induces an automorphism of C. Hence we have the homomorphism $\alpha : \mathrm{Aut}\,X_5 \longrightarrow \mathrm{Aut}\,C \cong \mathrm{PGL}(V)$. Since $\alpha|_{\mathrm{PGL}(V)}$ is an isomorphism, $\mathrm{Aut}\,X_5$ is isomorphic to $\mathrm{PGL}(V) \times \mathrm{Ker}\,\alpha$. Let g be an automorphism of X_5 which commutes with every element of $\mathrm{SL}(V)$. Since S^6V is an irreducible $\mathrm{SL}(V)$-module, g is the identity by Schur's lemma.   q.e.d.

Next we construct an equivariant embedding of X_5 into the Grassmann variety $G(2, S^4V)$. Let W be the 2-dimensional subspace of S^4V generated by $x^4 + y^4$ and x^2y^2 for some basis $\{x, y\}$ of V and Y the closure of the orbit $\mathrm{PGL}(V)\cdot[W]$ in $G(2, S^4V)$. Consider the morphism $J : G(2, S^4V) \longrightarrow \mathbf{P}_*(S^6V)$ for which

$$J([\mathbf{C}g + \mathbf{C}h]) = \det\begin{pmatrix} g_X & g_Y \\ h_X & h_Y \end{pmatrix},$$

where $\{X, Y\}$ is the dual basis of $\{x, y\}$. Then J is a $\mathrm{PGL}(V)$-equivariant morphism and sends $[W]$ to the point $\bar{f}, f = xy(x^4 - y^4)$. Hence J maps Y onto $X_5 \subset \mathbf{P}_*(S^6V)$. Define two $\mathrm{GL}(V)$-homomorphisms $\varphi : \wedge^2 S^4V \longrightarrow S^2V \otimes (\det V)^3$ and $j : \wedge^2 S^4V \longrightarrow S^6V \otimes \det V$ by

$$\varphi(g \wedge h) = \sum_{i,j,k=\pm 1} ijk(D_iD_jD_kg)(D_{-i}D_{-j}D_{-k}h) \otimes (X \wedge Y)^{-3}$$

and

$$j(g \wedge h) = \det\begin{pmatrix} D_1(g) & D_{-1}(g) \\ D_1(h) & D_{-1}(h) \end{pmatrix} \otimes (X \wedge Y)^{-1},$$

where $D_{\pm 1}$ are the derivations by X and Y. The $\mathrm{GL}(V)$-module $\wedge^2 S^4 V$ is decomposed into the direct sum of irreducible $\mathrm{GL}(V)$-submodules $\mathrm{Ker}\,\varphi$ and $\mathrm{Ker}\,j$. Since $\varphi(\wedge^2 W) = 0$, the Plücker coordinates of W lies in the linear subspace $P = \mathbf{P}_*(\mathrm{Ker}\,\varphi)$ of $\mathbf{P}_*(\wedge^2 S^4 V)$ and Y is contained in the intersection $G(2, S^4 V) \cap P$. The morphism J is the composite of the Plücker embedding $G(2, S^4 V) \subset \mathbf{P}_*(\wedge^2 S^4 V)$ and the projection $\mathbf{P}_*(j) : \mathbf{P}_*(\wedge^2 S^4 V) \cdots \to \mathbf{P}_*(S^6 V)$ from the linear subspace $\mathbf{P}_*(\mathrm{Ker}\,j)$. Since the restriction of $\mathbf{P}_*(j)$ to P is an isomorphism, J gives a $\mathrm{PGL}(V)$-equivariant isomorphism from the projective variety $Y \subset P$ onto $X_5 \subset \mathbf{P}_*(S^6 V)$.

Lemma 4.5. *Y coincides with the intersection of* $G(2, S^4 V)$ *and* P *in* $\mathbf{P}_*(\wedge^2 S^4 V)$.

Proof. Let Y' be the intersection of $G(2, S^4 V)$ and P and B (resp. B') the vector space consisting of quadratic forms on P which vanish on Y (resp. Y'). Both Y and Y' are intersections of quadratic hypersurfaces. Hence it suffices to show that $B = B'$. Since $G(2, S^4 V)$ does not contain P, B' is not zero. On the other hand, since $Y \subset P$ is isomorphic to $X_5 \subset \mathbf{P}^6$, B is an irreducible $\mathrm{SL}(V)$-module by Proposition 4.2. As we saw above, Y is contained $\mathrm{PGL}(V)$-equivariantly in Y' and hence B' is an $\mathrm{SL}(V)$-submodule of B. Hence B' coincides with B. q.e.d.

So we have constructed a $\mathrm{PGL}(V)$-equivariant embedding of X_5 into $G(2, S^4 V)$ and shown that X_5 coincides with the intersection of its linear hull and $G(2, S^4 V)$.

Proof of Theorem 4.1. There is a universal exact sequence

$$0 \longrightarrow \mathcal{E} \longrightarrow S^4 V \otimes \mathcal{O}_X \longrightarrow \mathcal{F} \longrightarrow 0$$

on $G(2, S^4 V)$, where $\mathcal{E}$ (resp. $\mathcal{F}$) is the universal sub- (resp. quotient) bundle and has rank 2 (resp. 3). Let S and S' be two members of the anticanonical linear system $|-K_X|$ on X_5. By the same arguments as in Sections 2 and 3, we have

(i) $H^i(S, sl(\mathcal{E})|_S) = 0$ for every i,

(ii) If S is general, then the vector bundle $\mathcal{E}|_S$ is stable with respect to $\mathcal{O}_S(1)$, and

(iii) If $\mathcal{E}|_S$ and $\mathcal{E}|_{S'}$ are stable with respect to $\mathcal{O}_S(1)$ and $\mathcal{O}_{S'}(1)$, respectively, and if S and S' are isomorphic as polarized surfaces, then there are isomorphisms $\alpha : \mathcal{E}|_S \longrightarrow \mathcal{E}|_{S'}$ and $\beta \in \mathrm{GL}(S^4 V)$ such that the diagram

$$
\begin{array}{ccc}
0 \longrightarrow & E|_S & \longrightarrow & S^4 V \otimes \mathcal{O}_S \\
& \alpha \downarrow \wr & & \downarrow \wr \; \beta \otimes 1 \\
0 \longrightarrow & E|_{S'} & \longrightarrow & S^4 V \otimes \mathcal{O}_{S'}
\end{array}
$$

is commutative. In particular, the automorphism $\bar{\beta}$ of $G(2, S^4V)$ induced by β maps S onto S' isomorphically.

Since X_5 is the intersection of $G(2, S^4V)$ and the linear span of S (resp. S'), the automorphism $\bar{\beta}$ maps X_5 onto itself. Hence, by Proposition 4.4, S and S' are equivalent under the action of $\mathrm{PGL}(V)$ on X_5. q.e.d.

§5. Fano 3-folds of genus 10

In this section we shall prove Theorem 0.9 in the case $g = 10$. The other cases $g = 7, 8$ and 9 are very similar.

Let V and V' be Fano 3-folds which are complete intersections of $X_{18} \subset \mathbf{P}^{13}$ and linear subspaces of codimension 2. By the Lefschetz theorem, both $\mathrm{Pic}\,V$ and $\mathrm{Pic}\,V'$ are generated by hyperplane sections. Let $\mathcal{U}$ be the universal subbundle of $\mathcal{C}_0 \otimes \mathcal{O}_{X_{18}}$ as in Section 1 and F and F' the restrictions of $\mathcal{U}$ to V and V', respectively.

Proposition 5.1. *Let $\varphi : V \xrightarrow{\sim} V'$ be an isomorphism. Then $\varphi^*(F')$ is isomorphic to F.*

Proof. Let S be the generic member of $|-K_V|$ and put $S' = \varphi(S)$. The Picard group of S is generated by the hyperplane section. The restrictions $E = F|_S$ and $E' = F'|_{S'}$ are stable vector bundles as we saw in the proof of Proposition 2.1. Hence F and F' are also stable vector bundles. Put $M = \mathcal{H}om_{\mathcal{O}_V}(F, \varphi^*F')$. By Step I in Section 2, there is an isomorphism $f_0 : E \xrightarrow{\sim} (\varphi|_S)^*E'$. Hence the restriction of M to S is isomorphic to $\mathcal{E}nd_{\mathcal{O}_S}(E)$. By Proposition 1.10, we have

$$H^1(S, M(n)|_S) \cong H^1(S, \mathcal{O}_S(n)) \oplus H^1(S, (sl\, E)(n)) = 0$$

for every integer n. Since $H^1(V, M(n))$ is zero, if n is sufficiently negative, we have by induction on n that $H^1(V, M(n))$ is zero for every n. In particular $H^1(V, M(-1))$ vanishes and hence the restriction map $H^0(V, M) \longrightarrow H^0(S, M|_S)$ is surjective. It follows that there is a nonzero homomorphism $f : F \longrightarrow \varphi^*F'$ such that $f|_S = f_0$. Since f_0 is an isomorphism, the cokernel of f has a support on a finite set. Since the Hilbert polynomials $\chi(F(n))$ and $\chi((\varphi^*F')(n))$ are same, the cokernel of f is zero and f is an isomorphism.

q.e.d.

By Proposition 5.1 and similar arguments as Step II-VII in Section 2, we have an isomorphism $\beta : F \xrightarrow{\sim} \varphi^*F'$ and an isomorphism $\gamma : \mathcal{C}_0 \longrightarrow \mathcal{C}_0$ such that the diagram

$$
\begin{array}{ccc}
F & \xrightarrow{\ \beta\ } & \varphi^*(F') \\
\cap & & \cap \\
\mathcal{C}_0 \otimes \mathcal{O}_V & \xrightarrow{\ \gamma \otimes 1\ } & \mathcal{C}_0 \otimes \mathcal{O}_V = \varphi^*(\mathcal{C}_0 \otimes \mathcal{O}_{V'})
\end{array}
$$

is commutative and such that $1 \oplus \gamma$ is an automorphism of the Cayley algebra C. Hence the automorphism of $X_{18} = G/P$ induced by $1 \oplus \gamma$ maps V onto V', which shows Theorem 0.9 in the case $g = 10$.

§6. Curves of genus ≤ 9

In this section we shall show the following:

Theorem 6.1. *The generic curve of genus ≤ 9 lie on a K3 surface.*

In the case $g \leq 6$, the generic curve is realized as a plane curve C of degree $d \leq 6$ with only ordinary double points. Take a general plane curve D of degree $6 - d$ and let S be the double covering of the plane with branch locus $C \cup D$. Then the minimal resolution $\tilde{S}$ of S is a K3 surface and contains a curve isomorphic to C.

In the case $6 \leq g \leq 9$, we shall show that the generic curve C of genus g can be embedded into $\mathbf{P}^5$ by the complete linear system of a line bundle L of degree $g + 4$ and that there is a K3 surface S which is a complete intersection of 3 quadratic hypersurfaces in $\mathbf{P}^5$ and which contains the image of C.

Let C be a curve of genus $6 \leq g \leq 9$ and D an effective divisor on C of degree $g - 6$. Put $L = \omega_C \otimes \mathcal{O}_C(-D)$. Then L is a line bundle of degree $g + 4$. If D is general, then $\dim H^0(C, L) = 6$. Since $\deg L^{\otimes 2} > \deg \omega_C$, we have $\dim H^0(C, L^{\otimes 2}) = 2(g + 4) + 1 - g = g + 9$.

Proposition 6.2. *If C and D are general, then we have*

(1) *L is very ample and $\dim H^0(C, L) = 6$,*

(2) *the natural map*

$$S^2 H^0(C, L) \longrightarrow H^0(C, L^{\otimes 2})$$

is surjective and its kernel V is of dimension $12 - g$, and

(3) *there are 3 quadratic hypersurfaces Q_1, Q_2 and Q_3 in $\mathbf{P}(H^0(C, L))$ which contains the image of C by $\Phi_{|L|}$ and such that the intersection $S = Q_1 \cap Q_2 \cap Q_3$ is a K3 surface.*

Proof. It suffices to show that there exists a pair of C and D which satisfies the conditions (1), (2) and (3). Let R be a smooth rational curve of degree $g - 4$ in $\mathbf{P}^5$ whose linear span $< R >$ has dimension $g - 4$. Since R is an intersection of quadratic hypersurfaces, the intersection of 3 general quadratic hypersurfaces Q_1, Q_2 and Q_3 which contain R is a smooth K3 surface. Let C_0 be the intersection of S and a general hyperplane H. We show that the pair of the generic member C of the complete linear system $|C_0 + R|$ on S and the divisor $D = R|_C$ satisfies the conditions (1), (2) and (3).

The intersection number $(C_0 \cdot R)$ is equal to $\deg R = g - 4 \geq 2$. Hence the linear system $|C_0 + R|$ has no base points. Therefore C is smooth and D is effective. The genus of C is equal to $(C_0 + R)^2/2 + 1 = g$ and the degree of D is equal to $(C_0 + R.R) = g - 6$. Since ω_C is isomorphic to $\mathcal{O}_C(C) \cong \mathcal{O}_C(C_0 + R)$, the line bundle $L = \omega_C(-D)$ is isomorphic to $\mathcal{O}_C(C_0)$, the restriction of the tautological line bundle of $\mathbf{P}^5$ to C. There is a natural exact sequence

$$0 \longrightarrow \mathcal{O}_S(-R) \longrightarrow \mathcal{O}_S(C_0) \longrightarrow \mathcal{O}_C(C_0) \longrightarrow 0.$$

Since $H^i(S, \mathcal{O}_S(-R)) = 0$ for $i = 0$ and 1, the restriction map $H^0(\mathbf{P}^5, \mathcal{O}_{\mathbf{P}}(1))$ $\overset{\sim}{\longrightarrow} H^0(S, \mathcal{O}_S(C_0)) \longrightarrow H^0(C, \mathcal{O}_C(C_0))$ is an isomorphism. Hence the morphism $\Phi_{|L|}$ is nothing but the inclusion map $C \hookrightarrow \mathbf{P}^5$ and (1) and (3) are obvious by our construction of C.

Claim. Let V_0 be the vector space of the quadratic forms on $\mathbf{P}^5$ which are identically zero on $C_0 \cup R$. Then the dimension of V_0 is at most $12 - g$.

Let $F_i = 0$ be the defining equation of the quadratic hypersurface Q_i for $i = 1, 2$ and 3 and $G = 0$ that of the hyperplane H. Let F be any quadratic form on $\mathbf{P}^5$ which is identically zero on $C_0 \cup R$. Since F is identically zero on C_0, F is equal to $a_1 F_1 + a_2 F_2 + a_3 F_3 + GG'$ for some constants a_1, a_2 and a_3 and linear form G'. Since F_1, F_2, F_3 and F are identically zero on R, so is GG'. Hence G' is identically zero on R. Therefore, the vector space V_0 is generated by F_1, F_2, F_3 and GG', G' being all linears from vanishing on $< R >$. Since $\dim < R >= g - 4$, we have $\dim V_0 \leq 3 + 5 - (g - 4) = 12 - g$.

Since C is a general deformation of $C_0 \cup R$, we have, by the claim, that the dimension of V is also at most $12 - g$. Since

$$\dim S^2 H^0(C, L) - \dim H^0(C, L^{\otimes 2}) = 21 - (g + 9) = 12 - g,$$

$H^0(C, L^{\otimes 2})$ is generated by $H^0(C, L)$ and V has exactly dimension $12 - g$.

$$\text{q.e.d.}$$

By the theorem and Corollaries 0.3 and 4.3, we have

Corollary 6.3. *The generic curve of genus $3 \leq g \leq 9$ is a complete intersection in a homogeneous space.*

References

[1] Borcea, C.: Smooth global complete intersections in certain compact homogeneous complex manifolds, J. Reine Angew. Math., **344** (1983), 65-70.

[2] Borel A. and F. Hirzebruch: Characteristic classes and homogeneous spaces I, Amer. J. Math., **80** (1958), 458-538: II, Amer. J. Math., **81** (1959), 315-382.

[3] Bott, R.: Homogeneous vector bundles, Ann. of Math., **66** (1957), 203-248.

[4] Bourbaki, N.: "Éléments de mathematique" Groupes et algébre de Lie, Chapitres 4,5, et 6, Hermann, Paris, 1968.

[5] Iskovskih, V.A.: Fano 3-folds I, Izv. Adak. Nauk SSSR Ser. Mat., **41** (1977), 516-562.

[6] Iskovskih, Fano 3-folds II, Izv. Akad. Nauk SSSR Ser. Mat., **42** (1978), 469-506.

[7] Kodaira, K. and D.C. Spencer: On deformation of complex analytic structures, I-II Ann of Math., **67** (1958), 328-466.

[8] Maruyama, M.: Openness of a family of torsion free sheaves, J. Math. Kyoto Univ., **16** (1976), 627-637.

[9] Moishezon, B.: Algebraic cohomology classes on algebraic manifolds, Izv. Akad. Nauk SSSR Ser. Mat., **31** (1967), 225-268.

[10] Mori, S. and S. Mukai: The uniruledness of the moduli space of curves of genus 11, ' Algebraic Geometry", Lecture Notes in Math.,n°1016, 334-353, Springer-Verlag, Berlin, Heidelberg, New York and Tokyo, 1983.

[11] Mukai, S. and H. Umemura: Minimal rational threefolds, "Algebraic Geometry", Lecture Notes in Math., n°1016, 490-518, Springer-Verlag, Berlin, Heidelberg, New York and Tokyo, 1983.

[12] Pijateckii-Shapiro, I and I.R. Shafarevic: A Torelli theorem for algebraic surfaces of type K3, Izv. Akad. Nauk. SSSR, Ser. Mat., **35** (1971), 503-572.

[13] Saint-Donat, B.: Projective models of K3-surfaces, Amer. J. Math., **96** (1974), 602-639.

Shigeru MUKAI

Department of Mathematics
Nagoya University
Furō-chō, Chikusa-ku
Nagoya, 464
Japan

Algebraic Geometry and Commutative Algebra
in Honor of Masayoshi NAGATA
pp. 379–404 (1987)

Threefolds Homeomorphic to
a Hyperquadric in $\mathbf{P}^4$

Iku NAKAMURA

§0. Introduction

The purpose of this article is to prove

(0.1) Theorem. *A compact complex threefold homeomorphic to a nonsingular hyperquadric $\mathbf{Q}^3$ in $\mathbf{P}^4$ is isomorphic to $\mathbf{Q}^3$ if $H^1(X, O_X) = 0$ and if there is a positive integer m such that $\dim H^0(X, -mK_X) > 1$.*

As its corollaries, we obtain

(0.2) Theorem. *A Moishezon threefold homeomorphic to $\mathbf{Q}^3$ is isomorphic to $\mathbf{Q}^3$ if its Kodaira dimension is less than three.*

A compact complex threefold is called a Moishezon threefold if it has three algebraically independent meromorphic functions on it.

(0.3) Theorem. *An arbitrary complex analytic (global) deformation of $\mathbf{Q}^3$ is isomorphic to $\mathbf{Q}^3$.*

We shall prove a stronger theorem (2.1) in arbitrary characteristic and apply this in complex case to derive (0.1). The above theorems in arbitrary dimension have been proved by Brieskorn [1] under the assumption that the manifold is kählerian. See also [2],[8],[10] for related results. When I completed the major parts of the present article, I received a preprint [14] of Peternell, in which he claims that he is able to prove the theorems (0.2) and (0.3) without assuming the condition on Kodaira dimension. See [11,(3.3)].

The main idea of the present article is the same as that of our previous work [11], in which we proved the similar theorems for complex projective space $\mathbf{P}^3$. However there arises a new problem that we have never seen in [11]. See (0.4), (3.6) and section 8.

Let X be a complex threefold with $H^1(X, O_X) = 0$, $\kappa(X, -K_X) \geq 1$ (see [6]), which is homeomorphic to a nonsingular hyperquadric $\mathbf{Q}^3$. Let L be the generator of $\operatorname{Pic} X$ $(\cong \mathbf{Z})$ with L^3 equal to two. Then $K_X = -3L$ by Brieskorn [1],

---

Received December 1, 1986.

Morrow [10] and [11,(1.1)]. In the same manner as in [11], we see that $\dim |L|$ is not less than four.

Let D and D' be an arbitrary pair of distinct members of $|L|$, ℓ the scheme-theoretic complete intersection $D \cap D'$ of D and D'. Then ℓ is a pure one dimensional connected closed analytic subspace of X containing $\mathrm{Bs}\,|L|$, the base locus of the linear system $|L|$. By studying ℓ and ℓ_{red} in detail, we eventually prove that the base locus $\mathrm{Bs}\,|L|$ is empty. Indeed, we are able to verify;

(0.4) Lemma. ℓ_{red} *is a connected* (*possibly reducible*) *curve whose irreducible components are nonsingular rational curves intersecting transversally and either*

(0.4.1) ℓ *is an irreducible nonsingular rational curve, or*

(0.4.2) ℓ *is "a double line" with* ℓ_{red} *irreducible nonsingular, or*

(0.4.3) ℓ *is "a double line" plus a nonsingular rational curve, or*

(0.4.4) ℓ *is reduced everywhere and is the union of two rational curves* (*"lines"*) *and a* (*possibly empty*) *chain of rational curves connecting the "lines", each component of the chain being algebraically equivalent to zero.*

It turns out after completing the proof of (0.1) that the case (0.4.3) is impossible and the chain in (0.4.4) is empty.

It follows from (0.4) that $\mathrm{Bs}\,|L|$ is empty so that the complete intersection $\ell = D \cap D'$ is irreducible nonsingular for a sufficiently general pair D and D', and that $\dim |L|$ is equal to four. Thus we have a bimeromorphic morphism f of X onto a (possibly singular) hyperquadric in $\mathbf{P}^4$ associated with the linear system $|L|$. It follows from $\mathrm{Pic}\,X \cong \mathbf{Z}$ and an elementary fact about singular hyperquadrics in $\mathbf{P}^4$ that the image $f(X)$ is nonsingular and that f is an isomorphism of X onto $\mathbf{Q}^3$.

The article is organized as follows. In section one, we recall elementary facts about algebraic two cycles on singular hyperquadrics in $\mathbf{P}^4$. In sections 2–8, we consider a threefold X with a line bundle L such that $\mathrm{Pic}\,X \cong \mathbf{Z}L$, $K_X = -3L$, L^3 is positive,$\kappa(X, L) \geq 1$ (see [6]). In section 2, we prove the vanishing of certain cohomology groups. We also prove $L^3 \geq 2$ and $h^0(X, L) \geq 5$.

In section 3, first we state without proof five lemmas (3.2)–(3.6) which are detailed forms of (0.4) and then by assuming these, prove that X is isomorphic to $\mathbf{Q}^3$.

In sections 4–8, we study a scheme-theoretic complete intersection $\ell = D \cap D'$ to prove the lemmas (3.2)–(3.6).

In section 9, we first give a slight improvement of a theorem in [11] and complete the proof of (0.1) by applying the results in sections 2–8.

Acknowledgement. We are very grateful to A. Fujiki and H. Watanabe for their encouragement and advices.

List of notations

Z	integers or the infinite cyclic group
C	complex numbers
X	a nonsingular threefold
$\kappa(X, L)$	L-dimension of X, L being a line bundle on X [6]
Bs $|L|$	the set of base points of the linear system $|L|$
$H^q(X, F)$	the q-th cohomology group of X with coefficients in a coherent sheaf F
$h^q(X, F)$	$\dim_\mathbf{C} H^q(X, F)$
$\chi(X, F)$	$\sum_{q \in \mathbf{Z}}(-1)^q h^q(X, F)$
O_X, O_X^*	the sheaf of germs over X of holomorphic (resp. nonvanishing holomorphic) functions
I_C, I_ℓ	the ideal sheaf in O_X defining C, resp. ℓ
Ω_X^p	the sheaf of germs over X of holomorphic p-forms
K_X	the canonical line bundle of X
$[D]$	the line bundle associated with a Cartier divisor D
b_q	the q-th Betti number (of X)
c_q	the q-th Chern class (of X)
$c_1(E)$	the first Chern class of a vector bundle E
$cl(C)$	the homology class of an irreducible curve C
$\mathbf{Q}^3, \mathbf{Q}_\nu^3$	hyperquadrics in $\mathbf{P}^4$, see (1.1)

§1. Hyperquadrics in $\mathbf{P}^4$

(1.1) We recall elementary facts about hyperquadrics in $\mathbf{P}^4$. Let x_i ($0 \leq i \leq 4$) be the homogeneous coordinate of $\mathbf{P}^4$, $F_\nu = \sum_{i=0}^{\nu+1} x_i^2$, $\mathbf{Q}_\nu^3$ a hypersurface defined by $F_\nu = 0$. The hypersurface $\mathbf{Q}_\nu^3$ ($\nu = 1, 2, 3$) is irreducible and $\mathbf{Q}^3$ ($:= \mathbf{Q}_3^3$) only is nonsingular.

The hypersurface $\mathbf{Q}_1^3$ contains a conic $q := \mathbf{Q}_1^3 \cap \{x_3 = x_4 = 0\}$ and a line $\ell := \{x_0 = x_1 = x_2 = 0\}$. Let U be a sufficiently small open neighborhood of ℓ in $\mathbf{Q}_1^3$. We may assume that $\mathbf{Q}_1^3 \setminus U$ (resp. U) is homotopic to q (resp. ℓ) and that ∂U, the boundary of U, is an S^3 - bundle over the conic q. By the Thom-Gysin sequence, we have,

$$(1.1.1) \qquad H_n(\partial U, \mathbf{Z}) = \begin{cases} \mathbf{Z} & n = 0, 2, 3, 5 \\ 0 & n = 1, 4 \end{cases}$$

In particular, $H_3(\partial U, \mathbf{Z}) \cong H_0(q, \mathbf{Z})$.

Also by the Mayer-Vietoris sequence of $\mathbf{Q}_1^3 = (\mathbf{Q}_1^3 \setminus U) \cup (\text{the closure of } U)$, we have,

$$(1.1.2) \qquad H_n(\mathbf{Q}_1^3, \mathbf{Z}) = \begin{cases} \mathbf{Z} & n = 0, 2, 4, 6 \\ 0 & n = 1, 3, 5 \end{cases}$$

By (1.1.1) and (1.1.2), we have,

$$(1.1.3) \qquad H_4(\mathbf{Q}_1^3, \mathbf{Z}) \cong H_3(\partial U, \mathbf{Z}) \cong H_0(q, \mathbf{Z}) \cong \mathbf{Z}.$$

(1.2) Lemma. *There is a Weil divisor on $\mathbf{Q}_1^3$ which is not an integral multiple of a hyperplane section H of $\mathbf{Q}_1^3$ in $H_4(\mathbf{Q}_1^3, \mathbf{Z})$.*

Proof. Let $a = [a_0, a_1, a_2]$ be a point of the conic q, $D_a =$ the closure of $\{[a_0, a_1, a_2, x_3, x_4] \in \mathbf{P}^4; x_3, x_4 \in \mathbf{C}\}$. Then by (1.1.3), $H = 2D_a$ in $H_4(\mathbf{Q}_1^3, \mathbf{Z})$.
                                                                        q.e.d.

(1.3) Lemma. *Let $\mathbf{Q}$ be a quadric surface $\mathbf{Q}_2^3 \cap \{x_4 = 0\}$ contained in $\mathbf{Q}_2^3$. Then $H_4(\mathbf{Q}_2^3, \mathbf{Z}) \cong H_2(\mathbf{Q}, \mathbf{Z})(\cong \mathbf{Z} \oplus \mathbf{Z})$ and $H_2(\mathbf{Q}, \mathbf{Z})$ is generated by fibers of two rulings via the isomorphism of $\mathbf{Q}$ with $\mathbf{P}^1 \times \mathbf{P}^1$.*

Proof. Similar to the above.   q.e.d.

(1.4) Remark. In arbitrary characteristic, any singular hyperquadric in $\mathbf{P}^4$ is a cone over a hyperplane section of it, whence it has a Weil divisor which is not (algebraically equivalent to) an integral multiple of a hyperplane section.

§2. Lemmas

Our first aim is to prove the following

(2.1) Theorem. *Let X be a compact complex threefold or a complete irreducible nonsingular algebraic threefold defined over an algebraically closed field of arbitrary characteristic, L a line bundle on X. Assume that $H^1(X, O_X) = 0$, $\operatorname{Pic} X \cong \mathbf{Z}L$, $L^3 > 0$, $K_X = -3L$, $\kappa(X, L) \geq 1$. Then $L^3 = 2$ and X is isomorphic to a nonsingular hyperquadric in $\mathbf{P}^4$.*

Compare [2], [8].

Sections 2–8 are devoted to proving (2.1). Throughout sections 2–8, we always assume that X is a compact complex threefold satisfying the conditions in (2.1). Our proof of (2.1) is completed in (3.8) by assuming (0.4), or more precisely, (3.2)–(3.6).

(2.2) Lemma. $H^3(X, O_X) = 0$ *and* $c_1 c_2 \geq 24$, $L^3 \geq 2$, $\chi(X, mL) \geq (m+1)(m+2)(2m+3)/6$.

Proof. We see $h^3(X, O_X) = 0$, $\chi(X, O_X) = 1 + h^2(X, O_X) \geq 1$ and $c_1 c_2 = 24\chi(X, O_X) \geq 24$, $\chi(X, mL) = \chi(X, O_X) + m(c_1^2 + c_2)L/12 + m^2 c_1 L^2/4 + m^3 L^3/6$. Assume $L^3 = 1$ to derive a contradiction. Let $c_1 c_2 = 24a$, $a \geq 1$. Hence $c_2 L = 8a$ by $L^3 = 1$. We also see that $\chi(X, L) = (5 + 5a)/3$, whence $1 + a = 0 \bmod 3$ and $a \geq 2$. Let $a = 3b + 2$, $b \geq 0$. Then $\chi(X, 2L) = 7b + (21/2)$, which is absurd. Consequently $L^3 \geq 2$ and $\chi(X, mL) \geq (m+1)(m+2)(2m+3)/6$ by $c_2 L = c_1 c_2/3 \geq 8$.
                                                                        q.e.d.

(2.3) Lemma. $h^0(X, L) \geq 5$.

Proof. The same proof as in [11,(1.5)] works by taking $d = 3$, $\chi(X, L) \geq 5$ instead of $d \geq 4$ and $\chi(X, L) \geq 4$.   *q.e.d.*

(2.4) Lemma. *Let D and D' be distinct members of $|L|$, $\ell = D \cap D'$ the scheme-theoretic intersection of D and D'. Then we have,*

(2.4.1) $H^q(X, -mL) = 0$ *for* $q = 0, 1, m > 0$; $q = 2, 0 \leq m \leq 3$; $q = 3, 0 \leq m \leq 2$,

(2.4.2) $H^q(D, -mL_D) = 0$ *for* $q = 0, m > 0$; $q = 1, 0 \leq m \leq 2$; $q = 2, m = 0, 1$,

(2.4.3) $H^0(\ell, -L_\ell) = 0$, $H^1(\ell, O_\ell) = 0$,

(2.4.4) $H^0(X, O_X) \cong H^0(D, O_D) \cong H^0(\ell, O_\ell) \cong \mathbf{C}$,

(2.4.5) $H^3(X, -3L) \cong H^2(D, -2L_D) \cong H^1(\ell, -L_\ell) \cong \mathbf{C}$.

Proof. The same as in [11,(1.7)] by using an exact sequence

$$0 \longrightarrow O_D(-L) \longrightarrow O_D \longrightarrow O_\ell \longrightarrow 0$$

[11,(1.4) and (1.6)].   *q.e.d.*

(2.5) Corollary. $H^2(X, O_X) = 0$ *and* $\chi(X, O_X) = 1$.

(2.6) Corollary. $\mathrm{Bs}\,|L| = \mathrm{Bs}\,|L_D| = \mathrm{Bs}\,|L_\ell|$.

§3. A complete intersection $\ell = D \cap D'$

Let X, L be the same as in section 2.

(3.1) Lemma. *Let D and D' be distinct members of the linear system $|L|$, $\ell := D \cap D'$ the complete intersection of D and D'. Let $\ell_{\mathrm{red}} = A_1 + \cdots + A_s$ be the decomposition of ℓ_{red} into irreducible components. Then*

(3.1.1) *each A_j is a nonsingular rational curve with $LA_j \leq 2$,*

(3.1.2) *if there is an irreducible component A_i with $LA_i = 2$, then $LA_j \leq 1$ for $j \neq i$.*

Proof. By (2.4.3), $H^1(\ell, O_\ell) = 0$. Hence $H^1(A_j, O_{A_j}) = 0$ for any j, whence A_j is a nonsingular rational curve. In view of (2.4.5), $h^1(\ell, -L_\ell) = 1$, whence $h^1(\ell_{\mathrm{red}}, -L_{\ell_{\mathrm{red}}}) \leq 1$. Therefore $\sum_{i=1}^s h^1(A_i, -L_{A_i}) = \sum_{i=1}^s h^0(A_i, O_{A_i}(-2 + LA_i)) \leq 1$. The assertions are therefore clear. See [11,(2.3)].   *q.e.d.*

In the subsequent sections 4-8, we shall prove the following five lemmas;

(3.2) Lemma. *Let $\ell = D \cap D'$ be the complete intersection in (3.1). Assume that there is an irreducible component C of ℓ_{red} with $LC \geq 2$. Then*

(3.2.1) *$LC = 2$ and ℓ is an irreducible nonsingular rational curve, isomorphic to C,*

(3.2.2) *$I_C/I_C^2 \cong O_C(-2) \oplus O_C(-2)$.*

(3.3) Lemma.   *Let $\ell = D \cap D'$ be the complete intersection in (3.1). Assume that there is an irreducible component C of ℓ_{red} with $LC = 1$ such that ℓ is nonreduced anywhere along C. Let I_ℓ (resp. I_C) be the ideal sheaf of O_X defining ℓ (resp. C). Then $I_\ell + I_C^2/I_C^2 \cong O_C$ or $O_C(-1)$. If $I_\ell + I_C^2/I_C^2 \cong O_C$, then*

(3.3.1) *ℓ_{red} is an irreducible nonsingular rational curve, isomorphic to C,*

(3.3.2) *ℓ is "a double line", to be precise, at any point p of C, the ideal sheaf I_ℓ (resp. I_C) is given by;*

$$I_\ell = O_{X,p}x + O_{X,p}y^2,$$
$$I_C = O_{X,p}x + O_{X,p}y$$

for suitable local parameters x and y at p,

(3.3.3) *$I_C \supset I_\ell \supset I_C^2$, $I_C/I_C^2 \cong O_C \oplus O_C(-1)$, $I_C/I_\ell \cong O_C(-1)$, $I_\ell/I_C^2 \cong O_C$.*

(3.4) Lemma.   *Let $\ell = D \cap D'$ be the complete intersection in (3.1). Assume that there is an irreducible component C of ℓ_{red} with $LC = 1$ such that ℓ is nonreduced anywhere along C. Assume that $I_\ell + I_C^2/I_C^2 \cong O_C(-1)$ and that if ℓ_{red} is reducible, then C meets an irreducible component C' of ℓ_{red} not contained in $\mathrm{Bs}\,|L|$. Then ℓ is a double line plus a nonsingular rational curve C'. To be more precise,*

(3.4.1) *ℓ_{red} is the union of C and C' with $LC = 1$, $LC' = 0$, the curve C intersecting C' transversally at a unique point p_0,*

(3.4.2) *the ideal sheaf I_ℓ (resp. I_C, $I_{C'}$) defining ℓ (resp. C, C') is given at p_0 by*

$$I_\ell = O_{X,p_0}x + O_{X,p_0}zy^2,$$
$$I_C = O_{X,p_0}x + O_{X,p_0}y,$$
$$I_{C'} = O_{X,p_0}x + O_{X,p_0}z,$$

for a local parameter system x,y and z at p_0 and except at p_0, ℓ is a double line along C in the sense of (3.3.2), and reduced along C',

(3.4.3) *$I_C/I_C^2 \cong O_C \oplus O_C(-1)$, $I_{C'}/I_{C'}^2 \cong O_{C'}(1) \oplus O_{C'}(1)$ or $O_{C'}(2) \oplus O_{C'}$.*

(3.5) Lemma.   *Let $\ell = D \cap D'$ be the complete intersection in (3.1). Assume that ℓ is reduced at a point of an irreducible component C_0 of ℓ_{red} with $LC_0 = 1$ and that C_0 intersects an irreducible component C' of ℓ_{red} not contained in $\mathrm{Bs}\,|L|$. Then,*

(3.5.1) *ℓ is reduced everywhere,*

(3.5.2) *there exist another irreducible component C_m of ℓ with $LC_m = 1$ and a chain of irreducible components C_j of ℓ with $LC_j = 0$ $(1 \leq j \leq m-1)$ such that ℓ is the union of C_j $(0 \leq j \leq m)$, the pair C_i and C_j $(i < j)$ intersect if and only if $i = j - 1$. If $i = j - 1$, then C_{j-1} and C_j intersect at a unique point*

p_j $(1 \leq j \leq m)$ *transversally, to be precise,* $\hat{O}_{\ell,p_j}$ $(:= the\ completion\ of\ O_{\ell,p_j})$ $\cong \mathbf{C}[[x,y,z]]/(x,yz)$, *for suitable local parameters* x,y,z *at* p_j,

$$(3.5.3) \quad I_C/I_C^2 = \begin{cases} O_C \oplus O_C(-1) & (C = C_0, C_m) \\ O_C(1) \oplus O_C(1)\ or\ O_C(2) \oplus O_C & (C = C_1, \ldots, C_{m-1}) \end{cases}$$

(3.6) Lemma. *Let* $\ell = D \cap D'$ *be the complete intersection in* (3.1). *Let* C *be an irreducible component of* ℓ_{red} *with* $LC = 1$. *If* ℓ_{red} *is reducible, then* C *intersects an irreducible component* C' *of* ℓ_{red} *not contained in* $\mathrm{Bs}\,|L|$.

From $(3.2) - (3.6)$, we infer the following

(3.7) Lemma. *The linear system* $|L|$ *is base point free and* $\dim |L| = 4, L^3 = 2$.

Proof by assuming (3.2)-(3.6). In view of (2.3), we are able to choose distinct members D and D' from $|L|$. Let $\ell = D \cap D'$ be the complete intersection. Let $\ell_{\mathrm{red}} = A_1 + \cdots + A_s$ be the decomposition into irreducible components. Then $cl(\ell) = n_1 cl(A_1) + \cdots + n_s cl(A_s) \in H_2(X, \mathbf{Z})$ for some $n_i > 0$ (see $[11,(2.1)]$). Since $L^3 = L\ell = n_1 LA_1 + \cdots + n_s LA_s$, there is at least a component A_i with $LA_i > 0$. We see that there are only three cases;

Case 1. ℓ_{red} contains an irreducible component C with $LC \geq 2$,

Case 2. ℓ_{red} contains no irreducible components C' with $LC' \geq 2$, but contains an irreducible component C with $LC = 1$ along which ℓ is nonreduced anywhere,

Case 3. ℓ_{red} contains no irreducible components C' with $LC' \geq 2$, but contains an irreducible component C_0 with $LC_0 = 1$ such that ℓ is reduced at a point of C_0.

Case 1. By (3.2), ℓ is isomorphic to C. By (2.6), $\mathrm{Bs}\,|L| = \mathrm{Bs}\,|L_\ell|$. Since $L^3 = L\ell = LC = 2$, we have $L_\ell = O_\ell(2)$, so that $|L_\ell|$ is base point free. Consequently $|L|$ is base point free and $h^0(X, L) = 2 + h^0(\ell, L_\ell) = 5$.

Case 2. First we assume that ℓ_{red} is irreducible. By (3.3) and (3.4), ℓ_{red} is isomorphic to C and $I_C/I_\ell \cong O_C(-1)$. Hence we have an exact sequence,

$$0 \to (I_C/I_\ell) \otimes L \to O_\ell(L) \to O_C(L) \to 0\ ,$$

whence $0 \to O_C \to O_\ell(L) \to O_C(1) \to 0$ is exact. It follows that

$$0 \to H^0(C, O_C) \to H^0(\ell, L_\ell) \to H^0(C, O_C(1))$$
$$\to H^1(C, O_C) \to H^1(\ell, L_\ell) \to H^1(C, O_C(1)) \to 0$$

is exact. Hence $|L|$ is base point free. Moreover $h^0(X, L) = 2 + h^0(\ell, L_\ell) = 5$. The intersection number $L^3 = L\ell = 2$ because $h^0(\ell, sL_\ell) = 2s + 1$. In this case, the proof of (3.7) is complete.

Next we consider the case where ℓ_{red} is reducible. Then by (3.4) and (3.6), ℓ is a double line plus a nonsingular rational curve C', whence $cl(\ell) = 2cl(C) + cl(C')$ and $L\ell = 2$. We define a subsheaf I_2 of I_C by $I_2 = (\{0\} \oplus O_C(-1)) + I_C^2$ via the isomorphism $I_C/I_C^2 \cong O_C \oplus O_C(-1)$. Let $p = C \cap C'$. We note that with the notations in (3.4), $I_{2,p}(:= \text{the stalk of } I_2 \text{ at } p) = O_{X,p}x + O_{X,p}y^2$. Then we have exact sequences;

$$0 \to O_\ell(L) \to O_{C'} \oplus (O_X/I_2)(L) \to \mathbf{C}^2 (\cong O_X/I_{C'} + I_2) \to 0 \,,$$

$$0 \to O_C(1) \to (O_X/I_2)(L) \to O_C(1) \to 0$$

because $I_C/I_2 \cong O_C$. We see that a subspace $H^0(O_{C'}) \oplus H^0((I_C/I_2)(L))$ of $H^0(O_{C'}) \oplus H^0((O_X/I_2)(L))$ is mapped onto $O_X/I_{C'} + I_2$ by the natural homomorphism. Therefore $h^0(X, L) = h^0(\ell, L_\ell) + 2 = 5, \mathrm{Bs}\,|L| = \mathrm{Bs}\,|L_\ell| = \emptyset$. This completes the proof of (3.7) in Case 2.

Case 3.   By (3.5) and (3.6), ℓ is reduced everywhere and $\ell = C_0 + \cdots + C_m$ with $LC_0 = LC_m = 1$, $LC_j = 0 (1 \le j \le m - 1)$. Then $L^3 = L\ell = L(C_0 + \cdots + C_m) = 2$. Consider an exact sequence,

$$0 \to O_\ell(L) \to O_{C_0}(1) \oplus O_{C_1} \oplus \cdots \oplus O_{C_{m-1}} \oplus O_{C_m}(1) \to \mathbf{C}^m \to 0.$$

It follows from this that $h^0(X, L) = 2 + h^0(\ell, L_\ell) = 5$, and that $|L|$ is base point free. Thus we complete the proof of (3.7).   q.e.d.

(3.8) Completion of the proof of (2.1) by assuming (3.2)-(3.6).   Let X be a compact complex threefold with a line bundle L satisfying the conditions in (2.1). By (3.7), we have a bimeromorphic morphism of X onto a hyperquadric in $\mathbf{P}^4$. The image $f(X)$ endowed with reduced structure is one of $Q_\nu^3 (\nu = 1, 2, 3)$. We note $\mathrm{Pic}\, X = \mathbf{Z}L = \mathbf{Z}[f^*H]$, where H is a hyperplane section of $f(X)$ and $[f^*H]$ is the line bundle associated with f^*H. If we are given a Weil divisor (an analytic two cycle) E of $f(X)$, then f^*E is a Cartier divisor of X and $E = f_*(f^*E)$ because f is bimeromorphic. Since $[f^*E]$ is an integral multiple of L, any Weil divisor of $f(X)$ is homologically (algebraically) equivalent to an integral multiple of H [3, Theorem 1.4]. Hence $f(X) \ne Q_1^3, Q_2^3$ in view of (1.2) and (1.3). We note that over an algebraically closed field of arbitrary characteristic, any singular hyperquadric in $\mathbf{P}^4$ has a Weil divisor which is not an integral multiple of a hyperplane section. Consequently $f(X) = Q^3$. Since f_* is an isomorphism of $\mathrm{Pic}\, X$ onto $\mathrm{Pic}\, Q^3 (= \mathbf{Z}[H])$, the exceptional set $(\det(\mathrm{Jac}\, f))$ of f is empty (see [11,(2.8)]). Therefore f is an isomorphism of X onto Q^3.
                              q.e.d.

Before closing this section, we prepare three lemmas for sections $4 - 8$.

(3.9) Lemma. *Let $\ell = D \cap D'$ be the complete intersection in* (3.1), *C an irreducible component of ℓ_{red}, I_C the ideal sheaf of O_X defining C, $c_1(I_C/I_C^2) = s \in H^2(C, \mathbf{Z})(\cong \mathbf{Z})$. Then $\chi(X, O_X/I_C^n) = n(n+1)(sn-s+3)/6$, $s = -3LC+2$.*

Proof. The first assrtion is clear from Riemann-Roch for $C = \mathbf{P}^1$. Next consider an exact sequence,

$$0 \to I_C/I_C^2 \to \Omega_X^1 \otimes O_C \to \Omega_C^1 \to 0.$$

Then we have $s = c_1(I_C/I_C^2) = K_X C + 2 = -3LC + 2$.　　　　q.e.d.

(3.10) Lemma. *Let ℓ and C be the same as in* (3.9). *Let $\phi : (I_\ell/I_\ell^2) \otimes O_C \to I_C/I_C^2$ be the natural homomorphism induced from the inclusion of I_ℓ into I_C. Then ϕ is injective everywhere on C if and only if ℓ is reduced at a point of C.*

Proof. We note that $(I_\ell/I_\ell^2) \otimes O_C \cong O_C(-L) \oplus O_C(-L)$ is locally free. Therefore the following conditions are equivalent to each other;

(3.10.1) ϕ is injective everywhere,
(3.10.2) ϕ is injective at a point q of C,
(3.10.3) $\mathrm{Coker}(\phi) = 0$ at a point p of C,
(3.10.4) $I_\ell + I_C^2 = I_C$ at a point p of C,
(3.10.5) $I_\ell = I_C$ at a point p of C.

Thus the assertion is clear.　　　　q.e.d.

(3.11) Lemma. *Let I and $I'(\neq O_X)$ be ideal sheaves of O_X. Suppose that $I \subset I'$ and $h^1(O_X/I) = 0$, $\dim \mathrm{Supp}(O_X/I) \leq 1$. Then $h^1(O_X/I') = 0$ and $\chi(O_X/I') \geq 1$.*

Clear.

§4. Proof of (3.2)

We apply a method of Mori [9, pp.167-170].

Assume that C is an irreducible component of ℓ_{red} with $LC \geq 2$. Then by (3.1.1), we have $LC = 2$. Then $(I_\ell/I_\ell^2) \otimes O_C \cong O_C(-2) \oplus O_C(-2)$. Since $C = \mathbf{P}^1$, by a theorem of Grothendieck, we express $I_C/I_C^2 = O_C(a) \oplus O_C(b), a \geq b$. By (3.9), $a + b = -4$.

(4.1) Lemma. $I_\ell \not\subset I_C^2$

Proof. Suppose $I_\ell \subset I_C^2$. Hence $h^1(O_X/I_C^2) = 0$ by (2.4.3) and (3.11). Hence $\chi(O_X/I_C^2) \geq 1$. However by (3.9), $\chi(O_X/I_C^2) = s + 3 = -1$ because $s = -4$. This is a contradiction.　　　　q.e.d.

In view of (4.1), we have a nontrivial natural homomorphism $\phi : (I_\ell/I_\ell^2) \otimes O_C \to I_C/I_C^2$. We shall prove

(4.2) Lemma. ϕ *is injective.*

Proof. Suppose not. Then both $\text{Ker}\,\phi$ and $\text{Im}\,\phi$ are torsion free sheaves of rank one, hence locally O_C-free. By a theorem of Grothendieck, we express $\text{Ker}\,\phi = O_C(c)$, $\text{Im}\,\phi = O_C(d)$ for some $c,d \in \mathbf{Z}$. Then we have an exact sequence,

$$0 \to O_C(c) \to O_C(-2) \oplus O_C(-2) \to O_C(d) \to 0.$$

Hence $c + d = -4$, $b \le c \le -2 \le d \le a$. Now we shall prove $b = d$ (hence $a = b = c = d = -2$). Assume $b < d$ to derive a contradiction. Then since $\text{Hom}_{O_C}(O_C(d), O_C(b)) = 0$, the sheaf $O_C(d)$ is contained in a direct summand $O_C(a)$ of I_C/I_C^2. Here we note that if $b < d$, then $b < a$ so that the subsheaf $O_C(a)$ in the splitting of I_C/I_C^2 is uniquely determined in I_C/I_C^2. Define a subsheaf I of I_C by $I = O_C(a) + I_C^2$. Then we see readily that $I_C \supset I \supset I_\ell$, $I/I_C^2 \cong O_C(a)$, $I_C/I = O_C(b)$. By $H^1(O_\ell) = H^1(O_X/I_\ell) = 0$, we have, $1 \le \chi(O_X/I) = \chi(O_X/I_C) + \chi(I_C/I) = 2 + b$, whence $b \ge -1$. This contradicts $b \le c \le -2$. Hence $a = b = c = d = -2$. Next we let $J = \text{Im}\,\phi + I_C^2 = I_\ell + I_C^2$. Then $J \supset I_\ell$, $I_C/J \cong O_C(-2)$. Therefore

$$1 \le \chi(O_X/J) = \chi(O_X/I_C) + \chi(O_C(-2)) = 0,$$

which is a contradiction. Hence ϕ is injective.   q.e.d.

(4.3) Completion of the proof of (3.2). By (4.2), we have the exact sequence,

$$0 \to (I_\ell/I_\ell^2) \otimes O_C \to I_C/I_C^2$$

Therefore $-2 \le a$, $-2 \le b$, whence $a = b = -2$. Hence $(I_\ell/I_\ell^2) \otimes O_C \cong I_C/I_C^2$. Let p be a point of X, $I_{\ell,p}$ (resp. $I_{C,p}$) the stalk of I_ℓ (resp. I_C) at p. Then $I_{\ell,p} + I_{C,p}^2 = I_{C,p}$ for any point p of C, whence $I_{\ell,p} = I_{C,p}$. This shows that ℓ is isomorphic to C anywhere on C. Since ℓ_{red} is connected by (2.4.4), ℓ is isomorphic to C. This completes the proof of (3.2).   q.e.d.

§5. Proof of (3.3)

(5.1) Lemma. *Assume that C is an irreducible component of ℓ_{red} with $LC = 1$. Then $I_C/I_C^2 \cong O_C \oplus O_C(-1)$.*

Proof. Since $C = \mathbf{P}^1$, we express $I_C/I_C^2 = O_C(a) \oplus O_C(b)$, $a \ge b$. Then by (3.9), $a + b = c_1(I_C/I_C^2) = -3LC + 2 = -1$, $a \ge 0$, $b \le -1$. We shall show $a = 0$, $b = -1$. We assume $b \le -2$ to derive a contradiction. Consider the natural homomorphism $\phi : (I_\ell/I_\ell^2) \otimes O_C \to I_C/I_C^2$. When $b \le -2$, $\text{Im}\,\phi$ is contained in $O_C(a) = O_C(a) \oplus \{0\}$. Let $I = O_C(a) + I_C^2$. Then $I_C \supset I \supset I_\ell$ and $I_C/I \cong O_C(b)$. Hence by (3.11),

$$1 \le \chi(O_X/I) = \chi(O_X/I_C) + \chi(I_C/I) = 2 + b.$$

This is a contradiction. Hence $a = 0$, $b = -1$. $\qquad$ q.e.d.

In what follows, we assume that C is an irreducible component of ℓ_{red} with $LC = 1$ along which ℓ is nonreduced anywhere.

(5.2) Lemma. $I_\ell \not\subset I_C^2$.

Proof. Assume $I_\ell \subset I_C^2$. Let $I = I_C I_\ell$. Then $I_\ell \supset I \supset I_\ell^2$, $I_C^3 \supset I$ and $I_\ell/I \cong (I_\ell/I_\ell^2) \otimes O_C \cong O_C(-1) \oplus O_C(-1)$. Therefore by (2.4), $h^0(O_X/I) = 1$, $h^1(O_X/I) = 0$. Consider the natural inclusion $\iota : I + I_C^4/I_C^4 \to I_C^3/I_C^4 \cong O_C \oplus O_C(-1) \oplus O_C(-2) \oplus O_C(-3)$. Then $\mathrm{Im}\,\iota$ is contained in $O_C \oplus O_C(-1) \oplus O_C(-2)$ because the following natural homomorphism is surjective,

$$(I_C/I_C^2) \otimes (I_\ell/I_\ell^2) \otimes O_C \to I_C I_\ell + I_C^4/I_C^4 = I + I_C^4/I_C^4 (\subset I_C^3/I_C^4)$$

and $(I_C/I_C^2) \otimes (I_\ell/I_\ell^2) \otimes O_C \cong O_C(-1) \oplus O_C(-1) \oplus O_C(-2) \oplus O_C(-2)$. Let $I' = O_C \oplus O_C(-1) \oplus O_C(-2) + I_C^4$. Then $I_C^3 \supset I' \supset I, I_C^3/I' \cong O_C(-3)$. By $h^1(O_X/I) = 0$ and (3.11), we have,

$$1 \le \chi(O_X/I') = \chi(O_X/I_C^3) + \chi(O_C(-3)) = 0$$

which is a contradiction. Hence $I_\ell \not\subset I_C^2$. $\qquad$ q.e.d.

(5.3) Completion of the proof of (3.3). We consider the natural homomorphism $\phi : (I_\ell/I_\ell^2) \otimes O_C \to I_C/I_C^2$. By (5.2), $\mathrm{Im}\,\phi$ is not zero. Since $\mathrm{Im}\,\phi(\cong I_\ell + I_C^2/I_C^2)$ is a subsheaf of a torsion free sheaf I_C/I_C^2, it is locally O_C-free. Since ℓ is nonreduced along C, $\mathrm{Im}\,\phi$ is of rank one by (3.10). Here we may set $\mathrm{Im}\,\phi = O_C(c)$ for some $c \in \mathbf{Z}$. Then $c = 0$ or -1 because $(I_\ell/I_\ell^2) \otimes O_C \cong O_C(-1) \oplus O_C(-1)$. In view of our assumption in (3.3), $\mathrm{Im}\,\phi \cong O_C$, $\mathrm{Ker}\,\phi \cong O_C(-2)$. Let $E = \mathrm{Coker}\,\phi \cong O_C(-1)$, $F = \mathrm{Im}\,\phi \cong O_C$. Then we may view $I_C/I_C^2 = E \oplus F$ because $H^1(C, E^\vee \otimes F) = 0$, $E^\vee$ being the dual of E. So we consider again the homomorphism ϕ as, $\phi : (I_\ell/I_\ell^2) \otimes O_C \to F \subset E \oplus F = I_C/I_C^2$. Let p be an arbitrary point of C. Then there are two generators x, y of $I_{C,p}$, and two generators f, g of $I_{\ell,p}$ such that $\phi(f) = x$, $\phi(g) = 0$, $x \bmod I_{C,p}^2$ (resp. $y \bmod I_{C,p}^2$) generates F (resp. E). Since $f = x \bmod I_{C,p}^2$, it is easy to see that f and y generates $I_{C,p}$ over $O_{X,p}$, so that we may take f instead of x. Then by deleting an $O_{X,p}$-multiple of x from g, we may assume $g = \beta y^m$ for some $\beta \in O_{X,p}$ and $m > 0$, the restriction of β to C being not identically zero. Thus we obtain local parameters x and $y \in I_{C,p}$ and $\beta \in O_{X,p}$, $m > 0$ such that

$$(5.3.1) \qquad I_{C,p} = O_{X,p}x + O_{X,p}y \quad I_{\ell,p} = O_{X,p}x + O_{X,p}\beta y^m$$

where the restriction β_C of β to C is not identically zero. The integer m is uniquely determined by the point p, but it is independent of the choice of $p \in C$. We note that $m \ge 2$ because $g \in I_{C,p}^2$. Let $\Phi = \{U_j\}$ be a sufficiently fine covering of an open neighborhood of C by Stein (or affine) open sets U_j. Then by (5.3.1), we have $x_j \in \Gamma(U_j, I_\ell)$, $y_j \in \Gamma(U_j, I_C)$, $\beta_j \in \Gamma(U_j, O_X)$ such that

$$(5.3.2) \qquad \begin{aligned} \Gamma(U_j, I_C) &= \Gamma(U_j, O_X)x_j + \Gamma(U_j, O_X)y_j \\ \Gamma(U_j, I_\ell) &= \Gamma(U_j, O_X)x_j + \Gamma(U_j, O_X)\beta_j y_j^m. \end{aligned}$$

Since $I_C/I_C^2 = F \oplus E \cong O_C \oplus O_C(-1)$, we may assume that

$$(5.3.3) \qquad x_j = x_k \bmod I_C^2, \quad y_j = \ell_{jk} y_k \bmod I_C^2,$$

where ℓ_{jk} stands for the one cocycle $L_C = O_C(1) \in H^1(C, O_C^*)$. Note that the second equation in (5.3.3) does make sense.

Hence $D_j \beta_j y_j^m$ and $D_k \beta_k y_k^m \in \operatorname{Ker} \phi(\cong O_C(-2))$ are identified if and only if (we may assume that)

$$(5.3.4) \qquad (D_j|_C) = \ell_{jk}^{-2}(D_k|_C).$$

This shows that

$$(5.3.5) \qquad (\beta_j|_C) = \ell_{jk}^{2-m}(\beta_k|_C).$$

In particular, $\beta_C := \{\beta_j|C; U_j \in \Phi\}$ is a nontrivial element of $H^0(C, O_C(2-m))$. This is possible only when $m = 2$ and β_C is a nonzero constant. Consequently ℓ_{red} is nonsingular anywhere on C, and it is isomorphic to C because it is connected by (2.4.4). Moreover ℓ is "a double line" in the sense that at any point p of C, there exist local parameters x, $y \in I_{C,p}$ such that

$$\begin{aligned} I_{C,p} &= O_{X,p}x + O_{X,p}y, \\ I_{\ell,p} &= O_{X,p}x + O_{X,p}y^2. \end{aligned}$$

This completes the proof of (3.3).   q.e.d.

§6. Proof of (3.4)

Let C be an irreducible component of ℓ_{red} with $LC = 1$, along which ℓ is nonreduced anywhere. By (5.3), there are two possibilities $\operatorname{Im}\phi \cong O_C$ or $O_C(-1)$. The case $\operatorname{Im}\phi \cong O_C$ was discussed in section 5. In this section, we shall discuss the case $\operatorname{Im}\phi \cong O_C(-1)$. We note that via the isomorphism $I_C/I_C^2 \cong O_C \oplus O_C(-1)$, the subsheaf $O_C = O_C \oplus \{0\}$ of I_C/I_C^2 is uniquely determined. First we prove

(6.1) Lemma. *Assume* $\operatorname{Im}\phi \cong O_C(-1)$. *Then* $\operatorname{Im}\phi$ *is not contained in* $O_C(= O_C \oplus \{0\}) \subset I_C/I_C^2$.

Proof. Assume $\operatorname{Im}\phi = O_C(-1) \subset O_C$ to derive a contradiction. Let p be an arbitrary point of C. Then there are two generators x, y of $I_{C,p}$ and two generators f, g of $I_{\ell,p}$ such that x mod $I_{C,p}^2$ (resp. y mod $I_{C,p}^2$) generates $O_C \oplus \{0\}$ (resp. $\{0\} \oplus O_C(-1)$) in $I_{C,p}/I_{C,p}^2$, and $\phi(f)$ generates Im ϕ, $\phi(g) = 0$, or equivalently $g \in I_{\ell,p} \cap I_{C,p}^2$. Since Im ϕ is contained in $O_C \oplus \{0\}$, we see that $f = \alpha x$ mod $I_{C,p}^2$ for some $\alpha \in O_{X,p}$. Thus we obtain,

$$(6.1.1) \qquad \begin{aligned} I_{C,p} &= O_{X,p} x + O_{X,p} y, \\ I_{\ell,p} &= O_{X,p} f + O_{X,p} g, \\ f &= \alpha x \text{ mod } I_{C,p}^2 \end{aligned}$$

where $\alpha \in O_{X,p}$, $g \in I_{\ell,p} \cap I_{C,p}^2$.

Let $\Phi = \{U_j\}$ be a sufficiently fine covering of an open neighborhood of C by Stein (or affine) open sets U_j. Then by (6.1.1), we have $x_j, y_j \in \Gamma(U_j, I_C)$, $f_j \in \Gamma(U_j, I_\ell)$, $g_j \in \Gamma(U_j, I_\ell \cap I_C^2)$, $\alpha_j \in \Gamma(U_j, O_X)$ such that

$$(6.1.2) \qquad \begin{aligned} \Gamma(U_j, I_C) &= \Gamma(U_j, O_X) x_j + \Gamma(U_j, O_X) y_j \\ \Gamma(U_j, I_\ell) &= \Gamma(U_j, O_X) f_j + \mathbf{Q}(U_j, O_X) g_j \\ f_j &= \alpha_j x_j \text{ mod } I_C^2, \end{aligned}$$

Moreover by the choice of the generators, we may assume

$$(6.1.3) \qquad \begin{aligned} f_j &= \ell_{jk} f_k && \text{mod} I_C I_\ell, \\ g_j &= \ell_{jk} g_k && \text{mod} I_C I_\ell, \\ x_j &= x_k && \text{mod} I_C^2, \\ y_j &= \ell_{jk} y_k && \text{mod} I_C^2 \end{aligned}$$

where ℓ_{jk} is the one cocycle $L_C = O_C(1) \in H^1(C, O_C^*)$.

Then one sees that $\alpha_C := \{\alpha_j|_C\}$ is a nontrivial element of $H^0(C, O_C(1))$. Hence α_C has a single zero at a point p_0 of C and it vanishes nowhere else.

If $\alpha|_C$ is nonvanishing at p in (6.1.1), then f and y generates $I_{C,p}$, so that we may take f instead of x and can normalize g into βy^m for some $\beta \in O_{X,p}$ and for some $m \geq 2$ so that the restriction of β to C is not identically zero. The integer m is independent of the choice of p.

If $\alpha|_C$ has a single zero at p in (6.1.1), then $z := \alpha$ forms a regular sequence at p together with the parameters x and y. Since $f = zx$ mod $I_{C,p}^2$ in (6.1.1), we may assume, by a suitable coordinate change, that $I_{C,p} = O_{X,p} x + O_{X,p} y$, $f = zx$ or $zx - y^s$ for some $s \geq 2$.

Therefore by taking a suitable refinement of Φ if necessary, we may assume that

$$(6.1.4) \qquad U_j \text{ contains } p_0 \text{ if and only if } j = 0,$$

$$(6.1.5) \qquad \begin{aligned} \Gamma(U_j, I_C) &= \Gamma(U_j, O_X) x_j + \Gamma(U_j, O_X) y_j, \\ \Gamma(U_j, I_\ell) &= \Gamma(U_j, O_X) x_j + \Gamma(U_j, O_X) g_j, \end{aligned}$$

for $j \neq 0$, $g_j = \beta_j y_j^m$, $\beta_j \in \Gamma(U_j, O_X)$, $m \geq 2$, $\beta_j|_C$ being not identically zero,

$$(6.1.6) \qquad \begin{aligned} \Gamma(U_0, I_C) &= \Gamma(U_0, O_X)x_0 + \Gamma(U_0, O_X)y_0 \\ \Gamma(U_0, I_\ell) &= \Gamma(U_0, O_X)f_0 + \Gamma(U_0, O_X)g_0 \\ f_0 &= z_0 x_0 \text{ or } z_0 x_0 - y_0^s \quad (s \geq 2) \end{aligned}$$

where x_0, y_0 and z_0 form a regular sequence everywhere on U_0, $g_0 \in I_C^2$, and moreover

(6.1.7) β_j $(j \neq 0)$ (resp. z_0) vanishes nowhere on U_{ij} $(:= U_i \cap U_j, \; i \neq j)$ (resp. on U_{0j}).

Now we define β_0 as follows. Let $x = x_0$, $y = y_0$, $z = z_0$.

Case 0. Assume $f_0 = zx$. Then the second generator g_0 of I_ℓ is normalized (mod f_0) into $g_0 = A_n(x,y)x + B_n(y,z)y^n$ for some $n \geq 2$, $A_n \in \Gamma(U_0, I_C)$, $B_n \in \Gamma(U_0, O_C)$, B_n being not identically zero on C. At a general point q of C sufficiently close to p, $I_{C,q}$ (resp. $I_{\ell,q}$) is generated by x and y (resp. x and y^m) by (6.1.7) because β does not vanish at q. It follows that $n \geq m$. We now define

$$\beta_0 = B_n(y,z)y^{n-m}.$$

Next we consider the case $f = zx - y^s$, $s \geq 2$.

Case 1. Assume $s > m$. We can choose f_j such that $f_j = x_j$ for $j \neq 0$ and $f_j = \ell_{jk}f_k \bmod I_C I_\ell$ for any j, k. We see

$$\begin{aligned} x &= (f_0 + y^s)/z = (\ell_{0j}/z)x_j \bmod I_C I_\ell, \\ y &= c_{0j}x_j + d_{0j}y_j \end{aligned}$$

for some c_{0j} and d_{0j} such that $c_{0j}|_C = 0$, $d_{0j}|_C = \ell_{0j}$. On the other hand,

$$g_0 = A_2(x,y)x + B_2(y,z)y^2$$

for some $A_2 \in \Gamma(U_0, I_C)$, $B_2 \in \Gamma(U_0, O_X)$. Since $\Gamma(U_{0j}, I_C I_\ell)$ is generated by x_j^2, $x_j y_j, y_j^{m+1}$, we have,

$$\begin{aligned} g_0 &= B_2(y,z)y^2 & \bmod I_C I_\ell \\ &= B_2(d_{0j}y_j, z)(d_{0j}y_j)^2 & \bmod(x_j^2, x_j y_j, y_j^{m+1}) \\ &= 0 & \bmod(x_j^2, x_j y_j, y_j^m) \end{aligned}$$

by (6.1.5), whence $B_2(d_{0j}y_j, z)$ is divisible by $(d_{0j}y_j)^{m-2}$. Hence $B_2(y,z)$ is divisible by y^{m-2}. So we define

$$\beta_0 = B_2(y,z)y^{-m+2}.$$

Case 2. Assume $s \le m$. Let $g_0 = \sum\limits_{\substack{\nu+\mu \ge 2 \\ \nu,\mu \ge 0}} A_{\nu\mu}(z) x^\nu y^\mu$. Then on U_{0j}, we see

$g_0 = \sum\limits_{\nu+\mu \ge 2} A_{\nu\mu}(z) y^{\nu s + \mu}/z^\nu \bmod I_C I_\ell$. By (6.1.7), we have $\sum\limits_{\nu s + \mu = k} A_{\nu\mu}/z^\nu = 0$

for $k < m$, and we define $\beta_0 = \sum\limits_{\substack{\nu s + \mu = m \\ \nu,\mu \ge 0}} A_{\nu\mu}/z^\nu$. By these definitions, we have,

$$
\begin{aligned}
\beta_0 y_0^m &= g_0 & \bmod I_C I_\ell \ \text{on} \ U_{0j} \\
&= \ell_{0j} g_j & \bmod I_C I_\ell \ \text{by (6.1.3)} \\
&= \ell_{0j} \beta_j y_j^m & \bmod I_C I_\ell \ \text{by (6.1.5)} \\
\beta_i y_i^m &= \ell_{ij} \beta_j y_j^m & \bmod I_C I_\ell \ \ (i, j \ne 0).
\end{aligned}
$$

Therefore we have $(\beta_i|_C) = \ell_{ij}^{1-m}(\beta_j|_C)$ for any i, j.

In Case 0 and Case 1, β_0 is holomorphic on U_0. In Case 2, $\beta_0 |_C$ is holomorphic except at p_0 and meromorphic at p_0, and it has a pole at p_0 of order at most $\nu_{\max} := \max\{\nu; \nu s + \mu = m, A_{\nu\mu} \ne 0\}$. Clearly $\nu_{\max} \le m - 2$ when $s \le m$. So $\beta_C := \{\beta_j|_C; U_j \in \Phi\}$ is a nontrivial element of $H^0(C, O_C(1 - m + \nu_{\max})) = 0$, which is a contradiction. Thus $\operatorname{Im} \phi$ is not contained in O_C.   q.e.d.

(6.2) Completion of the proof of (3.4). Let $E = \operatorname{Coker} \phi$, $F = \operatorname{Im} \phi$. Then by (6.1), $E \cong O_C$, $F \cong O_C(-1)$ and we may view $I_C/I_C^2 = E \oplus F$ because $H^1(C, E^\vee \otimes F) = 0$. So we consider again the homomorphism ϕ as, $\phi : (I_\ell/I_\ell^2) \otimes O_C \to F \subset E \oplus F = I_C/I_C^2$. Let p be an arbitrary point of C. Then there are two generators x, y of $I_{C,p}$, and two generators f, g of $I_{\ell,p}$ such that $\phi(f) = x$, $\phi(g) = 0$, and that $x \bmod I_{C,p}^2$ (resp. $y \bmod I_{C,p}^2$) generates F (resp. E). Since $f = x \bmod I_{C,p}^2$, we may take f instead of x. Then in the same way as in (5.3), by taking a sufficiently fine covering $\Phi = \{U_j\}$ of an open neighborhood of C by Stein open sets U_j, we have $x_j \in \Gamma(U_j, I_C), y_j \in \Gamma(U_j, I_C)$, $\beta_j \in \Gamma(U_j, O_X)$ such that x_j and y_j (resp. x_j and $\beta_j y_j^m$) generate $\Gamma(U_j, I_C)$ (resp. $\Gamma(U_j, I_\ell)$).

Since $I_C/I_C^2 = F \oplus E = O_C(-1) \oplus O_C$, we may assume

(6.2.1) $x_j = \ell_{jk} x_k \bmod I_C^2, \quad y_j = y_k \bmod I_C^2$

where ℓ_{jk} stands for the one cocycle $L_C \in H^1(C, O_C^*)$. Since $\operatorname{Ker} \phi$ is isomorphic to $O_C(-1)$, and it is generated locally by $\beta_j y_j^m$, we see that

(6.2.2) $(\beta_j|_C) = \ell_{jk}(\beta_k|_C).$

Therefore, $\beta_C := \{\beta_j|_C; U_j \in \Phi\}$ is a nontrivial element of $H^0(C, O_C(1))$ by (6.2.2). Consequently β_C has a single zero at a unique point $p_0 \in C$ and it vanishes nowhere else. Then $z := \beta_0$ forms a parameter system at p_0 with the parameters x and y. The curve C intersects a unique irreducible component C' of ℓ_{red} at p_0, but nowhere else. In particular, ℓ_{red} is reducible. Let $I_{C'}$ be the ideal sheaf of O_X defining C'. Then $I_{C',p_0} = O_{X,p_0} x + O_{X,p_0} z$. By the assumption in (3.4), we have $\delta := LC' \geq 0$. Let $I_{C'}/I_{C'}^2 \cong O_{C'}(a) \oplus O_{C'}(b)$, $a \geq b$, and $\phi_{C'} : (I_\ell/I_\ell^2) \otimes O_{C'} \to I_{C'}/I_{C'}^2$, the natural homomorphism. Then since ℓ is reduced generically along C', we have by (3.9) $a + b = -3\delta + 2$, and $\dim(\operatorname{Coker}\phi_{C'}) = a + b + 2\delta = -\delta + 2$. Since $\beta_0 = z$ is a local parameter, $\operatorname{Coker}\phi_{C',p_0} \cong \mathbf{C}[y]/(y^m)$, whence $m \leq -\delta + 2$. Hence $m = 2$, $\delta = 0$, $\operatorname{Coker}\phi_{C'} = \operatorname{Coker}\phi_{C',p_0} = \mathbf{C}[y]/(y^2)$. By $(I_\ell/I_\ell^2) \otimes O_{C'} \cong O_{C'} \oplus O_{C'}$, we have $a = b = 1$ or $a = 2, b = 0$. Moreover this shows that C' meets no irreducible component other than C. Thus the proof of (3.4) is complete.   q.e.d.

§7. Proof of (3.5)

Assume that there is an irreducible component C_0 of ℓ_{red} with $LC_0 = 1$ such that ℓ is reduced at a point of C_0. Assume moreover that C_0 intersects an irreducible component C_1 of ℓ_{red} not contained in $\mathrm{Bs}\,|L|$.

(7.1) Lemma. *Let $C = C_0$. We have,*

(7.1.1) C intersects the unique irreducible component C' of $\ell_{\mathrm{red}} - C$ (:= the closure of $\ell_{\mathrm{red}} \setminus C$) at a unique point p transversally, to be more precise, we can choose local parameters x,y and z at p such that

$$I_{C,p} = O_{X,p}x + O_{X,p}y,$$
$$I_{\ell,p} = O_{X,p}x + O_{X,p}zy,$$

(7.1.2) ℓ is reduced everywhere on C, and reduced generically along C'.

Proof. In view of (5.1), $I_C/I_C^2 \cong O_C \oplus O_C(-1)$. We consider the natural homomorphism $\phi : (I_\ell/I_\ell^2) \otimes O_C(\cong O_C(-1) \oplus O_C(-1)) \to I_C/I_C^2$. By (3.10), ϕ is injective and $\operatorname{Coker}\phi \cong O_C/O_C(-1) \cong \mathbf{C}$. Let p be $\operatorname{Supp}\operatorname{Coker}\phi$. Then in the same manner as in (5.3), we can find local parameters x,y,w and a germ $\beta \in O_{X,p}$ such that

$$I_{C,p} = O_{X,p}x + O_{X,p}y,$$
$$I_{\ell,p} = O_{X,p}x + O_{X,p}\beta(y,w)y^m$$

where $\beta(0,w)$ is not identically zero and $m \geq 1$. Since ϕ is injective, we have $m = 1$. Moreover we see that $\beta(0,w)$ has a single zero at p. Hence $\beta(y,w)$ forms a parameter system at p with x and y. So (7.1.1) is clear by setting $z = \beta(y,w)$. (7.1.2) is clear from (7.1.1).   q.e.d.

(7.2) Completion of the proof of (3.5). By the assumption, the irreducible component C_0 of ℓ_{red} with $LC_0 = 1$ intersects an irreducible component C_1 of ℓ_{red} which is not contained in $\mathrm{Bs}\,|L|$. Then $LC_1 = 0$ or 1. Assume $LC_1 = 0$. Let $I_{C_1}/I_{C_1}^2 = O_{C_1}(a) \oplus O_{C_1}(b)$, $a \geq b$. By (7.1.2), ℓ is reduced generically along C_1, whence the natural homomorphism $\phi_1 : (I_\ell/I_\ell^2) \otimes O_{C_1} \to I_{C_1}/I_{C_1}^2$

is injective. Hence $a \geq 0$, $b \geq 0$. Moreover $a + b = -3LC_1 + 2 = 2$. Therefore dim Coker $\phi_1 = 2$, $(a, b) = (1, 1)$ or $(2, 0)$. Since ϕ_1 is not surjective at $p_0 := C_0 \cap C_1$, there is a unique point p_1 of C_1, $p_1 \neq p_0$ such that ϕ_1 is not surjective at p_1. By the same argument as in (7.1), we can choose local parameters x, y, z at p_1 such that

$$
\begin{aligned}
I_{C_1, p_1} &= O_{X, p_1} x + O_{X, p_1} y, \\
I_{\ell, p_1} &= O_{X, p_1} x + O_{X, p_1} zy.
\end{aligned}
$$

Consequently there is the third component C_2 of ℓ_{red} intersecting C_1. Then C_2 is not contained in Bs $|L|$. Otherwise, C_1 is contained in Bs $|L|$ because $LC_1 = 0$. Hence $LC_2 = 0$ or 1. If $LC_2 = 0$, then we repeat the same argument as above and after a finite repetition of these steps, we eventually obtain C_m and a chain of rational curves $C_1, \ldots, C_{m-1}$ of ℓ_{red} such that $LC_j = 0 (1 \leq j \leq m - 1)$ and $LC_m = 1$, and the pair C_i and C_j $(i < j)$ intersect at a unique point p_j transversally iff $i = j - 1$. By the same argument as above no C_j $(0 \leq j \leq m)$ is contained in Bs $|L|$. Moreover by (5.1) and (7.1), $I_{C_m}/I_{C_m}^2 = O_{C_m} \oplus O_{C_m}(-1)$ and C_m intersects C_{m-1} only. By (2.4.4), ℓ is connected so that it is the union of $C_0, \ldots, C_m$. Hence ℓ is reduced everywhere. Thus the proof of (3.5) is complete.

§8. Proof of (3.6)

(8.1) Lemma. *Let C be an arbitrary irreducible component of ℓ_{red} with $LC = 1$. Then we have,*

(8.1.1) $I_C/I_C^2 \cong O_C \oplus O_C(-1)$,

(8.1.2) *C intersects $\ell_{\mathrm{red}} - C$ $(:=$ the closure of $\ell_{\mathrm{red}} \setminus C)$ at a unique point p transversally, to be more precise, we can choose local parameters x, y and z at p such that*

$$
\begin{aligned}
I_{C, p} &= O_{X, p} x + O_{X, p} y, \\
I_{\ell, p} &= O_{X, p} x + O_{X, p} zy^m,
\end{aligned}
$$

for some $m \geq 1$, (we call m the multiplicity of C in ℓ)

(8.1.3) *C is not contained in Bs $|L|$.*

Proof. The assertion (8.1.1) follows from (5.1). If ℓ is reduced at a point of C, then (8.1.2) follows in the same manner as in (7.1). Next we consider the case where ℓ is nonreduced along C. Consider the natural homomorphism $\phi : (I_\ell/I_\ell^2) \otimes O_C \to I_C/I_C^2$. Then by (3.3) and (6.1), Im $\phi = O_C(-1)$ and Im ϕ is not contained in $O_C \oplus \{0\}$. Let $E = \mathrm{Coker}\, \phi$, $F = \mathrm{Im}\, \phi$. Then we may view $I_C/I_C^2 = E \oplus F$ and consider the homomorphism ϕ as,

$$
\phi : (I_\ell/I_\ell^2) \otimes O_C \to F \subset E \oplus F = I_C/I_C^2.
$$

Then we are able to choose an open covering $\Phi = \{U_j\}$ of an open neighborhood of C and x_j, y_j and β_j satisfying (5.3.2). Here we may assume that

$$x_j = \ell_{jk} x_k \bmod I_C^2, \quad y_j = y_k \bmod I_C^2, \quad \beta_j|_C = \ell_{jk} \beta_k|_C$$

and that x_j (resp. y_j) generates F (resp. E). Hence $\beta_C := \{\beta_j|_C; U_j \in \Phi\}$ is a nontrivial element of $H^0(C, O_C(1))$, whence β_C has a single zero at a unique point p_0 of C. Then $x = x_j$, $y = y_j$ and $z = \beta_j$ at p_0 form a local parameter system at p_0. This completes the proof of (8.1.2).

Now we are able to construct a partial "normalization" of ℓ by using the expression (8.1.2) of I_C and I_ℓ as follows; With the notations in (8.1.2), we define an ideal subsheaf $I_{\ell'}$ of O_X by;

$$\begin{aligned}
I_{\ell',p} &= I_{\ell,p} & (p \in X \setminus C) \\
I_{\ell',p} &= O_{X,p} x + O_{X,p} z & (p = p_0) \\
I_{\ell',p} &= O_{X,p} & (p \in C \setminus \{p_0\})
\end{aligned}$$

where $I_{\ell',p}$ is the stalk of $I_{\ell'}$ at p. Let ℓ' be an analytic subspace of X with $\ell'_{\text{red}} = (\ell_{\text{red}} \setminus C) \cup p_0$, $O_{\ell'} = O_X/I_{\ell'}$, and $I_k = I_C^k + I_\ell$ $(1 \le k \le m)$. Then we have exact sequences;

(8.1.4) $\qquad 0 \to O_\ell \to O_X/I_{\ell'} \oplus O_X/I_m \to O_X/I_m + I_{\ell'} \to 0,$

(8.1.5) $\qquad 0 \to I_{k-1} \cap I_{\ell'}/I_k \cap I_{\ell'} \to I_k + I_{\ell'}/I_k \to I_{k-1} + I_{\ell'}/I_{k-1} \to 0$

(8.1.6) $\qquad 0 \to I_k + I_{\ell'}/I_k \to O_X/I_k \to O_X/I_k + I_{\ell'} \to 0$

We note that $O_X/I_k + I_{\ell'} = \mathbf{C}[y]/(y^k)$, $I_{k-1}/I_k \cong O_C$, $I_{k-1} \cap I_{\ell'}/I_k \cap I_{\ell'} = O_C(-p_0)$. Let $V_k = H^0((I_k + I_{\ell'}/I_k) \otimes O_X(L))$, $\eta_{i,j}$ the natural homomorphism of V_i into V_j for $i > j$. From (8.1.4)-(8.1.6) tensored with $O_X(L)$, we infer long exact sequences,

(8.1.7) $\qquad 0 \to H^0(O_\ell(L)) \to H^0((O_X/I_{\ell'})(L)) \oplus H^0((O_X/I_m)(L)) \to \mathbf{C}^m$

(8.1.8) $\qquad\qquad\qquad 0 \to H^0(O_C) \to V_k \to V_{k-1} \to 0$

(8.1.9) $\qquad 0 \to V_k \to H^0((O_X/I_k)(L)) \to O_X/I_k + I_{\ell'} (\cong \mathbf{C}^k)$

Then by (8.1.9), $V_k = \text{Ker}(H^0((O_X/I_k)(L)) \to O_X/I_k + I_{\ell'})$, whereas V_m is a subspace of $H^0(O_\ell(L))$ by (8.1.7). By (8.1.8), $\eta_{k,k-1}$ is surjective and $\dim V_k = \dim V_{k-1} + 1$, whence $\eta_{m,1} = \eta_{m,m-1}\eta_{m-1,m-2} \cdots \eta_{2,1} : V_m \to V_1$ is surjective. Since $V_1 = H^0(C, L_C - p_0)$ (:= elements of $H^0(C, L_C)$ vanishing at p_0) is a nontrivial subspace of $H^0(C, L_C)$, $C \setminus p_0$ is disjoint from $\text{Bs}|L_\ell|$ (= $\text{Bs}|L|$ by (2.6)). This completes the proof of (8.1.3).   q.e.d.

(8.2) Corollary. $\dim V_k = k$, $h^0((O_X/I_k)(L)) = 2k$.

Proof. By the above proof, $\dim V_k = \dim V_1 + k - 1$. From the exact sequence

$$0 \to (I_{k-1}/I_k)(L) \to (O_X/I_k)(L) \to (O_X/I_{k-1})(L) \to 0$$

and $(I_{k-1}/I_k)(L) \cong O_C(1)$, we infer $h^0((O_X/I_k)(L)) = h^0((O_X/I_{k-1})(L)) + 2$, whence the second assertion.   q.e.d.

(8.3) Proof of (3.6)– Start. Assume that there is an irreducible component C of ℓ_{red} with $LC = 1$ such that C intersects an irreducible component C' of ℓ_{red} not contained in $\mathrm{Bs}\,|L|$. Then by (3.2)-(3.5) and (3.7), $\mathrm{Bs}\,|L|$ is empty so that for any $\ell' = D'' \cap D'''$, D'', $D''' \in |L|$, any irreducible component C'' of ℓ'_{red} with $LC'' = 1$ intersects a component C''' of ℓ'_{red} not contained in $\mathrm{Bs}\,|L|$. Therefore it remains to consider the case where for any $\ell = D \cap D'$, D, $D' \in |L|$, any irreducible component C of ℓ_{red} with $LC = 1$ intersects a component of ℓ_{red} contained in $\mathrm{Bs}\,|L|$. Then C is not contained in $\mathrm{Bs}\,|L|$ and there is a unique irreducible component C' of ℓ_{red} intersecting C by (8.1). In what follows, we assume this to derive a contradiction in (8.10).

First we shall prove,

(8.4) Lemma. *Let C_j $(1 \le j \le s)$ be all the irreducible components of ℓ_{red} with $LC_j = 1$, B_j the unique irreducible component of ℓ_{red} contained in $\mathrm{Bs}\,|L|$ that C_j intersects. By choosing a general pair D and D', $B_1 = B_2 = \cdots = B_s$.*

Proof. We apply a variant of the argument in [11,(2.6)]. Assume the contrary. Then we can choose a one parameter family $D'_t(t \in \mathbf{P}^1)$ and a Zariski dense open subset U of $\mathbf{P}^1$ with the following properties;

$$(8.4.1) \qquad \begin{aligned} &\ell_{t,\mathrm{red}} = C_{1,t} + \cdots + C_{s(t),t} + B_1 + B_2 + \cdots, \quad t \in U, \\ &B_j \subset \mathrm{Bs}\,|L|, \quad B_1 \neq B_2 \end{aligned}$$

where $\ell_t = D \cap D'_t$,

$$(8.4.2) \qquad\qquad LC_{j,t} = 1, (1 \le j \le s(t)),$$

(8.4.3) $C_{1,t}$ (resp. $C_{2,t}$) intersects B_1 (resp. B_2) for any $t \in U$.

Let d (resp. d'_t) be the equation defining D (resp. D'_t), and define an analytic subset Z of $X \times \mathbf{P}^1$ by $Z = \{(x,t) \in X \times \mathbf{P}^1; d(x) = d'_t(x) = 0\}$. Let p_j be the j-th projection of $X \times \mathbf{P}^1$, Z_j all the irreducible components of Z_{red}, $g_j : Y_j \to Z_j$ the normalization of Z_j, $Y_j \overset{\pi_j}{\to} U_j \overset{h_j}{\to} \mathbf{P}^1$ the Stein factorization of $p_2 g_j$ $(1 \le j \le s)$. We may assume that $C_{j,t_j} = p_1 g_j(\pi_j^{-1}(u_j))$, $t_j = h_j(u_j)$ for some $u_j \in U_j$ $(j = 1,2)$. By (8.1), $p_1 g_j(\pi_j^{-1}(v))$ is irreducible nonsingular and intersects B_j only at one point when v moves in a Zariski dense open subset V_j of U_j. Then C_{1,t_1} and C_{2,t_2} intersect nowhere for general $u_1 \in V_1$, $u_2 \in V_2$. In fact, if $C_{1,t_1} \cap C_{2,t_2} = \{p,\ldots\} \neq \emptyset$ and if $p \neq C_{1,t_1} \cap B_1$, then D'_{t_2} contains C_{1,t_1} by $D'_{t_2} C_{1,t_1} = LC_{1,t_1} = 1$. Since D'_t is chosen general, this contradicts that C_{1,t_1} is not contained in $\mathrm{Bs}\,|L|$. Therefore we may assume that, $C_{1,t_1} \cap C_{2,t_2} = C_{1,t_1} \cap B_1 = C_{2,t_2} \cap B_2$. This shows that $C_{1,t}$ intersects $\ell_{t,\mathrm{red}} - C_{1,t}$ at the intersection $B_1 \cap B_2$ ($\neq \emptyset$) for general t. However this is impossible by (8.1.2). This proves that C_{1,t_1} and C_{2,t_2} intersect nowhere for general $u_1 \in V_1$, $u_2 \in V_2$. Hence the intersection of $p_1 g_1(Y_1)$ and $p_1 g_2(Y_2)$ is at most one dimensional. However $D = p_1 g_1(Y_1) = p_1 g_2(Y_2)$ because D is irreducible. This is a contradiction.   q.e.d.

By (8.4), all B_j are the same, say, $B_j = B$ for any j, by choosing a sufficiently general pair D, D' and $\ell = D \cap D'$. Let $n = -LB$, and let $\phi_B : (I_\ell/I_\ell^2) \otimes O_B \to I_B/I_B^2$ be the natural homomorphism. One sees $n \geq 0$ in view of (3.2) and (8.1).

(8.5) Lemma. *Let $\ell = D \cap D'$ for D, D' sufficiently general. Let m_j be the multiplicity of C_j in ℓ, $m := m_1 + \cdots + m_s$. Then $m = n + 2$, $n \geq 0$.*

Proof. By (8.1.2), ℓ is reduced generically along B, so that ϕ_B is injective by (3.10). Hence $\mathrm{Coker}\,\phi_B$ is finite. One sees that $\dim \mathrm{Coker}\,\phi_B = c_1(I_B/I_B^2) - c_1((I_\ell/I_\ell^2) \otimes O_B) = -LB + 2 = n + 2$, $\dim \mathrm{Coker}\,\phi_{B,p_j} = m_j$ at any intersection point $p_j := C_j \cap B$ by (8.1.2). Since p_j's are all distinct by (8.1.2), we have $m \leq n + 2$. Since ℓ is of multiplicity m_j generically along C_j and it is reduced generically along B, $cl(\ell)$ is equal to $m_1 cl(C_1) + \cdots + m_s cl(C_s) + cl(B) + cl(B')$ for some effective one cycle B' such that $\mathrm{Supp}\,B' \subset \mathrm{Bs}\,|L|$, $\mathrm{Supp}\,B' \not\supset B$. Then $LB' \leq 0$, because there is no irreducible component B'' of B'_{red} with $LB'' \geq 1$ by (3.2) and (8.1). Therefore we have by (2.2), $2 \leq L^3 = L\ell \leq L(m_1 C_1 + \cdots + m_s C_s + B) = m - n$. This proves $m \geq n + 2$, hence $m = n + 2$, $L^3 = 2$.   q.e.d.

(8.6) Corollary. *Let $\ell = D \cap D'$ for D, D' sufficiently general. The curve B intersects C_j ($1 \leq j \leq s$) only, ℓ is reduced everywhere along $B \setminus \cup_j(B \cap C_j)$, $\ell = m_1 C_1 + \cdots + m_s C_s + B$, $\mathrm{Bs}\,|L| = \mathrm{Bs}\,|L_\ell| = B$.*

(8.7) Lemma. *Let D'' and D''' be arbitrary members of $|L|$, $D'' \neq D'''$, $\ell' = D'' \cap D'''$. Then $\ell' = m_1' C_1' + \cdots + m_{s'}' C_{s'} + B$ for some m_j' and s' where $LC_j' = 1$, $m' := m_1' + \cdots + m_{s'}' = n + 2$, the structures of ℓ', C_j' and B at $C_j' \cap B$ are described in (8.1).*

Proof. The proof of (3.7) shows that $\mathrm{Bs}\,|L| = \emptyset$ in the cases (3.2)-(3.5). Since $\mathrm{Bs}\,|L| = B$ in our case, any irreducible component C_j' of ℓ'_{red} with $LC_j' = 1$ intersects B. By the above argument, we see $\ell' = m_1' C_1' + \cdots + m_{s'}' C_{s'} + B$ for some m_j' and s' where $LC_j' = 1$, $m' := m_1' + \cdots + m_{s'}' = n + 2$. The rest is clear from (8.1).   q.e.d.

(8.8) Lemma. $h^0(O_\ell(L)) = m$, *and* $n = -LB > 0$.

Proof. Let ℓ'' be an analytic subspace of X whose ideal in O_X is defined by $I_{\ell''} = I_\ell + \cap_{1 \leq j \leq s}(I_{C_j})^{m_j}$ where $m_j =$ multiplicity of C_j in ℓ. We easily see that

$$I_{\ell'',p} = \begin{cases} I_{\ell,p} & (p \in \ell_{\text{red}} \setminus B), \\ O_{X,p} + O_{X,p}y^{m_j} & (p = B \cap C_j), \\ O_{X,p} & (p \notin \cup_{1 \leq j \leq s} C_j). \end{cases}$$

Then there is an exact sequence

$$0 \to O_\ell(L) \to O_{\ell''}(L) \oplus O_B(L) \to \oplus_j \mathbf{C}^{m_j} \to 0.$$

Since the support of ℓ'' is the disjoint union of C_j $(1 \leq j \leq s)$ we see by (8.1.9) and (8.2) $h^0(O_{\ell''}(L)) = 2(m_1 + \cdots + m_s) = 2m$, and that the natural homomorphism $H^0(O_{\ell''}(L)) \to \oplus \mathbf{C}^{m_j}$ is surjective. Since $H^0(O_\ell(L))$ is mapped to zero in $H^0(O_B(L))$, we have $h^0(O_\ell(L)) = m, h^0(O_B(L)) = 0$. It follows from $O_B(L) = O_B(-n)$ that $n > 0$. q.e.d.

Let $h : Y \to X$ be the blowing-up of X with B center, $E = h^{-1}(B)_{\text{red}}$, $N = h^*L - [E]$, $\bar{D} = h^*D - E$, $\bar{D}' = h^*D' - E$, $\bar{C}_j =$ the proper transform of C_j $(1 \leq j \leq s)$. Then one checks

(8.9) Lemma. *For general D and D' in $|L|$, $\bar{C}_j$ is isomorphic to C_j and* $(\bar{D} \cap \bar{D}')_{\text{red}} = \cup_{j=1}^s \bar{C}_j$, $\bar{D}\bar{D}' = m_1\bar{C}_1 + \cdots + m_s\bar{C}_s$, $\bar{B}' := E \cap \bar{D}$ *is isomorphic to B.*

Proof. We note that a general member of $|L|$ is nonsingular along B. Indeed, assume that D and D' are singular at a point p of B, that is, $I_{D,p} \subset m_{X,p}^2$, $I_{D',p} \subset m_{X,p}^2$. Let $\ell = D \cap D'$. Then

$$\begin{aligned} T_{\ell,p} &= \text{Hom}(m_{\ell,p}/m_{\ell,p}^2, \mathbf{C}) \\ &= \text{Hom}(m_{X,p}/I_{D,p} + I_{D',p} + m_{X,p}^2, \mathbf{C}) \\ &= \text{Hom}(m_{X,p}/m_{X,p}^2, \mathbf{C}) \end{aligned}$$

whence $\dim T_{\ell,p} = 3$. However by (8.1.2), $\dim T_{\ell,p} \leq 2$, which is absurd. Let $q := B \cap C_j$, $a := m_j$. It suffices to consider the problem near q to prove (8.9). Then by (8.1.2),

$$\begin{aligned} I_{\ell,q} &= O_{X,q}x + O_{X,q}zy^a, \\ I_{B,q} &= O_{X,q}x + O_{X,q}z, \\ I_{C_j,q} &= O_{X,q}x + O_{X,q}y. \end{aligned}$$

We may assume without loss of generality that $I_{D,q} = O_{X,q}x$, $I_{D',q} = O_{X,q}(x + zy^a)$ because general D and D' are nonsigular along B. Now it is easy to check the assertions by a direct computation. q.e.d.

(8.10) Completion of the proof of (3.6). Since $\bar{C}_j$ is a movable part of $\bar{D} \cap \bar{D}'$, Bs $|N|$ consists of at most finitely many points, whence $N\bar{C}_j \geq 0$ for any j. Let $I_B/I_B^2 = O_B(a) \oplus O_B(b)$, $a \geq b, c = a - b$. Then by (3.9), $a + b = c + 2b = 3n + 2$ and E is a rational ruled surface Σ_c. By (8.9), $n > 0$. Let e (resp. f) be a section (resp. a fiber) of the ruling of Σ_c with $e^2 = c$, $f^2 = 0, ef = 1$. Let e_∞ be a section of Σ_c with $e^2 = -c$, $ee_\infty = 0$.

Let $\bar{B}' := E \cap \bar{D}$. We see $[E]_E = -e - bf$, $E\bar{B}' = E^2\bar{D} = E^2(h^*L - E) = -(2n+2)$, $E^3 = c_1(I_B/I_B^2) = 3n+2$, $N^3 = L^3 + 3h^*LE^2 - E^3 = 2 + 3n - (3n+2) = 0$. Consequently $m_1 N\bar{C}_1 + \cdots + m_s N\bar{C}_s = N\bar{D}\bar{D}' = N^3 = 0$, $N\bar{C}_j = 0$, and $|N|$ is base point free.

Let $\bar{B}' = pe + qf \in \text{Pic } E$. In view of (8.9), $\bar{B}'$ is isomorphic to B and $p = 1$. Since $[\bar{B}'] = [\bar{D}]_E = (h^*L - [E])_E = e + (b-n)f$, we have $q = b - n$. If $\bar{B}' = e_\infty$, then $q = -c$, $b = 2n+2$, $a - b = -n - 2 < 0$. This is a contradiction. Hence $\bar{B}' \neq e_\infty$, so that $q \geq 0$, $b \geq n$. Therefore $h^0(E, N \otimes O_E) = h^0(\Sigma_c, e + (b-n)f) = n+4$. We have by (8.8) $h^0(Y, N) = h^0(X, L) = h^0(\ell, L_\ell) + 2 = m + 2 = n + 4$, and by (8.1.2) $h^0(Y, N - E) = h^0(X, I_B^2 L) = 0$, whence we have a natural isomorphism $H^0(Y, N) \cong H^0(E, N \otimes O_E)$.

Let $g : Y \to \mathbf{P}^{n+3}$ $(n > 0)$ be the morphism associated with the linear system $|N|$, g_E the restriction of g to E. Any point y of Y is contained in some $\bar{C}_j$ of some $\bar{D} \cap \bar{D}'$ because $h^0(Y, N) \geq 5$. By $N\bar{C}_j = 0$, $\bar{C}_j$ is mapped to one point. Moreover $E\bar{C}_j = (N + E)\bar{C}_j = h^*L\bar{C}_j = L C_j = 1$, whence $g(y) = g(\bar{C}_j \cap E)$. Hence $g(Y) = g(E)$. We note that the linear system $|N \otimes O_E|$ defines an isomorphism g_E of E into $\mathbf{P}^{n+3}$ iff $b > n$. If $b = n$, then g_E is an isomorphism over $E \setminus e_\infty$.

We shall define a morphism ψ of Y onto E as follows. If $b > n$, then we define ψ to be the morphism g. In the general case, take an arbitrary point y of Y. Then choose two general members $\bar{D}$ and $\bar{D}'$ of $|N|$ such that $\bar{D}, \bar{D}'$ pass through y. The intersection $\bar{\ell}_{\text{red}} = (\bar{D} \cap \bar{D}')_{\text{red}}$ is the disjoint union of $\bar{C}_j$ $(j = 1, \dots, s)$ by (8.9). Hence there is a unique $\bar{C}_j$ passing through y. Since $E\bar{C}_j = 1, E$ intersects $\bar{C}_j$ at a unique point y' of E. We define $\psi(y) = y' = E \cap \bar{C}_j$. The point y' is independent of the choice of $\bar{D}$ and $\bar{D}'$. To show this, take $\bar{D}''$ and $\bar{D}'''$ of $|N|$ such that $\bar{D}''$ and $\bar{D}'''$ pass through y. Let $\bar{\ell}' = \bar{D}'' \cap \bar{D}'''$, $\bar{\ell}'_{\text{red}} = \bar{C}'_1 + \cdots + \bar{C}'_{s'}$. Then there is a unique $\bar{C}'_i$ passing through y. In fact, since $\bar{D}\bar{C}'_i = \bar{D}'\bar{C}'_i = N\bar{C}'_i = 0$, both $\bar{D}$ and $\bar{D}'$ contain $\bar{C}'_i$, whence $\bar{C}'_i$ is also the unique irreducible component of $\bar{\ell}_{\text{red}}$ passing through y. Hence $\bar{C}_j = \bar{C}'_i$. Therefore $\bar{C}_j \cap E = \bar{C}'_i \cap E$. It is easy to check that for any point y of E $(\cong \Sigma_c)$, there exist two members H and H' of $|N \otimes O_E|$ such that $H \cap H'$ is reduced and it contains y. By the isomorphism $H^0(Y, N) \cong H^0(E, N \otimes O_E)$, for a given y of Y, we can choose two members D'' and D''' of $|L|$ such that $\bar{D}'' \cap \bar{D}''' \cap E$ is reduced and $\bar{D}'', \bar{D}'''$ pass through y where $\bar{D}'' = h^*D'' - E$, $\bar{D}''' = h^*D''' - E$. Hence $D'' \cap D'''$ has $m'_j = 1$ for any j in (8.7). By using this it is easy to see that ψ is a morphism of Y into (indeed, onto) E. If $b = n$, the morphism ψ coincides with the morphism g on $Y \setminus g^{-1}(g(e_\infty))$. Note that $g(Y \setminus g^{-1}(g(e_\infty))) \cong E \setminus e_\infty$.

One also sees readily that any fiber of ψ endowed with reduced structure is

$\mathbf{P}^1$, one of the irreducible components of $\bar{D} \cap \bar{D}'$ for some $\bar{D}, \bar{D}' \in |N|$. Since both Y and E are smooth, ψ is flat and any general (scheme-theoretic) fiber of ψ is $\mathbf{P}^1$. Any fiber is mutually algebraically equivalent and $\psi^{-1}(x)E = 1$ for general $x \in E$. By the criterion of multiplicity one, we see that a fiber $\psi^{-1}(x)$ is generically reduced for any $x \in E$. Since any fiber is Gorenstein, it is therefore reduced everywhere, whence it is a nonsingular rational curve $\mathbf{P}^1$. Thus the morphism ψ gives a $\mathbf{P}^1$-bundle structure of Y over E with a section E. Therefore there is a rank two vector bundle F on E such that $Y \cong \mathbf{P}(F)$ and the following is exact;

$$0 \to O_E \to F \to \det F \to 0.$$

The surface $E(= \mathbf{P}(\det F))$ is embedded into Y by viewing $\det F$ as a quotient bundle of F as above. Then we have $\det F = -N_{E/Y} = -[E]_E = e + bf$. Since $H^1(E, -e - bf) = 0$, $F \cong \det F \oplus O_E$. Now it is easy to see that $b_2(Y)$ (or Picard number of Y) $= 3$. The threefold X is obtained from Y by contracting E to a curve B, whence $b_2(X)$ (or Picard number of X) $= 2$. This is a contradiction. This completes the proof of (3.6).

(8.11) Remark. The proofs in sections 2-8 work as well in arbitrary characteristic.

§9. Proof of (0.1)

(9.1) Theorem. *Let X be a compact complex threefold homeomorphic to $\mathbf{P}^3$ is isomorphic to $\mathbf{P}^3$ if $H^1(X, O_X) = 0$ and if $h^0(X, -mK_X) \geq 2$ for some positive integer m.*

Proof. Since $H^1(X, O_X) = 0$, we have a natural exact sequence $1 \to H^1(X, O_X^*) \to H^2(X, \mathbf{Z})(\cong \mathbf{Z}) \to H^2(X, O_X)$. Since $H^1(X, O_X^*)$ has a non-trivial element K_X and $H^2(X, O_X)$ is torsion free, $H^1(X, O_X^*)$ is mapped isomorphically onto $H^2(X, \mathbf{Z})$. Let L be the generator of $H^1(X, O_X^*)$ with $L^3 = 1$. Since the second Stiefel Whitney class $w_2(= c_1(X) \bmod 2)$ and rational Pontrjagin classes are topological invariants we see by [5,pp. 207-208]

$$c_1(X) = (2s + 4)L, \quad \chi(X, O_X) = (s + 1)(s + 2)(s + 3)/6$$

for an integer s. We see $h^3(X, O_X) = h^0(X, \Omega_X^3) \leq b_3 = 0$, $\chi(X, O_X) = 1 + h^2(X, O_X) \geq 1$. This shows $s \geq 0$, $K_X = -(2s + 4)L$, $2s + 4 \geq 4$. Thus all the assumptions in [11,(1.1)] are satisfied. Hence X is isomorphic to $\mathbf{P}^3$. q.e.d.

(9.2) Proof of (0.1). By the same argument as in (9.1), we see $\mathrm{Pic}\, X \cong H^2(X, \mathbf{Z}) \cong \mathbf{Z}$. Let L be a generator of $\mathrm{Pic}\, X$ with $L^3 = 2$. Then by [1, p.188] or [10, p.321], we have,

$$c_1(X) = (2s + 3)L, \quad \chi(X, O_X) = (s + 1)(s + 2)(2s + 3)/6$$

for an integer s. We see $\chi(X, O_X) \geq 1$, whence $s \geq 0$. If $s > 0$, then X is isomorphic to $\mathbf{P}^3$ by [11,(1.1)] which is a contradiction. Hence $s = 0$ and X is isomorphic to $\mathbf{Q}^3$ by (2.1).   q.e.d.

Theorems (0.2) and (0.3) are derived from (0.1). See [11],[12].

Appendix

As was pointed out by Fujita, the proof of (2.7) in [11] does not work in positive characteristic because it uses Bertini's theorem.

We shall give an alternative proof of [11,(1.1)] in arbitrary characteristic.

(A.1) Theorem. *Let X be a complete irreducible nonsingular algebraic threefold defined over an algebraically closed field of arbitrary characteristic. Assume* $\operatorname{Pic} X = \mathbf{Z}L$, $H^1(X, O_X) = 0$, $K_X = -dL$ $(d \geq 4)$, $L^3 > 0$, $\kappa(X, L) \geq 1$. *Then X is isomorphic to $\mathbf{P}^3$.*

For the proof of (A.1), in view of [11,(2.8) or (2.9)], it suffices to prove that a complete intersection $\ell = D \cap D'$ for any pair D and $D' \in |L|$ is a nonsingular rational curve.

In view of [11,(2.2)], $d = 4$, $K_X = -4L$. And by [11,(2.3)], there is a unique irreducible component A_1 of ℓ_{red} with $LA_1 = 1$. Here we write $C = A_1$ for simplicity.

First we show,

(A.2) Lemma. $I_\ell \not\subset I_C^2$.

Proof. Let I_ℓ (resp. I_C) be the ideal sheaf in O_X defining ℓ (resp. C). By Grothendieck's theorem, let $I_C/I_C^2 \cong O_C(a) \oplus O_C(b)$, $a \geq b$. We infer $c_1(I_C/I_C^2) + c_1(\Omega_C^1) = c_1(\Omega_X^1 \otimes O_C) = K_X C = -4LC = -4$. Hence $a + b = -2$. Hence $b \leq -1$. Assume $I_\ell \subset I_C^2$. We consider the natural homomorphism $\phi : (I_\ell/I_\ell^2) \otimes O_C \to I_C^2/I_C^3 (\cong O_C(2a) \oplus O_C(a+b) \oplus O_C(2b))$. Since $(I_\ell/I_\ell^2) \otimes O_C \cong O_C(-1) \oplus O_C(-1)$, $\operatorname{Im} \phi$ is contained in $O_C(2a)$. Let $I = O_C(2a) + I_C^3$. Then since by [11,(1.7.3)] $H^1(O_X/I_\ell) = 0$, we have $\chi(O_X/I) \geq 1$ by (3.11). It follows that $1 \leq \chi(O_X/I) = \chi(O_X/I_C) + \chi(I_C/I_C^2) + \chi(I_C^2/I) = 1 + 2b$, which is absurd.
                                                                                q.e.d.

(A.3) Lemma. $I_C/I_C^2 \cong O_C(-1) \oplus O_C(-1)$ *and the natural homomorphism* $\phi : (I_\ell/I_\ell^2) \otimes O_C \to I_C/I_C^2$ *is an isomorphism.*

Proof. By the proof of (A.2), we note $a + b = -2$, $b \leq -1 \leq a$. If $a > b$, then $\operatorname{Im} \phi$ is contained in $O_C(a)$. Let $I = O_C(a) + I_C^2$. Then $I_\ell \subset I \subset I_C$, $I_C/I = O_C(b)$. In the same manner as in (A.2), by [11,(1.7.3)] and (3.11), we have $1 \leq \chi(O_X/I) = \chi(O_X/I_C) + \chi(I_C/I) = 2 + b$. This is a contradiction. Hence $a = b = -1$. Assume ϕ is not injective. We note that by (A.2), ϕ is a nontrivial homomorphism. Since $\operatorname{Im} \phi \ (\cong I_\ell + I_C^2/I_C^2)$ is a rank one subsheaf of a torsion free sheaf I_C/I_C^2, it is locally O_C-free. Here we may set $\operatorname{Im} \phi = O_C(c)$ for some $c \in \mathbf{Z}$. Then $c = -1$ because $(I_\ell/I_\ell^2) \otimes O_C \cong O_C(-1) \oplus O_C(-1)$. Let $E = \operatorname{Coker} \phi \cong O_C(-1)$, $F = \operatorname{Im} \phi \cong O_C(-1)$. Then we may view $I_C/I_C^2 = E \oplus F$ because $H^1(C, E^\vee \otimes F) = 0$, $E^\vee$ being the dual of E. So we consider again the homomorphism ϕ as,

$$\phi : (I_\ell/I_\ell^2) \otimes O_C \to F \subset E \oplus F = I_C/I_C^2.$$

Let p be an arbitrary point of C. Then there are two generators x, y of $I_{C,p}$, and two generators f, g of $I_{\ell,p}$ such that $\phi(f) = x, \phi(g) = 0, x \bmod I_{C,p}^2$ (resp. $y \bmod I_{C,p}^2$) generates F (resp. E). In the same manner as in (5.3), we obtain local parameters x and $y \in I_{C,p}$ and $\beta \in O_{X,p}$, $m > 0$ such that

(A.3.1)
$$\begin{aligned} I_{C,p} &= O_{X,p}x + O_{X,p}y \\ I_{\ell,p} &= O_{X,p}x + O_{X,p}\beta y^m \end{aligned}$$

where the restriction $\beta|_C$ of β to C is not identically zero. We note that $m \geq 2$ because $g \in I_{C,p}^2$.

Let $\Phi = \{U_j\}$ be a sufficiently fine covering of an open neighborhood of C by Stein open sets U_j. Then by (A.3.1), we have $x_j \in \Gamma(U_j, I_\ell), y_j \in \Gamma(U_j, I_C), \beta_j \in \Gamma(U_j, O_X)$ such that x_j and y_j (resp.x_j and $\beta_j y_j^m$) generate $\Gamma(U_j, I_C)$ (resp. $\Gamma(U_j, I_\ell)$).

Since $(I_\ell/I_\ell^2) \otimes O_C \cong O_C(-1) \oplus O_C(-1)$, we may assume that

(A.3.2)
$$x_j = \ell_{jk} x_k \bmod I_C^2, \quad y_j = \ell_{jk} y_k \bmod I_C^2$$

where ℓ_{jk} stands for the one cocycle $L_C \in H^1(C, O_C^*)$.

Two elements $D_j \beta_j y_j^m$ and $D_k \beta_k y_k^m \in \operatorname{Ker} \phi (\cong O_C(-1))$ are identified if and only if (we may assume that)

(A.3.3)
$$(D_j|_C) = \ell_{jk}^{-1}(D_k|_C).$$

This shows that

(A.3.4)
$$(\beta_j|_C) = \ell_{jk}^{1-m}(\beta_k|_C).$$

In particular, $\beta_C := \{\beta_j|_C; U_j \in \Phi\}$ is a nontrivial element of $H^0(C, O_C(1-m))$. This is possible only when $m = 1$ and β_C is a nonzero constant. This contradicts $m \geq 2$. Consequently ϕ is injective whence it is an isomorphism. This completes the proof of (A.3).   q.e.d.

By (A.3), $I_{\ell,p} + I_{C,p}^2 = I_{C,p}$ whence $I_{\ell,p} = I_{C,p}$ for any point p of C. This shows that ℓ is nonsingular anywhere along C. Since ℓ is connected by [11,(1.7)], ℓ is isomorphic to C. Then it is easy to see that $\operatorname{Bs}|L| = \emptyset$ and that the morphism associated with $|L|$ is an isomorphism of X onto $\mathbf{P}^3$.   q.e.d.

References

[1] Brieskorn, E. : Ein Satz über die komplexen Quadriken, Math. Ann. 155 (1964), 184-193

[2] Fujita, T. : On the structure of polarized varieties with Δ-genera zero, J. Fac. Sci. Univ. of Tokyo, 22 (1975), 103-115

[3] Fulton, W. : Intersection Theory, Springer Verlag, 1984

[4] Hirzebruch, F. : Topological Methods in Algebraic Geometry, Springer Verlag, 1966

[5] Hirzebruch, F. and Kodaira, K. : On the complex projective spaces, J. Math. Pure Appl., 36 (1957), 201-216

[6] Iitaka, S. : On D-dimensions of algebraic varieties, J. Math. Soc. Japan, 23 (1971), 356-373

[7] — : Algebraic Geometry, Graduate Texts in Math., 76, Springer Verlag, 1981

[8] Kobayashi, S., Ochiai, T. : Characterizations of complex projective spaces and hyperquadrics, J. Math. Kyoto Univ., 13 (1973), 31-47

[9] Mori, S. : Threefolds whose canonical bundles are not numerically effective, Ann. of Math., 116 (1982), 133-176

[10] Morrow, J. : A survey of some results on complex Kähler manifolds, Global Analysis, Univ. of Tokyo Press, 1969, 315-324

[11] Nakamura, I. : Moishezon threefolds homeomorphic to $\mathbf{P}^3$, J. Math. Soc. Japan, 39(1987), 521-535

[12] — : Characterizations of $\mathbf{P}^3$ and hyperquadrics $\mathbf{Q}^3$ in $\mathbf{P}^4$, Proc. Japan Acad., 62, Ser. A, (1986), 230-233

[13] Peternell, T. : On the rigidity problem for the complex projective space, preprint

[14] — : Algebraic structures on certain 3-folds, Math. Ann. 274, (1986), 133-156

[15] Ueno, K. : Classification Theory of Algebraic Varieties and Compact Complex Spaces, Lecture Notes in Math., 439, Springer Verlag, 1975

Iku NAKAMURA

Department of Mathematics
Hokkaido University
Sapporo, 060
Japan